Library of
Davidson College

An Introduction to Astrophysical Hydrodynamics

An Introduction to Astrophysical Hydrodynamics

Steven N. Shore

GHRS Science Team
Computer Sciences Corporation
Goddard Space Flight Center
Greenbelt, Maryland

ACADEMIC PRESS, INC.
Harcourt Brace Jovanovich, Publishers
San Diego   New York   Boston
London   Sydney   Tokyo   Toronto

This book is printed on acid-free paper. ∞

Copyright © 1992 by ACADEMIC PRESS, INC.
All Rights Reserved.
No part of this publication may be reproduced or transmitted in any form or by any means, electronic or mechanical, including photocopy, recording, or any information storage and retrieval system, without permission in writing from the publisher.

Academic Press, Inc.
1250 Sixth Avenue, San Diego, California 92101-4311

United Kingdom Edition published by
Academic Press Limited
24–28 Oval Road, London NW1 7DX

Library of Congress Cataloging-in-Publication Data

Shore, Steven N.
　An introduction to astrophysical hydrodynamics / Steven N. Shore.
　　p.　cm.
　Includes bibliographical references and index.
　ISBN 0-12-640670-7
　1. Hydrodynamics.　2. Astrophysics.　I. Title.
　QB461.S446　1992
　523.01--dc20
　　　　　　　　　　　　　　　　　　　　　　　　　　　　91-28920
　　　　　　　　　　　　　　　　　　　　　　　　　　　　CIP

PRINTED IN THE UNITED STATES OF AMERICA
92　93　94　95　96　97　　MB　　9 8 7 6 5 4 3 2 1

To Lys Ann

*There's beauty in the bellow of the blast,
there's a grandeur in the growling of the gale.*

Contents

Preface   xi

Acknowledgments   xv

CHAPTER 1
The Equations of Fluid Motion   1
   1.1 Introductory Remarks   1
   1.2 Equations of Motion   2
   1.3 The Virial Theorem   12
   1.4 Energy Conservation   16
   1.5 Some Thermodynamics   18
   1.6 Conservative Form of the Fluid Equations   23
   1.7 Eulerian and Lagrangian Frames   23
   1.8 The Bernoulli Equation — The First Look   24
   Appendix: Langmuir Waves as Perturbations of the Distribution
      Function   31
   References   33

CHAPTER 2
Viscosity and Diffusion   34
   2.1 Introduction   34
   2.2 The Navier–Stokes Equation   35
   2.3 Dissipation and Viscous Coupling   45

2.4 Boundary Layers   46
2.5 Diffusion and Kinetic Theory   50
References   56

CHAPTER 3
Vorticity and Rotation   57
3.1 Introduction   57
3.2 Geostrophic Approximation   73
3.3 Viscous Effects: A Simple Example   86
3.5 Self-Gravitating Bodies   87
References   93

CHAPTER 4
Shocks   95
4.1 Introduction to Shock Phenomena   95
4.2 Generation of Shock Waves   96
4.3 The Rankine–Hugoniot Conditions   102
4.4 Some Additional Complications   110
4.5 A Quodlibet of Applications of Shocks to Astrophysical Problems   122
Appendix A: Details of the Spiral Shock Picture   141
Appendix B: Bending of Jets by a Supersonic Cross-Flow   144
Appendix C: Collisions of Galaxies with an Intracluster Gas   144
References   145

CHAPTER 5
Similarity Methods   148
5.1 Introduction   148
5.2 Similarity Solution for Static Configurations: Polytropes   159
5.3 Classical Gravitational Collapse   166
Appendix: Diffusion Equations   169
References   170

CHAPTER 6
Magnetic Fields in Astrophysics   171
6.1 Historical Introduction   171

6.2 The Basic Equations   174
6.3 Alfvén Waves   181
6.4 Magnetic Equilibrium: A Basic Example — Pinch Equilibrium   185
6.5 Force-Free Fields   188
6.6 Vorticity Analogy and Magnetic Helicity   191
6.7 Magnetic Dynamos   194
6.8 Magnetic Reconnection   209
Appendix: Rikitake's Toy Phenomenological Model   211
References   212

CHAPTER 7
Turbulence   214
7.1 Introduction   214
7.2 Astrophysical Environments   215
7.3 Incompressible Turbulence   216
7.4 Kolmogorov Theory: The Role of Dissipation   224
7.5 Time Dependence of the Turbulence Spectrum   227
7.6 The Transition to Turbulence   235
7.7 Compressible Turbulence: The Lighthill Process   237
7.8 More Physical Complications   240
7.9 Some Observational Signatures of Astrophysical Turbulence   247
Appendix A: Stochastic Functions and Their Application to Turbulence   248
Appendix B: Stochastic Differential Equations   249
Appendix C: Fractals   250
Appendix D: Nonlinear Maps and the Transition to Chaos   253
References   255

CHAPTER 8
Outflows and Accretion   257
8.1 Introduction   257
8.2 Historical Overview of Winds, Especially from Stars   257
8.3 The Isothermal Wind Problem   260
8.4 Driving Stellar Mass Loss   268

- 8.5 Magnetic Winds   274
- 8.6 Winds within Winds   282
- 8.7 Accretion Disks in Astrophysics   289
- 8.8 Spherical Accretion   305
- References   314

CHAPTER 9
Instabilities   318
- 9.1 Introduction   318
- 9.2 Waves   318
- 9.3 Convection   324
- 9.4 Pulsation as an Instability   343
- 9.5 Thermal Instability   353
- 9.6 A Selection of Important Instabilities   357
- 9.7 Exeunt   371
- References   371

CHAPTER 10
Diagnosis of Astrophysical Flows   373
- 10.1 Introduction   373
- 10.2 Radiative Transfer on the Cheap   374
- 10.3 A Sampling of Cosmical Flows   396
- References   415

Problems and Questions for Further Exploration   417

Appendix: Some Real Numbers   423

General Bibliography   428

Index   445

Preface

For he who'd make his fellow creatures wise should always gild the philosophic pill!
W. S. Gilbert, *Yeomen of the Guard*

The universe is the most wonderful hydrodynamics laboratory imaginable. Every kind of flow we observe in terrestrial environments occurs in the cosmos on a truly grand scale. Supersonic motion, magnetohydrodynamic flows, turbulence of every variety, and instabilities of perplexing richness all present themselves to the alert eye. It is the purpose of this book to introduce these processes to astronomers in the hope that they will find the tools necessary for understanding both their observations and the related theoretical literature.

Over the past decade the dramatic improvements in detectors and the instruments to which they are mated have opened a vast range of new phenomena to the scrutiny of observations. Not only can we determine velocities and observe structure, we can also get true flux measurements for objects fainter than ever before accessible. The resolution of spectrographs has always been high for optical observations; but now spacecraft, especially the Goddard High Resolution Spectrograph (GHRS) on the Hubble Space Telescope, have made it possible to obtain resolutions of order a few kilometers per second for the entire satellite ultraviolet range between 1200 and 3000 Å. Millimeter detectors have dramatically improved in sensitivity, and with new interferometers as well as large single dishes now available, both galactic and extragalactic observations of molecular clouds are possible at resolutions of a few hundreds of meters per second. In short, we are passing into a heroic epoch, one in which many of the hydrodynamic phenomena of the universe will lie open to observation.

In light of these advances, it appears to me to be a serious deficiency in the curriculum that so few courses on astrophysical hydrodynamics are offered. It is clear that astronomers know how intimately their work is related to fluid dynamics, but they have not had many resources to turn to when looking for an introduction to the subject written in their language and reflecting their concerns. I hope that this book helps to fill this lacuna. It is intended for the student with a physics background, with the usual complement of mathematics courses of a senior undergraduate or graduate level. However, no prior experience, other than mundane observation of the world at large, has been assumed. The astronomy is introduced in context. The material can be taught in a single semester, and the instructor and student should find the references sufficient so that a second semester seminar course could include deeper discussion of some of the topics in the various chapters.

The topics treated here are a selection of those frequently encountered both in the literature and in observations. Chapter 1 examines the basic conservation equations. It includes a section on the Bernoulli equation as an example of the simplest kind of solution for a flow. The effects of viscosity and boundary layers are introduced in Chapter 2. Chapter 3 addresses vorticity and rotation, including the mathematical devices and physical concepts required for magnetohydrodynamics (MHD) and turbulence theory. Chapter 4 deals with shock phenomena. Chapter 5 treats similarity solutions. In order to connect such solutions with a more familiar problem normally discussed in courses on stellar interiors, the polytropic sphere is presented as an example of a scaling solution. Chapter 6 addresses MHD and magnetic field generation. It is meant to serve as a general overview of the subject and for this reason does not deal explicitly with dilute plasmas. Chapter 7 treats turbulence, including the effects of self-gravitation and supersonic flows. Chapter 8 is an introduction to outflows and accretion phenomena. This chapter is, frankly, an experiment. I am trying here to show that many of the phenomena usually treated separately are really quite similar. Specifically, it is the theme of this chapter that in many ways accretion is just a wind with the direction of flow reversed. Chapter 9 deals with a range of astrophysically interesting instabilities. These are chosen to be representative rather than complete and are also intended to serve as a guide to the techniques for assessing the nature of instabilities in both the linear and nonlinear regimes. Finally, Chapter 10 presents some examples of the observational diagnosis of astrophysical flows. It introduces most of the tools needed for simple radiative transfer (the Sobolev approximation) and contains examples of some recent observational results for stellar winds and novae. While specific references have been included in each chapter, the general bibliography of monographs and conferences at the end of the book is intended as a supplement. I have tried to include as much as reasonable of the standard fluid dynamics corpus, especially from the engineering literature, but I have heavily spiced the selection with works on specific astrophysical applications.

A word now about the problems that follow Chapter 10. They are loosely grouped in the order of chapters, but there is crossover between sections. I hope this will help integrate some of the ideas and also lead to further study of specific astrophysical problems. I have found that the best way to learn the material is to re-derive the results and apply them to real situations rather than to just solve artificially posed problems.

Preface

I first formed the desire to write this book several years ago while teaching a course in astrophysics at Case Western Reserve University. The students in that course were expecting a traditional discussion of stellar physics and got a version of these lectures instead. Part of the MHD chapter was used in lectures in a graduate course on plasma physics at New Mexico Institute of Mining and Technology. Parts of the instabilities chapter and the accretion disk section of the wind–accretion chapter were delivered as a series of lectures in seminar series at the Very Large Array. And finally, the bulk of the winds and accretion sections of the book formed the basis of two GHRS seminar series on stellar winds at the Laboratory for Astronomy and Solar Physics at Goddard Space Flight Center.

And so now having given you, the reader, a view of my motives for producing this book and some background on its pedigree, I ask you only to turn a page or two and join me in what I hope will be some happy hours of contemplation of astrophysical hydrodynamics.

Steve Shore

Acknowledgments

While I have, by Heaven, I'll give, so damn your economy.
Sheridan, *School for Scandal*

There are many students and colleagues I feel under a deep obligation to acknowledge here. First, I want to thank Pierre Encrenaz and Françoise Praderie for their continuing kindness and interest in inviting me for many years to spend extended periods at the Ecole Normale Supérieure (Paris) and the Observatoire de Meudon. My continuing association with the group DEMIRM at Meudon has brought many stimulating hours of discussion, and I want especially to thank Patrick Boissé, Xavier Désert, Christophe Dupraz, Edith Falgarone, Michel Pérault, and Jean-Loup Puget for their kindness and interest in different aspects of this work. Their comments have been invaluable, as has been their friendship. Alain Omont and Bernard Lazareff kindly invited me to stay for some months at the Groupe d'Astrophysique at Grenoble, and I thank them for their hospitality. Federico Ferrini arranged several stays at Pisa and in addition has been a valued collaborator and close friend for many years. He also provided numerous incisive critiques for most of the chapters in this book. Marcello Rodonó and Santo Catalano were kind enough to arrange a visiting appointment at Catania, where some of the lectures in this book were delivered as a graduate short course.

I want to especially thank Ted LaRosa, my former student and deeply valued friend, for his indispensable help. He has been a constant source of discussion and encouragement and has provided detailed criticisms and suggestions for all of the chapters in this book, often forcing considerable rewriting and clarification of obscure points. I also thank Bob Hjellming, Jean Eilek, and Aileen O'Donoghue for discussions during VLA seminars on topics related to material in this book. Doug Brown, George Sonneborn, Sumner Starrfield, Tom Ake, Bruce Altner, and Sally Heap have been the finest co-workers anyone could ever hope for; they have borne my distractions as a result of this book like true and understanding friends. Albert Petschek and Dave Raymond, two of the finest physicists I have ever known or worked with, have been friends, colleagues, and wise guides when I was getting lost in the deeper waters of the subject. I also thank Xavier Désert, Christophe Dupraz, Ivan Hubeny, Richard Robinson, Ron Polidan, and Eli Dwek for comments on various chapters, Keith Feggans for his generous advice and assistance with software development and systems, and Dara Norman for introducing me to *Superpaint*, the program with which I have drawn most of the illustrations in this book.

I reserve special thanks for Tom Bolton, my advisor many years ago at Toronto and a constant source of scientific insight and personal counsel, and Kevin Prendergast, whose inspiration sparked my love of hydrodynamics.

My thanks also to the staff at Academic Press, especially Marvin Yelles, who first encouraged me to submit a prospectus of this work and has been a valued guide to the publishing process. My thanks also to Dean Irey and Steven Martin for their skillful and sensitive editorial treatment of this work.

Three acknowledgments are painful to record, but I trust the reader will understand. One is to my father, Frederick Shore, who died in 1988, just as this book was getting started. From him I learned the meaning of education, of service, and of hard work. I deeply regret that he never had a chance to hold in his hands the work you are now reading. The second is to Johannes Hardorp, my first teacher and colleague in astrophysics, who died in 1988. He welcomed me as a novice into his work and his life, and it was from him that I learned what riches lay in the field. Finally, I record here my great debt to Nick Sanduleak, whose death in 1990 cut short a long collaboration and golden friendship. He was the gentlest person I have ever known and the one most responsible for making me understand the deepest motivations and methods of a true observer of nature.

The most important acknowledgment of all is saved for the dedication. This book would never have happened without the prodding, wisdom, skill, and unfailing encouragement of Lys Ann Shore. There is not a portion that she has not read and not a slip of writing that her keen eye has not made some comment on. Whatever flaws still remain in this work are not for her lack of trying to make it a good read and a coherent and logical argument. There is no way that I'll ever be able to record how much I owe to her, but I hope she, and you, the reader, will be pleased with the final product.

I have received support from a number of organizations while writing portions of this work. Visiting appointments have been supported by grants from the CNRS (France), the CNR (Italy), and the French and Italian ministries of education. Some support has come from NASA and from the GHRS Science Team.

The final draft of this book was written during a visiting appointment at the ENS in February 1991.

CHAPTER 1

The Equations of Fluid Motion

> *Upon the word, accoutered as I was, I plunged*
> *in and bade him follow. So indeed he did.*
> Shakespeare, *Julius Caesar*

We live in an incredibly dynamical universe. Any chance observation bears witness to the enormous diversity of motions and interactions that dominate the structure of matter in astrophysical environments. It is the aim of this book to put some of this in context. We see that gravitation, the ultimate structuring force on the scales typical in astrophysical problems, plays a key role in understanding how matter evolves. But we also must reckon with magnetic fields, turbulence, and many of the same processes that dominate laboratory studies.

1.1 Introductory Remarks

When you think of a fluid, the idea of a structureless deformable *continuous* medium comes to mind. Other properties, ingrained in childhood, are also likely to occur to you; things like incompressibility, viscosity (if you were really precocious), and perhaps even the statement that a fluid will not support shear and will yield freely in the presence of an applied force. So what have these to do with an astrophysical context? More than these definitions and properties might lead you to suspect. Of course, you will likely think, since stars are gaseous (something we've known since the 1930s), that we should be dealing with kinetic theory and gas dynamics. The processes of rarefied media are likely to dominate our understanding of cosmic bodies. This isn't quite true. Since a star is composed of gas which is homogeneous (in *most* cases) and which acts collectively to create its own gravitational field, it mimics rather well the behavior of a fluid moving (or sitting) under gravity. The collision times are so short (or, put another way, *the mean free paths are so short compared with any scale lengths in the medium*) in the interior of the star that any disturbances can be washed out and the structure can be described as continuous. Naturally, this is, for the moment, only an assertion. We shall prove it in due time.

The most crucial point is that stars and all other cosmic matter can be treated as an ensemble object or system *when we have carefully chosen some scales of length and time*. In a nutshell, the reason for this book is that we can always, at some magnification of scale or some rate of clock ticking, apply a fluid approximation to the problems at hand. This book is meant to provide the machinery, both computational and conceptual, with which to begin treating dynamical and static problems posed by *fluids* in a nonterrestrial environment.

1.2 Equations of Motion

1.2.1 Distribution Functions

To begin with, we shall treat the case of a homogeneous medium of identical particles. Forget for the moment that this may be *too* restrictive an assumption. Imagine that this group of particles is characterized by a *global* velocity distribution. Also, assume that we can know this distribution function and that the positional information can eventually be derived for the particles as well. Let us start with a gas that consists of a collection of myriads of particles, all of identical mass. If we assume that these particles execute collective motions, we will be able to take ensemble averages and treat them as if they were a continuous medium. This is what we mean by a fluid in an astrophysical context. But before we can reach the stage of describing the matter as a classical substance, we need to consider the microscale phenomena and how to incorporate them into a macroscopic description of the motion and thermal properties. To do this, we begin with a statistical mechanics treatment and then generalize from there.

Let us say that there exists a possibly time-dependent distribution function f, which is a function of $\mathbf{x}$, the particle positions, and $\mathbf{v}$, their velocities, and which provides a complete description of the chance of any single particle having a specific position and velocity at any time. We assume that the particle motions are individually governed by any forces imposed externally by the medium and also by any interactions that the matter may have within itself. By this we mean that the particles see one another individually through short-range interactions and also collectively through bulk or ensemble interactions. A good example of the former is the electrostatic interaction between charged particles in a plasma, while the latter is exemplified by the integrated gravitational field resulting from the distribution of the whole mass of the material. Both feed back into the distribution function, both alter the microscale properties, and therefore both internal and external forces must be considered if we are to calculate the physical attributes of the medium in the large.

Some constraints can be placed on the form of the distribution function right from the start. For one thing, it should depend only on the magnitude of the velocity, not on its direction. Another way of saying this is that it should be symmetric with respect to the spatial and velocity components, that is, $f(\mathbf{x}, \mathbf{v}) = f(-\mathbf{x}, -\mathbf{v})$. Since the distribution function is assumed to be a measure of the probability that a particle will have a specific position and velocity, f must be integrable and normalizable. It need not be algebraic; for example, a delta function, $\delta(\mathbf{x})\delta(\mathbf{v})$, is allowable. Now the hyperspace we are dealing with is well known. It is the *phase space* of the ensemble, that collection of individual momenta and positions familiar from classical mechanics. We can picture this collection as a group of free particles all passing through a box in which we have placed an observer. This observer has no idea where these buggers came from or where they are headed, but can at least describe them in the vicinity. They will arrive in this corner of the world, interact (perhaps), and then exit. The overall result is that a complete distribution function can be specified and, if this observer isn't too egocentric, this function can even be generalized to describe all of spacetime containing these particles.

A few things are then clear. The distribution function must be scalar and depend only on scalar quantities. That is, it cannot depend on the placement of the observer within the ensemble of particles. It must depend only on scalar quantities, although these may themselves be combinations of many physical properties. If the distribution function is to be global, it must be characterized by some global parameter which is a constant for the system under study. The assertion that the distribution does not change on inversion suggests that it cannot be a pseudoscalar. So f must be positive everywhere. Given the fact that the distribution function is defined in terms of a probability, we wouldn't know how to interpret *negative* values for f. But this property of probabilities is very important for our considerations that follow.

If every particle in a gas has a position and a velocity, we might ask what the mean value of any quantity connected with this distribution is. For instance, we may wish to know what the average velocity is, or what the average distance is that a particle may be away from the statistical center of the distribution. These are *moments* of the ensemble. If we cannot observe the motion of every constituent component of a body, if we cannot distinguish the histories of the individual particle trajectories, we can still say something about the most probable values that any of the measurable quantities will have.

Let's examine this physical picture in terms of simple probabilities, reducing the distribution function to only one independent quantity. We must be careful to choose a meaningful physical attribute. For instance, position means something in an extended medium. But color probably doesn't. Even if the particles have different colors, masses, or whatever, we can always ignore these attributes until some quantity that we happen to be interested in requires taking them into consideration. Take, for example, the position of a particle. If the probability of being some distance from a fixed point in space, $x_0 = 0$, is $P(x)$, then the mean value for the displacement is $\langle x \rangle = \int x P(x)\,dx$. Now assume that we have a one-dimensional distribution, but one that extends over the range $[-\infty, \infty]$. Since the probability of being on the negative side or the positive side of the reference point is assumed to be the same, the mean value for the position must vanish; that is, on average the particle will be at the reference point. Another way of saying this is that the integrand consists of a symmetric and an antisymmetric part and therefore vanishes over the whole space. But this clearly doesn't make sense if the ensemble is extended. There must be some other way of treating the fact that many of the particles, although perhaps equally distributed on the two axes, may not be concentrated at the nominal zero point. We require a quantity that does not vanish on integrating over the whole ensemble, x^2. This is a measure of the dispersion of the particles in space, and unlike $\langle x \rangle$, $\langle x^2 \rangle$ is finite. Now we have both a symmetric function and a symmetric interval and the mean value therefore does not vanish.

This has been a rather long digression. It is, however, prompted by the need to place the process of taking moments in context.

1.2.2 Moments of the Distribution Function

Of all the quantities that you can think of as characterizing this gas, the most obvious ones are functions of velocity and density. This is just a product of our Newtonian bias. We will separate the equations for the velocity into two components. One is the mean velocity, which we shall write as V_i, and the other is the random motion, which is assumed to have a mean of zero. This velocity we shall call u_i. All of the moments will be taken assuming that the distribution function is taken over the random

1.2 Equations of Motion

velocities *only*. For instance, there are three quantities around which classical descriptions in physics revolve: the *number density*, $n(\mathbf{x})$, the *momentum flux*, $n(\mathbf{x})\mathbf{V}(\mathbf{x})$, and the *energy density*, $\frac{1}{2}n(\mathbf{x})\mathbf{V}\cdot\mathbf{V}$. You will notice that each of these is a function of some power of the velocity although each depends only on space and time, not on the internal velocity distribution of the particles.

It is then not hard to see how to generalize this process to create as large a collection of *moments* as we'd like. Now you see why that long digression was necessary. The principal reason for taking the various moments is to remove the individual velocity components from the picture, to average over that portion of phase space, and therefore to obtain mean physical quantities that characterize the macroscopic spatial distribution of the matter. To do this within the limits of the function $f(\mathbf{x},\mathbf{v})$, we proceed as follows.

If we integrate over the entire volume of phase space, we *must* recover the total number of particles in the system. That is,

$$N = \int_{-\infty}^{\infty} d\mathbf{v} \int_{-\infty}^{\infty} d\mathbf{x}\, f(\mathbf{x},\mathbf{v}) = \int d\mathbf{x}\, n(\mathbf{x}) \tag{1}$$

so that if we are interested in keeping free the information about the spatial variations, that is, those depending on $\mathbf{x}$, we should restrict our integration to only the velocity components. We now have the prescription for taking moments! Assume that we have components $v_i(x,t) = V_i + u_i$ and that we are free to choose any such components for examination. The subscript is then a dummy, so that we can multiply these together as $fv_i v_j \ldots v_n$, which we can then integrate over the normalizable distribution function in velocity space.

The various moments are averages over the random velocity distribution function. Since we cannot measure this in detail, we get rid of it via integration (which is equivalent to averaging). Statistically, all of the macroscopic properties of the medium are the expectation values of the distribution and its moments. Historically, it was an important step forward when it was realized that the proper treatment of thermodynamics, namely the statistical approach rather than the vaguer mean-value methods of the mid-nineteenth century, could also be taken over into dynamics of media composed of individual randomly moving particles. It is no accident that the theory of statistics paralleled closely (and was in turn paralleled by) the development of statistical mechanics. It provided a natural arena in which to display the ideas.

It is now time to begin taking the moments of the distribution. For example, the density is clearly (from the previous discussion and definition) the 0th moment. The momentum flux is the first, the energy the second. Note that *the moment need not be a scalar*—only $f(\mathbf{x}, \mathbf{v})$ was required to satisfy this condition. We can now clarify why an ensemble can be treated as a fluid and what it has to do with astrophysics. The choice is not forced on us by any *a priori* principles; it is just that we grow up with the voice of Newton ringing in our ears and it tells us how to evolve *these* particular quantities. We know that thermodynamics will enter at some stage in the evolution of the system and that therefore, since the kinetic basis of this subject is well established, we should try to build our equations to match those of the statistical approach. The thermokinetic approach gives us some insight into the evolution of the function f, and if the equations for the fluid are to have meaningful macroscopic form, we must be able to combine the two approaches. In order to do this, we must first write down some differential equation for the evolution of the distribution function and then see if the successive moments will give us what we are hoping for—a full-scale dynamical system for the observables.

Rather than present a full-blown derivation of the evolution equation based on the ergodic principle, let's proceed more heuristically. In the end it will get us to the same place, and this way is more of a *royal road* approach. For once, it is appropriate. Once we have arrived at the fluid equations, the distribution function will no longer play an important role. But it is important that we go through these steps, because stars in galaxies can be treated as if they were a fluid just as a gas can be, and the derivation of the stellar hydrodynamic equations involves much the same logic. The simplest, almost schematic, equation we can write down for the evolution of the distribution function is

$$\frac{\partial f}{\partial t} + v_j \frac{\partial f}{\partial x_j} + \dot{v}_j \frac{\partial f}{\partial v_j} = \left(\frac{df}{dt}\right)_c \qquad (2)$$

where we have included the possibility of reshuffling effects into the collisional term, $(df/dt)_c$. This term is really just a symbol, holding the place of a more complex, and more complete, expression that we will not need to write down. It describes the cumulative effect of collisional redistribution of particles among the different coarse-grained bins of the distribution function. As such, it is essential to the understanding of the microphysics but is averaged out when we take moments and isn't important in the macroscopic problems we will face here.

This equation is known as the *Boltzmann equation* and is the central one for the evolution of the distribution function. The collisional term as first described by Boltzmann is an integral term, which makes this highly nonlinear equation (when all effects are included properly) an integro-differential equation *for the distribution function, not for the dynamical*

1.2 Equations of Motion

variables as such. The velocities and accelerations are for the peculiar velocities, not the mean observables, and therefore must be included in any integration over the system. The right-hand side of the equation is a roulette machine of sorts. It shuffles particles around in the phase space by collisions of different *varieties* (Boltzmann's phrasing) and is, in effect, the cause of the entropy of the system. Note that this is the point of the famous H-theorem.[1] The definition of H and its evolution equation is through the right-hand side of Eq. (2).

Although neglect of the collision term may seem rash, since in a fluid we are presented with an inherently chaotic collisionally dominated system, we can begin to make some limited progress. One way of looking at it is this. If the system has reached equilibrium, there will likely be as many collisions that transfer particles into as out of a particular volume of phase space. Thus the term on the right-hand side has no net effect. If we then restrict our attention to *equilibrium* states, we will not be far from the mark when we neglect this collisional term. A new name is now applied to the equation, the *Vlasov* equation. This is a small price to pay for gaining a tractable system of equations. Actually, it is even more than that: the equation is now *exact*. That is, our equation has the form

$$df(\mathbf{q}) = \frac{\partial f}{\partial q_i} dq_i$$

where we have assumed that the distribution function is dependent on the variables $\mathbf{q}$ and that the variables are summed over repeated indices. This is called the *Einstein* convention, and it will be employed throughout this book to economize the notation.[2] *If* the equation vanishes, then the variables $\mathbf{q}$ are not linearly independent, and so the equation can be called exact and has *characteristic solutions*. The Vlasov equation (as distinct from the full Boltzmann equation) satisfies this condition easily. The characteristics exist and are derived from

$$dt = \frac{dx_i}{v_i} = \frac{dv_i}{\dot{v}_i} \qquad (3)$$

---

[1] I must add that this is not the Latin letter H, but rather the Greek capital eta. The theorem was meant to characterize entropy. Once you know this, it may place the whole process in a broader context.

[2] This convention is as follows. For a scalar product, $\mathbf{u} \cdot \mathbf{v} = \Sigma_i u_i v_i \to u_i v_i$. For a matrix product, $\Sigma_j A_{ij} x_j \to A_{ij} x_j$. The summation, unless explicitly stated, is over repeated indices. The trace is $\text{Tr}(A) = A_{ii}$, while A_{jj}, no summation, indicates a diagonal component.

and it is this system which defines the constants of the motion—the constants on which the distribution function will depend. The function f is therefore given by

$$f(x_i, v_i; t) = f\left(x_i - \int v_i \, dt, \tfrac{1}{2}v_i v_i - \int \dot{v}_i \, dx_i, v_i - \int \dot{v}_i \, dt\right) \tag{4}$$

where the last term is trivial, and the first term just describes translations along the particle trajectory and is also trivial. The middle term is the most interesting, since we see that this is the energy of the system and the one term which is connected with the moments of the equation. We now show that there is an easy, and historically interesting, way of arriving at the precise form of this function if the characteristics are known.

A Quick and Dirty Derivation of f

We have just said that f is a function of the characteristics $x_i - v_i t$, $v_i^2 - \dot{v}_i x_i$, and the trivial characteristic $\dot{v}_i t - v_i$. Is there anything else that we might use to specify the *form* of the function? In fact, we can do this by thinking for a moment about phase space. The distribution function for any cell in this space can be called f_i. For the space, the distribution function should be given by

$$f(\mathbf{x}, \mathbf{v}) = \prod_{i=1}^{N} f_i(x_i, v_i) \tag{5}$$

The function must depend on velocity, must be spatially homogeneous, and must be a scalar in the velocity in each of the spaces. In addition, it must be normalizable. One additional point should be noted:

$$f(\Gamma_1 + \Gamma_2) = f(\Gamma_1)f(\Gamma_2) \tag{6}$$

that is, if we have a phase space composed of two subunits, the total distribution function should be formed from the product of the individual functions. The simplest function which satisfies all of these restrictions is *an exponential in the energy—that is, in the square of the velocity*! It is one of the characteristics and thus is a conserved quantity in phase space. Therefore, the simplest distribution function has the form

$$f \, d\Gamma = \prod_i e^{-v_i^2/2\sigma_i^2} \, d\Gamma_i \tag{7}$$

where $d\Gamma_i \sim dv_i \, dx_i$. In other words, we get the Maxwellian velocity distribution. This derivation is the one first presented by Maxwell in his discussion of the rings of Saturn in his Adams Prize essay; he based it on some ideas by John Herschel concerning the normal distribution function in probability. In fact, it is the right choice for a thermal distribution and,

1.2 Equations of Motion

since we neglect here effects of degeneracy, the right one for us as well. The Vlasov equation simply makes it clear why the distribution function has the particular dependences it does. Or put differently, through the Vlasov equation we find the trajectories along which the information about the distribution function will propagate through the phase space.

1.2.3 Continuity and Momentum Equations

So far, we have learned only that there are three constants of the motion (for each dimension), a fact we can read off Eq. (4). Now, we take the first moment by multiplying Eq. (2), without the collision term, by v_i to obtain

$$v_i \frac{Df}{Dt} + v_i \dot{v}_j \frac{\partial f}{\partial v_j} = 0 \qquad (8)$$

using the convective derivative, defined by

$$\frac{D}{Dt} = \frac{\partial}{\partial t} + v_j \frac{\partial}{\partial x_j} \qquad (9)$$

We use the definitions of the moments and integrate by parts (remember: *in phase space, the velocities and accelerations are independent variables, and f does not depend on the accelerations*). Thus, we obtain

$$\rho \frac{DV_i}{Dt} = F_i \qquad (10)$$

where now F_i is the force density. Amazingly enough, we now have regained the equation of motion. What would have happened if we had just gone ahead and taken the 0th moment? Try it! The result is

$$\frac{Dn}{Dt} + n \frac{\partial V_j}{\partial x_j} = 0 \qquad (11)$$

In other words, we get the *continuity equation*! The point should then be clear that the evolution equations are those for the conservation conditions on the macroscopic observables. How did we get rid of the troublesome term

$$\int_{-\infty}^{\infty} \frac{\partial}{\partial v_j} (f v_i v_j \cdots) d\mathbf{v}?$$

By simply recalling that the function f is defined as integrable and normalizable. Consequently, it vanishes at $+\infty$ and $-\infty$. Thus, the terms which are integrated *under the derivative operator* vanish symmetrically. The term obtained from the derivatives with respect to $\dot{v}_j$ is also easily dealt with. It can be integrated by parts:

$$v_i \frac{\partial f}{\partial v_j} = \frac{\partial}{\partial v_j}(fv_i) - f\delta_{ij} \qquad (12)$$

Here we have used the fact that in the coordinate system we have chosen, v_i and v_j are orthonormal. This greatly simplifies the calculation, since then terms like $F_j \delta_{ij} \to F_i$ replace the higher-order terms.

A word on procedure is now in order. We assume that the distribution function is a function only of the random components of the velocity field. Consequently, we can use *either* **u** or **v** as the integration variable. The mean velocity doesn't need to enter the calculation.

Let us step back for a moment and examine one of the terms in the derivation. We have taken

$$v_i v_j \frac{\partial f}{\partial x_j} = \frac{\partial}{\partial x_j}(fv_i v_j) - f\frac{\partial}{\partial x_j}(v_i v_j)$$

On integrating over the peculiar velocities, we see that certain terms in the expansion will fall out. To see this, notice the expansion of the first derivative term:

$$fv_i v_j = f(V_i V_j + V_i u_j + V_j u_i + u_i u_j)$$

so that, since the integral is taken over the random velocities, we have

$$\int v_i v_j \frac{\partial f}{\partial x_j} d\mathbf{v} = n \frac{\partial}{\partial x_j}(V_i V_j + \langle u_i u_j \rangle) \qquad (13)$$

since the mean of the random component is $\langle u_i \rangle = 0$. There is, of course, no mean for the divergence, hence the second term in the expansion vanishes. This is just an expanded version of what we had seen before for the derivation of the equation of motion, but it makes clear a very important point connected with the Boltzmann equation.

To write the equations of motion and continuity in a more compact component form, we look at the individual components. Now we will revert, for a moment, to lower case representation for the velocity components, since we'll be using this through the rest of the book. We have derived for motion for a given by a potential field Φ:

$$\frac{\partial v_i}{\partial t} + v_j \frac{\partial v_i}{\partial x_j} = -\frac{1}{\rho}\frac{\partial p}{\partial x_i} - \frac{\partial \Phi}{\partial x_i} \qquad (14)$$

Again, we sum over repeated indices. The continuity equation is

$$\frac{\partial \rho}{\partial t} + \frac{\partial}{\partial x_j}\rho v_j = 0 \qquad (15)$$

1.2 Equations of Motion

Combining the two, we obtain a compact form for the equations:

$$\frac{\partial \rho v_i}{\partial t} = -\frac{\partial T_{ij}}{\partial x_j} - \rho \frac{\partial \Phi}{\partial x_i} \qquad (16)$$

We have introduced the tensor T_{ij}, called the stress tensor. It is symmetric (since one of the variables is a dummy for summation and the other is a free choice). It is defined by

$$T_{ij} \equiv \rho v_i v_j + p\delta_{ij} \qquad (17)$$

This way of writing the equations is especially useful. The primary difference among the various treatments of hydrodynamic problems is geometry. The choice of a coordinate system can dramatically alter both the formalism and results for the equations of motion and the conservation conditions. By resorting to the most general representation, we can get around a lot of the specific differences and concentrate on the most general results. This is an especially convenient form for considering the forces on a body. For a steady flow (one for which the time derivative vanishes), the net force is given by the integral of the stress over the area of the body $F_i = \int T_{ij} dS_j$.

The stress tensor contains the Bernoulli term (see Section 1.8), which, for steady-state flow, must be divergenceless. It is then correct to say that it is the divergence of this term which drives the evolution of the flow in non-steady-state conditions. In the discussion of accretion disks (Chapter 8) we will have recourse to this equation in its most useful form. For now, we merely note that the equation written in this form makes the generalization to different coordinate systems especially easy to understand, since we have a "machine" which will allow us to take the coordinate transformations into account in a natural way.

We began with the 0th moment. Here, we simply integrated the Boltzmann–Vlasov equation directly and obtained a term which depended on velocity. In the first moment, we have obtained second-order terms in the velocity. Clearly, by induction, with every higher moment the nth moment will always contain a term which is the mean of $n + 1$ components in the peculiar velocity. This is a real problem—and one which is at the core of all proposed solutions to the Boltzmann equation—the so-called closure problem. We have seen that the means are of the square of the peculiar velocity in the case of the first moment. This is the same as the velocity dispersion squared, so that the random motion is assumed to have a nonvanishing *mean square*. Now, if you think about a Gaussian for a moment, you will realize that it is a function which is characterized by three properties: it has a mean value of zero, a nonzero dispersion, and integrates to unity. No wonder the Boltzmann equation in its "Vlasovian" incarnation winds up having a Gaussian velocity distribution as its most direct solution. It is the simplest that preserves the stochastic properties of the distribution function.

Before we complete the system of conservation laws with the addition of the energy equation, we can already look at the virial theorem and its consequences. It is interesting to note that this theorem was originally defined for bodies for which only the motions, and not the energies, were known.

1.3 The Virial Theorem

Let us now stop taking the moments, short of having in hand the full three conservation conditions, and ask: what happens if we take the first two *spatial* moments? After all, phase space is a $6n$-dimensional hyperspace—three spatial and three velocity components. If we can take the velocity moments, why not also examine what happens in space?

A more precise way of stating our aim is this. Suppose, now that we have been able to average over the velocity portion of the distribution function, we seek truly global properties of the body in question. Are there any *generalized* constraints that the *entire* distribution of matter must obey, regardless of the individual components and their particular situation in space? Several examples immediately come to mind. For instance, for an isolated body the total energy is constant, and the shape of the body and all of its other properties must be consistent with this fact. This isn't a spatially dependent statement, it is holistic. As we shall shortly observe, a number of such statements apply to mechanical systems, fluid or otherwise, and these will be important in understanding some of the constraints that structure astrophysical flows. In effect, given the Vlasov equation we can take simultaneous moments of the form

$$\int_\mathbf{x} \int_\mathbf{v} x_i \cdots x_n v_i \cdots v_n (L_{\text{Vlasov}} f(\mathbf{x},\mathbf{v}))\, d\mathbf{v}\, d\mathbf{x} = 0 \qquad (18)$$

where the integrals are taken over all possible spatial and velocity components and values, and L_{Vlasov} is the linear operator on f [Eq. (2)]. By so doing, we have a function that will depend only on time, the last variable over which we have not taken any averages. And even this situation can be altered if we perform time averaging. Since the first velocity moment gives us the equation of motion, the first spatial moment will yield important information about the energy.

To show this, we take the first spatial moment of Eq. (14) with respect

1.3 The Virial Theorem

to x_j:

$$x_j \frac{DV_i}{Dt} - x_j G_i = 0 \qquad (19)$$

where now G_i is a generalized force.

Integrating over all x, we can derive a tensor (or in the case of the ith moment a scalar) equation for the energy of the system. This is most easily seen by recalling that $v_i = \dot{x}_i$ and so integration by parts and using the definition of the mass in terms of the density will give an equation for the kinetic and potential energies and the moment of inertia. If we were to work strictly with the scalar form of the equation and neglect any of the explicitly time-dependent terms, we would wind up with a familiar equation:

$$\frac{1}{2} \frac{d^2 I}{dt^2} = 2T + W \qquad (20)$$

which is the *virial theorem*. It is a constraint on the equilibrium state of the *whole* system, a statement of how the energy is redistributed to different parts of the system if any changes are made. Notice that the theorem provides a means for connecting bulk changes in the distribution of matter in the system with variations in the kinetic and potential energies. Here T and W are given by integrals over the mass of the system (found from the definition of the density used in the moment equations and noting that $\int \rho d\mathbf{x}$ is the mass of the system, $T = \int (\frac{1}{2} \dot{x}_i \dot{x}_i) dm$ is the kinetic energy, $W = \int x_i G_i dm$ is the potential energy, and $I = \int x_i x_i dm$ is the moment of inertia.

We should pause here for a moment and examine the consequences of this theorem for equilibrium configurations. This is, without a doubt, a most central result for astrophysics and therefore deserves further reflection. If we take a system which is in a state of gravitational equilibrium, as defined by stating that W is the gravitational potential, then the kinetic energy is easily found. If this configuration is a spherical particle distribution (see the discussion of the Lane–Emden equation in the chapter on similarity solutions), we can derive a velocity dispersion for the system using Eq. (20). In astrophysical environments, however—for instance the stars or galaxies in a cluster—we often measure only the velocity dispersion and *not* directly the total mass. By judicious application of the virial theorem, we can derive a mass. It should be noted that this is the mass required for the system to have such and such a velocity dispersion *in equilibrium, that is, after relaxation*. Recall that we stated that the moments are taken relative to the collisionless Boltzmann equation and therefore assume that the distribution function satisfies that equation. Such distribution functions are in equilibrium and therefore completely related. Put differently: In order for the virial theorem result to yield a physically useful number, rather than merely a constraint, we must *assume* that the medium is already behaving like a fluid as we have defined it in this chapter.

We now look in more detail at the meaning of the term we have called W. In the case of the momentum equation, we were able to show that there exists a term which is the pressure, as long as the medium has a velocity dispersion (in other words, as long as the medium is hot, since the temperature is the measure of the dispersion). Now we can make use of this explicitly. We separate out two terms in W, that which depends on external fields, which we shall now call simply Ω (to keep in line with the standard notation), and the pressure. The momentum equation has a term of the form:

$$\int x_i \frac{\partial p}{\partial x_i} dV = \int \frac{\partial x_i p}{\partial x_i} dV - 3 \int p\, dV = \int p n_i\, dS_i - 3 \int p\, dV \qquad (21)$$

where we have obtained the factor of 3 by taking the divergence of x_i (prove this for yourself) and the reduction of the first term to a surface integral by $\int \nabla \cdot \mathbf{Q}\, dV = \int \mathbf{Q} \cdot d\mathbf{S}$. The surface integral in Eq. (21) is assumed to vanish in the absence of surface tension or other surface-related terms. Do not forget from here on that *stars have no walls*. Now we specify that the body forces come from the gravitational potential, the energy of which is identified as Ω. Therefore, we have

$$2T + \Omega - 3 \int p\, dV = 0 \qquad (22)$$

for the virial theorem with pressure included explicitly. You have seen this last term before—it is simply the thermodynamic representation of the work done by the system under compression or expansion. In the absence of ordered motion, then, the equation relates the action of external forces (or as we shall see later of gravity) to the work done by the system. If there is a thermal and ordered motion as well, then the virial equation is the full form just derived.

The total energy for a system undergoing no external compression (or in other words in equilibrium) is

$$E = T + \Omega \qquad (23)$$

and therefore the binding energy, which is always negative, is given by

$$E = +\tfrac{1}{2}\Omega \qquad (24)$$

In order for the system to be bound and stable in the presence of a potential, the total energy of the system must be negative. It can be shown

1.3 The Virial Theorem

(and will be later) that the second variation in the energy is $\delta^2 E < 0$ and that the first variation is $\delta E = 0$ in order for hydrostatic equilibrium to hold. These are the most basic equations of structure applied to fluid configurations which are gravitationally bound and also the source of the so-called *negative specific heat* behavior of self-gravitating systems.

1.3.1 Higher-Order Virial Equations

We have one more digression to get to before we return to the process of moment taking. We should briefly examine some generalizations of the virial theorem to higher-order terms. If we instead take the moments in a more general way, allowing for the time dependence and also for the off-diagonal terms (terms of the form $x_i x_j$), we get a so-called higher-order virial theorem of the form

$$\frac{1}{2} \frac{d^2}{dt^2} I_{ij} = 2 T_{ij} + W_{ij} \qquad (25)$$

It should be obvious how we have generalized the result. The only reason for adding this is that the virial theorem applies to any dynamical system, regardless of the number of dimensions or the inherent symmetry of the system. Thus it should be possible to take the moments with respect to off-diagonal elements of nonspherical or nonsymmetric configurations and look for the *normal modes* by taking all time dependences to be periodic and varying like $e^{\omega t}$. Then we get an eigenvalue equation:

$$\tfrac{1}{2}\omega^2 I_{ij} + 2 T_{ij} + W_{ij} = 0 \qquad (26)$$

We shall make some use of this much later, in the discussion of stability problems, but it is useful to show it here for a start, since it is an immediate and clever generalization of the usual mechanical virial. It should be added, by the way, that even more generalized virial theorems are possible if moments of the virial itself are taken (Chandrasekhar 1966, 1967). That is, in analogy with the velocity moments, we can take higher-order moments of the spatial components as well—nothing but timidity stops us.[3]

---

[3] The equations of motion in tensor form provide the simplest way of seeing how to generalize the virial theorem. If the stress tensor is T_{ij}, then

$$x_i \left(\frac{\partial \rho v_i}{\partial t} + \frac{\partial}{\partial x_j} T_{ij} \right) = -x_i \rho \frac{\partial \Phi}{\partial x_i}$$

Therefore, we have

$$\frac{\partial^2}{\partial t^2} \int x_i x_i \, dM - 2 \int v_i v_i \, dM + \int T_{ij} x_j n_i \, dS - \int \delta_{ij} T_{ij} \, d\mathbf{x} = \int \Phi \rho \, d\mathbf{x}$$

Here $dM = \rho \, d\mathbf{x}$ is the mass. This result can be generalized by taking the products with off-diagonal components (x_j). Now the surface term is also explicitly included so you can see the effects of boundary conditions on the virial theorem result.

1.4 Energy Conservation

Once we have completed the job of getting the equations for the dynamics of the fluid, we must come to grips with the fact that any fluid which is hot and/or radiating or conducting heat will be subject to a tendency to cool. This is a simplistic way of saying it, perhaps, but the point is that any hot matter in an open world will tend to a state of lower temperature and somehow this should be expressed within the context of the equations with which we have been dealing. Of course, it should be clear that if it can be derived from the Boltzmann–Vlasov equation, we should be able to do it using moments. In fact, this is also more easily seen by considering the following point.

In the introductory part of this chapter, we wrote down the grammatical equivalents for the various moments, namely the density, momentum, and energy, and also defined the fluxes accordingly. Now the time has come to make use of the third one of these—the second moment. Take the moment of the Vlasov equation in such a way as to form a scalar. That is, take the moment of L_{Vlasov} with $v^2 = v_i v_i$. Therefore, we have

$$v_i v_i \frac{\partial f}{\partial t} + v_i v_i v_j \frac{\partial f}{\partial x_j} + v_i v_i \dot{v}_j \frac{\partial f}{\partial v_j} = 0 \qquad (27)$$

The velocity is again composed of mean and random components. Remember that we are going to take the integrals only over the stochastic components. First, if we separate these components out, it will be easier to see what's going on. The product $v^2 = V^2 + 2u_i V_i + u^2$, where u is the thermal component. The advective term becomes

$$v^2 v_j \frac{\partial f}{\partial x_j} = \frac{\partial}{\partial x_j}(V^2 + 2u_i V_i)(V_j + u_j)f - f\frac{\partial}{\partial x_j}(v^2 v_j) \qquad (28)$$

so that only the first term remains. But on taking the integral, we see that only the correlated terms remain: $\langle u_i u_j \rangle = \langle u^2 \rangle \delta_{ij}$. That is, for the divergence we have only the isotropic terms remaining. This is because we assume chaos and that the random components u are uncorrelated unless they are in the same direction. The last term becomes

$$v^2 \dot{v}_j \frac{\partial f}{\partial v_j} = \dot{v}_j \frac{\partial}{\partial v_j}(V^2 + 2u_i V_i + u^2) - \dot{v}_i(V_i = u_i)f \qquad (29)$$

Only the mean velocity contributes to the last equation. Since the force is presumed to be velocity independent, the integral over the distribution function vanishes for the first term. Collecting the final integrated values,

1.4 Energy Conservation

we find that

$$\frac{\partial}{\partial t}\left(\rho \frac{1}{2}v^2\right) + \frac{\partial}{\partial x_j}\left(\rho v_j\left[\frac{1}{2}v^2 + \frac{p}{\rho}\right]\right) + F_j v_j = 0 \qquad (30)$$

Now notice that the divergence appears again, but this time involving the quantity $\mathbf{F}_{\text{heat}} = \rho \mathbf{v}(\frac{1}{2}v^2 + \mathscr{E})$. Here $\mathscr{E}$ represents the internal thermal energy. This quantity, the energy flux, has the characteristic dimensions of energy per unit area per unit of time and is the rate of transport of the energy out of the region in question. If the body is not isolated, there will be an additional term for the work, since g_j is the acceleration and therefore this will provide a work term for the bulk changes in the medium. We now perform the same tricks as before, taking the partial derivatives and integrating by parts, to obtain

$$\frac{\partial \rho E}{\partial t} + \frac{\partial}{\partial x_j}(F_{\text{heat},j}) = W \qquad (31)$$

where W is the work done by the system and consists of all of the thermodynamic functions, which we will discuss in the next section. In anticipation of the results for the thermodynamic variables, let's look at another form for the energy conservation equation. The kinetic energy density is $\frac{1}{2}\rho v^2$ and the internal energy density is ρE. The time derivative of the sum of these two energies, $d\rho\mathscr{E}/dt$, is

$$\rho\frac{d\mathscr{E}}{dt} + \mathscr{E}\frac{d\rho}{dt} = \rho\frac{d\mathscr{E}}{dt} - \rho\mathscr{E}\frac{\partial v_j}{\partial x_j} \qquad (32)$$

where we have used the continuity equation, $\dot{\rho} = -\rho\nabla\cdot\mathbf{v}$, for the last step. Now:

$$\frac{d}{dt}\left(\frac{1}{2}v^2 + E\right) = v_j\frac{dv_j}{dt}\frac{dE}{dt} = -\frac{1}{\rho}v_j\frac{\partial p}{\partial v_j} + T\frac{dS}{dt} - \frac{p}{\rho}\frac{\partial v_j}{\partial x_j} \qquad (33)$$

We have used (in anticipation of the next section) the relation that

$$\frac{dE}{dt} = T\frac{dS}{dt} + \frac{p}{\rho^2}\frac{d\rho}{dt} \qquad (34)$$

where T is the temperature and S is the entropy. Collecting terms we therefore have

$$\frac{\partial}{\partial t}\rho\left(\frac{1}{2}v^2 + E\right) + \frac{\partial}{\partial x_j}\rho v_j\left(\frac{1}{2}v^2 + E + \frac{p}{\rho}\right) = \frac{d\epsilon}{dt} - \rho T\frac{dS}{dt} = \Lambda \qquad (35)$$

where for compactness we have written $\rho\mathscr{E} = \epsilon$ as the total energy density. All of the loss terms are on the right-hand side, grouped into a term $\Lambda(T, \rho)$

that encapsulates all of the radiative, viscous, and dissipative processes that are not included in the Vlasov equation. If this term vanishes, we obtain the normal form of the heat equation. If the medium is at rest so that v vanishes, then the energy derivative is the change in the internal energy, $\dot{E}$, which is a function of temperature, the heat flux is given by $-\kappa \nabla T$, the conductivity, and we obtain

$$\frac{\partial T}{\partial t} = \frac{\partial}{\partial x_j} \kappa \frac{\partial T}{\partial x_j} \quad (36)$$

which is the normal form for the heat conduction equation, where κ is the heat conductivity (and we have not assumed that it is constant throughout the medium).

1.5· Some Thermodynamics

We now examine the thermodynamic variables in order to complete our picture of the fluid and its physical description. Let's start with the old but familiar first law of thermodynamics. It states that any work done on a thermal system produces a change in the internal energy and also in the heat. It also relates the entropy, S, and the state variables, namely the pressure, p, temperature, T, and volume, V. Put more formally, the first law states that the quasi-exact derivative of the heat function is given by

$$d\mathcal{Q} = T\,dS = dE + p\,dV \quad (37)$$

Although you have certainly seen this many times before, it is useful here to note a few inter-relations among the variables which may come in handy later—especially because we see that this equation is essential to the definition of the equation of state. Let's write this last equation as

$$d\mathcal{Q} = \frac{dE}{dT} dT + p\,dV$$

The change in the internal energy with respect to the temperature is the specific heat at constant volume—at least if the two extensive variables are T and V. We also know, from the equation of state for a perfect gas, that $pV = \mathcal{R}T/\mu$, where $\mathcal{R}$ is the gas constant and μ is the mean molecular weight. Therefore, we get $d\mathcal{Q} = (c_v + \mathcal{R}/\mu)\,dT - V\,dp$. From the fact that here we have taken the variation of the heat function with respect to T and p, we obtain

$$c_p = c_v + \mathcal{R}/\mu \quad (38)$$

and it therefore follows that $c_p/c_v > 1$. In fact, if we take a somewhat different form for the equation of state, in the case of $E = E(V, T)$ as a

1.5 Some Thermodynamics

more general formalism, we get

$$c_p - c_v = \left(\left(\frac{\partial E}{\partial V}\right)_T + p\right)\left(\frac{\partial V}{\partial T}\right)_p \tag{39}$$

We define a new variable, with which you will doubtless become quite familiar in the chapters ahead, that is, the ratio of specific heats $\gamma = c_p/c_v$. Notice that by Eq. (39), γ is always ≥ 1. In terms of the density, the first law is written as

$$dE = T\,dS + \frac{p}{\rho^2}\,d\rho \tag{40}$$

where the specific volume is $V = 1/\rho$. The enthalpy is given by $H = E + p/\rho$, which is also called the work function or the heat function given by

$$dH = T\,dS + \frac{1}{\rho}\,dp \tag{41}$$

Thus you see that the heat flux is really an enthalpy flux, and the energy conservation equation has a term $\nabla \cdot \rho \mathbf{v}(\tfrac{1}{2}v^2 + H)$ for the flux divergence term.

1.5.1 More Virial Theorem Results

Let's now return to the virial theorem, for it has other manifestations that are really quite dramatic. Having defined γ not only helps in keeping the notation compact but also yields some insight into how the equation of state and the global equilibrium conditions interface. The internal kinetic energy is

$$\mathcal{T} = \tfrac{3}{2}(c_p - c_v)T = \tfrac{3}{2}(\gamma - 1)c_v T \tag{42}$$

where we have replaced $\mathcal{R}/\mu$ by $c_p - c_v$. The internal energy is $U = c_v T$, so that

$$3(\gamma - 1)U + \Omega = 0 \tag{43}$$

Now since the total energy is given by $U + \Omega$, we see that

$$E = -(3\gamma - 4)U = \frac{3\gamma - 4}{3(\gamma - 1)}\Omega \tag{44}$$

The total energy of the configuration vanishes if $\gamma = \tfrac{4}{3}$. Keep this in mind, because shortly you will need it. In one very important respect the virial theorem leads to a most interesting astrophysical consequence. If a self-gravitating body contracts, the total energy increases. This is the *negative specific heat* problem that was mentioned earlier [see Saslaw (1985) for an especially complete discussion of this phenomenon]. You can think of this as the process that leads to gravitational collapse; that is, as a self-gravitating body cools it is forced to contract. This contraction heats it up, increases its energy losses (because of the increase in T the surface loss terms go up), and therefore the contraction rate increases to compensate for the losses. If the equation of state should be independent of temperature, as it is for a degenerate gas, the energy losses do not produce a contraction—the reason why white dwarf stars are stable and main sequence stars are not, for instance. Thus, our attention must be directed toward the equation that connects the thermodynamic variables, specifically the density and temperature, to the pressure, the equation of state.

1.5.2 Equation of State for a Polytrope

We now show that it is possible to write the equation of state in terms of the density variable and that we have a simple way of connecting the thermodynamics with the hydrodynamic equations. Take the case of an adiabatic medium by using

$$d\mathcal{Q} = c_v dT + \frac{\mathcal{R}T}{\mu V} dV \tag{45}$$

where the universal gas constant is $\mathcal{R}$ and μ is the mean molecular weight. If we set $d\mathcal{Q} = 0$ then

$$\frac{dT}{T} - (\gamma - 1)\frac{d\rho}{\rho} = 0, \quad \frac{dp}{p} + \frac{\gamma}{\gamma - 1}\frac{dT}{T} = 0, \quad \frac{dp}{p} - \gamma\frac{d\rho}{\rho} = 0 \tag{46}$$

Note that these relations are written with the density rather than the volume (they are reciprocals). Notice that we are now able to obtain a closed-form solution for the pressure–density relation, the equation of state, provided all changes in the medium are adiabatic. This isn't too stringent a restriction, however, because if the state of the gas doesn't change, then the equation of state won't either. Put another way, if the gas doesn't change its specific heats and behaves more or less ideally, then the equation of state will remain valid. Again, there are no radiative processes which make the system inherently open and thus preclude these definitions from applying throughout the medium. A simple form for the equation of state which will be useful in later work is

$$p = K\rho^\gamma \tag{47}$$

where K is usually called the *entropy constant*. An equation of state of this form is called *polytropic*. It is one of the most generally used forms for the

1.5 Some Thermodynamics

equation of state, being grounded firmly in the thermodynamic state of the medium. For a perfect gas, one composed of identical particles with only translational degrees of freedom and no additional correlations in the distribution function, the ratio of specific heats is $\frac{5}{3}$. This comes from the fact that the kinetic energy for each free motion is $\frac{1}{2}kT$; if internal degrees of freedom are available, such as rotation or vibration, γ is reduced. It should be noted that we have actually to deal with three possible adiabatic exponents. The basic polytropic equation is $p = K\rho^{\Gamma_1}$. The temperature equation is $\ln p = [\Gamma_2/(\Gamma_2 - 1)] \ln T$ and $\ln T = (\Gamma_3 - 1) \ln \rho$. For a perfect gas, the three exponents are equal, but in astrophysical fluids, which are especially prone to multiple ionization states and even the effects of degeneracy, these need not be identical. For an ideal or perfect gas, $p = (\mathcal{R}/\mu)\rho T$ and $c_v = \frac{5}{3}$; for an adiabatic perfect gas all of the adiabatic indices are equal.

Since it is important to be able to derive the pressure laws for different types of fluids, and we have already examined the case of an ideal gas, let's look at one of the other kinds frequently encountered in astrophysical flows, a radiation-dominated gas. Remember that one distinguishing feature of astrophysical problems is the importance of radiation: in some cases, the energy density in the radiation can actually dominate over that of the matter, producing a much different type of gas than you will have encountered in the laboratory.

1.5.3 The Sound Speed

Here is the appropriate place to introduce a new dynamical variable, the sound speed. It is the speed with which a pressure disturbance propagates through a medium with an internal density ρ and pressure P. Because the pressure depends on the internal velocity dispersion of the gas, the sound speed is directly related to the temperature for a perfect gas. It is defined by

$$a_s = \left(\frac{\partial p}{\partial \rho}\right)^{1/2} = \left(\frac{\gamma p}{\rho}\right)^{1/2} \tag{48}$$

for a polytropic equation of state. We will make much use of this quantity throughout this book, but will reserve the derivation of its connection with the speed of sound until the chapter on instabilities.

1.5.4 The Equation of State for a Photon Gas

We have written out the equation for the entropy in a general way. Now, let us look at what happens for the adiabatic case in which we have an isotropic pressure resulting from a photon gas. We see that the entropy is an exact derivative, so that

$$\frac{1}{T}\frac{\partial}{\partial V}\left(\frac{\partial E}{\partial T}\right)_V = \frac{\partial}{\partial T}\left(\frac{1}{T}\left[p + \left(\frac{\partial E}{\partial V}\right)_T\right]\right) \qquad (49)$$

We now assert that the pressure is given by $\frac{1}{3}\epsilon(T)$, where $\epsilon(T)$ is the energy density, and that $E(T, V) = \epsilon(T)V$. Therefore, we have for the energy density

$$\frac{1}{T}\frac{d\epsilon}{dT} = -\frac{4\epsilon}{3T^2} + \frac{4}{3T}\frac{d\epsilon}{dT} \rightarrow \epsilon(T) = aT^4 \qquad (50)$$

If we then take the adiabatic equation for a polytrope that was derived a moment ago, we see that a radiation-dominated gas is one which has $\gamma = \frac{4}{3}$. This is a very soft equation of state, in fact the softest that is permitted for stability of a self-gravitating fluid, and is due to the relativistic nature of the "particles" in the gas. Therefore, we have a simple way of connecting the equation of state with the thermodynamic state of the system. For astrophysical problems, this is most useful. We generally must deal with radiation-dominated fluids, or at any rate ones in which the radiation cannot be completely neglected.

1.5.5 Virial Theorem for Self-Gravitating Bodies

In light of the dynamical arguments we have just been through and the thermodynamic relations we have defined, let's collect a few relations that are of importance for cosmic bodies. The one common thread that makes these especially "astrophysical" is that they involve self-gravity of the object. For a body of constant mass, the internal structure is given by $dM(r)/dr = 4\pi r^2 \rho(r)$. The gravitational self-energy of a mass M of radius R is

$$E_{\text{grav}} = -G\int_0^M \frac{M(r)\,dM(r)}{r} \qquad (51)$$

This becomes $-\frac{3}{5}GM^2/R$ for a uniform-density sphere. The total thermal energy content is

$$E_{\text{thermal}} = \int \frac{3}{2}kT(r)\,dM(r) \qquad (52)$$

for a perfect gas. And the energy evolution equation becomes

$$\frac{d}{dt}\left(\frac{3}{2}kT\right) - \frac{p}{\rho^2}\frac{d\rho}{dt} + \epsilon - \frac{1}{4\pi r^2 \rho}\frac{dL(r)}{dr} \qquad (53)$$

where $\dot{\epsilon}$ is the rate of energy generation and $L(r)$ is the radiative loss, also called the luminosity. Equation (53) is also the basic equation for energy generation in a stellar interior.

1.6 Conservative Form of the Fluid Equations

For a perfect gas, we can collect the fluid equations in vector form. The continuity equation is written as

$$\frac{\partial \rho}{\partial t} + \nabla \cdot \rho \mathbf{v} = 0 \tag{54}$$

The momentum conservation equations become

$$\frac{\partial \rho \mathbf{v}}{\partial t} + \nabla \cdot (p\mathbf{I} + \rho \mathbf{vv}) = -\rho \nabla \Phi + \eta \nabla^2 \mathbf{v} \tag{55}$$

where $\mathbf{I}$ is the unit matrix. The (scalar) energy conservation equation becomes

$$\frac{\partial \epsilon}{\partial t} + \nabla \cdot \rho \mathbf{v} \left(\frac{1}{2} v^2 + \frac{\gamma}{\gamma - 1} \frac{p}{\rho} + \Phi \right) = \rho \mathcal{L}(\rho, T) \tag{56}$$

where $\epsilon = \rho(\frac{1}{2}v^2 + E)$ is the total energy density, E is the internal energy, and $\mathcal{L}$ is a volumetric energy loss rate. The energy loss term takes into account all radiative and viscous loss terms (see Chapter 2 for the derivation of the Navier–Stokes term, which does not immediately follow from the Vlasov equation because we explicitly ignore the collisional term from the Boltzmann equation). These forms of the conservation equations are especially advantageous for our purposes because they show immediately the conditions that must be conserved within a flow. In addition, they have been shown to be convenient for numerical calculations.

1.7 Eulerian and Lagrangian Frames

This will be a short section. It assumes some experience on your part. Imagine that you are reading a map for someone who is driving and providing the directions for the route. Here individual style is important. Some, on giving directions, will say where to turn and where to go straight, giving distances between reference points and then locations of the critical places, as in "go 3 km north on Blvd. St. Michel and turn left at Blvd. St. Germain." Others would say, "continue for 3 minutes until Blvd. St. Germain comes up and turn left." This is essentially the difference between Eulerian and Lagrangian coordinates. One is in the frame of the external world, the other in the frame of the moving vehicle. Both are correct but, depending on the situation, both are not equally appropriate.

For a Lagrangian frame, the equations of motion take Newtonian form. That is, in this frame, because it is co-moving with the fluid, the time derivatives are ordinary and the advective term is absorbed. The coordinate **x** is a function of time, and velocities of external objects are referred to the frame of motion. In order to translate back into the stationary frame, a coordinate transformation has to be applied. Thus the Lagrangian forms for the fluid equations are

$$\frac{d\rho}{dt} + \rho \nabla \cdot \mathbf{v} = 0, \qquad \rho \frac{d\mathbf{v}}{dt} + \nabla p + \rho \nabla \Phi = 0 \qquad (57)$$

In many astrophysical problems, the equations are solved instead in a frame depending on the mass. That is, you choose a convenient mass fraction (if the total mass remains constant) and, instead of solving for the mass at a specific radius, you solve for the radius at which that mass is located. The difference is that in time-dependent problems (as we will see in Chapter 9) this often provides a more useful frame of reference.

1.8 The Bernoulli Equation—The First Look

Now that we have defined both the method and the details of the process for a single equation of state, one in which we have assumed that there is only a perfect gas, it should also be clear that this can be generalized to include the cases in which radiation pressure, or even degeneracy, might be important. However, these are beyond the scope of this discussion and more properly dealt with in a course on the physics of stellar interiors. For completeness, however, a discussion of degeneracy is included in the appendix to this chapter. We shall later have recourse to the case in which the medium is radiation dominated. It should be noted, however, that even in this case it is possible to write down the equation of state in this form. It is useful to note that for a perfect gas, for which there are three degrees of freedom for the particles, the value of $\gamma = \frac{5}{3}$. For any gas in which there are more degrees of freedom (like a diatomic or polyatomic gas, in which internal states are available), the value of γ will decrease. Any ionization will also decrease its value. In Chapter 9 we shall discuss how this relates the thermal pool to the stability of a region.

1.8 The Bernoulli Equation—The First Look

Since we now have in hand an equation for the pressure as a function of the density, we can write down an additional description of a flow. Take the simple case of an adiabatic compressible fluid. The fluid is assumed to have no internal vorticity. This is because the advection term is

$$\mathbf{v} \cdot \nabla \mathbf{v} = -(\nabla \times \mathbf{v}) \times \mathbf{v} + \tfrac{1}{2}\nabla v^2 \tag{58}$$

and we assume that $\nabla \times \mathbf{v}$, the *vorticity*, vanishes (see the chapter on rotation and vorticity). This also means that $\mathbf{v} = \nabla \phi$, where ϕ is called the velocity potential (since the curl of this quantity vanishes identically everywhere). Such motion is said to define a streamline. Then by using Eq. (14) we get for the equation of motion

$$v_j \frac{\partial v_i}{\partial x_j} + \gamma \rho^{\gamma-1} \frac{\partial \rho}{\partial x_i} = 0 \tag{59}$$

if there are no other body forces acting on the fluid. For a one-dimensional flow this becomes

$$\frac{1}{2} v^2 + \frac{\gamma}{\gamma-1} \frac{p}{\rho} = \text{constant} \tag{60}$$

which is the *definition* of Bernoulli flow. Named in honor of D. Bernoulli, this equation is one of the most basic tools for the analysis of flows. It states that if the internal velocity of the flow goes up, the internal pressure must drop. Therefore, the flow becomes more constricted if the velocity field within it increases. Another consequence of this equation is that

$$\frac{\partial \phi}{\partial t} + \frac{1}{2}(\nabla \phi)^2 + \frac{\gamma}{\gamma-1} \frac{p}{\rho} = \text{constant} \tag{61}$$

which is the equation for a streamline ϕ. This particular parametrization of the velocity in terms of a potential field is especially useful for inviscid flows, because the streamlines define the motion of the fluid and permit the use of analytical techniques (like complex analysis and conformal representations) for solving the flow equations. We will not deal with these methods, however [but see Batchelor (1967), in the General Bibliography, for a very thorough discussion of their applications].

Instead, let's look at some of the implications of the Bernoulli equation. Think about the problem of liquid flowing down the side of a pitcher, and the fact that it will always stick to the walls, and you will have a good feeling (albeit a wet one) for what is implied by this equation. In turn, it is the equation which is determined along a streamline to be a universal—it defines the characteristics of the flow and must be conserved along any streamline, by definition. Therefore, in the case of any fluid within a pressure-bounding medium, the flow is confined by the internal velocity and pressure against expansion into the background gas, and this equation also serves as the boundary condition for the flow. The enthalpy is defined by

$$H = c_p T + \frac{1}{2} v^2 = \frac{\gamma}{\gamma - 1} \frac{p}{\rho} + \frac{1}{2} v^2 \tag{62}$$

which is essentially the same as the integration of the stress tensor for a one-dimensional flow. We can change the form of the equation for the one-dimensional system to

$$\frac{d\rho}{dv} = -\frac{\rho v}{a_s^2} \tag{63}$$

Here we have represented the internal energy by $a_s^2/(\gamma - 1)$; a_s is the sound speed. With this form of the equation, we obtain

$$\frac{dJ}{dv} = \rho \left(1 - \frac{v^2}{a_s^2} \right) \tag{64}$$

for a simple flow for which $J = \rho v$. This equation gives the mass flux as a function of velocity. Clearly this function has a maximum at the sonic point and decreases for higher velocities. Let us assume that $J = J_\star$ at this point. We also assume that there is some point in the flow at which the velocity is zero. We can call this point x_0. Here we have for the flow an entropy S_0 and an enthalpy at x_0 of $H_0 = E_0$. The internal energy is given by $w \to c_p T = a_s^2/(\gamma - 1)$. This results from the assumption of a perfect gas and the relation $c_p - c_v = \mathcal{R}/\mu$. Thus, for adiabatic flow, we obtain $S_\star = S_0$, and $E_\star + \frac{1}{2} v_\star^2 = E_0$, and that the velocity at the maximum is the *local* sound speed $a_{s,\star}$. The maximum velocity for the flow is therefore

$$v_\star = a_{s,0} \left(\frac{2}{\gamma - 1} \right)^{1/2} \tag{65}$$

and we find that the sound speed at the maximum is

$$\frac{a_{s,\star}^2}{\gamma - 1} + \frac{1}{2} a_{s,\star}^2 = \frac{a_{s,0}^2}{\gamma - 1} \to a_{s,\star} = a_{s,0}^2 \left(\frac{2}{\gamma + 1} \right)^{1/2} \tag{66}$$

From this, we can obtain all of the thermodynamic variables as a function of velocity and therefore as a function of position in the flow. This is the result for an accelerating flow which transits the sonic point. It is a flow for which the cross section is not included. We now look at what happens if we include the effect of an aperture of variable size through which the flow is directed.

1.8.1 The de Laval Nozzle: Bernoulli Flow with Confinement

Suppose we have, in addition to the Bernoulli equation, the continuity equation in a steady-state approximation (that is, the solution to the equation without time dependence):

$$C = \rho v \cdot \Sigma \tag{67}$$

where Σ is the area through which the flow is moving. We then take the area to be a function of position in the flow such that, for a one-dimensional flow, we have $\Sigma = f(x)$. Now we have for a compressible fluid the equation

$$\rho v \frac{dv}{dx} = -\frac{dp}{dx} \tag{68}$$

in the absence of external forces. We make the approximation for the equation of state that the velocity dispersion is a constant and that therefore we can write $P = \rho a_s^2$ where a_s is the sound speed. For now assume that it is a constant. Then, by the equation of continuity we have

$$v \frac{dv}{dx} = a_s^2 \left(\frac{d \ln v}{dx} + \frac{d \ln \Sigma}{dx} \right) \tag{69}$$

We therefore see that there is a solution of the form

$$\tfrac{1}{2} v^2 - a_s^2 \ln v - a_s^2 \ln \Sigma = \text{constant} \tag{70}$$

so that as the flow is constricted into a narrower aperture it must move faster. This solution for the fluid flow has some remarkable consequences. In the case of a polytropic gas, where the density is a power law, we get

$$v \frac{dv}{dx} = -\frac{\gamma}{\gamma - 1} \frac{d p}{dx} \tag{71}$$

We assume that the ratio of specific heats remains constant throughout the flow. An alternative form for the equation of motion is

$$v \frac{dv}{dx} = -a_s^2 \frac{d \ln \rho}{dx} - \frac{da_s^2}{dx}$$

so that we can remove the density by using the continuity equation:

$$(v^2 - a_s^2) \frac{1}{v} \frac{dv}{dx} = -\frac{da_s^2}{dx} + a_s^2 \frac{d \ln \Sigma}{dx} \tag{72}$$

A remarkable result has just been obtained. We can relate the gradient of the sound speed to the velocity gradient via the gradient in the cross section. In fact, we have the result that when the sound speed is reached in the flow, the equation for the sound speed gives

$$v = a_s \rightarrow 2\left(\frac{d \ln a_s}{dx}\right)_\star = \left(\frac{d \ln \Sigma}{dx}\right)_\star \qquad (73)$$

We will have many recourses to the Bernoulli equation in the discussions that follow. This introduction has been meant only to whet your appetite.

1.8.2 Jets: Introduction

Accretion disks are fine examples of astrophysics in two dimensions. They impose axial symmetry on their environment and introduce angular momentum into any material that happens to escape from their surface. It may not therefore be surprising that jets, axisymmetric and essentially one-dimensional flows, are one of the most ubiquitous phenomena in astrophysics. They are one-dimensional because the momentum is carried primarily *along* the axis of the flow. And they are axisymmetric because the azimuthal coordinate does not enter into the structure—well, it is not entirely unimportant. Because a jet is free to move in three dimensions, even if it doesn't in the unperturbed state, there are free modes along the surface and within the body of the jet that make the azimuthal coordinate very important. The engineering literature has contained work on jets since the first decades of the century. There was considerable interest in this means of propulsion, first in naval architecture and later in aerodynamics, as an alternative to propellers. Several reasons seem to have dominated. One was the problem of stability. Another was the efficiency of the momentum transfer possible with jet propulsion.

The astrophysical importance of jetlike flows was also recognized quite early. Curtis (1918)[4] described the *visual* observations of M87, the central elliptical galaxy in the Virgo cluster, as a jet. This was elaborated by Minkowski and Baade in the 1940s. The galaxy associated with the quasar 3C 273 also has an optical jet that was noted in the first observations in the early 1960s. Plume and jetlike phenomena were invoked to explain doublelobed radio sources, the first important attempt being by Blandford and Rees (1974). Much elaboration of the model, though, was spurred by a remarkable discovery: SS 433. This source was first observed by the Ariel V and SAS-3 x-ray satellites and quickly identified as a radio source. Its optical identification soon provoked astonishment (I don't use that term lightly). Optical spectra showed that the lines of the Balmer series and

---

[4] The references to jets are so numerous, and occur in so many places throughout this book, that they are included as a separate section in the general bibliography and will not be repeated at the ends of the chapters.

1.8 The Bernoulli Equation—The First Look

of neutral helium display time-variable velocities with an amplitude of $0.26c$. The lines are narrow, never show P Cygni structure, and while variable otherwise act as a reliable clock (Margon 1984). Interferometric radio observations are also completely consistent with a precessing jet that occasionally belches out blobs (Hjellming and Johnston 1986). The discovery of bipolar outflows from some planetary nebulae and observations of jetlike structures connected with sites of active star formation have made jets and their related phenomenology among the most familiar of all hydrodynamic phenomena in astronomy. There is too much written about the phenomena of radio jets to go into here, and this is generally specialized to radio observations of near- and ultrarelativistic flows for nonneutral plasmas. There is enough to do just to understand ordinary sluggish Newtonian flows. That is the purpose of this section.

For the purposes of our discussion, a jet will be defined as a confined fluid structure, bounded by stable fluid and along which there is an essentially one-dimensional flow. This is an important definition to keep in mind because observationally, by definition, a collimated structure is called a jet. A lot often gets assumed. First, that there is a flow. This can be observed directly in one class of sources, namely bipolar flows and emission line jets seen in active star-forming regions of molecular clouds, and less directly in another, specifically SS 433. But in the more spectacular cases, radio galaxies, the direct evidence is lacking and more indirect arguments and analogies play the dominant role. There will not be much discussion of large-scale radio jets here, however. Observations point to the bulk motion in these jets as being relativistic. Because such problems would require more expansion than we have space available for, all but the most schematic properties of these will be ignored.

1.8.3 Basic Physics

An expanding jet is a fully three-dimensional object, at least in this world.[5] Jets can be treated as either, but the consequences of a two-dimensional treatment are quite severe when comparing the calculated structure to real astrophysical objects. A two-dimensional jet is generated by many different physical conditions. For instance, a fluid issuing from a rectangular slit will enter the surrounding medium in a plane. It is an axisymmetric flow in the sense that it has a central axis with the maximum

---

[5] George Abbott, whose book *Flatland* has been the source of much inspiration to generations of geometers and physicists, would certainly have understood the difference between saying that a jet is imbedded in a three-dimensional world and saying that it is a three-dimensional object.

velocity and a finite momentum. A fluid leaving a wall-confined region also bears many of the same characteristics.

1.8.4 Subsonic Jets

Laminar jets present a classical laboratory example of the basic physics of axisymmetric unidirectional flows. For this reason, let's look at how the analysis of such objects proceeds and then generalize to more astrophysically interesting problems.

A two-dimensional jet is generated when a flow emerges from a slit or when it leaves the edge of a wall or plate. It also resembles a wake in many of its characteristics. Although such a jet is not a very good approximation of an astrophysical entity, it provides an example of closed-form solutions and serves to illustrate the basic physical problem. Also, in a sense, the axial jet is also two-dimensional, but the differences are very important between planar and cylindrical (or solid-angle) jets. Before generalizing to a truly axial jet, let's take a look at the planar one. Call the velocity component parallel to the x direction along the jet u. The momentum per unit volume is ρu carried by a symmetric jet (one symmetric on reflection about the midplane) and the rate of transport of momentum across a surface is defined by $\mathcal{M} = \int \rho v^2 \, dS$. For a two-dimensional jet, dS is replaced by dy, the line transverse to the jet. For an incompressible fluid $\nabla \cdot \mathbf{v} = 0$. The transport of momentum is constant along the direction of propagation. To see this take

$$\frac{d\mathcal{M}}{dx} = 2\rho \int_{-\infty}^{\infty} u \frac{\partial u}{\partial x} dy \tag{74}$$

and substitute the continuity equation. Since $\partial u/\partial y$ and v both change sign across the jet axis, the integral vanishes. Therefore $d\mathcal{M}/dx = 0$. The next assumption we can make is that the jet has a simple velocity law. The spreading away from the axis is assumed to be self-similar and to depend only on y/δ, where δ is the width; thus $u(x, y) = u_0(x) f(y/\delta)$. Next, and this is merely for simplicity, we take the axial speed u_0 to vary as x^a and δ to vary as x^b. From the equation of motion, this implies that $U^2 x^{-1} \sim \nu U \delta^{-2}$ so that $a - 1 + 2b = 0$. Then from the definition of $\mathcal{M}$, it follows that $2a + b = 0$. This gives a scaling law of the form $a = -\frac{1}{3}$ and $b = \frac{2}{3}$ so the opening angle of the jet is constant with a half-angle of approximately 33°.

The jet expands by plowing up material ahead of it, some of which is "absorbed" and some of which is deflected. Its surface exerts a force on the medium $\rho v_j^2 A$, where A is the surface area of the head. The medium in turn is acting to slow it down, so the velocity of the head into the external medium simply scales as $v_h = (\rho_j/\rho_0)^{1/2} v_j$. It is very important to note, though, that this estimates only the rate of momentum transfer and neglects some very important dynamical effects.

We see that a jet carries momentum of a magnitude $\dot{M}u_1$ and a bulk kinetic energy $\frac{1}{2}\dot{M}(u_1^2 + \sigma^2)$, where σ is the random component estimated by the spectral line widths or by random motions observed directly within the jet. It has a force of $\rho_j u_1^2 A_j$ on the background gas and stagnates when the background pressure, $\rho_0 a_{s,0}^2$ is of the same order. Therefore, to ensure that the jet not stall, the Mach number for the jet must be approximately $M_j \gtrsim (\rho_0/\rho_j)^{1/2}$. Notice that for an overdense jet any supersonic flow, and even subsonic flows, will not be stopped simply by ram pressure. But for an underdense jet (a so-called *light jet*) the flow must be supersonic in order to ensure continued propagation.

Appendix: Langmuir Waves as Perturbations of the Distribution Function

We will not deal very much with particle kinetics or plasmas in this book. But it is impossible to pass up the opportunity to illustrate an example of how the distribution function can be used, with moment methods and the Vlasov equation so fresh in your mind. The other reason for this appendix is that it is the best way to introduce an important quantity, the plasma frequency, and also to illustrate how perturbations on the distribution function result in dispersion relations. All of these ideas will be important later on. The best arena for studying the uses of the Vlasov equation is in the context of dilute gases, in which the individual particle kinetics are important. Recall that the point of the Vlasov equation is that it permits the study of what collections of weakly interacting particles will do in the presence of background fields.

Consider the Vlasov equation including the effects of charged particles. Here the force comes from the generated E field. We must supplement the equation for the distribution function by using a field equation, in this case the Poisson equation:

$$\nabla \cdot \mathbf{E} = 4\pi\rho \tag{75}$$

where ρ is the charge density. This is the *net* charge, that is, the excess of the charge from a zero value. We can consider that the field gives rise to a charge separation and that if the plasma initially has some velocity distribution function, which is the same for the ions and electrons, the application of a background field will produce a change in the two velocity distributions so that the electrons will be "heated" relative to the ions.

In this way, the electrons understand that they are living in the same neighborhood, and their response to the imposed field is collective. It is through this response, governed by the Poisson equation, that the distribution function evolves. Otherwise the particles would respond individually and we would simply have a collection of collisionless orbits. But here, because once the distribution of velocities is changed in the gas it also feeds back into the long-range forces and the accelerations, there is a mechanism for changing the distribution function of the entire plasma. However, the electrons are not free to move, and they must drag the ions along as they are accelerated. This is because the plasma, unless there are constraints which prevent the free motion of the ions, will try to cancel the applied field and also to heal the rift between the electrons and the ions.

Since we have $\rho = e(n_i - n_e)$, and the two densities are moments of the respective distribution functions, we have

$$\frac{\partial}{\partial t}f_s + u\frac{\partial}{\partial x}f_s + \frac{q_s}{m_s}E\frac{\partial}{\partial u}f_s = 0 \tag{76}$$

assuming that we have two species (s being i for the ions and e for the electrons). Further, $q_i = e$ and $q_e = -e$. We assume that we have a one-dimensional electric field placed across the medium, but that the distribution function is fully three-dimensional. Now assume that n_0 is the density of the undisturbed medium, so that

$$f_s(\mathbf{x}, \mathbf{v}, t) = f_0(\mathbf{v}) + f_{1s}(\mathbf{x}, \mathbf{v}, t) \tag{77}$$

and that f_0 is the unperturbed distribution function. Further, assume that the electric field is a small perturbation. Perhaps it is a wave incident on the medium, or somebody turning on the potential across the plasma between two plates. Assume finally that all spatial and temporal terms are given by the Fourier amplitudes and that they vary with space as k and with time as ω where these are the wave number and frequency, respectively.

$$ikE(\omega) = 4\pi e n_0 \int_{-\infty}^{\infty} du\, dv\, dw (\phi_{1i} - \phi_{1e}) \tag{78}$$

Here we have assumed that

$$\phi_{1s} = \int f_{1s}(\mathbf{x}, \mathbf{v}, t) e^{i\omega t - kx}\, dt\, dx \tag{79}$$

The transform of the Vlasov equation becomes

$$(-i\omega + iku)\phi_{1s} + \frac{q_s}{m_s}E(\omega)\frac{df_{0s}}{du} = 0 \tag{80}$$

so that we can solve for ϕ_{1s} to obtain

$$\phi_{1s} = i\frac{q_s/m_s}{\omega/k - u}\frac{df_{0s}}{du} \tag{81}$$

Substituting this into the Poisson equation, we derive a simple relation for the *dispersion relation* for a wave in this plasma, strictly in terms of the kinetic distribution function:

$$-\frac{4\pi n_0 e^2}{k^2}\int d\mathbf{v}\,\frac{dg_0}{du}\left(\frac{1}{\omega/k - u}\right) = 1 \tag{82}$$

where we have abbreviated $f_{0i}/m_i - f_{0e}/m_e$ with $g(u)$. Since nature is constructed in such a way (unless we deal with a positron–electron plasma) that $m_e \ll m_i$, we can assume that the electrons dominate and therefore neglect the term involving f_{0i}.

We now define the plasma frequency, a characteristic frequency for the oscillation of the separated charges, as

$$\omega_p^2 = \frac{4\pi e^2 n_0}{m_e} \tag{83}$$

so that we can rewrite Eq. (82) as

$$1 + \frac{\omega_p^2}{k^2}\int d\mathbf{v}\,\frac{df_{0e}(u)/du}{\omega/k - u} = 0 \tag{84}$$

In order to avoid the pole at $\omega/k = u$, we assume that the phase velocity of the waves is large compared with the velocity of the electrons. First we integrate by parts to remove the derivative of the distribution function from under the integral in Eq. (83). Now expand the denominator as a power series:

$$\frac{1}{(\omega/k - u)^2} \approx \left(\frac{k}{\omega}\right)^2 \left(1 + 2\frac{ku}{\omega} + 3\left(\frac{ku}{\omega}\right)^2 + \cdots\right) \tag{85}$$

The integral of the even powers does not vanish, because of the symmetry of f_{0e}, so that, assuming that f_{0e} is normalized and that the second moment of the distribution function is σ_e, we get

$$1 - \frac{\omega_p^2}{\omega^2} - 3\frac{k^2\sigma_e^2}{\omega^4} = 0 \tag{86}$$

which is a quadratic in ω^2. The final dispersion relation, which relates k to ω, is

$$\omega^2 = \omega_p^2 + 3\sigma_e^2 k^2 \tag{87}$$

which is also known as the *Langmuir dispersion relation*. Note that it implies that the group velocity of the waves is different from the phase speed, such that the two are reciprocals of each other.

The reason for dwelling on this in the chapter on the Vlasov equation is that the methods used in the derivative of this equation are really quite general. With the exception of the specific equation for the electrostatic field E, we could have used the Poisson equation for a gravitational field. In this case, we would have replaced E by $\nabla\Phi$, the gradient of the gravitational potential, and used only the mass of the *single species* of particle rather than the charge and the ion–electron pairs. The same method is used for a galaxy consisting of collisionless stars as for the plasma with which we have been dealing. Both are possibly unstable, and both have a characteristic dispersion relation for propagating disturbances. In the case of the self-gravitating medium, we shall return to the stability problem in Chapter 9. For the moment, keep it in mind as something coming down the road.

References

Chandrasekhar, S. (1939). *Introduction to the Study of Stellar Structure*. Chicago: The University of Chicago Press; reprinted by Dover.
Chandrasekhar, S. (1966). Higher order virial equations. In *Lectures in Theoretical Physics*, Vol. 6 (W. Britten ed.). New York: Gordon & Breach, p. 1.
Chandrasekhar, S. (1967). *Ellipsoidal Figures in Equilibrium*. New Haven, CT: Yale University Press; reprinted by Dover.
Collins, G. W. II. (1983). *The Virial Theorem and Its Applications to Astrophysics*. Tucson, AZ: Pachart Press.
Cox, J., and Giuli, R. (1968). *Principles of Stellar Structure*, Vol. 1. New York: Gordon & Breach.
Landau, L. D., and Lifshitz, E. M. (1978). *Statistical Physics*, Vol. I. Oxford: Pergamon.
Saslaw, W. C. (1985). *Gravitational Physics of Stellar and Galactic Systems*. Cambridge: Cambridge University Press.

CHAPTER 2

Viscosity and Diffusion

> *No one means all he says, and yet very few say all they mean, for words are slippery and thought is viscous.*
> Henry Adams, *The Education of Henry Adams*

2.1 Introduction

The action of internal friction is the most important dissipation mechanism in fluids. It acts on all scales, from molecular to megaparsec sizes, and powers most of the internal energy conversion processes. In the laboratory it is often an annoyance, making flows difficult to control and unstable. In the cosmos, it appears that viscous action, whether by molecular or turbulent processes, shapes most of the hydrodynamic structure we see. Here we will look at a general approach to what viscosity is, and how it acts, in a rather traditional fashion. The generalizations will have to wait for now. But once you have a feel for what viscous coupling is about, it should be possible to anticipate some applications even before we reach them.

The particle jumps and momentum coupling within a continuous medium can have many causes. On the kinetic level, they result from collisions. On a macroscopic level, they may be due to turbulent eddies transporting material through the fluid or gas. Collisions of waves within a fluid can simulate a turbulence, indeed power it, and these will produce a macroscopic diffusion. Because such waves can transport momentum as well as mass, they can be thought of as a source of friction, just as the kinetic scale can produce a similar effect. This friction is called the *viscosity* and gives rise to most of the interesting effects in fluids. For instance, Feynman referred to two kinds of fluid-mechanical treatments, those concerned with "dry water" being ones which neglect viscosity (so-called ideal fluids) and those concerned with "wet water" ones which include such effects. In fact, it is viscosity that produces the phenomenon of "wetting," the adhesion of a fluid to a surface, so the labels are most apt.

This chapter is divided into two broad sections. The first is devoted to the classical treatment of viscosity. It introduces the extension to the equations of motion affected by the introduction of long-range coupling within the fluid caused by viscosity. The Navier–Stokes equation is the goal of the discussion, and we will examine some of its consequences. Many more applications will occur in later chapters. The second part is an introduction to the spatial version of the Fokker–Planck formalism that leads to the diffusion equation. This is included for two reasons. One is that it makes the origin of the viscous terms a bit clearer and leads to an equation that looks much like a continuity equation, in fact one that contains diffusion. The second reason is more pedagogical in its intent. The diffusion equation is a stochastic realization of a Chapman–Kolmogorov process. Frequently in astrophysics we are faced with random processes that demand statistical interpretation. But too frequently, we still prefer to attack these problems in deterministic terms. The treatment in the second half is meant to introduce some of the methods that have wide dynamical application and also connect the kinetic approach with the continuous hydrodynamic methods of the rest of the book.

2.2 The Navier–Stokes Equation

The kinetic approach shows how we can understand the origin of internal dissipation and coupling in a gas, but it doesn't achieve our aim of a dynamical equation for the fluid including viscosity. To do this, we have to resort to the representation of the fluid as a continuous medium and examine the flows in the context of deformations introduced by the viscous forces. In short, we require the Navier–Stokes equation.

2.2.1 Historical Introduction

The approach to continuous media that leads to the Navier–Stokes equation is interesting and worth diverting to for just a moment. One of the most debated subjects at the end of the eighteenth century and through the first half of the nineteenth century was the proper treatment of mechanics for continuous media, especially for elastic solids. It was known that solids can support many fluidlike motions, in particular that they can shear and can also support various wave modes. When a fluid is cooled so that it is almost solid, for instance, molten metal, it will behave like a plastic, moving differentially in the presence of walls and developing and supporting shearing. At the time, electromagnetic phenomena, especially light, were also of considerable interest. The analogy between waves and light, forcefully demonstrated by Young through the phenomenon of interference, reinforced the belief that if light was transported through space, it must be through some medium capable of sustaining deformation and vibration.

Here an interesting set of questions arose, which it appeared could be answered through analogy with the mechanical properties of fluids. Real fluids were known to show the property of viscosity, so would the same apply to the medium supporting light waves? That is, did the ether, the hypothetical medium responsible for the transmission of such waves, have any internal friction? Also, since bodies immersed in an electric field developed charge, would it be possible to treat the electrical phenomena observed in the laboratory by using fluid analogies for the properties of the ether? While of course now it is not fashionable, or necessary, to consider such questions, and relativity disposes of the need to consider them at all, they were nevertheless important in formulating many of the basic concepts of electromagnetic theory and as such spurred much of the work of the last century in fluid mechanics. For our purposes, they are important as examples often overlooked of analogous reasoning, so fundamental to astrophysical understanding, and can be useful as a stimulation to imagination. We still use the same kind of "as if" epistemology in approaching cosmic environments where we do not have a complete handle on the processes involved but can observe what appear to be phenomenological behaviors similar to those noted in the laboratory. The roots of much of astrophysical thought lie deep within the soil of the last century, and since many of the problems of astrophysics currently revolve around large-scale fluid motion, it seems appropriate to pause to consider the forebears of this mode of scientific exploration.

2.2.2 Strain, Shear, and Stress

With that aside, let's now consider a medium undergoing strain, shear, and stress. A stress is a force; strain and shear are displacements. A strain is either a distension or compression in the direction in which a force is applied and can be thought of as essentially one-dimensional. A shear is a differential displacement of adjacent regions. That is, if we take a point in space, x, and imagine that a differential force has been applied to two points at $x \pm \delta x$, then these points will be displaced in directions which are not necessarily the same. The effect is to act along the plane. The discovery that the stresses on a body decompose naturally into components acting normal to, and along, a body's surface is due to Cauchy. There is an

2.2 The Navier–Stokes Equation

enormous literature on the research concerning continuous media during the last century. Relative to the central position, such media can rotate. The presence of a shear is this rotation, which is the evidence that there are planar components to the stress.

Geometry

To make this more specific, take a position in a fluid or other continuous elastic medium and deform the medium by an amount ξ from its original location. Then in three dimensions the distance between the original and final locations of the reference point has increased by a quantity $ds^2 = (dx_i + d\xi_i)^2$. Expanding this out, we need the change in the displacement with respect to the original coordinate system, which is given by

$$d\xi_i = \frac{\partial \xi_i}{\partial x_j} dx_j \tag{1}$$

This means that the displacement is

$$ds^2 = dx_i^2 + 2\frac{\partial \xi_i}{\partial x_j} dx_i dx_j + \frac{\partial \xi_i}{\partial x_j} \frac{\partial \xi_i}{\partial x_k} dx_j dx_k \tag{2}$$

The last term is second order and we will drop it. Now using the symmetry of the displacements, we rewrite this as

$$\frac{\partial \xi_i}{\partial x_j} dx_i dx_j = \left(\frac{\partial \xi_i}{\partial x_j} + \frac{\partial \xi_j}{\partial x_i}\right) dx_i dx_j = S_{ij} dx_i dx_j \tag{3}$$

The displacement law therefore turns into the following expression:

$$ds^2 = dx_i^2 + S_{ij} dx_i dx_j \tag{4}$$

where S_{ij} is the *shear*. The shear is a symmetric tensor which expresses the deformation due to displacement in a plane, something like displacement in two directions simultaneously. Now, take the time derivative to get V_i. The term S_{ij} can be decomposed into symmetric and antisymmetric parts:

$$\frac{\partial v_i}{\partial x_j} = \frac{1}{2}\left(\frac{\partial v_i}{\partial x_j} + \frac{\partial v_j}{\partial x_i}\right) + \frac{1}{2}\left(\frac{\partial v_i}{\partial x_j} - \frac{\partial v_j}{\partial x_i}\right) \tag{5}$$

The second term is familiar. It is the *curl* of the velocity and represents local circulation. If we take a rigid body and push on it along one of its surfaces, it rotates. This does not deform the object, and it is not of any interest for us. The first term is the important one, the shear. Now the stress we already know about is the pressure, the forces normal to the fluid's surface (that is, for any body immersed within it). We know also from the equations of conservation that T_{ij} is symmetric. So we will concentrate on the term that represents deformation, the shear.

2.2.3 Viscous Stress

The effects of viscosity must depend on the relative motion of two parcels of matter. There must be shear. Even if one of them is stationary, the interaction between these parcels is such as to exchange momentum and to bring them into uniform motion. In the process, since they cannot be in isolation in a fluid, the viscous action disseminates momentum, and energy, throughout the fluid and tries to bring it to a state of rigid motion. You can look at it another way. For two parcels of gas to exchange momentum means that a shear must be involved. We are interested in the gradient of the momentum along a direction perpendicular to the motion of the fluid, for instance, $\partial v_x/\partial y$. The differential force that results from this has a phenomenological coefficient η to describe the coupling with the rest of the medium.

Some symmetries can be placed on the stress tensor even before we write it down. For an ideal fluid the diagonal components must be the pressure and all of the diagonal components must be equal. The reasoning here is simple and is due to Pascal and Torricelli. The pressure acts normally to the walls of a container, regardless of the shape of the container, when a body is immersed in a fluid. Thus the diagonal components of the stress tensor are written as

$$T_{ii} = p\delta_{ij} \tag{6}$$

We also know that the stress tensor is symmetric because, from the tensor form of the equations of motion, we see that $a_i = -\partial T_{ij}/\partial x_j$ and since i and j are dummy indices they can be interchanged. Therefore there are six independent components of the stress tensor (which actually become only four because of the diagonal components in an isotropic medium like a fluid). Since the shear terms are due to the off-diagonal components, we can assume that the shear stress is linearly related to the shearing strain. Just arguing from dimensional considerations, we would expect to arrive for T_{ij}, the off-diagonal part of the stress tensor, at something like $T'_{ij} \sim -\mu S_{ij}$ for the shearing term. Here μ acts like a kind of spring constant measuring the displacement in the fluid as a result of the action of the stress on the medium. It is assumed to be a constant, which can be measured directly from the flow or calculated from first principles in a limited number of cases. Recall that the shear-induced stress, since it represents a frictional force, must retard the motion of the fluid and as such must give rise to a dissipation as well as a momentum transfer.

2.2 The Navier–Stokes Equation

The force per unit area is given by the normal component of the stress tensor through that area, $f_i = -T_{ij}n_j$, and the force is the integral over the area:

$$F_i = -\int T_{ij} n_j \, dS = -\int \frac{\partial T_{ij}}{\partial x_j} \, dV \tag{7}$$

We can separate out the diagonal but still symmetric parts of the stress tensor, thereby writing the equations of motion in the form

$$\frac{\partial}{\partial t} \rho v_i + \frac{\partial}{\partial x_j} T_{ij} = \rho a_i \tag{8}$$

and using the continuity equation we arrive at

$$\rho \frac{dv_i}{dt} = -\frac{\partial p}{\partial x_i} + \frac{\partial T'_{ij}}{\partial x_j} + \rho a_i \tag{9}$$

Here a_i is the external and/or potential field acceleration and T'_{ij} represents the off-diagonal parts of the stress tensor. Since the total stress is given by $T_{ij} = p\delta_{ij} - T'_{ij}$, the whole right-hand side can be replaced by the divergence of the total stress tensor T'_{ij}.[1]

Let's look in more detail at the derivation of T'_{ij}. We know one very important fact from experience: friction produces drag. This is a force that acts *along* the surface of an immersed body. Because it is along each surface, rather than normal to it, friction *must* be represented by off-diagonal components in the stress tensor. We can expect that any representation of the fluid's tendency to resistance to motion through viscosity will therefore depend on the shear (see Fig. 2.1). For an inviscid fluid, the stress tensor is simply the one we have derived from kinetic theory, namely $T_{ii} = 3p$ (the trace of the stress tensor). As we have already remarked, the stress tensor is symmetric with respect to $i \leftrightarrow j$, and we suspect that the same will be true for any additional forces. We also know that shear will contribute a component of the force. For a static medium stress and strain are linearly related by a constant.

Now we assert the same thing for a shearing medium (assuming that shear is the dynamical manifestation of strain) so that we initially guess that

$$T'_{ij} = C_{ijkl} \frac{\partial v_k}{\partial x_l} \tag{10}$$

---

[1] Remember that this is a divergence, so the form of the operator changes depending on the coordinate system used even though the form of the equation does not change.

Davidson College Library

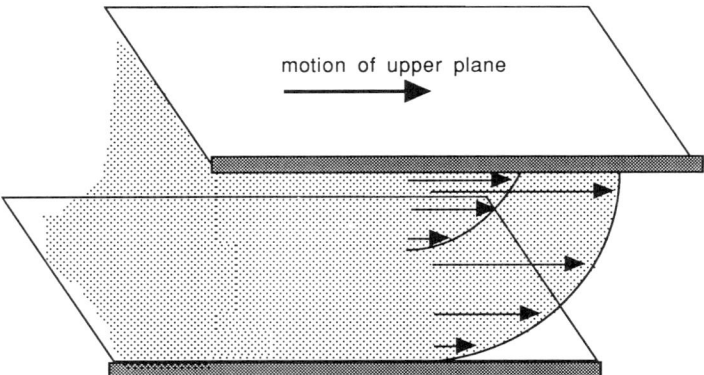

Figure 2.1 Shear in a viscous fluid between parallel moving plates.

where the C_{ijkl} terms are all constants (and there are lots of them). But we also can limit the number of components. First, we know that $C_{ijkl} = C_{jikl}$. Then, writing the symmetric and antisymmetric parts of the derivative and throwing away the antisymmetric (which is just rotation about each of the three axes), we also know that $C_{ijkl} = C_{ijlk}$. So the coefficients are quite reduced in number. But there is a more direct way still of getting the final form for the viscous term. For a completely symmetric tensor, we can find the coefficients using permutations of the combinations of Kronecker deltas in a general way:

$$C_{ijkl} = c_0 \delta_{ij} \delta_{kl} + c_1 \delta_{ik} \delta_{jl} + c_2 \delta_{il} \delta_{jk} \tag{11}$$

There are now only three constants and we have included all permitted permutations of the indices. So now we can write

$$T'_{ij} = c_0 \delta_{ij} \frac{\partial v_k}{\partial x_k} + c_1 \frac{\partial v_i}{\partial x_j} + c_2 \frac{\partial v_j}{\partial x_i} \tag{12}$$

The first term contains the divergence of the velocity, so this term must represent whole-body effects. The second and third terms are obviously the components of the shear. Now the viscous terms should not be diagonal. They should vanish when the trace is taken because the trace of the total stress tensor must be the pressure. That is, $T_{ii} = 3p$ so $T'_{ii} = 0$, from which it follows that $3c_0 + c_1 + c_2 = 0$ so that $3c_0 + 2c_1 = 0$. Thus, we have the full representation for the stress tensor:

$$T_{ij} = p\delta_{ij} - 2c_0 \left[\sigma_{ij} - \frac{1}{3} \frac{\partial v_k}{\partial x_k} \delta_{ij} \right] \tag{13}$$

2.2 The Navier–Stokes Equation

and we now identify c_0 with η, the coefficient of viscosity. When we resubstitute Eq. (13) into the equation for conservation of momentum, we arrive at last at the Navier–Stokes equation in vector form:

$$\frac{\partial \mathbf{v}}{\partial t} + \mathbf{v} \cdot \nabla \mathbf{v} = -\nabla p + \mathbf{f} + \nu \left(\nabla^2 \mathbf{v} - \frac{2}{3} \nabla \nabla \cdot \mathbf{v} \right) \tag{14}$$

where ν is called the kinematic viscosity. The second Navier–Stokes term is a result of compressibility; for a compressible fluid it is absent but for a highly compressible one it will play a role and must be included. In component form, Eq. (14) becomes

$$\rho \left(\frac{\partial}{\partial t} + v_j \frac{\partial}{\partial x_j} \right) v_i = -\frac{\partial T_{ij}}{\partial x_i} = -\frac{\partial p}{\partial x_i} = \mu \frac{\partial^2}{\partial x_j \partial x_j} v_i \tag{15}$$

Notice that in the absence of this viscous term, there are no length or time scales in the equation of motion. That is, the motion occurs on characteristics along which there are simple streamlines and there is no typical time scale or length scale on which the momentum is changed. Once a viscous coefficient is introduced there is no longer a completely scale-free environment in which to work. Should motions exceed a specific rate, or should there be a scale of length small enough, viscosity will act on that scale. As we will see shortly, dissipation will play a role, so that the fluid motion ceases to be reversible.

Now let's try an alternative approach. We know that the generalized equation for the energy is

$$\rho \frac{d}{dt} \frac{1}{2} v^2 = v_i \frac{\partial p}{\partial x_i} - v_i \frac{\partial T'_{ij}}{\partial x_j} + \rho v_i a_i \tag{16}$$

for the kinetic energy of the flow. We will return to this equation presently when discussing dissipation, but we already can use it to place some constraints on any form for the shear stress. We can break the problem of deforming the fluid into two parts, one due to shear and the other due to the compression of the body:

$$T'_{ij} = b_1 \left(\frac{\partial v_i}{\partial x_j} + \frac{\partial v_j}{\partial x_i} \right) + b_2 \frac{\partial v_k}{\partial x_k} \delta_{ij} \tag{17}$$

where the first term is due to the distortion of the body due to shear and the second is due to changes in the volume of the fluid. Now define *two* coefficients of viscosity, $\eta = b_1$ and $\zeta = \frac{2}{3} b_1 + b_2$, using the same argument that the trace of the viscous stress vanishes. Then

$$T'_{ij} = \eta \left(\frac{\partial v_i}{\partial x_j} + \frac{\partial v_j}{\partial x_i} - \frac{2}{3} \frac{\partial v_k}{\partial x_k} \delta_{ij} \right) + \zeta \frac{\partial v_k}{\partial x_k} \delta_{ij} \tag{18}$$

For strictly incompressible flow, you see that ζ is unimportant since $\nabla \cdot \mathbf{v} = 0$ and therefore there is only one coefficient of viscosity. Recall that $\nu = \eta/\rho$, which is the coefficient calculated from kinetic theory. We will soon see that the energy due to the stress is $T'_{ij} \partial v_i/\partial x_j$, so that

$$T'_{ij}\frac{\partial v_i}{\partial x_j} = \frac{1}{2}\eta\left(\frac{\partial v_i}{\partial x_j} + \frac{\partial v_j}{\partial x_i} - \frac{2}{3}\frac{\partial v_k}{\partial x_k}\delta_{ij}\right)^2 + \left(\frac{2}{3}\eta - \zeta\right)\left(\frac{\partial v_k}{\partial x_k}\right)^2 \tag{19}$$

which must be positive. While we are still dealing with the two coefficients of viscosity, let's explore one nice illustration of what was implied by J. Herschel's comment[2] that the viscous state is between that of a fluid and a solid.

Recall that a perfect fluid does not resist shear. But the essential physical mechanism that we have indicated gives rise to viscous stress is shear. Therefore, we might suspect that a viscous fluid will be able to support both shear and compressional waves. To see this, take a medium initially at rest and with uniform density ρ. Then the equation for the perturbation in the motion is

$$\rho\frac{\partial \mathbf{v}}{\partial t} = -\nabla p + \left(\eta + \frac{4}{3}\zeta\right)\nabla\nabla \cdot \mathbf{v} - \eta\nabla \times \nabla \times \mathbf{v} \tag{20}$$

where we have expanded $\nabla^2 \mathbf{v}$. Assuming that the velocity is a small perturbation, we have linearized Eq. (15) and taken ρ to be constant. We have also assumed that $p = \rho a_s^2$. Notice that we have two different terms in the equation of motion that contain the velocity. One is the divergence, which represents longitudinal waves; the other is the curl, which gives transverse waves. Take the time derivative of both sides of Eq. (20) and then form the divergence to obtain

$$\rho\frac{\partial^2}{\partial t^2}\nabla \cdot \mathbf{v} = -a_s^2\rho\nabla^2\nabla \cdot \mathbf{v} + \left(\eta + \frac{4}{3}\zeta\right)\nabla^2\nabla \cdot \mathbf{v} \tag{21}$$

which is a wave equation for $\nabla \cdot \mathbf{v}$. This wave has a phase speed $v_p^2 = a_s^2 + (\eta + \frac{4}{3}\zeta)/\rho$. We have used the continuity equation to remove the $\partial \rho/\partial t$ term.[3] Now taking the curl in place of the divergence, we obtain

$$\frac{\partial^2}{\partial t^2}\nabla \times \mathbf{v} = -a_s^2\nabla^2\nabla \times \mathbf{v} + \zeta\nabla^2\nabla \times \mathbf{v} \tag{22}$$

---

[2] From the definition of *viscous* in the Oxford English Dictionary.
[3] The divergence of the last term in Eq. (21) vanishes identically.

2.2 The Navier–Stokes Equation

This is the wave equation for a transverse or *shear* wave. It has a phase speed of $v_p^2 = a_s^2 + \zeta/\rho$.[4] For a perfect gas, since we eliminate the $\boldsymbol{\nabla}\boldsymbol{\nabla}\cdot\mathbf{v}$ term from the Navier–Stokes equation, we do not have the shear waves. Notice that the two phase speeds are different and that the wave will therefore be polarized. You will meet with this problem again when we discuss Alfvén waves, in the sense that the medium becomes anisotropic in the presence of a magnetic field and the difference in the viscous coefficients is like the propagation of a magnetosonic wave.

2.2.4 The Reynolds Number

If we start with the Navier–Stokes equation, we see that we can form a dimensionless number from the scales of length and time. The equation of motion, with or without viscosity, is ultimately a force equation. It therefore has the dimensions MLT^{-2}, or MU^2L^{-1}. The viscous coefficient must have the dimensions of ML^2T^{-1}; that is how the Navier–Stokes equation was originally constructed. The advection term is of order U^2L^{-1} and the time derivative adds a term of order UT^{-1}. We don't know the time scale, but we can remove it anyway by taking $T \sim LU^{-1}$. Then if we divide both sides of the Navier–Stokes equation by U^2L^{-1}, we arrive at a dimensionless form for Eq. (14):

$$\left(\frac{\partial}{\partial t'} + \mathbf{v}'\cdot\boldsymbol{\nabla}'\right)\mathbf{v}' = \mathbf{f}' + \frac{\nu}{UL}\nabla'^2\mathbf{v} \tag{23}$$

where all primed symbols denote dimensionless quantities. We could have done this by focusing only on the Laplacian and its coefficient because this is the only one involving the viscosity. We have obtained a pure number by combining velocity, length, and the viscosity in the form $L^2T^{-1}\nu^{-1}$. The combination of parameters of the fluid is called the *Reynolds* number, after O. Reynolds, whose investigations of the transition to turbulence were the

---

[4] Were we to take the equation of motion for an *elastic solid*, rather than a fluid, and look at solenoidal or potential *displacements*, rather than the velocity, we would arrive at a similar conclusion. There are two waves in an anisotropic elastic medium. One is due to shear and the other is compressional. These two waves are familiar from earthquakes, where they are called P and S waves, respectively (P from primary because these waves travel faster and arrive first, and S for either secondary or shear). While normal astrophysical media will not display these distinct propagation modes, there are several types of cosmic bodies that do. First, obviously, planets show these. The S waves do not propagate through the fluid cores and form S shadow zones. The same effects may happen in neutron stars, where the crustal adjustments might produce something akin to quakes. But this is a rather far reach from the more prosaic environments we'll be dealing with throughout this book.

pioneering effort to understand the role of viscosity in the phenomenon (Rott 1990). It describes the rate of transport of momentum by viscosity versus advection and is formally given by

$$\mathrm{Re} = \frac{UL}{\nu} \tag{24}$$

where, just to remind you, $\nu = \eta/\rho$. It is important to note that L is any typical scale of length in the system. Thus viscosity dominates over a length scale L when the Reynolds number is very small. However, there is always some length scale over which the viscosity will be important, provided the fluid is not ideal, and therefore there is some minimal length below which, for any arbitrary velocity U, we must abandon our ideal fluid approximation.

Reynolds' hydrodynamic studies, conducted in the second half of the nineteenth century, were in connection with channel flow. He found that dye tracers moved smoothly in a fluid flowing in a flat, narrow laboratory channel when the characteristic number now named after him was small. That is, laminar flow is a characteristic of small Reynolds number, when viscosity dominates. However, the flow underwent a striking transition for large values of the number. The dye marking the channel flow became turbid and mixed rapidly. There was no evidence of a precursor stage to this change. It had an onset very much like a phase transition. Reynolds argued that the causative agent was the generation of vorticity by shear in the fluid which on some scale was causing dissipation and that the onset of turbulence must be associated with the action of a viscosity. We shall return to this point in due time when discussing both instabilities and turbulence.

Whether deriving the effects of viscosity phenomenologically, as in the Navier–Stokes equation, or from kinetic theory, as in the Fokker–Planck equation, we assume that the scale on which dissipation takes place is microscopic. To analyze the flows, however, does not really require any detailed knowledge of the underlying physical mechanism for viscosity. At the "engineering" level of astrophysical fluids, we can turn the problem around and ask what viscosity we require to produce the observed flow. Ultimately, however, we have to ask where the viscosity comes from. Whatever we take as the dissipative mechanisms for the flow, they always act on a scale smaller than we can directly observe. In this way, any chaotic, dissipative motions can play the same role as viscosity, on some length or time scale. This means we have to change our magnification when viewing different parts of the medium. In fact, we have to ask different questions depending on the scale. On some level, for some range of velocities and lengths, any fluid is inviscid. On some other, it is very likely

2.3 Dissipation and Viscous Coupling

it will be like molasses. Or more likely, on some scale the description of fluid motion may break down altogether and we will have to resort to a kinetic description.

2.3 Dissipation and Viscous Coupling

The *leitmotif* of this chapter is that viscosity is intrinsically dissipative of momentum and energy. To see the consequences of viscous coupling of the fluid to its environment, or to internal dissipation, we need to find how the energy losses depend on the stresses within the fluid. We assume that the flow is driven by the stress T_{ij} which contains the viscosity. We will generalize this discussion, without regard to the details of T_{ij}, because the only condition we need to place on the stress is that it is symmetric in its indices. Therefore we have

$$\rho \frac{dv_i}{dt} = -\frac{\partial T_{ij}}{\partial x_j} \tag{25}$$

Now if we take the scalar product with **v** we obtain the equation for the work performed by viscous stresses. This can be written as

$$\rho v_i \frac{dv_i}{dt} = -\frac{\partial}{\partial x_j}(v_i T_{ij}) + T_{ij} \frac{\partial v_i}{\partial x_j} \tag{26}$$

Now notice that we can expand out the shearing term (the second term on the right):

$$\frac{\partial v_i}{\partial x_j} = \frac{1}{2}\left(\frac{\partial v_i}{\partial x_j} + \frac{\partial v_j}{\partial x_i}\right) + \frac{1}{2}\left(\frac{\partial v_i}{\partial x_j} - \frac{\partial v_j}{\partial x_i}\right) \tag{27}$$

The internal energy E is what we are after. The energy equation is

$$\frac{\partial}{\partial t}\left(\frac{1}{2}\rho v^2\right) + \frac{\partial}{\partial x_j}\rho v_j\left(\frac{1}{2}v^2 + \frac{p}{\rho}\right) = -\frac{\partial}{\partial x_j} v_i T'_{ij} + \frac{\partial}{\partial x_k} \kappa \frac{\partial T}{\partial x_k} \tag{28}$$

where T is the temperature and κ is the heat conduction coefficient. We have a complete equation for the change in the *internal* energy given by

$$\frac{dE}{dt} = T'_{ij} S_{ij} \tag{29}$$

for the rate of energy dissipation. This is the increase in the internal energy *at the expense of the kinetic energy of the flow*. Now we will identify the S_{ij} with the shear tensor, σ_{ij}. In other words, the stress times the shear is the source of the dissipation of energy. The dissipation rate is given by a scalar equation, the scalar product of tensors. So it will look the same even if we go to non-Cartesian coordinates. We have a way of dissipating energy. An application of viscous stresses in astrophysical flows is the powering of accretion disk emission. The stresses are taken as proportional to the shear so that $\dot{E} \sim -\Omega'^2$ where Ω' is the radial gradient in the angular frequency of a circulating flow. We will discuss this in greater detail in the winds and accretion chapter.

Actually, what is happening is that the flow is losing bulk kinetic energy to random motions. It isn't precisely correct to say that the energy is lost. It is simply lost to the overall flow, but shows up in the pressure and in the entropy—the medium heats up. The effect of this heating is displayed in many astrophysical situations. For example, in an accretion disk, this is the source of the internal heating that supports the disk vertically against collapse. In prominences, it is the source for radiation and also for the expansion of the magnetic structure. For radio jets, it may be the source for the electron heating, the agent that feeds the turbulence thought to be responsible for accelerating the electrons to the relativistic energies required for the observed synchroton emission. In short, although the net effect of this dissipation in a viscous fluid is to slow the medium down and to bring it to a state of rigid motion, it also provides many ancillary signatures of its presence. These are fundamental for gaining an understanding of the viscous agent.

The radiative losses that result in most astrophysical flows serve to dissipate the bulk kinetic energy. If thermal diffusion is important, that is, if the medium is optically thick, then the rate of dissipation is ultimately given by the surface and volume emissivity of the fluid. We will see that in the case of accretion disks, especially, the input of turbulent energy is responsible for the emissivity of the disk. In this particular case, the fact that the velocity field can be specified in advance permits the calculation of the continuum spectrum, but this is actually an unusual case. More frequently, if the viscous dissipation is large, we must calculate the emission self-consistently, taking into account the effect of the change in internal energy on the kinetic energy of the bulk flow. We will meet this again in Chapter 7.

2.4 Boundary Layers

2.4.1 Introduction

A bounding surface has several effects of a viscous flow. Examples of these you know well. When looking at a stream, you've likely noticed that

2.4 Boundary Layers

gyres, little vortices, are formed in the immediate vicinity of the bank and that the waves generated by disturbances in the interior of the flow seem to propagate almost at normal incidence to the bank. The blades of a fan get dusty even though they are pushing the air over them at a high rate. And a propeller of a boat gets encrusted. All of these indicate that some of the fluid, and particulate matter that it transports, is sticking to a surface. It must therefore be comoving with the surface, and this is the region called the *boundary layer*.

The essential feature of a boundary layer is that it is a region of locally low Reynolds number. Here the velocity, U, is large but L, the scale length, is small. Therefore the viscous terms dominate the flow and the region moves almost as a harmonic flow with the surface. A short distance outside this region, however, the fluid quickly reverts to its nearly inviscid state and the advective derivative again dominates. Take x as the direction of the mean fluid motion and y perpendicular to it. Assume that there is a wall at constant y, a plane. Then another way of stating the condition for forming a boundary layer along the wall is that $\tau = \mu(\partial v_x/\partial y)$, the viscous stress, dominates the motion there. This means that the motion in y is driven by the pressure gradients in x, the velocity gradients in x, and the viscous stress. The connection between the motion in x and y is provided by the continuity equation. The most accessible case for analysis is therefore the incompressible fluid problem, since there we have potential flow with the continuity equation describing only the velocity and not the density. We will start with this and generalize from there.

2.4.2 Laminar Plane Boundary Layers

The equations for an incompressible boundary layer assume that only the shear terms contribute to the momentum flux. The motion normal to the wall is circulatory, and since the flow is incompressible there can be no true mass accumulation there. The pressure gradient therefore vanishes and only the shear stress remains to drive the circulation. The equation of motion is therefore

$$u\frac{\partial u}{\partial x} + v\frac{\partial u}{\partial y} = \nu\frac{\partial^2 u}{\partial y^2} \tag{30}$$

supplemented by the continuity equation

$$\frac{\partial u}{\partial x} + \frac{\partial v}{\partial y} = 0 \tag{31}$$

The flow in the main channel is uniform, so within the flow along x there is no characteristic length. But because of the viscosity, there is one for the y

direction given by $\delta = (\nu x/U)^{1/2}$. We choose to write the flow in terms of a y coordinate scaled to the thickness of the boundary layer. The ratio y/δ is dimensionless. Taking a hint from this, and from the interest in rendering the flow equations scale-free, we create a new variable:

$$\eta \equiv y/\sqrt{x} \tag{32}$$

Now we are ready to solve the boundary layer structure. Note that if we take $u = Uf(\eta)$, we will be able to obtain a scaled form for v because of the continuity equation. In effect, we are asking for the representation of the flow that allows the Reynolds number to remain constant as we move away from the walls. We assume that the flow is stationary; this removes the need to include the time. This version of a boundary layer also ignores one of the more important features of viscous flows, the entrainment of material from quiescent parts of the environment, but it is illustrative anyway.

Now transform the derivatives as follows:

$$\frac{\partial}{\partial x} = -\frac{1}{2}\frac{\eta}{x}\frac{d}{d\eta}, \qquad \frac{\partial}{\partial y} = \frac{\eta}{y}\frac{d}{d\eta} \tag{33}$$

Because the flow satisfies $\nabla \cdot \mathbf{v} = 0$ (it is incompressible) the velocity is given by a solenoidal potential:

$$u = \frac{\partial \psi}{\partial y}, \qquad v = -\frac{\partial \psi}{\partial x} \tag{34}$$

and we can then take $\psi = (\nu U x)^{1/2} F(\eta)$. You see that this one is forced by the requirement that the equations be independent of x or y individually. To illustrate the derivation of the velocity components, take v. Here:

$$v = -\frac{\partial}{\partial x} x^{1/2} F(\eta) = -\frac{1}{2} x^{-1/2} F + \frac{1}{2}\eta x^{-1/2}\frac{dF}{d\eta} \tag{35}$$

For u, we get the simpler form $u = dF/d\eta = F'$. So for the left-hand side of the boundary layer equation we have

$$u\frac{\partial u}{\partial x} + v\frac{\partial u}{\partial y} = -\frac{1}{2}\eta x^{-1} F'F'' + \frac{1}{2}\eta y^{-1} x^{-1/2}(\eta F' - F) F'' = -\frac{1}{2x} FF'' \tag{36}$$

Taking the transformed derivatives of u for the diffusion term yields F''', so the final equation for the boundary layer transforms into

$$\tfrac{1}{2} FF'' + F''' = 0$$

This is the *Blasius* problem. Notice that we have a function that will yield the velocity components, but we do not have u and v directly. We must

2.4 Boundary Layers

take derivatives [see Eq. (34)]. Since at the walls the normal velocity, v, vanishes, $F(0)$ and $F'(0)$ vanish at $\eta = 0$. This can be at the wall or at infinite distance along the wall. Finally, $F'(\eta) \to 1$ as $\eta \to \infty$, a statement that the boundary layer is finite in width and that the viscous stress vanishes as we move toward the midstream. Equation (37) has no analytical solution; its importance is that it is self-similar.

This solution has been discussed many time in many places. It has taken on particular astronomical relevance in the past few years because of the problem of siting large optical interferometric telescopes. We will divert the flow of the discussion for just a moment to consider the problem of flow over terrain as an application of the boundary layer picture.

2.4.3 Boundary Layers and Astronomical Seeing

We will discuss the problem of turbulence at length in its own chapter, but here we will discuss a simplified model for flow over terrain. We assume that the shear is large enough that over some scale the flow is turbulent. This is actually possible because we really don't need to know what turbulence is in order to discuss some of its effects, especially in practical applications. All we require is a phenomenological theory that allows us to scale from one environment to another in a consistent way. The problem of boundary layers in flow over a mountaintop has been discussed as much in the halls of observatories as by atmospheric physicists in the past few years. Turbulence is generated when an otherwise smooth (*laminar*) flow encounters an obstacle and feels a drag force. Vorticity is created in the boundary layer, and this being injected into the flow causes random spatial and temporal fluctuations in the temperature and density of the fluid. If this fluid happens to be air and we are attempting to make astronomical observations through it, the variations of the thermal and dynamical state of the air produce "seeing." This is the stochastic fluctuation of the size and brightness of images viewed through this turbulent flow. The assumption is that the shear is given by the change in the wind speed over a characteristic distance when the air flows over some rough terrain. A good way to represent the shear in the equations of motion is therefore to scale on u, the horizontal component. We can assume that the boundary layer grows with distance so that the stress can be represented as being inversely proportional to the distance above the surface, y. The advective speed is U and the turbulent flow produces a viscosity $U\lambda$, so that if the shear is from the surface by

$$\frac{\partial u}{\partial y} \approx \frac{K}{y} \tag{38}$$

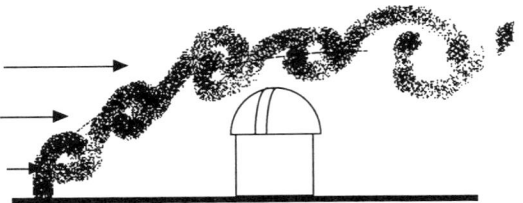

Figure 2.2 The von Kármán boundary layer over a plane surface.

the velocity is approximated by $u(x, y) = K \ln y$ + constant. This is often called the *von Kármán* boundary layer profile (see Fig. 2.2). The constant K depends on the viscosity and the speed of the mean flow, but the important thing is that the velocity profile has a universal functional form.

2.4.4 Turbulent Boundary Layers

Even in the absence of more details about turbulence, we can estimate its effects on the structure of a boundary layer. The viscosity coefficient has the dimensions $L^2 T^{-1}$, so we can approximate by noting that the shear provides a scale length through $V/(\partial V/\partial z)$ where $\hat{z}$ is normal to the bounding surface and $V \to V_\infty$ is the stream flow in the $\hat{x}$ direction. Then the Reynolds number is given by $\text{Re} \sim VL^2/(\langle v_t^2 \rangle (V - V_\infty))$ where the derivative has been replaced by a scale length. At a specified value of the Reynolds number, the Reynolds stress given by the average of the turbulent energy $\langle v_t^2 \rangle$ causes the heating in the layer and also the dissipation. If we have a prescription for connecting the energy density in the turbulence with that of the mean flow, we can calculate the viscous coefficient. In general, however, we don't have this information and must rely on some generalized turbulence theory instead. Such theories are important for providing the physical connection between vorticity, typical of boundary layers (especially in turbulent transitions), and shear stress. We shall postpone further discussion of this until the chapter on turbulence (and also the section on accretion disks in the wind and accretion chapter).

2.5 Diffusion and Kinetic Theory

2.5.1 Introduction

To discuss viscosity for a normal fluid is simpler than treating it in an astrophysical context. Normal liquids are obviously wet, obviously wetting, and obviously dense and incompressible. A kinetic treatment for them is

2.5 Diffusion and Kinetic Theory

not only hard, it is inappropriate at the level of normal experience. It takes a great deal of effort to calculate the origin of viscosity in a terrestrial liquid, because its structure at the microscopic level is dominated by short-range correlation effects and detailed quantum mechanical state-counting problems. But for a cosmic plasma, the problem is a bit easier. First of all, the densities are generally pretty low once you get outside a stellar interior. The main effect of viscosity is that it redistributes momentum, especially angular momentum, throughout a fluid. The main source for the viscous action is not collisions, as normally encountered in a gas, but turbulence or other large-scale chaotic motions that can be treated *as if they are a viscosity*. But because the formalism is similar for kinetic and *pseudokinetic* treatments, namely the Fokker–Planck equation, we will examine it *before* starting on the more traditional treatment of the viscosity problem. Then we will look at the continuous medium of a fluid and see how the Navier–Stokes equation provides the proper treatment of viscosity in the equations of motion.

2.5.2 Background Justification

In the case of dilute fluids, such as the interstellar medium or even intergalactic space, a good first guess would be that the medium can be treated as inviscid. The reason is simple: the distance over which the particles are capable of moving is large in comparison with the scale of the density or pressure gradients. In the case of the interstellar medium in disk galaxies, for example, the typical densities are 0.01 cm^{-3} for the electrons and of order 0.1–1 cm^{-3} for the mass overall. The typical velocity is of order the sound speed, which is about 10 km s^{-1}. Therefore for a cross section of order 10^{-18} cm^{-2} the mean collision time is $\tau = (n\sigma v)^{-1} \approx 10^5$ years, short compared to the Hubble time but of the order of 10^{-2} to 10^{-3} of a galactic rotation period. The mean free path is of order 1 pc, small in comparison with the scale lengths in the galaxy. The medium consequently *behaves as if it is viscous*; there will be ample chance for collisions to redistribute momentum gained by a parcel of gas throughout the medium on time scales which are comparatively short. This is, for instance, considerably different from the time scale for stars to collide. The gas must be treated as if it is viscous, while the single-particle approximation works exceedingly well for stars.

In the intergalactic medium, where the densities are of order 10^{-6} cm^{-3}, viscosity is considerably reduced and therefore there would seem to be less reason to consider its effects in the case of this extremely dilute medium. But one of the most amazing, and still poorly understood, facts is that the interstellar medium, the interplanetary medium (specifically the solar wind), and a whole slew of astrophysical environments do not behave as if they were inviscid. Dissipation appears to be the rule, whether due to molecular viscosity or through the agency of turbulence. The observed structures of radio and optical jets, and fluctuations detected in cometary comas and tails, and a wide range of other observations point to the role of viscosity in structuring cosmic fluids. For stars, we cannot exclude the effects of this internal friction. Molecular viscosity plays a role in the redistribution of angular momentum and, as we shall see in the discussion of convection in Chapter 9, also in the mechanical redistribution of energy. We will also see, in this section, that there is another form of viscosity in stars. This is the "radiative" viscosity, which is also to be included in the generation of circulation currents.

2.5.3 Derivation of the Fokker–Planck Equation

Consider an ensemble of particles, characterized by a distribution function in velocity and a spatial density. We shall, for the moment, concentrate on the latter in order to extend the treatment of the continuity equation.

Normally, the Fokker–Planck equation is a means for deriving the distribution function. The two coordinates are the velocity and space, in addition to the time. But the same formalism can be used to determine the density as a function of just space and time, if these are the random variables. The general theory connects the Fokker–Planck formalism with Langevin equations and stochastic equations. These are not often discussed in the fluid context for astrophysical applications. Instead they usually appear in discussions of stellar dynamics (as in Spitzer 1987).

In the continuity equation for a compressible medium, it is assumed that the velocity of a particle is due to some mean flow. That is, every particle in the gas has both a mean motion and some random deviation from it. The latter is derived from the distribution function. Now if we look instead at what would happen in an ensemble of particles, each of which random-walks through space, we would be able to imagine what the effects of random encounters between particles would be like and what effect they would have on density and pressure fluctuations in a medium. One characteristic of a random walk, or a *Markov process*, is that all steps are independent of the history of the particle's motion. At any step at some time t, the probability of being at a location $x(t)$ is $P_x(t) = \Pr\{X = x(t) | X = y(s)\}$, which is to say that it depends only on what the position was at the previous step at time s and not on any of the transitions which have previously taken place. In order to compute the number of particles at some point in space, that is, the number density (or mass density if they all have the same mass), it is necessary to consider what the distribution of the

2.5 Diffusion and Kinetic Theory

particles was in space at some time prior to the instant of interest and then weight the probability that they would have jumped some distance from their original location, in some random direction, arriving at the spot at which we have taken the sample. Should this be a random walk, then we know something about the distribution of step sizes. The large jumps are not very likely, and the larger the jump the less likely it will be. For collisions, this is equivalent to the assertion that close encounters between particles are very rare and that each deflection, due to interaction among the constituents, is on average small or even zero.

The probability of jumping some distance Δx is normalized, $\int_{-\infty}^{\infty} P(x; \Delta x)\, d\Delta x = 1$, and we can compute the probability from the detailed solution of the phase space trajectories of the interacting particles. We thus have a propagator problem. The question is how to get particles located at some position $x - \Delta x$ at some time $t - \Delta t$ to a position x at time t. If we assume that we know the spatial density $n(x - \Delta x, t - \Delta t)$ at some earlier time, and we know $G(x; \Delta x)$ at that time, which is the propagator, then we can solve for the density distribution at the next future time:

$$n(x, t) = \int_{-\infty}^{\infty} n(x - \Delta x, t - \Delta t) P(x; \Delta x)\, d\Delta x \tag{39}$$

Since we are interested in the evolution of the distribution function for only very small time steps, take Δt to be very small. However, it should also be noted that the time step is in *our* control. It is not a distributed random variable like Δx. It has no P associated with it. Also, and this is a particularly important assumption, the process by which particles jump from one place in space to another is stochastically stationary, so the probability distribution is independent of time. Finally, Δx is a true random variable and not a function of position. This is an important assumption of the Markov process. It assumes that there are no attractive forces from some large-scale field which could be erecting barriers to the jumping process. In a homogeneous medium, this is very likely to be true, but in some solid-state systems or viscoelastic media it will not apply. This does not mean that we do not have a directed Markov process, that there could be a uniform field across the medium which causes the probability of one direction of a jump to exceed another direction, but only that this is independent of the position of the particle in the medium.

We are interested in the probability of moving a particle from a position x_0 at some time in the past to a position $x = x_0 + \Delta x$ at some time Δt later. Because we are interested in the equations of dynamics, and we already know that the mean value of the displacement may vanish but the second moment does not, we take the first-order expansion in time but go to second order in space. Note again that Δx is a random variable that records the response of a particle distribution to randomly acting forces. We have no *direct* knowledge of the magnitude of the spatial step, Δx, that the particle will take during that time. Thus, we want to integrate over all possible step sizes to remove this unknown (and indeed *unknowable*) variable from the problem. But we want to know how the spatial density distribution changes in time because of these forces.

We therefore assume that we have only small intervals of time over which we seek information about the evolution of n. We expand the density to first order in time but second order in space. This is because we are interested in the immediate temporal history for steps of Δt, which we have control over. Therefore, we integrate over the propagator, that is, the probability, for a step of size Δx:

$$n(x - \Delta x, t - \Delta t) = n(x, t) - \frac{\partial n}{\partial x}\Delta x - \frac{\partial n}{\partial t}\Delta t + \frac{1}{2}\frac{\partial^2 n}{\partial x^2}\Delta x^2 \quad (40)$$

Now the propagator (the probability function) must also be expanded. Thus P becomes

$$P(x - \Delta x; \Delta x) = P(x; \Delta x) - \frac{\partial P}{\partial x}\Delta x + \frac{1}{2}\frac{\partial^2 P}{\partial x^2}\Delta x^2 \quad (41)$$

and retain terms up to second order in Δx and only to first order in Δt:

$$\frac{\partial n}{\partial t}\Delta t = -\frac{\partial}{\partial x}\left\{\int_{-\infty}^{\infty} P(x; \Delta x)\Delta x n(x, t)\, d\Delta x\right\}$$
$$+ \frac{1}{2}\frac{\partial^2}{\partial x^2}\left\{\int_{-\infty}^{\infty} \Delta x^2 P(x; \Delta x) n(x, t)\, d\Delta x\right\} \quad (42)$$

The left side of the equation is the temporal evolution of the density [we have made use of the fact that $\int d\Delta x\, \Delta t\, P(x; \Delta x) = \Delta t$], while the right side is the rate of drift of particles through the action of the collisions. In a random walk, the absence of a background field produces jumps with equal probability of being in either the positive or negative direction. Think of the case of a true coin being flipped. The probabilities of heads and tails are equal but this does not mean that large runs of heads and tails are impossible. It simply means that, on average, there is an expected outcome of zero if we call heads, say, $+1$ and tails -1. The fact that long runs can occur, however, means that there can be a drift of the "payoff" for a while. In the case of a physical process, this means that the mean position of the particle will remain fixed in space but that the distance of a particle from the center

2.5 Diffusion and Kinetic Theory

will steadily increase with the number of steps taken. The steps in each of the coordinates are strictly independent of those in every other direction, so that $\langle \Delta x \rangle \equiv \int d\Delta x\, P(x; \Delta x)\, \Delta x = 0$ while $\langle \Delta x^2 \rangle \geq 0$. This slow, *diffusive*, drift is the characteristic of a random walk as first discussed by Einstein in the famous series of papers on Brownian motion in 1902. The equation of motion for this random walk is

$$\frac{\partial n}{\partial t} = \eta \frac{\partial^2 n}{\partial x^2} \qquad (43)$$

where $\eta \equiv \langle \Delta x^2 / \Delta t \rangle$ in the limit of vanishingly small Δt. Here η is the diffusion coefficient, which can be identified as the result of the internal collisional processes in a gas or fluid. The modified continuity equation is therefore

$$\frac{\partial n}{\partial t} = -\frac{\partial J}{\partial x} \qquad (44)$$

where J is the current:

$$J = \frac{\langle \Delta x \rangle}{\Delta t} n(x,t) - \frac{1}{2} \frac{\langle \Delta x^2 \rangle}{\Delta t} \frac{\partial n}{\partial x} \qquad (45)$$

The generalization of this equation to three dimensions is quite straightforward (and left as an exercise). Notice that we have two terms now, a divergence of the mass flux and a term which depends on the gradient of the density, which drives the diffusion of particles through the collisional term. At this point, it is tempting to say something like "we have one equation which shows that a spatial gradient of a scalar variable can drive a current stochastically, so why not try this for a vector like the velocity?" For instance, we could say that all random processes in a medium add a term of the form $\eta_n \nabla Q$ to any evolution equation, whether Q is a scalar or a tensor. But this is merely an assertion and one which we have to examine in more detail. The Fokker–Planck equation was, in fact, first derived for the random walk of particles in *phase space*, actually in velocity, due to collisions and was a way of treating the evolution of the velocity distribution function in time. We deal with this stochastic picture here also because macroscopic effects like turbulent viscosity are well described by this approach. So even though originally formulated to handle a microscopic phase-space description of an ensemble, the Fokker–Planck equation has many added uses.

References

Chandrasekhar, S. (1943). Stochastic problems in physics and astronomy. *Rev. Mod. Phys.*, **15**, 1.
Lifshitz, E. M., and Pittaevskii, L. P. (1981). *Physical Kinetics*. Oxford: Pergamon Press.
Rott, N. (1990). Note on the history of the reynolds number. *Annu. Rev. Fluid Mech.*, **22**, 1.
Spitzer, L., Jr. (1987). *Dynamical Evolution of Globular Clusters*. Princeton, N. J.: Princeton University Press.

CHAPTER 3

Vorticity and Rotation

And by turning, turning, we come 'round right.
Shaker hymn

3.1 Introduction

The study of rotation of cosmic bodies is as old as astronomy and was one of its initial inspirations. Nearly all celestial bodies exhibit some form of rotational or circulatory motion. The Earth's rotation structures its atmospheric and oceanic circulation patterns. The latitudinally dependent rotation of the Jovian planets dominates the structure of their atmospheres. The Sun rotates, albeit slowly, and in so doing drives its magnetic dynamo. And the interaction between the Sun and the solar wind produces both the spindown of the Sun and many of the well-known effects observed in cometary tails. Stars rotate, as determined from the Doppler broadening of their spectral lines. Material exchanged between the components of a close binary system forms an accretion disk that slowly collects onto the mass-gaining star. And in the nuclei of active galaxies, the excess angular momentum of material accreting onto the central "engine" probably causes it to form a toroid or accretion disk.

Fluids support rotation. It is the quintessential property of the fluid state that *the medium does not resist shearing*. You know that if the fluid is viscous the shear generates some redistribution and dissipation, so this statement isn't strictly true, but we will look at the consequences of internal effects as well. For the moment we will work with the equations of motion for an incompressible fluid, introducing the basic ideas of vorticity and the coordinate systems which are natural to the problem, and then we will expand this to problems of astrophysical interest. Some applications, specifically those related to accretion disks, will be postponed until the chapter on outflows and accretion.

3.1.1 Coordinate Transformations

Rotating systems posses a natural symmetry, centered on the rotation axis, and it is best to begin immediately by writing the equations of motion in such a frame. This approach has other pedagogical benefits because the formal structure of the equations of motion is invariant under coordinate transformations. We can most directly make explicit use of this property in what follows. We begin with the equations of motion, this time keeping track of the components and unit vectors for a non-Cartesian coordinate system by writing $\mathbf{u}$ as $u_i \hat{e}_i$ where u_i is the vector component and $\hat{e}_i$ is the unit vector. Now take the momentum equation of the form

$$\rho \left(\frac{\partial(v_i \hat{e}_i)}{\partial t} + v_j \frac{\partial}{\partial x_j}(v_i \hat{e}_i) \right) = F_i \hat{e}_i \tag{1}$$

where F_i is the component of the force density. Normally if we are dealing with rectilinear coordinates, the unit vectors are all constants and independent of the other coordinates. Therefore, the simple form of the equation with which we have been dealing thus far results—terms of the form $\hat{e}_{i,j}$, where the comma plus subscript indicates the partial derivative with respect to the jth coordinate, don't occur. In cylindrical coordinates, however, this is no longer true. To see this, consider the line element for the axisymmetric coordinate system (r, ϕ, z), $dl^2 = dx_i dx_i = dr^2 + r^2 d\phi^2 + dz^2$ Converting the coordinates to Cartesian shows that there is a mixture between the r and ϕ coordinates. In fact,

$$\frac{\partial \hat{r}}{\partial \phi} = \hat{\phi}, \qquad \frac{\partial \hat{\phi}}{\partial \phi} = -\hat{r} \tag{2}$$

All other derivatives vanish. To see this, try a graphical approach. Draw a vector tangentially to a circle (in the ϕ direction) and rotate it about the center, keeping track of the direction in which the length of the vector changes. For *spherical* coordinates, it follows that all of the radial derivatives vanish as they do for the cylindrical case, but there are two angles and a line element that goes as $ds^2 = dr^2 + r^2 d\theta^2 + r^2 \sin^2\theta d\phi^2$ and the nonvanishing derivatives of the unit vectors become

$$\frac{\partial \hat{r}}{\partial \theta} = \hat{\theta}, \qquad \frac{1}{\sin\theta} \frac{\partial \hat{r}}{\partial \phi} = \hat{\phi} \tag{3}$$

$$\frac{\partial \hat{\theta}}{\partial \theta} = -\hat{r}, \qquad \frac{1}{\sin\theta} \frac{\partial \hat{\theta}}{\partial \phi} = \frac{\cos\theta}{\sin\theta} \hat{\phi} \tag{4}$$

$$\frac{1}{\sin\theta} \frac{\partial \hat{\phi}}{\partial \phi} = -\hat{r} - \frac{\cos\theta}{\sin\theta} \hat{\theta} \tag{5}$$

3.1 Introduction

For Cartesian coordinates the derivatives all vanish, as we have asserted. With these, you should be able to derive the appropriate operators.

We now expand the velocity derivative as follows:

$$\frac{\partial}{\partial x_j} v_i \hat{e}_i = \hat{e}_i \frac{\partial v_i}{\partial x_j} + v_i \frac{\partial \hat{e}_i}{\partial x_j}$$

Using the cylindrical coordinate system, we see that this term *mixes the radial and angular coordinates*. Taking the derivatives of the unit vectors transforms the terms in the preceding equation into

$$v_r \frac{\partial v_r}{\partial r} \hat{r} + v_\phi v_r \hat{\phi}, \quad \left(v_r \frac{\partial v_\phi}{\partial r} + v_\phi \frac{\partial v_\phi}{r \partial \phi} \right) \hat{\phi} - \frac{v_\phi^2}{r} \hat{r}$$

For most of what follows, we *assume* that the velocity is independent of ϕ, that is, axisymmetric, so that the second term in the last equation vanishes. Spiral waves are thus explicitly excluded, as are any periodic waves which are not strictly radial. We also can *assume* that the flow is constant on cylinders, so that angular momentum is conserved on cylinders; this will be shown to be a consequence of the Taylor–Proudman theorem. Collecting terms by their unit vectors, we have as the time-dependent equations of motion the following:

$$\frac{\partial v_r}{\partial t} + v_r \frac{\partial v_r}{\partial r} - \frac{v_\phi^2}{r} = \frac{F_r}{\rho} \tag{6}$$

$$\frac{\partial v_\phi}{\partial t} + v_r \frac{\partial v_\phi}{\partial r} + \frac{v_r v_\phi}{r} = \frac{\partial v_\phi}{\partial t} + \frac{v_r}{r} \frac{\partial}{\partial r}(r v_\phi) = \frac{F_\phi}{\rho} \tag{7}$$

For the moment, we also ignore the $\hat{z}$ component. Equation (7) describes the equation for conservation of angular momentum, where the right side of the equation is the torque. Also, notice the centrifugal term in Eq. (6).

We can now examine the Laplacian in component form. It is written as

$$\nabla^2 \mathbf{v} = \frac{\partial}{\partial x_j} \left(\frac{\partial (v_i \hat{e}_i)}{\partial x_j} \right) = \frac{\partial^2 v_i}{\partial x_j^2} \hat{e}_i + 2 \frac{\partial \hat{e}_i}{\partial x_j} \frac{\partial v_i}{\partial x_j} + v_i \frac{\partial^2 \hat{e}_i}{\partial x_j^2} \tag{8}$$

Notice that here the unit vectors have to be operated on separately from the vector components, so the transformation is not just a double application of the gradient operator to u_i alone. It is straightforward to see that we can obtain the full form of the Laplacian for *any* coordinate system. Let us look again at the cylindrical coordinate system with which we have been working. The *second* derivatives of the unit vectors $\hat{r}$ and $\hat{\phi}$ are $-\hat{r}$ and $-\hat{\phi}$, respectively. Then the components for the Laplacian are

$$\nabla^2 v_r \hat{r} = \hat{r} \left[\nabla^2 v_r - \frac{2}{r^2} \frac{\partial v_\phi}{\partial \phi} - \frac{v_r}{r^2} \right], \quad \nabla^2 v_\phi \hat{\phi} = \hat{\phi} \left[\nabla^2 v_\phi + \frac{2}{r} \frac{\partial v_r}{\partial r} - \frac{v_\phi}{r^2} \right] \tag{9}$$

Again the effect of the non-Cartesian coordinates is to mix the components. By the way, we have been dropping the $\hat{z}$ components throughout this discussion. But since this coordinate is effectively independent of the others, it seemed an appropriate thing to do without loss of generality.

3.1.2 The Equations of Motion

We now consider the angular momentum equation explicitly. If there is no torque, and if the system is static, then the equation has the solution $v_\phi \sim 1/r$. This can be substituted into the radial equation to provide the usual equation for a uniform rotation rate $\omega = $ constant so that the centrifugal term in the radial equation is simply j^2/r^3. The point is that in the case of constant specific angular momentum, $j = $ constant,

$$\frac{Dv_r}{Dt} = -\frac{1}{\rho}\frac{\partial \Psi}{\partial r} \tag{10}$$

Here Ψ is the *effective* potential, including the rotational energy (given in terms of the specific angular momentum, j), $\Psi = \Phi_{\text{grav}} - \frac{1}{2}r^2\omega^2$. This equation gives a set of bound states (as is usually the case in rotational systems or bound states in central fields in classical mechanics).

There is one more conservation condition that we have not yet written down, and this is one which will have very central importance in our deliberations from here on. The continuity equation for an incompressible fluid is $\nabla \cdot \mathbf{v} = 0$. Now, we know a solution to this equation, without any additional conditions coming into consideration. It is that $\mathbf{v} = \nabla \times \boldsymbol{\psi}$ where $\boldsymbol{\psi}$ is a vector potential. That is, a divergenceless flow always can be represented by a solenoidal vector potential. Actually, you can see this essentially linguistically, and it is in fact the origin of the terms for the operators. We can now define a new quantity, which in fact we have already used, namely the *vorticity*:

$$\boldsymbol{\omega} = \nabla \times \mathbf{v} \tag{11}$$

which has the dimensions of the frequency. Therefore, we can write the equation of motion as

$$\frac{\partial \mathbf{v}}{\partial t} + \mathbf{v} \times \boldsymbol{\omega} = -\nabla\left(\frac{p}{\rho} + \frac{1}{2}v^2\right) - \nabla\Phi$$

Taking the curl of both sides and noting that the curl of a gradient always vanishes, we have

$$\frac{\partial \boldsymbol{\omega}}{\partial t} + \nabla \times (\mathbf{v} \times \boldsymbol{\omega}) = 0 \tag{12}$$

3.1 Introduction

This happens only if the fluid is incompressible or if the condition $\nabla\rho \times \nabla p = 0$ is met. We shall come upon this condition frequently from here on, so note it well. This equation is an "induction equation," of sorts, an analog of the Faraday induction equation in the Maxwell description of the electromagnetic field. Actually, it is really the case that the equations were found in the inverse order—the hydrodynamics came prior to the electrodynamics. The equation for conservation of vorticity results from the fact that, for a system that has no divergence and has (prove this yourself) no divergence of the vorticity, then

$$\frac{D\boldsymbol{\omega}}{Dt} = \boldsymbol{\omega} \cdot \nabla \mathbf{v}$$

We have made use of the fact, for an incompressible fluid, $\nabla \cdot \mathbf{v} = 0$.

Now in the case of a compressible medium, the continuity equation must (as we have already discussed) be solved explicitly. The nonvanishing of the density fluctuation in time and space means that there are potential source and sink terms for the angular momentum and that the vorticity as such is no longer conserved. The reason is that the increase in the local density at constant angular frequency can increase the total angular momentum in a closed region and thus couple the density field to the velocity field. As we shall see later, this is the same problem that one encounters in the case of the difference between two- and three-dimensional flows. But, for the moment, let's continue with the more easily solved case of incompressible flow.

If the vortex is stretched by a velocity field, its strength changes. Specifically, if there is a mean velocity component that is a uniform flow the vorticity is increased by stretching. (See Fig. 3.1.) Consider, for example, the case of an updraft in a vorticial flow. Assume that there is an initial

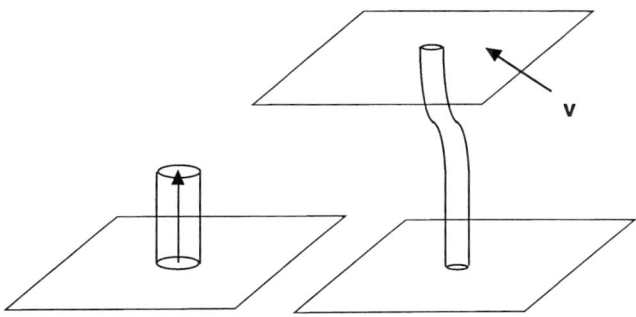

Figure 3.1 Vortex stretching.

whirlpool which is stretched downward or upward. Then the vorticity is increased. This last point can be observed quite easily in a simple experiment. Take a cup of coffee and pour in some very cold cream. As the cream sinks (see Chapter 9 for some additional consequences of this experiment) the rotation rate of the cream is seen to increase dramatically! As the cream heats up and becomes buoyant, the rate of rotation of the central column is seen to decrease, since the line is contracted, resulting in a lowering of the required vorticity. Similar observations can easily be made in a bathtub. To analyze this situation further, we consider the equation of motion with viscosity included:

$$\frac{D\boldsymbol{\omega}}{Dt} = \boldsymbol{\omega} \cdot \nabla \mathbf{v} + \nu \nabla^2 \boldsymbol{\omega} \tag{13}$$

and take the scalar product with the vorticity ω. Integrating over the volume of the fluid, we obtain

$$\frac{d}{dt} \int \frac{1}{2} \omega^2 \, dV = \int \boldsymbol{\omega} \cdot (\boldsymbol{\omega} \cdot \nabla \mathbf{v} + \nu \nabla^2 \boldsymbol{\omega}) \, dV \tag{14}$$

Two important quantities have now appeared. The first is the *enstrophy*, defined by $\frac{1}{2} \int \omega^2 \, dV$. This is like a total energy; it represents the quantity of energy tied up in the fluid in the vorticity field. The second quantity is the *helicity*, defined by $\int \mathbf{v} \cdot \boldsymbol{\omega} \, dV$, but this is more implicit and we will study it shortly. This is a measure of the anisotropy of the vorticity distribution with respect to the mean flow of the medium. Now identify the components of the vorticity and velocity, writing this last equation as

$$\frac{d}{dt} \int \frac{1}{2} \omega_i \omega_j \, dV = \int \omega_i \omega_j \frac{\partial v_i}{\partial x_j} dV - \nu \int \left(\frac{\partial \omega_i}{\partial x_j}\right)^2 dV + \frac{1}{2} \nu \int \frac{\partial (\omega_i \omega_j)}{\partial x_j} dS_j \tag{15}$$

where dS is the surface element normal to the flow. The components are represented in Cartesian coordinates for simplicity. Notice that the critical term in this last equation is the one which stretches the vortex lines—it should be noted that the other terms are either strictly positive or vanishing. It is very important to keep in mind that the vortex amplification is possible due to the convergence of the field lines. If the velocity increases along the direction of the vortices, then the rate of rotation increases. The quantity that describes this property is the helicity.

The equation for the helicity is derived as follows. Call $\mathcal{H}$ the helicity density, $\mathbf{v} \cdot \boldsymbol{\omega}$. Since it is scalar, you may suspect that it will be transported by the potential vorticity in the same way that the entropy (the potential temperature) is (see below). Take the scalar product $\boldsymbol{\omega} \cdot \dot{\mathbf{v}} + \mathbf{v} \cdot \dot{\boldsymbol{\omega}} = \dot{\mathcal{H}}$ to

3.1 Introduction

obtain the evolution equation:

$$\frac{\partial \mathcal{H}}{\partial t} + \nabla \cdot \mathcal{H}\mathbf{v} = -\omega \cdot \left(\frac{1}{\rho}\nabla p + \nabla \frac{1}{2}v^2\right) + \mathbf{v} \cdot \left(\frac{1}{\rho^2}\nabla \rho \times \nabla p\right)$$

$$-\omega \cdot \mathbf{g} + \nu \nabla^2 \mathcal{H} - 2(\nabla \mathbf{v}) \cdot (\nabla \omega) \tag{16}$$

This is a continuity equation for the helicity density. The terms on the right-hand side represent creation and annihilation terms. The stretching and combination of vortices by the mean flow are responsible for generation; the viscosity is a dissipative term. For an inviscid barotropic fluid (read: "incomprehensible" as well) that is moving adiabatically, the volume integral of Eq. (16) gives

$$\frac{\partial}{\partial t}\int \mathcal{H}\, dV = 0 \tag{17}$$

so that the helicity is, under these specific conditions, a strictly conserved quantity in the flow.

Vortices are especially interesting because they can be studied topologically. A whole subdiscipline, topological fluid dynamics, focuses on this aspect, especially in the context of turbulence. As an illustration of the qualitative reasoning in this business, think of a single vortex imbedded in a background medium. Assume that the viscosity does not vanish in either the vortex or the surrounding fluid. As it propagates, due to shear, the vortex becomes unstable and develops nonaxisymmetric deformations. These form rolls and waves around the periphery of the ring. Eventually, as the amplitude of nonplanar motion increases, coupling to the environment can cause some of these kinks to break off and recombine, spawning new vortices of smaller radius. These can collide, or recombine, and eventually the entire ring dissipates into turbulent uniform fluid.

3.1.3 Working in the Rotating Frame

On the corner of a Second Empire apartment building on Rue Vaugirard in Paris, a plaque records that "in the stairwell of this building, Foucault demonstrated the rotation of the Earth." A sculpted pendulum accompanies the inscription. The Coriolis force, demonstrated by the pendulum experiment, is absolutely vital to an understanding of the treatment of rotating coordinate systems. For the case of slow rotation, and of slow motion within the frame, it may in fact be the dominant force.

We need to consider the frame in which rotation problems are posed. If we are interested in the appearance of the body, as in the case of starspots or binary system phenomenology, then the external (stationary) vantage point may be the best one for describing the behavior of the system. In this case, we are essentially dissecting the surface structure, using the rotation merely as a means by which the body presents its various surface features to the observer. For the dynamical problems, however, working in this frame is absurdly hard. Instead, we put ourselves in the corotating frame and examine the dynamics there. All time derivatives in this frame also contain the motion of the frame itself, and therefore the flow acquires a new force, $2\mathbf{\Omega} \times \mathbf{v}$, the Coriolis force. It depends on motion toward or away from the rotation axis, and this is because of conservation of angular momentum. Any motion on the plane normal to the rotation axis produces some circulation. Note, for instance, that any motion on the surface of a sphere not parallel to the equator has some component toward or away from the rotational axis. If we look at the equations for the vorticity, we obtain

$$\frac{d}{dt}\boldsymbol{\omega} + \boldsymbol{\nabla} \times (2\Omega\hat{\mathbf{z}} + \boldsymbol{\omega}) \times \mathbf{v} = \frac{1}{\rho^2}\boldsymbol{\nabla}\rho \times \boldsymbol{\nabla}p + \boldsymbol{\nabla} \times \frac{1}{\rho}\mathbf{f} \qquad (18)$$

where we have now explicitly included the Coriolis term in the equations prior to taking the curl; also, for a change, we have *not* assumed incompressibility. If the rotation rate Ω is independent of time (although it may depend on position), we can bring it under the time derivative by defining a new quantity, $\boldsymbol{\varpi} \equiv 2\boldsymbol{\Omega} + \boldsymbol{\omega}$, called the absolute vorticity (also called ω_a in some texts):

$$\frac{d}{dt}\frac{\boldsymbol{\varpi}}{\rho} = \frac{\boldsymbol{\varpi}}{\rho}\cdot\boldsymbol{\nabla}\mathbf{v} + \frac{1}{\rho^3}\boldsymbol{\nabla}\rho \times \boldsymbol{\nabla}p + \frac{1}{\rho}\boldsymbol{\nabla} \times \frac{1}{\rho}\mathbf{f} \qquad (19)$$

To get Eq. (19) requires using $d\rho/dt = -\rho\boldsymbol{\nabla}\cdot\mathbf{v}$. The rotation of the frame adds to any vorticity in the velocity field; any motion within a rotating fluid serves as a potential source for vorticity. For now, define $\chi \equiv \boldsymbol{\varpi}/\rho$. We won't need this for long, but it will help.

As it stands, Eq. (19) provides a means for examining the transport of thermal properties of the fluid as well as the momenta. All thermodynamic variables are scalars, since they concern the energy of the medium. Therefore, let's look at some quantity Q and examine how its redistribution may be altered in a rotating compared with a stationary frame. Suppose in the fluid there is a gradient $\boldsymbol{\nabla}Q$ in this quantity. You know instances of this, like the pressure. We may be interested in how fluid motions alter this gradient and redistribute Q through the medium. Alternatively, we may want to know something about the conditions under which the gradient is

3.1 Introduction

large enough that it drives the motion, rather than being transported like a *passive* scalar. Now write

$$\chi \cdot \frac{d}{dt}\nabla Q = \frac{d}{dt}(\chi \cdot \nabla Q) - \nabla Q \cdot \frac{d\chi}{dt}$$

$$= \frac{d}{dt}(\chi \cdot \nabla Q) - \nabla Q \cdot (\chi \cdot \nabla \mathbf{v}) + \frac{1}{\rho}\nabla Q \cdot (\text{remaining terms}) \quad (20)$$

where the last term will not concern us for the moment. Notice that the term $\nabla Q \cdot (\chi \cdot \nabla \mathbf{v})$ can be written in component form as

$$\chi_k \frac{\partial Q}{\partial x_i} \frac{\partial v_i}{\partial x_k}$$

which makes it easier to see that it becomes $\chi \cdot \nabla(\mathbf{v} \cdot \nabla Q) = \chi \cdot \nabla \dot{Q}$. Notice that the last transformation used the advective derivative. If we now go back to our original variables, we can define a new quantity

$$\Pi = \frac{\boldsymbol{\varpi}}{\rho} \cdot \nabla Q \quad (21)$$

called the *potential vorticity* by geophysical fluid types. This quantity is constant if the medium is barotropic and there are no torques; otherwise it is conserved. The equation of conservation of potential vorticity is, finally,

$$\frac{d}{dt}\Pi = \frac{\boldsymbol{\varpi}}{\rho} \cdot \nabla \dot{Q} + \frac{1}{\rho^3}\nabla Q \cdot (\nabla \rho \times \nabla p) + \frac{1}{\rho}\nabla Q \cdot \left(\nabla \times \frac{1}{\rho}\mathbf{f}\right) \quad (22)$$

If Q is a conserved quantity, then transportation of potential vorticity is effected by departures from the barotropic state (the second term on the right) and/or external driving torques. So you can already see that under circumstances where the driving forces, like gravity, are not in the same direction as the rotation axis, there will be trouble.

What should we choose for the quantity Q? Actually, any scalar quantity is fine. For instance, we can look at the entropy S. This is a conserved quantity for adiabatic motion. We have used $\ln(p\rho^{-\gamma})$ for entropy; a representation used in meteorological literature is called the potential temperature, $\theta = T(p/p_0)^{(\gamma-1)/\gamma}$ and the entropy is given by $S = c_p \ln \theta$. Then it can be said that potential vorticity transports potential temperature, an essential part of front generation and the formation of large-scale weather systems (Hoskins et al. 1985).

3.1.4 Taylor–Proudman Theorem

This theorem, related to the circulation, is best demonstrated for incompressible flows. But since incompressibility and constant density mean the same thing, as long as we have a layer thin enough or there are no density effects due to radiation or a very low γ, this is not a bad approximation. In order to prove this theorem, it is necessary to look at the equation for the vorticity (the induction-like equation) and expand it to obtain $\mathbf{v} \cdot \nabla \omega = 0$. Take the curl of the equation of motion to obtain an evolution equation for the vorticity as before, but now assume that ω is conserved. Then

$$2\mathbf{\Omega} \cdot \nabla \mathbf{v} - 2\mathbf{\Omega} \nabla \cdot \mathbf{v} + \frac{1}{\rho^2} \nabla \rho \times \nabla p = 0 \qquad (23)$$

The second term vanishes for an incompressible fluid, and if the pressure depends only on the density, as it does for a polytrope, the third term also vanishes. This leaves

$$\mathbf{\Omega} \cdot \nabla \mathbf{v} = 0 \qquad (24)$$

That is to say, the fluid motion is constant along a vertical column defined by the rotation axis. (See Fig. 3.2.) This result, known as the *Taylor–Proudman theorem*, is basic to an understanding of geostrophic motion, and says that the motion is defined as a function of the plane coordinates

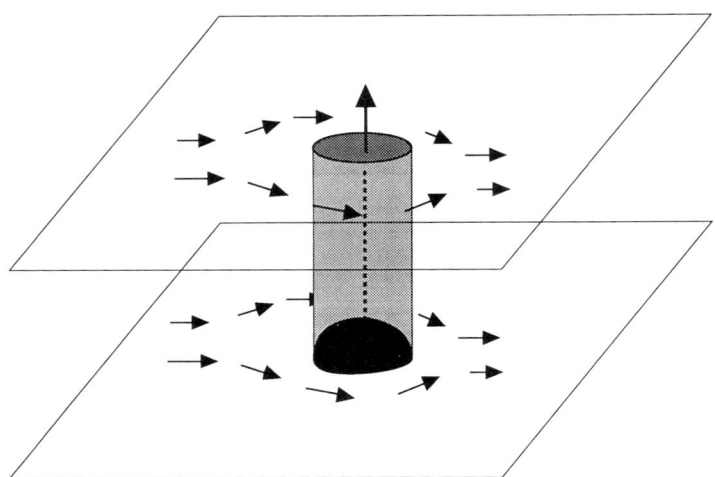

Figure 3.2 Vortex pinning on a barrier forming a Taylor column.

(x, y) rather than the coordinates perpendicular to the plane. It is one of the conditions which leads to the statement that there are tornado alleys and that there is a stability to vortex lines in a rotating superfluid—oddly enough, both problems with considerable astrophysical application. In the presence of an external force this will also be true provided the force is given by a gradient of a potential. As an example of the application, note that if we have a cylinder with gravity acting along the axis, then $\mathbf{g} = g\hat{\mathbf{z}}$ has no curl, and the theorem holds. We could also have started with the geostrophic equations. In this case, the equations of motion are given by $2\mathbf{\Omega} \times \mathbf{v} = -(1/\rho)\nabla p$. We take the curl of both sides and assume the barotropic condition and incompressibility and we obtain the same result.

A remarkable consequence of this theorem is that some convection patterns are constrained to lie along isorotational surfaces. Consider a rotating spherical annulus. If the fluid is incompressible and gravitation acts only as $\mathbf{g} = g\hat{\mathbf{r}}$, then the motion of any buoyant fluid will obey the Taylor–Proudman theorem and meridional motion will be established. This means that the equivalent of Hadley circulation or "banana cells" are the product of convective motions in a thin shell. Such motion has been directly simulated by Hart, Glatzmeier, and Toomre (1986) using a space shuttle–based experiment. Gravity was simulated for a silicone fluid shell using electrical potentials and the experiment of inducing convection in a rotating hemispherical fluid annulus was conducted in essentially zero-gravity conditions. The circulatory pattern observed was essentially that expected from the Taylor–Proudman theorem, in addition to a very rich ensemble of higher-order modes that we will discuss in the chapter on instabilities.

3.1.5 The Barotropic Condition

We have made use of a very important property of rotating fluids in deriving the Taylor–Proudman result. For an incompressible medium, the density is just a constant that we can ignore (or rather absorb into the pressure). But for a compressible fluid, the usual astrophysical situation, we have actually made an assumption in taking the curl of the two sides of the geostrophic equation:

$$\nabla \rho \times \nabla p = 0 \qquad (25)$$

This is the *barotropic* condition. It states that the surfaces of constant pressure are also ones of constant density. Alternatively, it says that the thermal state of the gas depends only on the density (obviously satisfied by a simple polytrope for a nonrotating star). We will meet later with some physical conditions that can violate this constraint, but you should keep in mind that such an assumption is tacit in any derivation of the Taylor–Proudman theorem when applied to gaseous rotating bodies.

3.1.6 Circulation and Kelvin's Theorem

We now state the *Kelvin circulation theorem*. Define a quantity

$$C \equiv \int \boldsymbol{\omega} \cdot d\mathbf{S} \tag{26}$$

which is the vorticity passing normally through an area dS of fluid. This is, by Stokes' theorem, the same as

$$C(t) = \int_{\text{contour}} \mathbf{v} \cdot d\mathbf{l} \tag{27}$$

This quantity, first defined by Kelvin, is called the *circulation*; it is also written as Γ in some discussions. Since the fluid velocity and the vorticity may both be functions of time, we suspect that there is also an evolution equation for the circulation. Take the time derivative, substitute the Navier–Stokes equation in the absence of body forces, and you will get

$$\frac{dC}{dt} = \int \frac{D\mathbf{v}}{Dt} \cdot d\mathbf{l} + \int \mathbf{v} \cdot (d\mathbf{l} \cdot \nabla \mathbf{v}) = \nu \int (\nabla^2 \mathbf{v}) \cdot d\mathbf{l} \tag{28}$$

The only term from the equations of motion that remains is the viscosity. All of the others are the result of gradients (potential fields, like gravitation) and therefore the line integrals vanish for these. The incompressibility of the fluid is used to transform $\nabla^2 \mathbf{v}$ into $-\nabla \times \boldsymbol{\omega}$, thus providing the dissipative term from the viscosity:

$$\frac{dC}{dt} + \nu \int (\nabla \times \boldsymbol{\omega}) \cdot d\mathbf{l} = 0 \tag{29}$$

For an *inviscid* fluid, $C(t)$ is conserved exactly. But an essential feature of viscosity is that it redistributes momentum and dissipates energy and therefore the circulation of a medium must be altered when it is not inviscidly flowing. We shall see more of this later. Note, by the way, that in a rotating frame there is no flow that does not have some circulation. The same form of Kelvin's theorem applies in this case, with the exception that the vorticity is replaced by $\boldsymbol{\omega} + 2\boldsymbol{\Omega}$.

The most important consequence of the Kelvin theorem is this. An inviscid fluid initially at rest has no circulation. It therefore *never* has any, regardless of what disturbances it may be subjected to. So if you move a

3.1 Introduction

body through the medium in such a way that the motion is reversible (not dissipative), the flow around the body will be antisymmetric. No permanent net angular momentum can be transferred to the medium by this perturbation. You have often encountered such an effect, when moving oars through water or when moving your hand through a bath initially at rest. Counterrotating vortices are generated on either side of your hand, and while these may persist for some time, and even separate, they do not represent a net increase in the angular momentum of the fluid. It may be hard to realize this as you watch these vortices seem to take on lives of their own, but there is no net spin-up on the medium. These are also called *von Kármán* vortices, since they are also formed in the wake of an airfoil because of vorticity shedding.

There is an interesting use of the circulation theorem in the case of airfoils, which can serve to illustrate many of the effects of flows around asymmetric blunt objects. Before proceeding, it should be said that stars accreting matter out of the stellar wind of a companion in a binary system, or galaxies producing wakes in the intergalactic medium of a cluster, can also be thought of in these terms. So it is not entirely inappropriate to use the analogy here. If the density is constant in a flow, then the deviation of the material around an object produces a pressure differential provided the body is asymmetric. (See Fig. 3.3.) This will happen even in the absence of a gravitational field to provide the asymmetry. Specifically, recalling the Bernoulli equation, the pressure is decreased when the flow velocity is increased along a streamline. Therefore, in the case of an airfoil, there is a pressure deficit on the upper side of the wing, $\Delta P = \frac{1}{2}\rho(v_{top}^2 - v_{bottom}^2) \approx \rho U \Delta u$, where U is the mean flow speed and we assume small deviations in the velocity result from the body. Then, since the flow is incompressible, we have for the lift force the integration over the span of the

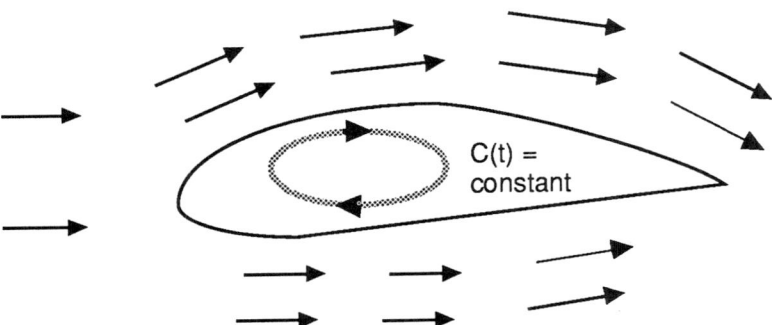

Figure 3.3 Flow over an airfoil and Kelvin's theorem.

wing in order to produce the force per unit length:

$$f = \rho U \int \mathbf{v} \cdot d\mathbf{l} \equiv \rho U C \tag{30}$$

where we have now identified the integral as the circulation. The circulation results, in the rest frame of the wing, from the velocity difference on the top and bottom. Thus, an increase in the circulation is equivalent to an increase in the lift.

There is another interesting consequence, one that may have more astrophysical application, and that is wake formation. Let's look at the case of a blob, or blunt body, moving in a stream. The flow is initially in one direction. The body forms an obstacle that diverts the flow around it on either side. If the fluid is not inviscid, a boundary layer is formed that separates some distance behind the blob. This viscous boundary forms a bisymmetric flow about the blob and the wake therefore feels a pressure deficit on separation and some ambient fluid is entrained in the wake that is formed. This wake is vortical, and it is here that the lift analogy is important.

If there is initially no vorticity, no viscosity, and the flow maintains constant density (or equivalently, the fluid is incompressible), then the flow takes the form of a streamline satisfying $\nabla^2 \psi = 0$. For a sphere of radius a in a flow having the form $v_\infty \cos \theta =$ constant (unidirectional), the streamlines have the form

$$\psi(r, \theta) = -\frac{1}{2} v_\infty r^2 \left(1 - \frac{a^3}{r^3}\right) \sin^2 \theta \tag{31}$$

and the velocity is given by $\mathbf{v} = \nabla \psi$ because we have been able to assume that the flow is irrotational ($\nabla \times \mathbf{v} = 0$) and divergenceless. This flow is perfectly symmetrical and as such can be taken as a model for inviscid flow. It forms no wake, injects no vorticity into the flow. It is relative to this model that viscosity can be understood as introducing a symmetry-breaking effect into the motion. If the flow is at least a little viscous, then a wake forms. This is a leeward circulatory flow, some of which is directed back toward the blob. This means that material from the stream is transported back toward the body. The pressure deficit that results is responsible for drag, but for our case there is another effect: material brought into the wake may penetrate the blob and entrain and mix.

Physically, this process is observed in a variety of bodies, notably starting plumes and vortex rings, and also in the rising thermals of clouds and explosions. In the case of a rising cloud from a nuclear explosion, the heated gases from the fireball rise and entrain material both from the ground (hence the fallback often observed from the heavier particles dripping out of the rising entrained flow) and from the environment. The entrained material is cooler than the blast and mixes with it. In the case of a more quiescent rising thermal, the precursor to a cumulus cloud, the entrained fluid is colder air and this lowers the buoyancy of the thermal. In the case of an astrophysical environment, a blob in convective flow, or a blob formed in a stellar wind, will develop vorticity on moving through a background gas. The entrainment may be responsible for more efficiently mixing material and limiting the rise of the blob than the normal ideas of mixing might indicate. At any rate, in an incompressible fluid, the entrainment is not penetrative because only velocity and vorticity can be transported within the fluid. Once compressibility effects are felt, it is likely that mixing is more efficient.

3.1.7 Vortex Dynamics and Line Stretching

We note that the *Taylor–Proudman* theorem also has important implications for vortex dynamics. A vortex line is essentially pinned to any object which happens to produce a change in the shape of the line. If we take a barrier, like a rock, and place it in the way of a vortex line, the line will pin to the obstacle. The motion of the blockage will then drag the line with it, with the result that the line will appear to be originating from the region. Kelvin's circulation theorem can also be written in terms of the vorticity and the area. Remember that, in this form, it states that the vorticity passing through an area of fluids is conserved. So if the area is changed, say decreased, the vorticity must increase. From the potential vorticity equation you can see that this means that the gradient in some scalar transported variable is reduced when ϖ is increased, at least in the absence of dissipation.

3.1.8 Magnus Force

In the case of the flow between two cylinders which are rotating, or in the case of a single rotating cylinder, this is the cause of the *Magnus force*. (See Fig. 3.4.) For a single rotating cylinder this is the lift that results from the fact that rotation orthogonal to the flow breaks the symmetry of the flow and generates the analogy of the airfoil. For the flow between two cylinders, if the fluid is forced between them at high velocity, it is compressed and therefore must travel faster, resulting in a reduction of pressure and an attractive force between the cylinders. In either case, this force acts perpendicular to the symmetry axis of the cylinder.

There is a very interesting application of this motion in an extreme astrophysical environment which can be mentioned here. It is that the

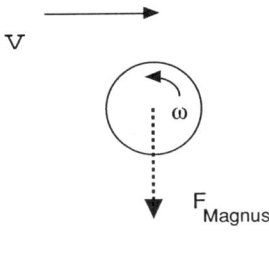

Figure 3.4 The Magnus effect for a rotating disk in a flow.

superfluid, argued to exist in the mantle of a neutron star (the deep envelope is so dense that it behaves like a quantum fluid and therefore the term envelope seems somehow out of place), quantizes due to the rotation of the neutron star. This means that the angular momentum on a vortex line takes on only discrete values and these behave like particles in the background moving fluid. The behavior of such a system is well known from laboratory studies of ^{4}He, so it is only the fact that neutrons and proton pairs form the fluid that is different. The motion of the vortex lines in the presence of a circulating fluid is therefore like a set of cylinders which pin to the two surfaces which bound the superfluid, namely the core and the underside of the crust. The pinning is due to the Taylor–Proudman theorem. If the rotation of the star is large, the lines will feel a tendency to drift in their pinning sites due to the Magnus force and therefore will strain the crust at the point of contact. This provides a mechanism for the storage of rotational energy in the vortices and a large potential energy to be tapped in the event that the sites should become depinned. Currently, this is one explanation offered for the production of the "glitch" event, the discontinuous rotational spin-up of the surface of a pulsar.

There is an additional place in which the Magnus force may play a role. In the case of a jet emerging from the core of a radio galaxy, the flow appears to the background interstellar medium and intergalactic medium as a highly collimated structure with a large inertia due to internal flows. If this appears difficult to picture, imagine a chimney instead with a plume of exhaust rising rapidly above it (List 1982). The flow of a background wind will be deviated by the rising jet and will therefore, if there is a helicity in the jet (as there might be), produce a Magnus force deviation of the plume. This enhances the deviation of the jet from the vertical, and the increase in the area of the plume in the direction of the flow also increases the entrainment (a topic to be discussed in a later chapter). Thus the deviation begins the process of destruction of the jet and also the eventual mixing with the surrounding medium. Whether this plays a role in the structuring of radio source jets is still an open question, but the analogy with other fluid flows may be a useful illustration.

3.2 Geostrophic Approximation

Although the term derives from ocean and atmospheric dynamics, the *geostrophic approximation* has astrophysical uses. It is a means for treating driven flows within a slowly rotating frame. The motions are assumed to be sufficiently slow as well that the only accelerations are due to pressure and gravitational gradients and the Coriolis force. In this chapter we will merely touch on this subject. It would otherwise occupy the entire discussion, so central is this to the study of atmospheric dynamics and geophysical flows. It is essential that the rotating layer be in vertical pressure balance, so that there are initially no driven motions along the vertical direction. The one assumption which we require is that the existence of pressure gradients in a plane is balanced by the Coriolis acceleration and that all of the other net velocities are very slow. That is, we assume that $\mathbf{v} \cdot \nabla \mathbf{v}$ is very small. Then the steady-state equation of motion *in the plane* is given by

$$2\rho \mathbf{\Omega} \times \mathbf{v} = -\nabla p - \rho \nabla \Phi \tag{32}$$

Therefore, given a way of calculating the pressure gradients, the velocities can be found. If there is a temperature gradient along the surface, the pressure will not be uniform and therefore a flow can be set up which will transport energy and momentum in an attempt to balance out the net driving resulting from the gradient. Any additional accelerations are provided by gravity (in general), and these equations are supplemented with the continuity equation for the density and the equation of state for the pressure. Vertical gradients in temperature and/or pressure are responsible for driving shear. This is seen from the fact that the flows obey not only momentum conservation but also mass conservation. Any vertical gradients in the driving forces must translate, through the vorticity and velocity divergence, to a motion in the (x, y) plane.

3.2.1 Rotation in an Incompressible Shallow Layer

To illustrate how such a calculation proceeds, take the case of a fluid layer in the *shallow-water* approximation. This is essentially an incompressible fluid, but it also holds when the depth of the layer in question is small compared with the scale height. Vertical hydrostatic equilibrium gives a pressure that depends only on the depth of the layer and the perturbation in this pressure, δp, is given by $\rho_0 g \eta$, where $\eta = \delta z$. Now because of the incompressibility assumption, the height of the layer is a sort of density variable. For this reason the continuity equation takes the form

$$\frac{\partial \eta}{\partial t} + \nabla_H \cdot (z_0 \mathbf{v}) = 0 \tag{33}$$

where ∇_H is the horizontal divergence. There is no motion induced in the $\hat{\mathbf{z}}$ direction. The perturbed geostrophic equations are

$$\frac{\partial u}{\partial t} + fv = -g\frac{\partial \eta}{\partial x}, \quad \frac{\partial v}{\partial t} - fu = -g\frac{\partial \eta}{\partial y} \tag{34}$$

where $f = 2\Omega_0$ is the Coriolis parameter. In effect, we close the equations through the assumption that the vertical displacement is equivalent to a pressure. Take the divergence and the curl of Eq. (34) separately and combine to get

$$\frac{\partial}{\partial t}(\nabla \times \mathbf{v}) + f \nabla \times (\hat{\mathbf{z}} \times \mathbf{v}) = 0, \quad \frac{\partial}{\partial t}(\mathbf{v} \cdot \nabla \ln z_0) - \frac{\partial^2 \eta}{\partial t^2} + f \nabla \hat{\mathbf{z}} \times \mathbf{v} = -g \nabla^2 \eta \tag{35}$$

These provide the final equation for the vertical displacement:

$$\frac{\partial}{\partial t}\left[\left(\frac{\partial^2}{\partial t^2} + f^2\right)\eta - \nabla \cdot gz_0 \nabla \eta\right] - gf\frac{\partial(z_0, \eta)}{\partial(x, y)} = 0 \tag{36}$$

where the last term is the Jacobian; this term vanishes if z_0 is constant.

Binaries actually provide the justification for talking about the shallow-layer problem in an astrophysical context. One reason for introducing this form of the equations is that they have actually been modeled in a laboratory simulation. The inspiration was a discussion by Prendergast (1960) applying the geostrophic approximation to the problem of mass transfer in slowly rotating binary star systems (see also Shu and Lubow 1981). Imagine the case of a binary system in which there is a gradient in the potential in the orbital plane. (See Fig. 3.5.) The equation of motion is solved using the three-body gravitational potential, also called a Roche potential:

$$\Phi_{\text{Roche}}(x, y) = -G\left(\frac{m_1}{|\mathbf{r} - \mathbf{r}_1|} + \frac{m_2}{|\mathbf{r} - \mathbf{r}_2|}\right) - \frac{1}{2}\Omega^2 r^2 \tag{37}$$

3.2 Geostrophic Approximation

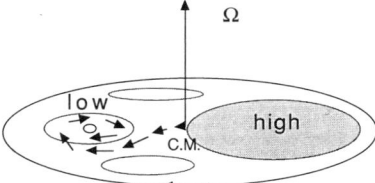

Figure 3.5 Flow in a binary star system as an example of geostrophic flow. The mass loser is represented as a local high and the mass gainer and its accretion disk form the local low.

so that there is a force which is produced by the fact that the particle feels the two stars, m_1 and m_2, at any point in its orbit, and this is also a centrifugal acceleration which results from the rotation of the coordinate system at a rate Ω. If the motions of the particle are slow, that is, if there are no large net accelerations in the system, it is possible to solve for the velocity at every point in the flow due to the pressure gradient which is set up by this acceleration. Unfortunately, the Roche potential does not provide a natural coordinate system in which to solve the geostrophic equations globally. The potential does, however, lend itself nicely to local analysis, which we will discuss more in the section on accretion disks and mass transfer in binary star systems. The simulation proceeded by modeling the depth of a rotating tank according to the Roche potential in three dimensions and then filling the model with a layer of water. As far as I know, this has not been repeated elsewhere but makes a wonderful illustration of how an analytical treatment can be realized in a laboratory experiment.

Motion on a Sphere: The Full Equations

The equations of motion in a thin layer on the surface of a sphere are a little different from those we have been using so far, so let's take a moment to examine them. There is one sense in which the motion is governed by the rotating frame. Any parcel in motion normal to the surface experiences a change in its angular momentum. This means that it carries a vorticity with it simply by initially being on the surface. Any motion along the normal direction involves motion away from the rotation axis and therefore every parcel of fluid experiences a torque. The motion in the radial direction is governed by the geostrophic term

$$2\rho\Omega v_\phi \sin\theta = -\frac{\partial p}{\partial r} \tag{38}$$

while the two on the surface of the sphere are given by

$$\frac{Dv_\theta}{Dt} - 2\Omega v_\phi \cos\theta = \frac{\partial v_\theta}{\partial t} + \frac{v_\theta}{r}\frac{\partial v_\theta}{\partial \theta} + \frac{v_\phi^2 \cot\theta}{r} - 2\Omega v_\phi \cos\theta$$

$$= -\frac{1}{\rho r}\frac{\partial p}{\partial \theta} - \frac{1}{r}\frac{\partial \Phi}{\partial \theta} \tag{39}$$

$$\frac{Dv_\phi}{Dt} - 2\Omega v_\theta \cos\theta = \frac{\partial v_\phi}{\partial t} + \frac{v_\theta}{r}\frac{\partial v_\phi}{\partial \theta} + \frac{1}{r\sin\theta}\frac{\partial v_\phi}{\partial \phi} + \frac{v_\theta v_\phi}{r}\cot\theta$$

$$= -\frac{1}{\rho r \sin\theta}\frac{\partial p}{\partial \phi} - \frac{1}{r\sin\theta}\frac{\partial \Phi}{\partial \phi} \tag{40}$$

Motion on a sphere is fundamentally different from that for a plane because of the possibility of crossing the equator. The Coriolis term reverses sign between opposite hemispheres.

Approximations

These equations are sufficiently complex that some simplification is necessary. Let's really oversimplify them at first. Assume that we take a small enough section of the sphere so that the tangential plane can be taken as a constant inclination to the surface. (See Fig. 3.6.) Then assume that the three directions are *up*, that is, normal to the surface, *along y*, and *along x*. The latter two can be taken as northward and eastward if you

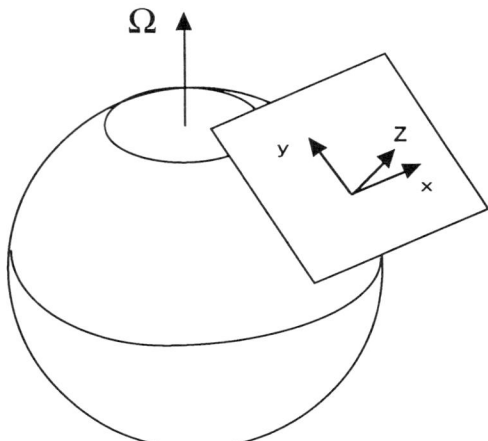

Figure 3.6 Geostrophic coordinate system.

3.2 Geostrophic Approximation

want, but it is better to think of one (y) as being essentially meridional and the other (x) as essentially azimuthal. Take the plane to be at some latitude away from the equator. Then the motion along y is actually at some inclination to the plane. The vertical motion is therefore $v \tan \lambda$, where λ is the inclination of the plane to the surface. Take the vertical gradient to be over one pressure scale height, h. This gradient is substituted in the continuity equation $\partial w/\partial z$; now $\nabla \cdot \mathbf{v}$ therefore resolves into

$$\frac{\partial u}{\partial x} + \frac{\partial v}{\partial y} + \frac{v \tan \lambda}{h} = 0 \tag{41}$$

This is the critical step. We assume that the motion in $\hat{z}$ is due entirely to shear in v and therefore we can eliminate the pressure term by taking $\nabla \times \nabla p/\rho$ and assuming barotropicity. The equations of motion simplify to

$$\frac{\partial u}{\partial t} + 2\Omega v = -\frac{1}{\rho}\frac{\partial p}{\partial x}, \qquad \frac{\partial v}{\partial t} - 2\Omega u = -\frac{1}{\rho}\frac{\partial p}{\partial y} \tag{42}$$

Now consider that for a thin shell we can assume that there is no dependence of the velocity field on y. Assume that the flow is only in the $\hat{x}$ direction initially. Then from the absolute vorticity, we know that $\tilde{\omega} = 0$ so that

$$\frac{d}{dt}(\hat{z} \cdot \nabla \times \mathbf{v}) = -2v \frac{\partial}{\partial y} \Omega \cdot \hat{z} \tag{43}$$

where we have assumed that the vertical lies in the $\hat{z}$ direction but that this direction is not parallel to the rotation axis except at the poles; Ω is not a function of time. Now for the thin shell, we can assume that the angle between Ω and $\hat{z}$ varies slowly with θ for some middle region of either hemisphere. We also take $(1/r)\partial/\partial \theta \to \partial/\partial y$ and $r \to R$, where R is a constant and the radius of the sphere. Therefore, the Coriolis force is a function of θ and we take a first-order expansion so that taking $\beta = \partial(2\Omega\hat{\omega} \times \hat{z})/\partial y \approx 2\Omega \cos \theta / R$, we can approximate Eq. (43) by

$$\frac{d\zeta}{dt} = -\beta v \tag{44}$$

Here we use the conventional notation $\zeta \equiv \hat{z} \cdot \nabla \times \mathbf{v}$, the vertical component of the vorticity. This remarkable equation is the basis for studying the perturbations in the rotating frame and in meteorology is called the *β-plane approximation*. Notice that we have no gravity in this problem. Once we make the approximation that the layer is thin and geostrophic, all large-scale flows are formed by the pressure gradients, which in turn reflect the local gravity. The small-scale perturbations are entirely dominated by the

Coriolis force. Equation (44) states that the change in the vorticity with time is due to the fact that the Coriolis acceleration is actually a function of latitude and so looks like a sloping lower surface for a vortex line. Let's complete the discussion by assuming that the flow is initially only in the $\hat{x}$ direction. Then

$$\frac{\partial \zeta}{\partial t} + u_0 \frac{\partial \zeta}{\partial x} + \beta v = 0 \rightarrow \frac{\partial^2 v}{\partial t \partial x} + u_0 \frac{\partial^2 v}{\partial x^2} + \beta v = 0 \qquad (45)$$

so that using the fact that the phase speed is ω/k and that k is a characteristic wave number, we get for the phase speed of the wave $v_p = u_0[1 - \beta/(k^2 u_0)]$ or

$$v_p = u_0 \left(1 - \frac{\Omega R \sin \theta}{2\pi^2 u_0} \right) \qquad (46)$$

which is the phase speed for a Rossby wave. Notice that u_0 is the flow along the lines of constant latitude, in the rotating frame, and that the wave moves more slowly than the local flow. It is in this way that Rossby (1939) was able to derive the basic properties for the transport of vorticity and generation of large-scale fronts in terrestrial weather systems. Any disturbance in the local flow in a rotating frame generates waves of the Rossby type. You might also notice that even as $u_0 \rightarrow 0$ the characteristic phase speed for the wave simply approaches the local acceleration $\beta R^2/4\pi^2$.

3.2.2 The Rossby Number

In the equations of motion in a rotating frame, we are presented dimensionally with two different forces. The inertial term varies as U^2/L, where U and L are some characteristic speed and length scale, respectively. The Coriolis term is ΩU. The ratio Ro = $U/\Omega L$ is a dimensionless parameter which, if small, signals that the rotation dominates the motion. This quantity is called the *Rossby number*. It was first introduced by C. G. Rossby (1939) in a study of flows in the Earth's atmosphere, where buoyancy and rotation are extremely important. Small Rossby number flows are called *geostrophic*. Any pressure gradients are balanced by the Coriolis term, and vorticity is the primary means of momentum transfer between fluid parcels. The problem is that we don't know what U is. We know that if there is a characteristic velocity, or a characteristic length and time scale, we will get a value for Ro. But we have no *a priori* idea what to take.

The Rossby number has been employed in recent years in the study of magnetic activity in cool stars. Stellar dynamos are governed as well by the effects of buoyancy, a role played by convection, and by rotation. While

3.2 Geostrophic Approximation

this is not the place for the detailed examination of such correlations (see Hartmann and Noyes 1987), some power of the quantity $\Omega\tau_c$, where here Ω is the stellar rotation frequency and τ_c is the convective eddy turnover time, has been used as a scaling factor for the intensity of chromospheric emission in late-type stars. This is not precisely the Rossby number as first introduced (actually it is its inverse), but it is one available way of quantifying the relation between the properties of the stellar envelope (τ_c) and the large-scale structuring of the convective pattern by rotation.

3.2.3 The Ekman Layer

It is now appropriate to introduce two new ideas, the Ekman layer (or rotationally generated boundary layer) and the idea of scaling within a rotating frame in the context of geostrophic balance. The first provides a new dimensionless number for the effects of viscosity within a rotating frame; the second provides new conditions for the appropriate time scales for shear and advection within slowly rotating frames.

The Ekman Number

First, let's write the equation of motion in a rotating frame, in which the motion is steady state and the rotation is slow. The latter condition is imposed to ensure that the centrifugal forces can be neglected. Let's also include viscous forces. For a normal fluid, we have (in rectilinear coordinates)

$$2\Omega \times \mathbf{v} = \nu \nabla^2 \mathbf{v} \qquad (47)$$

where the Coriolis term is the only external acceleration. Take the two coordinates (x, y) and assume that the fluid velocity (u, v) depends only on the depth z; then the Laplacian ∇^2 is replaced by d^2/dz^2. The equations for the velocity are now

$$\frac{2\Omega}{\nu} v = \frac{d^2 u}{dz^2}, \qquad -\frac{2\Omega}{\nu} u = \frac{d^2 v}{dz^2} \qquad (48)$$

If there is a characteristic length for the $\hat{z}$ direction, call it l, we can scale Eq. (48) so as to render it dimensionless. This provides us with a new dimensionless number, $E = \nu/\Omega l^2$, which is called the *Ekman number*. Multiply the second equation by i and combine the two to get

$$\frac{d^2}{dz^2}(u + iv) = \frac{2\Omega l^2}{\nu}(u + iv)$$

This has a solution of the form $u + iv = C \exp[(1 + i)kz]$ and C is a constant. The wave number, k, scales with the Ekman nuumber, by $E^{1/2}$.

Here we find that the value of C is provided by the free surface condition: the motion at the surface is determined by whatever forcing is taking place there. We will follow the traditional choice and assume that the upper surface is being forced by a shear stress, so that at the top of the layer

$$\left.\frac{du}{dz}\right|_{(0)} = \tau, \qquad \left.\frac{dv}{dz}\right|_{(0)} = 0$$

These boundary conditions assume that we have a unidirectional flow across the surface (call it wind) and that the motion we seek to find is the response of the fluid to this constant stress. We also assume that at depth, that is, as $z \to -\infty$, the velocity vanishes (both u and v vanish identically). This gives an amplitude to the motion of $C = (1-i)\tau/(2k)$ [recall that $1 + i = \sqrt{2}\exp(i\pi/4)$]. The condition that the effect of the surface stress asymptotically vanishes with depth yields a remarkable type of motion:

$$u = \frac{\tau}{2^{1/2}k}e^{kz}\cos\left(kz - \frac{\pi}{4}\right), \qquad v = \frac{\tau}{2^{1/2}k}e^{kz}\sin\left(kz - \frac{\pi}{4}\right) \qquad (49)$$

This velocity field is called the *Ekman spiral*. You see that there is a phase shift between them that depends on z. At $z = 0$ we get a motion which is at 45° to the direction of forcing, and the force is dissipated on a scale of $E^{1/2}$. This was of great importance in understanding the motion of ice floes under wind stress and other aspects of polar research and was one of the reasons for the predominance of the Bergen school of fluid mechanics in this research.

Now what does the Ekman number actually provide? It compares the rate of geostrophic transport with the viscous dissipation. We could find it by asking, what are the comparative magnitudes of the various forces in a rotating frame to which a fluid is subject? The Coriolis force is approximated by ΩU, the viscosity by $U\nu/L^2$, the pressure gradient by a_s^2/L and the buoyancy by $\delta\rho g/\rho_0$. So the contrast of viscous and Coriolis forces is shown by E, and this scales the time over which viscosity alone will distribute momentum through the medium. This is independent of the Rayleigh criterion or any baroclinic effects. If buoyancy is responsible for the redistribution, then Ro measures the relative role of Coriolis acceleration in the momentum transport. For astrophysical bodies, the molecular viscosity is usually trivially small. But in hot stars, the radiative viscosity (found by assuming a value for the Prandtl number and scaling between the heat conductivity and the viscosity) may be large and E will therefore actually enter the equations of motion. Usually, buoyancy dominates over viscosity, as in stellar convection zones and turbulent regions of rotating molecular clouds, and therefore the Rossby number is a more appropriate estimator of conditions.

Ekman Pumping

Astrophysically, the Ekman spiral in this form is not terribly important. The viscous forces in most systems are too low to be really important, or the effective depths are too shallow. The time scale for the circulation generated by Ekman pumping is longer than most of the radiatively driven time scales. It does play a role, however, in your experience and in geophysical problems. The point is that in a rotating environment there is an angular momentum which is being dissipated by the shear in the layer, and in the rotating frame this is expressed by the motion of the forced object at a specified angle to the direction of forcing. In a cosmic object, this produces a deviation of the flow due to the depth dependence of the shear. Imagine the case of tidal forcing of a rotating star in a binary system, in which this kind of flow is certainly important. The structure of the envelope of the star is dominated by the geostrophic flow, and therefore there will be a shear which is generated in the circulation currents that are set up and driven by the tides.

Several examples from daily life well illustrate the importance of this layer in the structuring of fluids. Take, for instance, the flow of water in a cup of tea (Fig. 3.7). First, spin-up takes place on a time scale $t_{\text{spinup}} \sim E$ and the boundary layer thickness scales as $E^{-1/2}$. As any of you who frequent fortune-tellers know, the reading of tea leaves is a finely honed art. What we will now discuss is how to analyze the results for fun and hydrodynamic profit. Consider the flow in the cup. The fluid is forced outward, against the walls of the cup, once the medium is set to rotating. The centrifugal acceleration increases at the top compared with the base, since there is drag at the bottom of the cup. This will cause a net circulation to be set up, which will be outward from the center at the top and inward along the bottom, rising at the center (since water is incompressible, remember?).

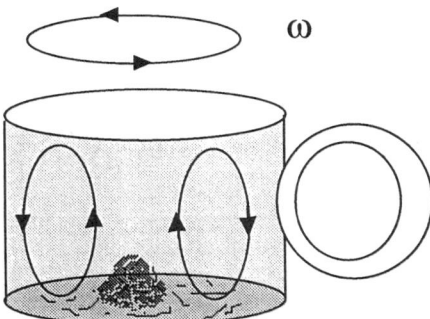

Figure 3.7 The circulation pattern in Ekman pumping in a teacup.

Therefore, any flotsam which happens to be around will be dragged with the flow and concentrated in the center. You have seen this result many times—but haven't capitalized on it—as the formation of a pile of tea leaves in the center of the cup. The form is strictly determined by the flow. If you slow down the flow by introducing a counterrevolution in the cup, you will find that the leaves form a ring. This is an example of what happens if the effective dissipation is rapidly changed; the change in the rotation has the same effect as changing the viscosity within the boundary layer. Also, if you take the cup at rest and spin up the walls instead of the fluid, you will be able to see the pumping at work as the boundary layer acceleration forces the tea leaves outward from the center.

3.3 Rayleigh Stability Criterion

In a rotating fluid, there is a condition of stability which must be satisfied. We want to be sure, in the case of a differentially rotating medium, that if we displace a fluid element, it will return to the same point. That is, it will have a low enough angular momentum relative to the background that it will drift inward again.

3.3.1 General Argument

We can look at the problem first using the Keplerian laws. We know that for a gravitation field of a point mass, the angular velocity satisfies Kepler's harmonic law, $\Omega^2 r^3 =$ constant, so that the angular momentum (specific angular momentum per unit mass) is given by $j = r^2\Omega \sim r^{1/2}$. If we displace a fluid element outward in this field, the centrifugal force it feels is $F_{\text{cent}} \sim j^2/r^3$, so we are assured that this force will always decrease as we go outward. Therefore, the fluid will find itself in a region of higher angular momentum if it is displaced outward and will drift back.

Assume that our circulating fluid moves in an effective potential $\Phi = \frac{1}{2}r^2\Omega(r)^2 = \frac{1}{2}r^{-2}j(r)^2$. Notice that there is no gravitational force here. We want to know under what conditions the gradient in the angular momentum alone can be a restoring force that keeps the body in a stable orbit. This will have an important consequence—it specifies properties of the gravitational field—but that will come later. For now, we look at a parcel of fluid if we displace it by some δr from the axis. Then the equation of motion for a small perturbation is

$$\frac{\partial^2 \delta r}{\partial t^2} = -\left[\frac{1}{r^3}\frac{\partial j(r)^2}{\partial r}\right]_0 \delta r \qquad (50)$$

3.3 Rayleigh Stability Criterion

If we work on the surface of a sphere rather than in a plane, the r variable is replaced (in conventional notation) by ϖ, the distance from the rotation axis. So if the angular momentum increases as we go outward, simple harmonic motion is possible. The generalized statement of this law is the so-called *Rayleigh criterion*: if in a rotating fluid the *Rayleigh frequency* (also called the Rayleigh discriminant) ever satisfies the condition

$$f_R^2 = -\frac{1}{r^3}\frac{\partial(r^2\Omega)^2}{\partial r} < 0 \tag{51}$$

then the fluid is *unstable* to redistribution of angular momentum. In addition, to ensure stability the angular momentum must have no dependence on displacement along the axis:

$$\frac{\partial}{\partial z} r^2 \Omega = 0 \tag{52}$$

that is, the rotation must be stable on cylinders.[1]

The Keplerian orbit is a stable pattern of circulation. However, if a body loses or gains angular momentum, it will move inward or outward, respectively. This is the reason that, in the case of the solar system, the planets stay in the orbits in which they have been inserted, but the reason that the Poynting–Robertson effect (the radiative drag that particles which scatter sunlight feel) causes the dust in the system to drift inward toward the Sun. The presence of viscosity in the system guarantees that the particles will drift inward as they lose angular momentum (there can be torques in such systems) and therefore the central object will ultimately accrete. In the next section, we shall have recourse to more application of viscosity to the rotating-fluid case, but it is useful to keep in mind that the Rayleigh criterion also plays a role in the structure of meridional circulation in stars which have molecular weight gradients (the so-called μ-barrier) and in the accretion rate in stellar and galactic disks around compact objects.

3.3.2 Epicyclic Frequency

Large-scale circulation is typical of many bodies. A large class of investigation has involved the stability of galactic disks, especially in light of

---

[1] Another form for the argument makes the physics more transparent. If we displace a parcel from r to $r+dr$, then the centrifugal acceleration at this new position is $j^2/(r+dr)^3 \approx (j^2/r^3)(1-3dr/r)$. Now if the angular frequency at this new location is $\omega + d\omega$, then the centrifugal acceleration is for corotation is $(\omega + d\omega)^2(r+dr) = \omega^2 r(1 + 2\,d\omega/\omega + dr/r)$. For stability we therefore get $2\,d\omega/\omega + 4\,dr/r > 0$ as before.

the generation of and stability of spiral patterns. That the Galaxy is differentially rotating has been known since the work of J. Oort in 1927. But the generalized equations of motion in such systems have been the evolving work of the past 50 years. The first derivations of stability criteria for astrophysical disks were done for the case of spiral galaxies. These have two interesting features. The stars in a galaxy behave as if they form a fluid only because there is a velocity dispersion and there are long-range forces, namely gravity, which stimulate some of the collisional effects that you see in a gas. However, unlike a perfect fluid, where the pressure tensor is isotropic, stars in a galaxy need not have isotropized their random motions. In fact, in general, there are substantial differences between the two velocity dispersions within the plane of a spiral disk galaxy and between the disk and vertical component of the motion.

Consider, however, that in the absence of viscosity, and with only gravity present, the stars in a galaxy will still have to obey the laws of motion. It should therefore be possible to approximate their motion as that of a fluid and to use this to look at the effects of perturbations in a differentially rotating fluid system with gravity. First, consider the effective potential. In a rotating system with gravity, there are two components. (See Fig. 3.8.) The first is due to gravity; the second is the result of the frame (we have already discussed this above). First, consider the case of an orbit with an angular frequency $\Omega(r)$ where the frequency is taken as a function of distance (this means that the disk circulation is not necessarily at a constant angular rate). Start with a locally circular orbit. Then $\Omega = \Omega_0$ locally at $r = r_0$. Such an orbit is *defined* by the fact that the gravitational acceleration is balanced by the centrifugal acceleration in the effective

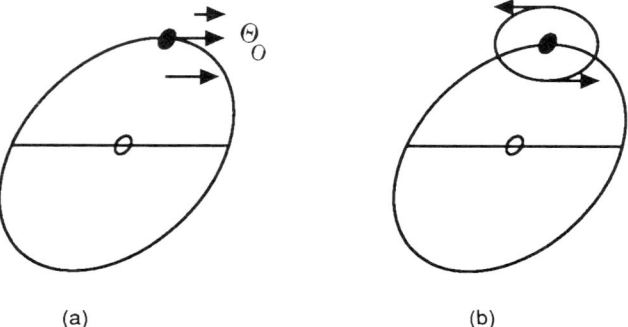

Figure 3.8 The meaning of the epicyclic frequency in a differentially rotating flow. (a) External observer's frame: differential rotation; (b) corotating frame: epicycles.

3.3 Rayleigh Stability Criterion

potential. Thus the magnitude of the orbital frequency is given by:

$$\Omega(r_0) = -\left(\frac{1}{r}\frac{\partial \Psi}{\partial r}\right)^{1/2}_{r_0} \tag{53}$$

This depends only on the mass distribution. It can also be inverted to determine $M(r)$, the mass within a specific radius. Now we need to examine the motion in the rotating frame. Consider small departures from strict centrifugal balance. This means that we consider that only the gradient in the angular momentum is responsible for maintaining a body in stable motion. It sounds worse than it is, because for the linearized treatment we will use here, you will only need to keep a few terms in the perturbation expansion. Here take $r = r_0 + \delta r$ and $\phi = \Omega_0 t + \delta \phi$, where we assume that Ω_0 is a reference frequency for the orbit. Now the radial component of the motion is given by

$$\frac{d^2 \delta r}{dt^2} - 2\Omega r_0 \frac{d\delta \phi}{dt} = -\frac{\partial \delta \Psi}{\partial r} \tag{54}$$

and from the condition that the angular momentum is conserved, $d(r^2 \dot{\phi})/dt = 0$, we get

$$r_0^2 \frac{d\delta \phi}{dt} + 2r_0 \Omega_0 \delta r = 0 \tag{55}$$

Take the potential and expand it to second order (the equilibrium is given by the first-order term, which vanishes in the corotating frame): $\Psi \approx \Psi_0 + \frac{1}{2}\Psi_0'' \delta r^2$. This is just a harmonic oscillator and a characteristic frequency for stability emerges:

$$\omega_{\pm} = \pm\left(1 + \frac{\Psi_0''}{3\Omega_0^2}\right)^{1/2} \tag{56}$$

As in the standard case of a harmonic oscillator, the natural frequency is proportional to $(\Psi'')^{1/2}$. To continue, the stability of fluid motion, under inviscid collisionless conditions, requires that angular momentum remain constant and that the gradient in the effective potential remain zero for all displacements:

$$3\Omega^2 + \frac{\partial}{\partial r}(r^2\Omega) \geq 0 \rightarrow \frac{\partial}{\partial r}(r^4\Omega^2) \geq 0 \tag{57}$$

so that if $\Omega \sim r^{-\alpha}$ then $\alpha \geq -2$. We know one law that certainly conforms to this, namely Kepler's law. The motion in a central point source gravitational field has $\alpha = -\frac{3}{2}$, so this is stable by the conditions we have just derived. That is, the motion resulting from any perturbation will lead to strictly periodic motion. And you see that we have recovered the Rayleigh criterion for stability of the motion. The epicyclic frequency contains the key to the stability.

Now one of the problems is that there may be more than one characteristic speed for the perturbations, and this is where the fluidlike behavior of the system comes into play. The change in the gravitational field, caused by the perturbations of the stellar orbits, feeds back into the acceleration and therefore into the density of stars at any part of the disk. If this perturbation propagates at some speed through the disk, it is called a wave. On the other hand, perturbations may form standing patterns in the disk, or modes, and these may be stationary. The difference between the picture we have developed for the stars alone and that required for a full disk of stars *plus gas* is that the gas is dissipative, compressible, and will be responsible for altering the flow of stars. But we will return to this later in this chapter.

3.4 Viscous Effects: A Simple Example

To underscore the effects of viscosity on circulation, a problem that we will address in more detail in the wind and accretion chapter, we now examine a simple example. Begin with the vorticity evolution equation:

$$\frac{d\boldsymbol{\omega}}{dt} = \boldsymbol{\omega} \cdot \nabla \mathbf{v} + \nu \nabla^2 \boldsymbol{\omega} \tag{58}$$

and assume that we confine our attention to planar flows (and assume that there are no motions in the $\hat{\mathbf{z}}$ direction. Then, since $\boldsymbol{\omega}$ is parallel to $\hat{\mathbf{z}}$, we get

$$\frac{\partial \omega}{\partial t} = \nu \frac{1}{r} \frac{\partial}{\partial r} \frac{\partial \omega}{\partial r} \tag{59}$$

which is a simple diffusion equation for the vorticity. The shear distributes the vorticity throughout the fluid. Therefore,

$$\omega = \frac{C}{(\pi \nu t)^{1/2}} e^{-r^2/4\nu t} \tag{60}$$

so that the vorticity at a fixed position initially decays like $t^{-1/2}$. Notice that this means that with increasing r, the vorticity is transferred outward. Thus the primary effect of viscosity is to cause a slow loss of vorticity within a circulating disk. For astrophysical accretion disks, this is the primary mode of driving accretion onto a compact central body. The viscosity adds a torque to the system that couples the more slowly circulating material in the outer part of the disk with the more rapidly moving material in the inner part.

3.5 Self-Gravitating Bodies

Consider a star in radiative equilibrium. That is, we assume that the body is generating its own energy by some means and that this energy flows out of the star on the thermal time scale. Also, we need to assume that the body maintains mechanical equilibrium on at least the radiative time scale. The problem of thermal balance in a spherical star is hard, but at least solutions to the problem exist. If the star is rotating, as we shall now see, the problem becomes very much more complicated. It isn't just that the mathematics becomes harder, that's daunting enough. Instead, the physical process of radiative transfer becomes more difficult.

3.5.1 Baroclinicity

If a star doesn't rotate, it will be spherical (Marks 1977). All this changes when rotation appears. To begin with, instead of being one-dimensional (a sphere depends only on the radial coordinate), the potential depends on the latitude. The effective potential is not the same as the gravitational potential because the star now has some centrifugal support. And the temperature and pressure may vary over the surface. For a sphere, assuming that the star is in equilibrium, we have

$$\frac{d}{dt}\mathbf{v} = -\frac{1}{\rho}\boldsymbol{\nabla} p - \boldsymbol{\nabla}\Psi \tag{61}$$

so that taking the curl of the two sides, we obtain

$$\frac{d\omega}{dt} = \frac{1}{\rho^2}\boldsymbol{\nabla}\rho \times \boldsymbol{\nabla} p \tag{62}$$

For a nonrotating body $\boldsymbol{\nabla} p \times \boldsymbol{\nabla}\rho = 0$, the star is barotropic, and the pressure is given by $p(\rho)$. A fundamental corollary to this is that the temperature is constant on equipotentials. Naturally, in the case of a sphere it is easy to see that this will be so. But on examination of the problem in the case of nonspherical stars, it is also easy to see why this should *not* be so. The condition for rotational stability requires that the angular momentum increase going away from the axis. If gravity is involved, this needs to be modified. (See Fig. 3.9.)

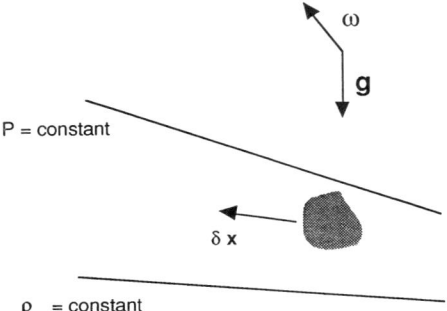

Figure 3.9 Physical basis for the baroclinic instability.

Rotation modifies both the magnitude and direction of the local gravitational acceleration compared with a spherical body. For a sphere, isobars and isotherms line up and a parcel displaced at constant pressure doesn't find itself buoyant relative to its surroundings. In a rotating body, displacement along such surfaces cuts through those of temperature and density, and buoyancy comes into play. There is a restoring force if the angular momentum distribution is stable, but this may not be enough. It is best to say that, to be certainly stable against axisymmetric perturbations, the angular momentum must satisfy the Rayleigh criterion on isobars, $(\partial [r^2 \Omega)^2]/\partial r)_p > 0$. We know that the gradient of the effective potential produces a local effective gravity $\mathbf{g}_{\text{eff}} = \mathbf{g} + \varpi \Omega^2 \hat{\varpi}$, whose direction is inclined to $\mathbf{g}$. Here ϖ is the distance from the axis (so as not to confuse r, the distance from the center, with the distance on a sphere from the rotation axis), and the direction parallel to Ω is $\hat{z}$. The hydrostatic equilibrium is given by $-\nabla p + \rho \mathbf{g}_{\text{eff}} = 0$. Notice that we're neglecting Coriolis terms and conservation of angular momentum for now, assuming that the perturbation is not axisymmetric. Taking the curl of both terms and noting that $\mathbf{g}$ is a potential field, then $\nabla \cdot \mathbf{g} = 0$ and we derive

$$\rho \frac{\partial}{\partial z} \varpi \Omega^2 = g_{\text{eff}} \hat{\mathbf{s}} \cdot \nabla \rho \qquad (63)$$

where $\hat{\mathbf{s}}$ is a unit vector along surfaces of constant effective gravity. Displace a fluid blob by a small amount along the body's surface to study its motion. Its subsequent behavior depends on the direction. If displaced away from the axis (that is, toward the rotational equator) along surfaces of constant pressure, it finds itself in a denser medium than the one it came

3.5 Self-Gravitating Bodies

from. It feels a buoyant restoring force, and a zonal circulation pattern is established where the parcel receives energy in the equatorward parts of its motion and releases energy as it moves poleward.

When we include the radiative losses from the surface of the body, the reason for surface temperature variations becomes more apparent. A sphere presents the same surface area for each cone of constant solid angle connecting its center and surface. A spheroid doesn't. Since the flux is integrated through both the surface area and the solid angle, a constant solid angle intercepts a greater surface area at the equator than at the pole. Thus, in order to be in radiative equilibrium, the equatorial flux needs to be lower than that at the pole. Otherwise there would be a variation in the total energy over the body. You see, for a steady-state condition, we still must satisfy the requirement that the divergence of the flux is globally zero and this would be unbalanced if the flux were constant. For the flux to vary requires either that the temperature is no longer constant on equipotential surfaces, those on which the pressure is defined, or that the flux is actually variable because of some other mechanical means of flux transport, or both. We shall in fact see that *both* is the best choice.

3.5.2 von Zeipl's Theorem for Rotating Self-Gravitating Bodies

The equation of hydrostatic equilibrium is

$$\nabla p = -\rho \nabla \Psi \tag{64}$$

where Ψ is the effective potential. For a rotating body, this is given by

$$\Psi = \Phi_{\text{grav}} - \tfrac{1}{2} r^2 \sin^2 \partial \Omega^2 \tag{65}$$

where Ω is the rotational frequency. The distance from the axis is $\varpi = r(\theta) \sin \theta$, although for our purposes we will assume that there are only very small departures from spherical symmetry. Now we write the energy flux as

$$\mathbf{F} = -\frac{4ac}{3\kappa\rho} T^3 \nabla T \tag{66}$$

where T is the local temperature and κ is the opacity. We can rewrite the flux as $\mathbf{F} = K\nabla T$ in analogy with the usual conductive law for heat. Now take the divergence of the flux and equate it to the energy generation within the star:

$$\epsilon = \nabla \cdot (K \nabla T) = \nabla K \cdot \nabla T + K \nabla^2 T \tag{67}$$

We can assert that, under ideal conditions, $K = K(\Psi)$ alone. Then the equation for radiative equilibrium becomes

$$\epsilon = (\nabla\Psi)^2 \frac{dK}{d\Psi} + K\nabla^2\Psi \sim \left(1 - \frac{\Omega^2}{2\pi G\rho}\right) \tag{68}$$

Here we have use of Poisson's equation, duly modified for the effective potential, to account for the Laplacian. As the surface is approached and the density vanishes, it becomes increasingly difficult for the star to maintain radiative equilibrium. Its only solution is to establish a temperature gradient along the equipotentials, making the fluid now baroclinic. The result we have just obtained is also known as *von Zeipl's theorem*, having first been stated in 1924 for general rotating bodies in radiative equilibrium. A new dimensionless parameter has now appeared in the problem,

$$\lambda_\Omega = \frac{\Omega^2}{2\pi G\rho} \tag{69}$$

If instead of the local density we choose the mean density $\langle\rho\rangle \sim MR^{-3}$, where M and R are the stellar mass and radius, respectively, then $\lambda_\Omega \sim \Omega^2 R^3/GM$. The parameter λ_Ω is extremely important because, for a rigid rotator, it represents the perturbation parameter required for the expansion of the equations of motion and for all estimates of the importance of rotational distortion.

3.5.3 Circulation Currents

The fact that the temperature cannot remain constant along equipotential surfaces has a very important consequence. Circulation currents must be established relatively rapidly within the envelope for the purpose of redistributing heat throughout the optically thick part of the stellar envelope. These currents were first discussed by Eddington and Vogt and later elaborated by Sweet, Mestel, Zahn, and Tassoul (see Tassoul 1978). Assume that the star departs only a little from being spherical. The flux therefore decomposes into two parts, one spherical and the other perturbed. The spherical flux is divergenceless, while the perturbation has the magnitude $K\nabla T'$. We have already found that the relevant parameter for determining the effect of rotation is λ_Ω. Since the thermal time scale depends on the gravitational potential, one would expect that any circulation will redistribute heat on a time scale of order $\lambda^{-1} t_{KH}$.

The rate of energy generation per unit mass for a sphere is given by $(1/\rho)\nabla \cdot K\nabla T$, which is approximately L/M. For a distorted body, this is

3.5 Self-Gravitating Bodies

changed by the alteration in the shape of the star. The nonspherical contribution to the divergence is the deviation of the star from a spherical surface. Since this is given by the ratio of the rotational to the gravitational potential per unit mass, the net result is that

$$\frac{1}{\rho}\nabla \cdot (\nabla T)_d \approx \lambda_\Omega \frac{L}{M} \tag{70}$$

To see what the characteristic speeds might be for any redistribution of energy, we need some other estimate of what would be an appropriate time for a spherical body to lose its thermal energy by radiation. This estimate is provided by the Kelvin–Helmholtz time, $t_{KH} = E_{\text{thermal}}/L_{\text{rad}}$. By the virial theorem, the thermal energy is directly proportional to the total gravitational energy, so we obtain as an estimate for the timescale of

$$t_{KH} = \frac{GM^2}{LR} \approx 3.1 \times 10^7 \text{ yr} \frac{m^2}{lr} \tag{71}$$

where in the scaled time all units are solar. This really is not a very long time for bodies the size of stars. The transport of heat through the interior of an adiabatic body is described by

$$\rho T \frac{dS}{dt} \to \rho T \mathbf{v} \cdot \nabla S = \nabla \cdot (K \nabla T)_{\text{distort}} \tag{72}$$

so that replacing the time derivative for the entropy with $\mathbf{v} \cdot \nabla S$, assuming that the spherical part of the flux is balanced by the rate of energy generation, we obtain an estimate of the speed for the thermal currents:

$$v \approx \frac{LR^2}{GM^2} \frac{R^3 \Omega^2}{GM} (|\nabla S|)^{-1} \tag{73}$$

We need some way to estimate the entropy gradient. This can be done using the deviation of the temperature gradient from the adiabatic one. We can write the temperature perturbation as

$$\frac{\delta T}{T} = \Delta \left(\frac{\partial \ln \delta T}{\partial \ln p} \right) \frac{d \ln p}{dr} \delta r \equiv (\nabla - \nabla_{\text{Ad}}) \frac{\delta r}{H_p} \tag{74}$$

which gives an expression for the entropy gradient of the form

$$\nabla S = c_p T_0 \Delta (\nabla - \nabla_{\text{Ad}}) \approx \mathcal{R} H_p^{-1} (\nabla - \nabla_{\text{Ad}}) \tag{75}$$

So we have the rate of entropy transport and can therefore relate this to the stellar flux. This gives a characteristic time scale for the circulation currents:

$$v_{ES} \sim \frac{LR^2}{GM^2} \lambda_\Omega (\nabla - \nabla_{Ad})^{-1} \tag{76}$$

Notice that in order to get the currents to move at reasonable speeds the temperature gradient must be superadiabatic. Dividing by the stellar radius gives

$$t_{ES} = \lambda_\Omega^{-1} t_{KH} \tag{77}$$

This is called both the *Eddington–Vogt* and *Eddington–Sweet* time scale; it estimates the characteristic rate of redistribution of the heat by thermal currents in the absence of any viscosity and neglecting mass transport. Some of the flow is also being driven by the potential gradients that now lie along, as well as normal to, the equipotentials (just as in tidal acceleration). This estimate of the current is much too short because it neglects all of the dissipative effects that dominate the deep flow. But it at least provides a fastest time scale on which the circulation would be expected to redistribute the energy. (See Fig. 3.10.)

We have not, however, included the effects of the distortion terms in the dynamics. Since the surfaces are no longer spherical, a gravitational

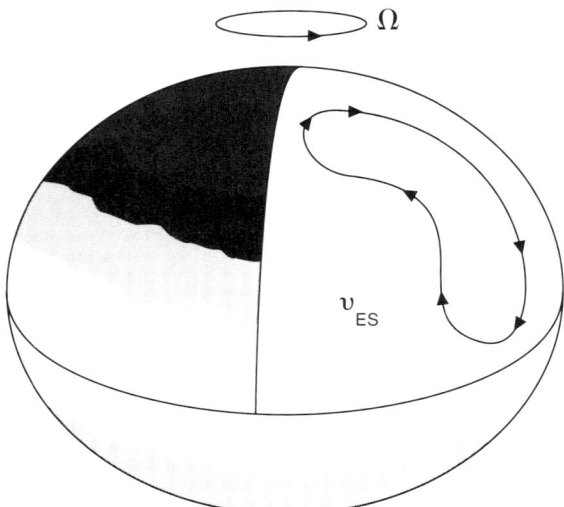

Figure 3.10 Eddington–Sweet meridional circulation in a rapidly rotating star, consistent with von Zeipl's law. The gradient in the surface brightness is a schematic representation of the flux differential between the pole and equator.

gradient appears *along* the equipotentials and this drives some of the mixing. It has a magnitude of order $\lambda_\Omega g_s$, so an estimate of the speed of the currents being driven by the distortion is $(\lambda \Omega g_s R)^{1/2}$. Comparing this with the Eddington–Sweet estimate shows that the thermal currents are more important for the mixing than the distortion. Accretion disks that are supported by radiation pressure and are self-gravitating may show some of the same problems. As you will note in the chapter on outflows and accretion, accretion disks have an integral energy generation mechanism: they convert viscous shear dissipation into radiative energy. This is transferred vertically in a rotating frame. But if the circulation is not Keplerian and the disk is sufficiently optically thick, it starts to look much like a star of the sort that von Zeipl had in mind, a fat torus, and the circulation theorem also applies to these.

References

Abt, H. A. (ed). (1970). *Stellar Rotation: IAU Colloq. 10*. Dordrecht, The Netherlands: Reidel.
Achinson, D., and Hide, R. (1973) Hydromagnetics of rotating fluids. *Rep. Prog. Phys.* **36**, 159.
Greenspan, H. P. (1968). *Theory of Rotating Fluids*. Cambridge: Cambridge University Press.
Hart, J. E. (1979). Baroclinic and barotropic instabilities. *Annu. Rev. Fluid Mech.* **11**, 147.
Hart, J. E., Glatzmaier, G., and Toomre, J. (1986). Space-laboratory and numerical simulation of thermal convection in a rotating hemispherical shell with radial gravity. *J. Fluid Mech.* **173**, 519.
Hartmann, L. W., and Noyes, R. W. (1987). Rotation and magnetic activity in main sequence stars. *Annu. Rev. Astron. Astrophys.* **25**, 271.
Hoskins, B. J., McIntyre, M. E., and Robertson, A. W. (1985). On the use and significance of isentropic potential vorticity maps. *Quart J. R. Met. Soc.* **111**, 877.
Houghton, J. T. (1977). *The Physics of Atmospheres*. Cambridge: Cambridge University Press.
Knobloch, E., and Spruit, H. C. (1982). Stability of differential rotation in stars. *Astron. Astrophys.* **113**, 261. [See also Spruit, H. C., and Knobloch, E. (1984). Baroclinic instabilities in stars. *Astron. Astrophys.* **132**, 89.]
List, E. J. (1982). Turbulent jets and plumes. *Ann. Rev. Fluid Mech.* **14**, 189.
Marks, D. W. (1977). On the spherical symmetry of static stars in general relatively. *Astrophys. J.* **211**, 266.
Mestel, L. (1965). Rotation and circulation in stars. In *Stellar Structure: Stars and Stelllar Systems*, Vol. 7 (p. 465). L. Aller and D. McLaughlin (ed.) Chicago: The University of Chicago Press.
Pedlosky, J. (1979). *Geophysical Fluid Dynamics*. Berlin: Springer-Verlag.
Prendergast, K. (1960). The motion of gas streams in close binary systems. *Astrophys. J.* **132**, 164. [See also Biermann, P. (1971). A simple gasdynamical model of mass exchange in a close binary system. *Astron. Astrophys.* **10**, 205.]

Putterman, S. (1972). The phenomenology of vortices in superfluid helium. *Phys. Rep.* **4C**, 67.

Randers, G. S. (1941). Large-scale motion in stars. *Astrophys. J.* **95**, 454.

Rossby, C. G. (1939). Relation between variations in the intensity of the zonal circulation of the atmosphere and the displacements of the semi-permanent centers of action. *J. Mar. Res.* **2**, 38. [Reprinted in Saltzman, B. (1962). *Selected Papers on the Theorey of Thermal Convection*. New York: Dover.]

Shu, F. H., and Lubow, S. H. (1981). Mass, angular momentum, and energy transfer in close binary stars. *Ann. Rev. Astron. Ap.*, **19**, 277.

Tassoul, J. (1978). *Theory of Rotating Stars*. Princeton: Princeton University Press. von Zeipl, H. (1924). On the radiative equilibrium of rotating gaseous masses. *Mon. Not. R. Astron. Soc.* **84**, 665. [See also Vogt, H. (1925). *Astron. Nachr*, **223**, 229; Eddington, A. S. (1925), *Observatory* **48**, 73; Sweet, P. A. (1950), *Mon. Not. R. Astron. Soc.* **110**, 548.]

Wasiutynski, J. (1946). Hydrodynamics and structure of stars and planets. *Astrophys. Norveg.* **4**, 1.

Zahn, J. -P. (1983). Instability and mixing processes in upper main sequence stars. In *Astrophysical Processes in Upper Main Sequence Stars: 13th Saas-Fee Course*. B. Hauck and A. Maeder (eds.) Geneva: Geneva Observatory.

CHAPTER 4

Shocks

> *Nobody ever heard the bullet that killed him.*[1]
> Th. von Kármán, *Aerodynamics*

4.1 Introduction to Shock Phenomena

The idea that flow must be steady state, which is what we have been using for most of this course, is a carryover from the last century. Since most fluid mechanical problems are posed in an equilibrium condition, for fairly slow flows or in the case of viscosity dominating the structure of the medium, it is normal to consider reversible transformations which are analogous to the Newtonian conditions in mechanics. In the case of strong explosive phenomena or supersonic flows, however, which as you know are disturbances in the medium to which the gas cannot quasi-statically accommodate itself, the situation is quite different.

The origin of shock wave theory can be traced back to B. Riemann's work on the propagation of finite acoustic disturbances and on the theory of characteristics in the 1860s. The laboratory study of the phenomena associated with shocks began in the 1880s with the work of E. Mach, who looked into the question of supersonic flows. Specifically, Mach was intrigued with the problem of motion of bodies at speeds faster than sound, especially ballistics. He developed the visualization techniques still employed for such studies, specifically the Mach–Zehnder interferometer and spark schlieren, and was responsible for the first investigations of shock reflection (Merzkirk 1970; Blackmore 1972). In addition, Mach was one of

---

[1] This is equally true for the crack of a whip in air, but the ballistic example is probably the best one for illustrating most of the discussion in this chapter. You will find much reward in contemplating the many physical implications of this conventional wisdom!

the first to note the internal reflections in supersonic expanding jets that now are referred to as Mach disks (1888) and to note the existence of bow shocks in front of blunt bodies (1886). L. Prandtl, in the 1940s, began the study of supersonic boundary layers, again in connection with the design of projectiles. The subject did not, on the other hand, gain currency or notoriety until the Manhattan project. During this period, and in the analysis of explosive phenomenology in general, it became necessary to face up to problems which had long been avoided—namely, what are the details of the propagation of strong discontinuities in a medium and what are the thermodynamic consequences of such disturbances? One of the most important innovations to come out of the bomb studies was the detailed examination of the treatment of radiating shocks; another important result was the examination of reflection of spherical shocks (Glasstone 1977).

Shock waves are almost the normal, rather than the exceptional, type of astrophysical fluid flow. Part of the reason for this is how far most of the places in the universe, accessible to our observations, are from thermal and mechanical equilibrium. The time scales for dissipating energy are often long. The energy sources for flows are often remote from them and from the environment the flows interact with. The escape velocity from most cosmic bodies far exceeds the sound speed, so any material escaping into the interstellar medium at large will be a loose cannon, whanging about the medium supersonically, eventually dissipating its momentum and energy and mixing. We will begin with a simple analysis of why discontinuities are expected to occur in some kinds of flows. Then we will look at the case of a strong pressure disturbance and finally examine some additional physical processes governing the structure and evolution of the discontinuity. One point to recall is that in the interstellar medium, and in many stellar phenomena as well, the mean free path is sufficiently small that the viscosity may be neglected to first order. We will thus begin with a discussion of inviscid shocks and later include the effects of dissipation on the front's dynamics.

4.2 Generation of Shock Waves

So far, we have merely asserted that shocks are possible. Since they represent discontinuities in the flow, they must be quite nonlinear. You might expect, therefore, that they arise as a result of an instability which causes the flow to turn nonlinear in its dependence on velocity. You would not be wrong. To see how this happens, let's return to the case of sound waves, discussed earlier in the book.

4.2 Generation of Shock Waves

4.2.1 Steepening of Sound Waves

Suppose that we take a sound wave to solve both the continuity and dynamical equations and don't specify that the disturbance is small. Then when the density increases the velocity goes up. Let us show this.

$$\frac{\partial \rho}{\partial t} + \frac{\partial}{\partial \rho}(\rho v)\frac{\partial \rho}{\partial x} = 0 \tag{1}$$

$$\frac{\partial v}{\partial t} + \left(v + \frac{1}{\rho}\left(\frac{\partial p}{\partial v}\right)\right)\frac{\partial v}{\partial x} = 0 \tag{2}$$

This you have seen *many* times before in different guises. Notice that we now have two coupled nonlinear *hyperbolic* equations and so can use the method of characteristics to specify the solutions, knowing only the initial conditions. We first make use of the fact that

$$\frac{\partial y/\partial t}{\partial x/\partial t} = -\frac{\partial y}{\partial x}$$

(which you should prove for yourself) to get

$$\left(\frac{\partial x}{\partial t}\right)_\rho = \frac{d(\rho v)}{d\rho} \tag{3}$$

and also

$$\left(\frac{\partial x}{\partial t}\right)_v = v + \frac{1}{\rho}\frac{dp}{dv} \tag{4}$$

Make use of the equation

$$\frac{dp}{dv} = \frac{dp}{d\rho}\frac{d\rho}{dv}$$

to obtain a solution which says that the two time derivatives above are equal:

$$dv = \pm\frac{a_s d\rho}{\rho} \tag{5}$$

We have used the fact that $p = a_s^2 \rho$. From the characteristics, we then see that the phase velocity of the wave is a function of the density at any point in the profile and that therefore the wave will not be able to retain its shape as a function of time. The velocity depends on the density in a very specific way. If the density increases, so does the velocity of that part of the wave.

If it decreases, it slows down—*the Matthew effect strikes again.*[2] We can notice that the formalism can be more thoroughly developed by using the characteristics of the equations of motion and continuity, but this will be deferred for a moment.

What effect does this have? One way of looking at it is that as the wave amplitude grows, the trough of the disturbance progressively lags behind compared with the peak. The wave front must therefore steepen, and the resultant profile will be one which will inevitably form either a shock or a crest. Surfers are among the most famous nonlinear hydrodynamicists, but the same phenomenology occurs in many ways in astronomical situations. For example, in the case of sound waves propagating into the outer tenuous atmosphere of a star and generating local disturbances, the waves will turn into shocks. (See Fig. 4.1.) They will therefore deposit a considerable portion of their energy locally, weaken, and dissipate. Again, like waves on a beach, the wave will become multivalued at some stage and must break. In the case we are discussing here, the wave must form a shock. The irreversibility of the thermodynamic transformation which the gas undergoes on passage through the shock ensures that the energy of the wave will be dumped into the gas, heating it up and producing all of the attendant effects of shock passage that we have already discussed.

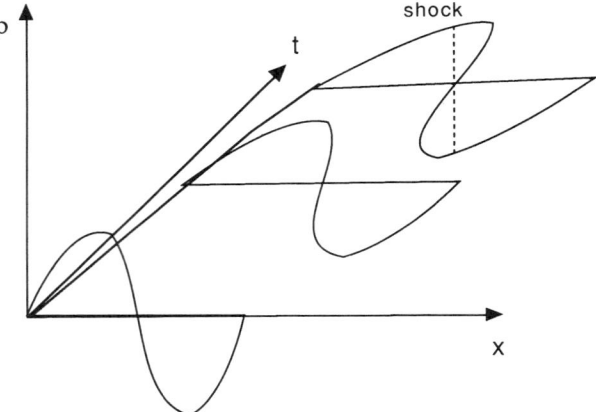

Figure 4.1 Schematic of the steepening of a finite amplitude sound wave to form a shock. The shock occurs following the appearance of a double-valued condition for the density or pressure.

---

[2] The Matthew effect, so named by the sociologist of science R. K. Merton, derives from the biblical quotation: "For whosoever hath, to him shall be given ... but whosoever hath not, from him shall be taken away even that which he hath" (Matt. 13:12).

4.2 Generation of Shock Waves

4.2.2 Riemann Invariants

The equations of motion lead to conservation conditions as we have already seen. These are especially useful when the flow is supersonic, because here the initial conditions set the stage for the flow in a way not encountered for subsonic flow. Essentially, supersonic flow is ballistic. Since a pressure disturbance can make itself known only through waves which propagate at the speed of sound, and by definition the flow is faster than that, the existence of shocks in a flow is the consequence of the breakdown of hydrostatic equilibrium. Thus, if something is conserved in the flow, it must be conserved at each point along the trajectory of the flow.

Let's put this in a different way. Suppose that a surface sends out a series of sprays each of which is a pencil flow and each of which is supersonic. These flows separate from one another. (See Fig. 4.2.) The rate of separation is faster than the sound speed, so that each evolves independently of the other. Whatever the conditions imposed at the outset, at the nozzle, those will be the ones that prevail throughout the history of the flow. Always keep this in mind: once the flow is faster than the sound speed, anything that happens near the source is frozen into the flow,

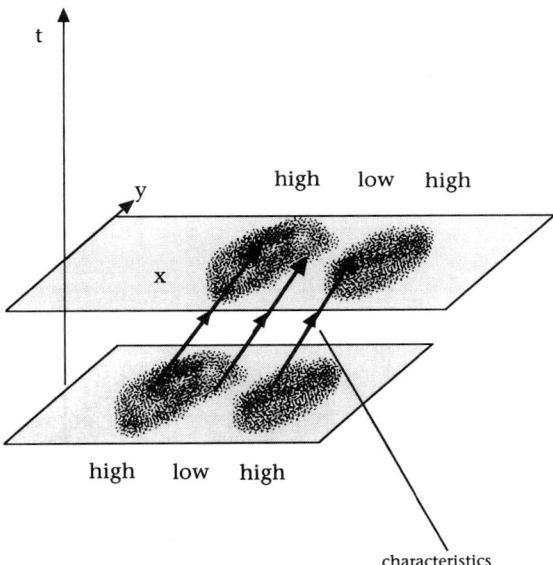

Figure 4.2 The propagation of characteristics in a fluid flow. The density fluctuations are marked in this space–time diagram.

and anything that happens upstream cannot make itself known to the flow until it is hit directly by the material. If there is a pressure jump at some point in the flow, the fluid will encounter it as a discontinuity. If this is at a point in the flow and the flow is along a surface or a three-dimensional flow, then the rest of the fluid will know of the existence of this disturbance only by the shock that radiates away from the site at some angle which depends on the flow speed of the incoming fluid.

To see this, let us go back to the fluid equations. The continuity equation is

$$\frac{\partial \rho}{\partial t} + \frac{\partial \rho u}{\partial x} = 0 \tag{6}$$

and the equation of motion is

$$\frac{\partial u}{\partial t} + u \frac{\partial u}{\partial x} = -\frac{1}{\rho} \frac{\partial p}{\partial x} \tag{7}$$

Now assume, as we have many times already, that $p = p(\rho)$ and write $\Phi \equiv \partial p / \partial \rho$ and $\Lambda \equiv \ln \rho$ to shorten the notation. The equations are then

$$\frac{\partial u}{\partial t} + u \frac{\partial u}{\partial x} = -\Phi \frac{\partial \Lambda}{\partial x} \tag{8}$$

$$\frac{\partial \Lambda}{\partial t} + u \frac{\partial \Lambda}{\partial x} = -\frac{\partial u}{\partial x} \tag{9}$$

If we multiply the equation of continuity by $\Phi^{1/2}$, we can combine the two equations to get

$$\frac{\partial u}{\partial t} + (u - \Phi^{1/2}) \frac{\partial u}{\partial x} = \Phi^{1/2} \left[\frac{\partial \Lambda}{\partial t} + (u - \Phi^{1/2}) \frac{\partial \Lambda}{\partial x} \right] \tag{10}$$

A similar result is obtained if we multiply by $-\Phi^{1/2}$, so we see that there is a new propagation condition, namely that the disturbance moves at a speed

$$U_\pm = u \pm \Phi^{1/2} \tag{11}$$

The equation of state enters into the propagation condition through $\Phi = \partial p / \partial \rho$. Depending on how the pressure changes with density, we will get acceleration or deceleration of the disturbance compared with a constant speed of sound. This can be understood as a condition that determines what happens when the pressure disturbance passes through a medium. If the sound speed is independent of density, the disturbance moves at constant speed. But suppose there is a change in the sound speed with change in density. Then as the medium is compressed the sound speed

4.2 Generation of Shock Waves

increases, since Φ is positive and the wave speeds up. The condition on the wave front is that

$$\left(\frac{dx}{dt}\right)_\pm = u \pm \left(\frac{\partial p}{\partial \rho}\right)^{1/2} \tag{12}$$

which are the trajectories of the fluid parcels. These are the *Riemann invariants*, so named because they were introduced by Riemann in his 1860 paper on the propagation of sound disturbances of finite amplitude. The Riemann invariants are also written as

$$\frac{dx}{dt} = v \pm \int \frac{dp}{\rho a_s} \tag{13}$$

where now the sound speed is a function of density through the equation of state.

What are we to make of these quantities? They are the trajectories along which the conserved quantities of the flow are transported through the fluid. If we have a discontinuity in any of these quantities, it will follow a trajectory given by the Riemann invariants. This is a most important point. The lines in the (x, t) plane are also called the *characteristics*, a term originally applied to the conservation lines which result for hyperbolic partial differential equations. Suppose we take u as the speed of a piston going into a gas. Then u is given as an externally imposed condition on the flow. We are free to change it, and it is controlled by physical properties at the source of the flow—a boundary or initial condition. It is set upstream and the flow cannot adjust downstream to its value, if it is faster than the sound speed. Then let us assume that the equation of state is $p = K\rho^\gamma$, where as usual γ is the ratio of the specific heats. Then

$$\left(\frac{dx}{dt}\right)_\pm = u \pm \frac{2a_s}{\gamma - 1} \tag{14}$$

and therefore the speed of propagation of the fluid is

$$U_\pm = a_{s,0}\left(1 \pm \frac{\gamma - 1}{\gamma + 1}\left[\left(\frac{\rho}{\rho_0}\right)^{(\gamma-1)/2} - 1\right]\right) \tag{15}$$

where the zero subscript refers to the undisturbed fluid. Notice that this also means that as the density increases, and as the density contrast increases, so does the propagation speed of the disturbance. This is the condition that is likely to form a shock wave. If the speed increases as the density increases, and decreases as the density decreases, then at some point the dense material overtakes the lower-density material. The result is the pileup that signals the initiation of a shock.

4.3 The Rankine–Hugoniot Conditions

The equations of evolution of a flow are, if viscosity is neglected, hyperbolic. They therefore admit characteristic solutions, streamlines along which the physical properties of the medium are constants. These spacetime trajectories are the ones along which the flow variables evolve, having started out at some initial condition at time t_0. Now we can begin our calculation by asserting that there are three conserved quantities that we know of: mass flux, momentum flux, and enthalpy flux. All three, if there are no explicit time dependences in the problem, are independently conserved.

4.3.1 The Conserved Quantities

The moment equations give the following conserved quantities. The mass flux $\rho \mathbf{u} \cdot \mathbf{n}$ is constant through a surface in the fluid, anywhere in the fluid. The same is true for $(\mathbf{p} + \rho \mathbf{uu}) \cdot \mathbf{n}$ for the momentum flux. Finally, the energy flux is constant across any surface $\rho \mathbf{u} \cdot \mathbf{n}(\frac{1}{2}u^2 + h + \phi)$, where h is the enthalpy. These come from the divergence conditions for a time-independent problem and were described by Rankine (1870) and Hugoniot (1889), starting from very different assumptions (Rankine was mainly interested in the thermodynamics of wave propagation; Hugoniot was interested in nonlinear sound waves following work by Poisson earlier in the century).

If we imagine that we have a discontinuity Σ across which the quantities are conserved, and if we denote the condition that $Q_{\Sigma_-} = Q_{\Sigma_+}$ by $[Q]_\Sigma = 0$, for some quantity Q, then

$$[\rho v]_\Sigma = 0, \qquad [\rho v^2 + p]_\Sigma = 0, \qquad \left[\frac{1}{2}v^2 + \frac{\gamma}{\gamma - 1}\frac{p}{\rho}\right]_\Sigma = 0 \qquad (16)$$

These are for one-dimensional propagation but can easily be generalized to spherical or cylindrical fronts by altering the structure of the surface and using only the normal component of the velocity. These are the conditions on which the magnitude of the change in the flow variables depends. If we are given a discontinuity in any one of these three basic variables (ρ, v, p), the other two can be given in turn. Look at it this way, in terms of a simple laboratory experiment.

4.3.2 Planar Shocks

Imagine that we have a long cylinder (as long as the budget will permit) with a thin membrane placed within it at some point, call it P, and

4.3 The Rankine–Hugoniot Conditions

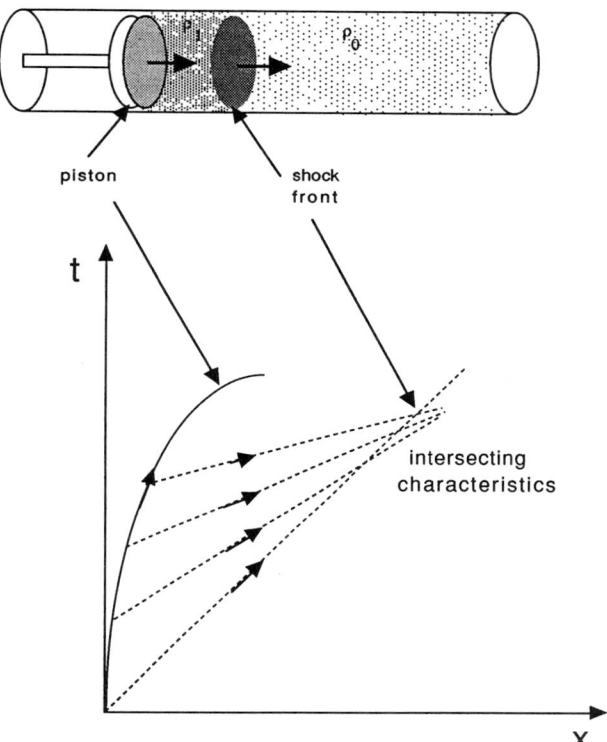

Figure 4.3 Formation of a shock ahead of a piston moving into a shock tube. The characteristics formed at the piston intersect at some distance ahead of the piston.

with some gas of interest in the right-hand side of the tube (Fig. 4.3). In the left side, as it were behind the membrane, we start to pump in an inert gas like argon or helium. We contrive the experimental setup to be one which reaches critical overpressure on the left side of the membrane at some controlled instant. At this point, the membrane ruptures and the inert gas floods out into the interesting gas. Collisional excitations will follow, the gas will glow, and the setup will act as a source for thermal excitation. Now what is the temperature which will be reached? The answer depends on the overpressure. We have a discontinuity moving outward into the tube which is, in fact, a shock. The apparatus is therefore called a *shock tube* and is part of the laboratory arsenal of every spectroscopist. The added feature of such a device is that there can be many passes of the front through the gas, reflected from the end of the tube, and so a compounding of the initial effect can be achieved. The density, velocity, and pressure being all specified at the front, its velocity can be determined and therefore the thermal energy input.

To see how to do this, both in the lab and in the sky, we consider the moment of rupture. If we sit on the front, we see the gas on the right-hand side rushing at us at a velocity V. Call this velocity v_1. Behind the shock, it undergoes a deceleration in the frame of the front and so a compression results. To see this, return to Eq. (16). If 2 denotes the postshock gas and 1 indicates the preshocked material, we get from the continuity equation

$$v_2 = \frac{\rho_1}{\rho_2} v_1 \qquad (17)$$

The front does not gain mass; what goes through the front emerges on the other side. Then, since momentum conservation gives

$$\rho_1 v_1^2 + p_1 = \rho_2 v_2^2 + p_2$$

we can eliminate one variable. Let's eliminate the velocity, for an example. Define a new variable $\Gamma \equiv \gamma/(\gamma - 1)$. By the momentum flux condition we have

$$J^2 = \frac{p_2 - p_1}{\rho_2 - \rho_1} \rho_1 \rho_2 \qquad (18)$$

Solving this equation gives us what is known as the *shock adiabat*:

$$\rho_1 v_1^2 + p_1 = \rho_2 \frac{\rho_1^2}{\rho_2^2} v_1^2 + p_2$$

where J is the mass flux, which is a constant across Σ. Since $\rho = 1/$(specific volume), we have a thermodynamic relation between p and ρ. The only parameter we need specify is J. But we know this in advance, since we have put ourselves on the front moving with a velocity V, and we know the initial density of the gas into which the shock front is moving.

We now use the enthalpy condition, the third equation [Eq. (17)], to get

$$v_1^2 \left(1 - \frac{\rho_1^2}{\rho_2^2}\right) = \frac{2\Gamma}{\rho_1 \rho_2} (p_2 \rho_1 - p_1 \rho_2) \qquad (19)$$

or, simplifying,

$$\frac{\rho_2}{\rho_1} = -\frac{(1 - 2\Gamma) p_2 - p_1}{p_2 - (1 - 2\Gamma) p_1} = \frac{(\gamma + 1) p_2 + (\gamma - 1) p_1}{(\gamma - 1) p_2 + (\gamma + 1) p_1} \qquad (20)$$

4.3 The Rankine–Hugoniot Conditions

This is the core result contained in the *Rankine–Hugoniot conditions*. Given a density or pressure jump and the ratio of specific heats (which is assumed to be constant for the moment), we can derive *algebraic* relations for the change in the other thermodynamic variables across the front. Clearly, if ρ_2 is much greater than ρ_1, we have a limit on the magnitude of a strong shock.

In the case of a strong shock, with $p_2 \gg p_1$, we get the limit on the magnitude of the compression to be $\rho_2/\rho_1 \to (\gamma+1)/(\gamma-1)$. For a perfect gas we know that $\gamma = \frac{5}{3}$, so the maximum compression we would expect in the strong shock limit is of the order of 4. Notice that the compression is greater for the strong shock limit in a lower-γ medium, like radiation-dominated gases, since the medium is intrinsically more compressible. Now recall that the density jump is inversely related to the velocity jump and the problem is done. In fact, if we can measure the velocity across the shock, we have everything else specified. In the shock tube case we have already thought of, the pressure jump is given; therefore we know the velocity jump from the Rankine–Hugoniot conditions.

One of the things that we can do next is take a specific equation of state. For instance, one which is a perfect gas comes to mind. Then the enthalpy condition becomes an equation for the temperature T, and so the three variables are v, ρ, and T. There is nothing special about these equations. The derivation of the shock equations in this simple form, exploited by Courant, von Neumann, and their collaborators, pioneered the calculation of supersonic phenomena. These conditions represent an example of *matching conditions* at the boundary of the moving region, which are taken in Lagrangian coordinates. The only thing that remains (and this is delayed until the final chapter) for such fronts is to find an equation which specifies the evolution of the density, pressure, and velocity after the passage of the material through the front.

What can one do, astrophysically, with the conditions as they currently stand? Actually, quite a bit. For one thing, we can ask whether any such conditions as we have discussed in the shock tube *gedankenexperiment* ever occur under cosmic conditions. Consider the case of a flare on the solar surface. It is a very high energy phenomenon, consisting of the release of large stores of energy in the form of radiation and energetic particles from a small region in a very short time. The medium cannot possibly respond slowly to such input (the flare may contain a sizable portion of the total radiative budget of the Sun) and so it will expand. Not into a vacuum, but into the solar corona and the low-density interplanetary medium. We therefore have a blast wave, which is a shock, that moves outward into the medium. If we wish to figure out what the conditions at the time of the energy release were, we need some way of getting from the observed velocity jumps and density jumps to a measure of the input energy, momentum, etc.

A more familiar example is that of a supernova blast. Ignore, for a moment, the initial instant of the explosion and concentrate on the later stages, the ones we can observationally probe in some detail. We can measure the expansion velocity directly, by observing the line shifts in the spectrum, but one of the questions we are interested in is the production of heavy metals during such events and how the matter gets mixed into the ambient diffuse medium. Knowing what the Rankine–Hugoniot conditions are, and the velocity jumps, we can get enough plasma diagnostic input to calculate abundances from the observed emission lines.

A final example is the H II region. This one we can look at qualitatively here and examine in more detail at the close of this chapter. Imagine that we have a star sitting in the center of a diffuse gaseous medium. Take the star to be sufficiently hot that radiation from it ionizes the gas. The photons will ionize as much of the material as they can, the rate of ionization being balanced only by the recombination rate for purposes of maintaining charge balance. If the central star continues to shine long enough, enough energy can be pumped into the material so that it will begin to heat up and expand. That is, the initially static state cannot be preserved forever. The result is that a front, across which there is a discontinuity in both the pressure and ionization, will begin to move outward into the interstellar medium (ISM). We say that there is an *ionization shock* because there is a discontinuity in the enthalpy across the ionization region. The ionization acts to change the enthalpy (and the entropy) of the gas, and so there is a weak shock. At the expanding front there is a complex structure that we will reveal shortly, consisting of a compressional and a rarefactional shock, but the simplest picture to think of is a tenuous, high-temperature, ionized region moving out into a cool, higher-density gas.

4.3.3 Another Form for the Rankine–Hugoniot Conditions

We have, in deriving the jump conditions, started with the variables most natural to the conservation equations. The final equations are expressed in terms of the velocity, density, and pressure. What would happen if, instead, we were interested in the ratio of densities on either side of a shock going at *Mach 2*? Remember, we are sitting on Σ, the shock front, and looking at the gas slam into us, so this is a reasonable question. It turns out that it is possible to reduce the problem to one which depends only on

4.3 The Rankine–Hugoniot Conditions

M_1, the Mach number of the preshock flow, and all ratios of post- to preshock values can be expressed in terms of this one variable. There is a caveat here: this representation is straightforward only if the ratio of specific heats, γ, is a constant (that is, there are no changes in the equation of state across the front). If you are going to apply this kind of approach, be sure of the assumptions before you start.

Use the sound speed to relate the state variables, $a_{s,i}^2 = \gamma p_i/\rho_i$, for side i of the front. The equations for the momentum and enthalpy flux are therefore rewritten as

$$\left[\rho\left(v^2 + \frac{a_s^2}{\gamma}\right)\right]_\Sigma = 0 \tag{21}$$

$$\left[\frac{1}{2}v^2 + \frac{1}{(\gamma-1)}a_s^2\right]_\Sigma = 0 \tag{22}$$

We now define $M_1 \equiv v_1/a_{s,1}$ to be the Mach number for the preshock flow. It does not matter that the sound speed may not be the isothermal value—we will simply remove it by this expedient. Notice that the mass flux equation is not altered by the change in variables. If you solve these two equations for $a_{s,2}^2/a_{s,1}^2$, you will ultimately get [3]

$$\frac{\rho_2}{\rho_1} = \frac{(\gamma+1)M_1^2}{(\gamma-1)M_1^2+2} = \frac{4M_1^2}{M_1^2+3} = \frac{v_1}{v_2} \tag{23}$$

You already know that the strong shock limit is one in which $M_1 \gg 1$, so that in this limit we recover the previous results for the Rankine–Hugoniot conditions. The second version of the equation assumes $\gamma = \frac{5}{3}$.

We can also derive the pressure ratio by noting that, given the ratio of sound speeds, we have

$$\frac{a_{s,2}^2}{a_{s,1}^2} = \frac{p_2\rho_1}{p_1\rho_2} = \frac{p_2}{p_1}\left(\frac{\rho_2 M_1}{\rho_1 M_2}\right)^2 \tag{24}$$

Then use

$$\frac{a_{s,2}^2}{a_{s,1}^2} = \frac{1}{2}\left(1 - \frac{\rho_1^2}{\rho_2^2}\right)(\gamma-1)M_1^2 - 1 \tag{25}$$

to eliminate the density term. Alternately, we could have used the formula we derived in the previous section for the density ratio as a function of

---

[3] Be forewarned that the algebra is a bit tricky, so be careful.

pressure to derive the pressure ratio as a function of the incident Mach number. The same is true for the ratio of velocities on either side of the shock, which is in fact a way of getting M_2 as a function of M_1. Keep in mind that it is indeed possible to derive them and that often it is the Mach number and not the velocity about which we know the most. Finally, in addition to the density condition we have derived above, we collect the jump conditions written in terms of the upstream Mach number. For convenience, we also include the form with $\gamma = \frac{5}{3}$. For the pressure, we have

$$\frac{p_2}{p_1} = \frac{2\gamma M_1^2 - (\gamma - 1)}{\gamma + 1} = \frac{1}{4}(5M_1^2 - 1) \tag{26}$$

and for the temperature (this is also obtained from the sound speed) we have

$$\frac{T_2}{T_1} = \frac{[2\gamma M_1^2 - (\gamma - 1)][(\gamma - 1)M_1^2 + 2]}{[(\gamma + 1)M_1]^2} \tag{27}$$

Finally, using the condition that $\rho_2/\rho_1 = v_1/v_2$, the downstream Mach number becomes

$$M_2^2 = \frac{(\gamma - 1)M_1^2 + 2}{2\gamma M_1^2 - (\gamma - 1)} = \frac{M_1^2 + 3}{5M_1^2 - 1} \tag{28}$$

Why would we want to look at the problem this way? For one thing, we often know only the velocity of the shock and the Mach number. Take the case, for example, of a pulsating variable star. We know the effective temperature of the atmosphere and, making the assumption of local thermodynamic equilibrium (LTE), we can assume that we know the sound speed. We can, by the light curve and spectroscopy, determine the velocity of the pulsationally generated wave in the atmosphere and can therefore specify the Mach number for the disturbance. Now we generally do not have the slightest idea what the compression ratio is, nor do we have a real handle on any of the other variables which might normally be needed for the Rankine–Hugoniot conditions. However, having in hand M_1, we can calculate all of the important diagnostic parameters for the shock and therefore predict what the environment after shock passage will be like. This will aid us in determining the effect of the shock on both the atmosphere dynamics and the diagnostics.

Now notice that the energy flux is constant for a steady-state flow with no radiative losses. This means that

$$\tfrac{1}{2}v^2 + c_p T = c_p T_0 \tag{29}$$

4.3 The Rankine–Hugoniot Conditions

Here T_0 is the temperature of the reservoir at large distance from the front. Since the temperature for a perfect gas is directly related to the sound speed, we have

$$\frac{1}{2}v^2 + \frac{a_s^2}{\gamma - 1} = \frac{a_{s,0}^2}{\gamma - 1} \qquad (30)$$

which means that

$$\frac{a_s}{a_{s,0}} = \left(1 + \frac{\gamma - 1}{2} M^2\right)^{-1/2} \qquad (31)$$

so that the Mach number is lower *after* the passage through a shock and therefore the temperature increase causes the flow after the shock to be subsonic. An important point is that the entropy of a compressive shock must increase. Let us look at this point in more detail. The entropy change is given by

$$s_2 - s_1 = c_p \ln \frac{T_2}{T_1} - \mathcal{R} \ln \frac{p_2}{p_1} \qquad (32)$$

which can also be expressed as

$$\frac{s_2 - s_1}{\mathcal{R}} = \ln\left[\left(\frac{p_2}{p_1}\right)\left(\frac{\rho_2}{\rho_1}\right)^{-\gamma}\right]^{\gamma - 1} \qquad (33)$$

For $M \approx 1$, we can expand the terms depending on $M^2 - 1$ as terms of first order to obtain

$$\frac{s_2 - s_1}{\mathcal{R}} \approx \frac{2\gamma}{3(\gamma - 1)^2}(M^2 - 1)^3 \qquad (34)$$

From this it appears the entropy increase after the passage of the shock with M greater than unity is a very strong function of the Mach number of the flow.

Notice that all of our discussion has been in the frame of the shock, so that the actual velocity of the matter after passage through the front will be $V_\Sigma - v_2$, where V_Σ is the speed of the front. Although a very simple point, it is essential that you remember this. For observational purposes, in general, we sit outside the shock and watch the material go by, front and all. This Eulerian framework is very different from the one used for the calculation of the postshock conditions, which are usually Lagrangian-type calculations. So a word to the wise: always keep in mind where you stand!

4.4 Some Additional Complications

The formalism we have been discussing is so flexible that it seems unreasonable to ignore possible generalizations. For instance, in the case that a magnetic field is present in the medium, the conservation condition for magnetic flux introduces an equation into the Rankine–Hugoniot conditions which is analogous to the mass flux condition and which can then be incorporated into the shock structural equations. In addition, since the magnetic field contributes a pressure and energy to the medium, the full set of jump relations will be altered. Another physical complication of astrophysical importance is the problem of oblique and interacting shocks. The two are really parts of the same problem, so we will discuss them together. The obliquity of a shock front leads to an interesting set of effects, which are analogies with geometric optics, and really highlights the nature of the *characteristics* or *Riemann invariants* which are the basis for much of the dynamical and numerical work on shock structure.

4.4.1 Reflected Planar Shocks

Normal Incidence

One property of the shocks in our *gedankenexperiment*—that of the multiple passes of the shock through the gas due to its reflection from the walls of the tube—should also be considered here. Multiple shocks may occur in astrophysical environments, and their properties are quite interesting. Obviously, each passage of the shock through the medium produces additional heating of the gas. The front, however, dissipates some energy with each pass. It is important to note, though, that the effects of successive shocks cannot be simply additive. You might think of this as the case of a region being subjected to multiple shocks, not just from reflection but from continuous generation of more fronts to pass through the gas. It is left as an exercise for you to derive the result, but *you can show* that for the second pass

$$\frac{p_3}{p_2} = \frac{(3\gamma - 1)p_2 - (\gamma - 1)p_1}{(\gamma - 1)p_2 + (\gamma + 1)p_1} \tag{35}$$

again under the condition that γ does not change as a result of the shock passage through the gas. If the shocks are not normally incident, however, the flow will be more complicated. This point will be treated again in the discussion of oblique shocks, but keep it in the back of your mind for now.

4.4 Some Additional Complications

4.4.2 Oblique Shocks

Stars don't have corners, but they are finite-sized bodies. So are galaxies. If there is a hydrodynamic phenomenon that takes place on the same scale, it will also be a truism that the phenomenon generally will not behave as a one-dimensional effect. In the case of flows of stellar winds in the vicinity of a companion, for example, the wind may see a barrier to free passage presented by the companion or its atmosphere. In the case of gaseous flows in a binary system of stars, or in a nonaxisymmetric galactic potential, it may be neccessary to forsake the simple picture of planar one-dimensional shocks and treat at least the two-dimensional case.

We can consider the problem as an analogy to the one in classical optics. Imagine that a wave transits between two media of different refractive index—or, put in more colloquial terms, the speed of light across the interface changes value. The variation in the velocity of the group will be sufficient to alter the overall direction of the wave, noting that conservation and jump conditions must be fulfilled. We will consider the case, again, of a front. (See Fig. 4.4.) Call it Σ. Now take Σ and slam it at an angle ϕ into a gaseous medium. If there is an intrinsic direction to this medium (say a preexistent flow with a given preferential direction), then the shock will have components parallel and perpendicular to the flow. We appeal to the Rankine–Hugoniot conditions once again and assert that the normal flow will undergo a discontinuous jump, while the flow *parallel* to the front will be unaltered (no jump). This latter statement is equivalent to saying that $[\mathbf{v} \times \hat{\mathbf{n}}]_\Sigma = 0$. We knew this anyway, since we have been neglecting vorticity in our streamline treatment of the flow. Let us see what this corresponds to formally and then see how it can be applied to astrophysically interesting scenarios.

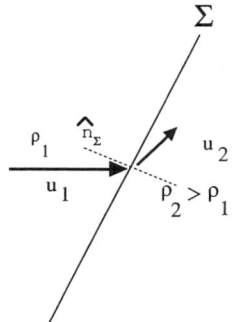

Figure 4.4 Oblique planar shock front.

The flow into the front will have velocity component $v_1 \sin \phi$ and $v_1 \cos \phi$. Here $\hat{x}$ is the direction normal to shock front Σ. The Rankine–Hugoniot conditions become

$$\rho_1 v_1 \sin \phi = \rho_2 v_{2,\perp}$$
$$\rho_1 (v_1 \sin \phi)^2 + p_1 = \rho_2 v_{2,\perp}^2 + p_2$$
$$\frac{1}{2}(v_1 \sin \phi)^2 + \frac{\gamma}{(\gamma-1)} \frac{p_1}{\rho_1} = \frac{1}{2} v_{2,\perp}^2 + \frac{\gamma}{(\gamma-1)} \frac{p_2}{\rho_2} \qquad (36)$$

We add the auxiliary condition that the velocity is unchanged in the tangential direction so the $v_1 \cos \phi = v_{2,\|}$. This is the full complement of jump conditions, which will then allow a direct solution of the shock jump equations as a function of oblique angle. Before we proceed with the algebra, let's examine an intuitively useful picture of what we might expect to happen. We know that in the case of a compressive shock, we will have the velocity slower (in the frame moving with Σ) after passage of the material through the front. Consequently, we would expect that, in a general way, v_1 is greater than v_2 in the normal direction. However, in the tangential direction, the component is unaltered. Therefore, the shock refracts *outward*, that is, away from the normal, and the flow becomes more parallel to the front. In the case of a *rarefaction shock*, the opposite occurs. This is the opposite, as well, of the optical case, where the decrease in the speed of light in a refractive medium causes convergence of the wave toward the normal. We will see that this has some remarkable consequences for astrophysical problems, especially related to the interstellar medium and the formation of jet structures.

In all of the equations, the velocity normal to the front is replaced, on the incoming side, by $v_1 \sin \phi$, while on the outgoing side it becomes $v_{2,\perp}$. For the deviation of the flow from the normal direction, we therefore can define an angle χ such that $v_{2,\perp}/v_{2,\|} = \tan \chi$. Having done this, solving for the Rankine–Hugoniot conditions, we note that there is a simple relation between the angle ϕ of the front to the incoming flow and χ, such that $\chi < \phi$ in general. Therefore, in the frame of the flow, the shock will not be parallel to the front, but it will also no longer be parallel to the original flow. To see this, we need only note the following. The perpendicular component of the velocity, postshock, is given by

$$v_{2,\perp} v_1 \sin \phi = \frac{p_2 - p_1}{\rho_2 - \rho_1} \qquad (37)$$

4.4 Some Additional Complications

which is the analog of the condition for planar shocks. We can reduce the Rankine–Hugoniot conditions for the oblique shock to the form

$$\frac{v_1 \sin \phi}{v_{2,\perp}} = \frac{(\gamma-1)p_2 + (\gamma-1)p_1}{(\gamma-1)p_2 + (\gamma-1)p_1} = \Lambda^{-1} \tag{38}$$

With this in hand, the velocity ratio is given by

$$\frac{v_{2,\perp}}{v_{2,\parallel}} = \tan \chi = \Lambda \tan \phi \tag{39}$$

Thus, in general, $\xi < \phi$ for the shock.[4] Now using the fact that the preshock Mach number is defined in terms of the component normal to the front,[5] the ratio of pressures across the front is given by

$$\frac{p_2}{p_1} = \frac{\gamma[2M_1^2 \sin^2 \phi - 1] + 1}{\gamma + 1} \tag{40}$$

Then having in hand the pressure jump across the shock, we can write

$$\tan \chi = \frac{2 \cot \phi (M_1^2 \sin \phi - 1)}{2 + M_1^2(\gamma + \cos 2\phi)} \tag{41}$$

It is also possible to write down the general expression for the postshock Mach number for an oblique shock as

$$M_2^2 = \frac{2 + (\gamma-1)M_1^2}{2\gamma M_1^2 \sin^2 \phi - (\gamma-1)} + \frac{2M_1^2 \cos^2 \phi}{2(\gamma-1)M_1^2 \sin^2 \phi} \tag{42}$$

This is a general expression for the jump conditions across an oblique shock, and it reduces (as it should) to the standard conditions for a normal-incidence planar front. It is also important that there is a maximum deviation angle for the refraction of the shock, $\chi_{\max}$.

---

[4] Here you should be careful, though, because of the way in which χ is defined.

[5] It's worth outlining here an alternative approach to this derivation. Take the incident velocity normal to the front to be $v_1 \sin \phi$ and the postshock flow to be $v_{2x} \sin \phi - v_{2y} \cos \phi$. The flow along the front is described by $v_1 \cos \phi$ and $v_{2x} \cos \phi + v_{2y} \sin \phi$. Therefore, we can use the ratio of densities with the modified Mach number, $M_1 \sin \phi$ to determine the ratio of the normal components, from which the rest of the Rankine–Hugoniot relations follow (Landau and Lifshitz 1987). The shock refraction angle can also be derived in a direct manner from these equations.

Shocks at Curved Fronts

Consider what happens if a flow collides with a shock that has a finite radius of curvature. This is similar to what happens if we take a uniform flow and slam it into a bow wave. Take the two directions, normal and parallel to the front, to have unit vectors $\hat{n}$ and $\hat{t}$. These are defined locally, at each point along the shock front. Note that here t refers to the *transverse* component, not the time. The velocity derivatives are

$$\frac{\partial \mathbf{v}}{\partial n} = \frac{\partial v_n}{\partial n}\hat{n} + \frac{\partial v_t}{\partial n}\hat{t} \tag{43}$$

$$\frac{\partial \mathbf{v}}{\partial t} = \left(\frac{\partial v_n}{\partial t} - \frac{v_t}{R}\right)\hat{n} + \left(\frac{\partial v_t}{\partial t} + \frac{v_n}{R}\right)\hat{t} \tag{44}$$

Therefore, for uniform flow, we have

$$\frac{\partial \rho_1}{\partial n} = 0, \quad \frac{\partial p_1}{\partial n} = 0, \quad \frac{\partial \rho_1}{\partial t} = 0, \quad \frac{\partial p_1}{\partial t} = 0, \quad \frac{\partial v_{n,1}}{\partial n} = 0$$
$$\frac{\partial v_{t,1}}{\partial n} = 0, \quad \frac{\partial v_{n,1}}{\partial t} = \frac{v_{t,1}}{R}, \quad \frac{\partial \rho_1}{\partial n} = -\frac{v_{n,1}}{R} \tag{45}$$

Notice that the transverse flow has an acceleration term that depends on the radius of curvature of the shock front. Now we can find the postshock flow properties by taking the derivatives of Eqs. (40–42) and using the fact that the Mach number for the incident flow is given by

$$M^2 = \left(\frac{\gamma p}{\rho}\right)^{-1} |\mathbf{u}|^2 \cos^2 \theta \tag{46}$$

where θ is the inclination of the shock front. Then the derivatives of the Mach number depend on $\partial \theta/\partial t = -1/R$ and $\partial \theta/\partial n = 0$ and we arrive at

$$\frac{\partial \rho_2}{\partial t} = \frac{4(\gamma+1)M^2 \tan \theta}{(\gamma-1)M^2 + 2} \frac{\rho_1}{R} \tag{47}$$

which is the gradient along the front of the postshock density. Notice that this gradient has a steep dependence on θ. The transverse pressure gradient is given by

$$\frac{\partial p_2}{\partial t} = \frac{4\gamma M^2 \tan \theta}{\gamma+1} \frac{p_1}{R} \tag{48}$$

and the transverse gradient in the normal component is found directly from the density jump condition. The reason you would want these may not have been clear at the start, but consider the postshock flow for a

4.4 Some Additional Complications

moment. The equations of motion for the flow after it has passed through the shock front decompose into a new coordinate system—parallel and normal to the front. Since the solution to these equations is set at the shock front, the gradients suffice to continue the calculation into the postshock region. For instance, for the normal component of the momentum, we have

$$\frac{\partial}{\partial n}\rho_2 v_{2,n} + \rho_2 \frac{v_n}{R} + \frac{\partial}{\partial t}\rho_2 v_{2,t} = 0$$

$$\rho_2 \left\{ v_{2,n}\frac{\partial v_{2,n}}{\partial n} + v_{2,t}\frac{\partial v_{2,n}}{\partial n} - \frac{v_{2,t}^2}{R} \right\} = -\frac{\partial p_2}{\partial n}$$

$$\rho_2 \left\{ v_{2,n}\frac{\partial v_{2,t}}{\partial n} + v_{2,t}\frac{\partial v_{2,t}}{\partial n} + \frac{v_{2,n}v_{2,t}}{R} \right\} = -\frac{\partial p_2}{\partial t}$$

$$\frac{\gamma}{\gamma-1}\left\{ \frac{\partial}{\partial n}p_2 v_{2,n} - \frac{\partial}{\partial t}p_2 v_{2,t} \right\} - \frac{1}{2}\rho_2 \left[v_{2,n}\frac{\partial}{\partial n} + v_{2,t}\frac{\partial}{\partial t} \right](v_{2,n}^2 - v_{2,t}^2) = 0$$
(49)

These equations, with the appropriate substitutions from the preshock conditions we derived a moment ago, provide the normal and transverse components for the gradients of all of the relevant physical quantities across the shock front. The radius of curvature also enters, so we require some additional information when applying these equations to real bodies, like clouds or spiral arms.[6]

Qualitative Phenomena of Oblique Reflected Shocks

Take a planar shock and imagine that it strikes a wall or other surface at some angle. A good example of such an effect would be a very small explosion of a star near a very large molecular cloud. Since we're at liberty to invent scenarios here, imagine that the cloud is sufficiently rigid that it simply reflects the incident shock. Now what happens? When a shock strikes a surface, it reflects. But the material into which it moves is also still supersonic, and it is unable to adjust to the second banging it gets from the reflected shock. The result is that the shock forms a triple point, with a contact surface that connects the two shocked regions. The condition for the formation of the *Mach stem* is that the reflection be at an angle greater than χ_{max}. For nearly normal incidence, the shock looks like a normally reflected wave and the angle of incidence is the same as the reflected angle. (See Fig. 4.5.)

---

[6] The final step, the derivation of the normal derivatives, can be completed in your copious spare time with the materials at hand.

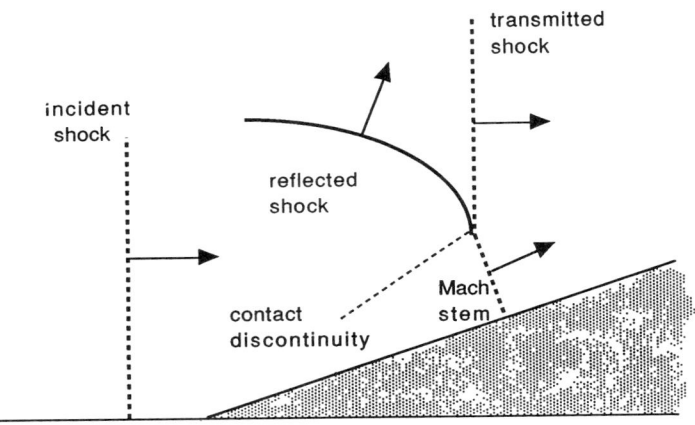

Figure 4.5 The interaction of a planar shock with an oblique barrier, indicating the formation of the Mach stem and reflected shock. The incident shock is shown at left.

Some Remarks on Applications of Oblique Shocks

One application of oblique shocks in astronomical problems is the interaction of either blast waves or ionization fronts with interstellar clouds. Since these are two- or three-dimensional problems, they are terribly cumbersome numerical experiments, often taking the largest available machines long periods to integrate. Recent work includes the evaporation of globules (Sandford et al. 1984; Tenorio-Tagle and Różycka 1983, 1984), the interaction of supernovae with clouds (Chevalier and Theys 1974), and one of the applications which has caught the fancy of many who work on grand design spiral galaxies—spiral nonlinear density wave shocks. It is this latter work which we will now discuss, since it is one of the few which we might even hope to observe.

4.4.3 Magnetic Shocks

A magnetic field enters the pressure and energy of the medium and acts like an additional, very compressible, fluid. The effects of the field can be included through the Alfvén velocity, v_A, which is given by $B/\sqrt{(4\pi\rho)}$, where B is the field strength. The energy density of the field is simply $B^2/8\pi$, while the pressure is ρv_A^2. The effect of the shock is to compress the field and therefore increase the local energy density and pressure, with a corresponding increase in the internal pressure of the gas. The presence of the field also changes the amount of compression of the gas and, since the

4.4 Some Additional Complications

behavior is that of a gas which does not obey a perfect equation of state, alters the temperature–density structure of the shock. Finally, and most important, the magnetic field allows the medium ahead of the pressure shock to know of its approach. This is because the speed of magnetosonic waves can exceed that of the shock front, and these waves prepare the state of the gas. In effect, as soon as a magnetic field is strong in the medium, it is possible to lessen drastically the effects of a pressure shock. We will only outline the problem here; more details are to be found in McKee (1987) and Draine (1980, 1981).

Because magnetic fields introduce an anisotropy into the medium, we need to choose a direction of shock incidence. This is because the Maxwell stress does not vanish for the off-diagnonal components (just like the Reynolds stress, for example) and therefore both parallel and perpendicular jump conditions have to be included. To simplify the problem we will assume that there are no electric fields (so that $[(\mathbf{v} \times \mathbf{B}) \times \hat{\mathbf{n}}]_\Sigma = 0$) and that the normal component of the magnetic field is conserved $[\mathbf{B} \cdot \mathbf{n}]_\Sigma = 0$. Then the modified momentum jump condition is written as

$$\left[p + \rho v^2 + \frac{B^2}{8\pi}\right]_\Sigma = 0 \tag{50}$$

and the energy jump condition becomes

$$\left[\frac{1}{2}v^2 + \frac{\gamma}{\gamma-1}\frac{p}{\rho} + \frac{B^2}{8\pi\rho} + w\right]_\Sigma = 0 \tag{51}$$

Here w is the internal energy. Notice that the last term is equivalent to $\frac{1}{2}v_A^2$. Besides altering the momentum equation and adding a term to the energy equation, there is an addition to the shock equations when a magnetic field is included. The flux-freezing condition in a fully ionized, perfectly conducting plasma, $\mathbf{B}/\rho =$ constant (see the magnetic fields chapter for more details). This provides a supplementary condition:

$$\left[\frac{\mathbf{B} \cdot \mathbf{n}}{\rho}\right]_\Sigma = 0 \tag{52}$$

that connects the changes in the magnetic field strength directly with changes in the velocity and density. Now define the quantity ratio of gas to magnetic field pressure to be $\beta = p/(B^2/8\pi)$. The density jump condition then becomes

$$2(\gamma-2)\left(\frac{\rho_2}{\rho_1}\right)^2 - [(\gamma-1)M_1^2 + 2\gamma(1+\beta_1)]\left(\frac{\rho_2}{\rho_1}\right) + (\gamma+1)M_1^2 = 0 \tag{53}$$

The difference between this idealized treatment and reality, especially in dilute environments like the interstellar medium, is that different phases of the fluid may feel the magnetic effects differently. Observationally, the effects of magnetic shocks are most clearly seen in radio maps of supernova remnants. The compressed field is quite clearly seen at the boundary of the shock, where the density of trapped energetic particles also increases and therefore the surface brightness is increased. Polarization maps of remnants have demonstrated that the field is also turbulent but compressed parallel to the front.

4.4.4 Supersonic Jets

Supersonic jets are an amazingly widespread phenomenon in astrophysics.[7] They present a physical problem that is more complicated than subsonic jet flows because of the existence of internal and external shocks. The key physical point to keep in mind is that because the flow cannot hydrostatically adjust to changes in the environment, or to internal changes, supersonic jets are highly dissipative, often turbulent, and shock dominated. The first, and most dramatic, difference between supersonic and subsonic jets is seen at the orifice: the expansion of the supersonic jet into a stationary medium is preceded by a bow shock. The matter moving through this surface is accelerated in the laboratory frame and is overtaken by the front of the material jet. The compressed region is called the *working surface*, the compression zone. A shock propagates into the jet from this front, forming a contact discontinuity. Keep in mind that any change that takes place in the head of the jet will not propagate downstream. The flow is supersonic and therefore must form internal shocks. Any deviation of the flow produces internal structures, equivalent to oblique shocks, that heat and compress the jet.

The expansion is due to the deflection of the jet by the oblique shock that develops at the nozzle. Since the flow is a rarefaction, the Mach number is higher after than before the flow and the shock overexpands [see Sanders (1979) and Königl (1982) for more discussion]. You can see where the deflection of the internal flow comes from by considering a rarefaction front within the axial flow and the angle that you can calculate for χ from the equations we have just derived. A subsonic jet expanding into a vacuum can do so in a self-similar manner. A supersonic jet doesn't have this option. Shocks develop at the orifice from which the jet issues as rar-

---

[7]The references for this chapter are located in the general bibliography. The literature on jets is so extensive that it is more appropriate to link all of the various topics together in a single list.

efaction fans. As the shock expands, it may become underpressured with respect to its environment and stall. The jet streams into the stalled matter, forming internal shocks that divert the flow by refracting it back into itself. A compressional shock is then formed internal to the jet and the matter expands outward from there again. The effect just described is very clearly seen in the films of rocket launches, especially when the engines on the space shuttle or the Ariane rocket start up. They appear as diamonds, apparently suspended within the flow. Called *Mach disks*, they are the refracted shocks from the reconfinement of the expanding flow coming out of the nozzle.

The front surface of a supersonic jet is preceded into its surroundings by a bow shock. The flow across this surface has a velocity toward the jet of $3v_{BS}/4$ and therefore encounters the head of the jet supersonically. This collision region is called the *working surface* and is the region where most of the dissipation occurs. The pressure equilibrium between the bow-shocked matter and the jet gives the rate of advance of this surface, which is approximately $\rho_J(v_J - v_{WS})^2 \approx \rho_0 v_{WS}^2$. The jet velocity is given by the initial conditions, plus any entrainment that may have occurred during its expansion. The luminosity of the different regions depends both on the energy dissipation there, ρv_i^3, where i characterizes the different parts of the jet, and on the surface area A_i of that part of the jet. Extragalactic radio jets appear to show these structures, although their interpretation in very dilute, charged, magnetized plasmas is far from obvious. They appear as localized bright regions seen in centimeter wavelength images. But what complicates matters in the case of these jets is that synchrotron radiation, the nonthermal emission mechanism responsible for emission at radio wavelengths, depends on the local conditions within the jet and on the acceleration of particles to relativistic speeds.

4.4.5 Radiative Phenomena in Shocks

Astrophysical shocks have an especially interesting feature: if they occur in an optically thin medium, the radiation of energy by atomic and molecular species can significantly alter the thermal structure of the post-shock medium. It is important to note this, because it allows for the diagnosis of the conditions in the gas, both the thermal and dynamical structure, as the result of a shock passage. First, we need to consider the thermal loss mechanisms and then show how these enter into the calculation of the postshock environment.

The primary source of radiative losses is line emission. Permitted lines have large transition probabilities and often, because of collisional excitation, can be populated after the shock has passed through the medium. For temperatures less than 10^4 K, the primary losses are due to weak lines of atomic and molecular species. However, once a temperature of about 10,000 K is reached, the Hα line is excited. This line, which cascades to the Lyman α (Lyα) line, is usually optically thin. Coming as it does from the most abundant atomic species, it is intense and produces significant losses of energy which increase with decreasing temperature. In fact, there is a local maximum in the cooling function, which we shall call Λ, at this temperature primarily because of this line. The Lyα line is often suppressed in emission because of the optical depth in the medium; it is a ground state transition and consequently often thick and ineffective without many scatterings, so the time scale for thermal loss is increased by the number of scatterings. A simple rule of thumb is that the cooling time increases as the number of scatterings in the medium increases, which is proportional to $\tau^{1/2}$, where τ is the optical depth. Unless large velocity gradients are present, the photons from the ground state Lyman series will be trapped within the gas and only slowly leak out; the Balmer series can immediately escape the medium.

The cooling function for the postshocked gas is determined by the elemental abundances and speed of the shock. The stronger the shock, the higher the Mach number in terms of the preshocked gas values, the stronger the emission. The energy of the postshocked gas decreases as it flows away from the front and radiates. This can be expressed in the frame of the shock as a change in the thermal parameters for the gas as a function of distance:

$$\frac{d}{dx}\left[\rho v\left\{\frac{1}{2}v^2 + \frac{\gamma}{\gamma-1}\frac{p}{\rho} + \mathcal{U}\right\} + \frac{B^2}{4\pi}v\right] + n^2\Lambda = 0 \qquad (54)$$

Here $\mathcal{U}$ is the specific internal energy of any quantum states. The magnetic field is included just for good measure so that you can see how it changes the relaxation of the thermodynamic variables.

Lines sometimes appear that would not normally be observed from photoexcited regions. The difference between a shocked layer and an H II region at the same temperature is essentially one of density. In a shock, the density jump is compressive. In an H II region, it is just the opposite. So collisionally excited ionic lines are the signature of a shocked environment.

The temperature and density in the postshocked region determine the emission spectrum from the gas. If the excitation is great enough, ultraviolet continuum emission may result. This is the *precursor*, so called because it stands ahead of the shock, depending on the optical depth of the surrounding gas. The effect of this radiation may be merely to excite the preshocked gas, or it may actually ionize it *before* the front sweeps it up. In strong explosions in air, for example for nuclear weapons, the formation of

4.4 Some Additional Complications

the fireball is the result of the precursor. It ionizes the surrounding medium to a very large distance. But because it is an impulsive emission, resulting from the first microseconds of the explosion, it rapidly recombines. The shock isn't hot enough after the first few seconds to maintain the ionization. At this point, the fireball clears and the material blast wave breaks through. In fact, in such an explosion what you see is two competing effects. One is the fireball recombining and shrinking from its maximum size (which is established "instantaneously" because the time scale is so small for terrestrial explosions) and the expansion of the blast wave. The debris follows the blast.

A good example of a precursor is the circumstellar emission observed for SN 1987A in the Large Magellanic Cloud. Soon after the optical maximum, after the ultraviolet from the blast had faded, narrow emission lines appeared in the UV spectra. These were from intercombination and forbidden transitions from fairly ionized species like N III and C III. It is possible that even more highly ionized species, coronal lines of the iron peak elements, have been observed. These would be the product of the precursor, the initial UV pulse, propagating into the surrounding medium ahead of the blast wave. In effect, the precursor and its associated radiative effects can be thought of as a rapidly evolving H II region, one that runs out of energy as a result of powering the ionized region in which it is expanding.

Other effects are easily observed from strong shocks. For magnetic shock waves, because of the additional pressure from the magnetic field, the pressure and density jumps can be offset. The shock may be supersonic, but the magnetoacoustic speed may be great enough that the jump is continuous. In this case, there is enhanced emission because of the increase in the density and energy in the postshocked gas, but the gas shows the enhanced emission over a large region of space (if it is spatially resolved). For shocks in low-density environments, the statistical equations must be solved to determine the emergent intensities.

The run of temperature and pressure behind the front is more complicated for astrophysical plasmas than for most laboratory ones. The medium is very tenuous and collisional equilibrium is not quickly established. Since LTE doesn't hold, the radiation may come from states whose populations differ dramatically from those encountered in terrestrial shocks. The temperature, for example, does not necessarily depend directly on the rate of energy loss. If collisions are responsible for the deexcitation and subsequent losses, then this assumption (LTE) is OK. But if the gas is losing energy primarily through recombination lines, the electrons may be much hotter than the cooling rate would indicate. Also, if the collisions are infrequent enough, two temperature instabilities may come into play. Another important effect in astrophysical shocks is charge exchange. This is the process by which an ion exchanges an electron with a stationary atom, thereby neutralizing but combining in an excited state. The subsequent loss of energy is ultimately a sink for the material and an additional cooling agent. In short, there are many avenues by which the gas can lose energy radiatively, and each shock environment affects the rate of loss and the thermal profile of the postshocked gas differently.

4.4.6 Collisionless Shocks

We will only briefly mention this problem. There is a vast and growing literature on the subject because it is quite frequently encountered in space plasmas. The term collisionless shocks is a sort of misnomer, since they do not involve the pressure conditions that we have been using normally to define such phenomena. However, there are certain astrophysical conditions where gas pressure plays no direct role, but other variables undergo discontinuous changes. Take for instance the case of a strongly magnetized plasma, one in which the magnetic energy density exceeds the thermal energy. If there is an abrupt change in the direction and/or strength of the field, particles will not be able to respond to this adiabatically and will behave as if a shock wave has passed through their vicinity. Let's examine further what this implies for the medium.

Take the case of a low-density electron gas. The critical parameter $\beta = 4\pi p_e/B^2$ is the ratio of the thermal to magnetic pressure through the gas and will be assumed to be small. The speed of the disturbance is measured in terms of the Alfvén speed (see magnetic fields chapter) v_A which depends on both the field strength and the local gas density. The addition of a magnetic field means, as we have discussed before, that the normal component of the magnetic field is continuous $[\mathbf{B} \cdot \mathbf{n}]_\Sigma = 0$ and that, for a steady-state problem, $[\mathbf{E} \times \mathbf{n}]_\Sigma = 0$ where $\mathbf{E} = -(1/c)\mathbf{v} \times \mathbf{B}$.

4.5 A Quodlibet of Applications of Shocks to Astrophysical Problems

In this section, we will examine a few problems that illustrate shock calculations. Not only are they not exhaustive, they are barely scratching the surface of what can be, or has been, done in the field. However, a reason for including the particular examples in the collection is that they are representative of classes of such problems. Also, they can be studied

4.5 A Quodlibet of Applications of Shocks to Astrophysical Problems

more or less analytically. While numerical calculations are now the norm for the whole range of hydrodynamic problems, quick estimates and evaluations of the underlying physics are often what is needed when confronting observations. So let's proceed.

4.5.1 Density Wave Shocks: Traffic

It is late at night. You are driving down an interstate highway, looking at the pattern of the traffic (and hopefully also the road). One thing will be quite distinct about the distribution—it is clumpy. There will be waves of cars, with a characteristic spacing between the lumps. (See Fig. 4.6.) The interaction of drivers, in the presence of an imposed speed limit, will cause the flow to show a wave packet not unlike a density wave. We now put a toll booth in the way. This acts like a shock—there is a density enhancement on the approaching side (recall that you have always before been

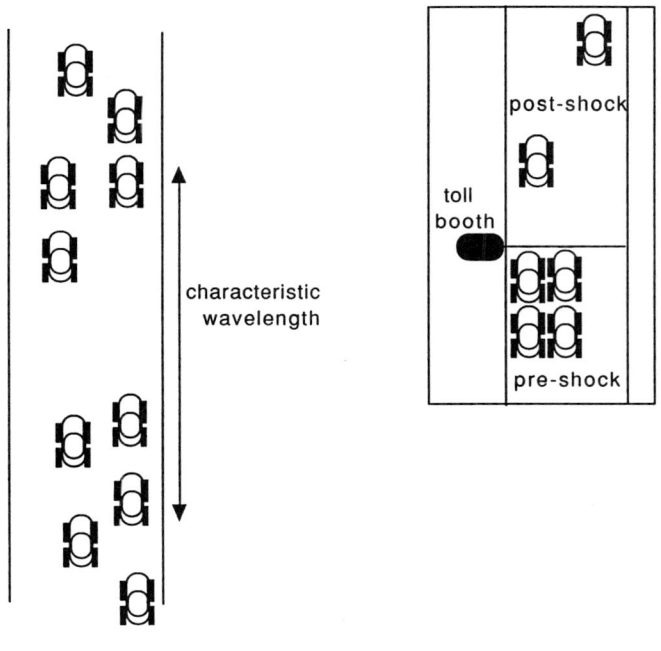

Figure 4.6 Traffic as a density wave problem (see also Prigogine and Herman 1977).

sitting on the front; here you are sitting in the flow). The density is enhanced because cars are an (approximately) incompressible medium, and therefore the decrease in the velocity produces an increase in the density (a compressional shock). On the lee side of the booths, the flow rapidly accelerates away leaving a rarefaction wave, which eventually settles down into the same density wave pattern that was typical of the upstream side of the flow before the toll. This is as close as you will likely come (considering the time scales for orbiting the center of the galaxy) to experiencing the effects of a density wave shock in a galaxy. A delightful book on this subject is by Prigogine and Herman (1971), who use a fluid approach to the flow equations for traffic.

4.5.2 Density Wave Shocks: Spiral Arm Shocks in Galaxies

Now let us examine how this analogy might be applied to galactic hydrodynamics. If we have a perturbation in the stellar component of the galactic disk in a flattened galaxy, there will be a perturbation in the gravitational potential via the Poisson equation. Since the gas flows "through" the stellar disk, it will be forced to respond to the change in the potential by accelerating as it enters the gravitational well generated by the stars, and it slows down as it climbs out. Roberts (1969) was the first to show that the full nonlinear equations for the flow possess shock solutions at the density wave. That is to say that given the local sound speed, which is the same as the stellar velocity distribution, the flow velocity accelerates to supersonic velocities within the density enhancement. We can now examine the equations for this flow, to show what the procedure for the calculation of such shocks is, and then continue our discussion of the more general applications of the approach. First, let's start off with a very schematic approach and then see how it can be generalized to the full nonlinear solution.

Begin with the equations of motion in a frame rotating with the pattern speed of the density wave. (See Fig. 4.7.) This means that, by some means, we have been able to determine Ω_p, the frequency of the wave. Assume a thin disk of gas and stars, and that the stars are solely responsible for the support of the wave. Further assume that the motion is slow compared with the rotation speed of the galaxy and therefore that we can use a local analysis in which the geostrophic approximation nearly holds. For stationary flow, the continuity equation reads as follows:

$$\frac{\partial \Sigma u}{\partial x} + \frac{\partial \Sigma v}{\partial y} = 0 \tag{55}$$

4.5 A Quodlibet of Applications of Shocks to Astrophysical Problems

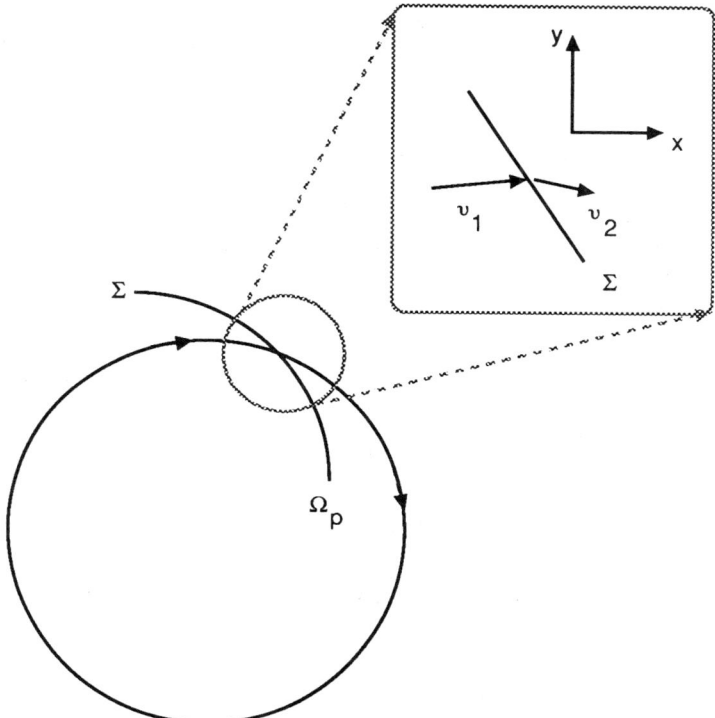

Figure 4.7 Coordinate system for spiral shock calculation in a spiral galaxy (see text for details).

and the equations of motion are

$$u\frac{\partial u}{\partial x} + v\frac{\partial u}{\partial y} + fv = -\frac{a_s^2}{\Sigma}\frac{\partial \Sigma}{\partial x} - \frac{\partial \Phi}{\partial x}$$

$$u\frac{\partial v}{\partial x} + v\frac{\partial v}{\partial y} - fu = -\frac{a_s^2}{\Sigma}\frac{\partial \Sigma}{\partial y} - \frac{\partial \Phi}{\partial y} \qquad (56)$$

Assume that the flow is parallel to the $\hat{x}$ direction and that there is no dependence of the motion on y other than there being a β-plane approximation for this motion. This value of β is given by $\partial \Omega_p/\partial y$ evaluated at $r = r_0$. The y direction is taken radially so that at constant radius the β approximation means that the Coriolis perturbation is the epicyclic frequency κ. Now assume that the gravitational potential is expanded as

$$\Phi(x,y) \approx \Phi_0 + \frac{1}{2}\frac{\partial^2 \Phi}{\partial x^2}x^2 + \frac{1}{2}\frac{\partial^2 \Phi}{\partial y^2}y^2$$

since we assume that at the spiral arm is a local density maximum. Note that the approximation we can make is that the density is constant along the arm; therefore the second term vanishes. We assume for simplicity that the arm is a planar shock lying along $\hat{y}$. The problem is a bit more complicated, as we will see, once we allow for the precise form of the spiral perturbation. In particular, the arm has a tilt i to the local circular streamlines so that it is really oblique to the flow u. But for the moment, let's continue with the analysis to see what happens in the schematic equations before studying the detailed solution.

The equation for u reduces to the approximate form

$$u\frac{\partial u}{\partial x} + fv = \frac{a_s^2}{\Sigma}\frac{\partial \Sigma}{\partial x} - \left(\frac{\partial^2 \Phi}{\partial x^2}\right)_0 x \tag{57}$$

The continuity equation is approximately

$$\Sigma\frac{\partial u}{\partial x} + u\frac{\partial \Sigma}{\partial x} = 0 \tag{58}$$

and is used to remove the pressure term from Eq. (57); thus the flow is described approximately by

$$\frac{1}{u}(u^2 - a_s^2)\frac{\partial u}{\partial x} = fv - \Phi_0'' x, \qquad \frac{\partial v}{\partial x} = -f \tag{59}$$

for the x component of the flow, perpendicular to the local arm. Therefore, the flow becomes sonic at $x = 0$, that is, at the spiral arm, and v is continuous across the arm.

You will see this again in the case of a stellar wind and also for the flow at the inner Lagrangian point of a Roche-distorted close binary system. However, here the condition on the flow is that the velocity does not start out small. In fact, the flow is highly supersonic due to the motion of the pattern through the surrounding gas, and therefore we have the condition for the formation of a shock. The gas is decelerated at the spiral arm because of the change in the sign of the gravitational force due to the perturbation there, and the information cannot propagate upstream to inform the material coming in that it needs to adjust to the change in the velocity. This means that the flow will plow into material near the arm and compress. The normal component of the flow, $u = \mathbf{v} \cdot \hat{\mathbf{x}}$ in our notation, then must satisfy the Rankine–Hugoniot conditions at the arm while the para-shock component, v, is continuous. The flow refracts away from the normal on passing through the shock and the streamlines are therefore distorted toward the arm if the shock front is oblique.

4.5 A Quodlibet of Applications of Shocks to Astrophysical Problems

For a trailing spiral this means that the gas is diverted toward the galactic center; the local density of tracers, the molecular clouds, increases; and the flow departs from strictly circular orbits. In the case of a very tightly wound spiral or a stellar bar, this has a drastic effect. Material is channeled toward the center of the galaxy, where it can serve as fuel for any accreter that resides there. Such flows are often implicated in the supply of material to central black holes in active galactic nuclei. Whatever the ultimate decision on the existence of stable density wave patterns in galaxies, it is certain that the flow we have just described is a necessary consequence of the stars being able to set up even a local version of the density wave. The amplitude of the velocity perturbation is directly proportional to the amplitude of the gravitational perturbation in this geostrophic-type picture. The details of the more complete solution are found in Appendix A.

4.5.3 Blunt Body Flows: Collisions of Shocks with Clouds

Blunt body shock theory is an outstanding example of the application of an engineering problem to astrophysical situations. The interaction of a shock with a blunt object was first developed for supersonic aircraft design in an attempt to optimize stability and minimize drag at transsonic and hypersonic speeds. The solution was to make the nose of the aircraft as sharp as possible and to make the attacking side of the wings as thin as possible. Much of the subsequent work in this field has been dominated by this design plan.

Unfortunately, most cosmic objects are spherical, not shaped like airfoils or cones. This is important. It means that the collision of a shock with such an obstacle will necessarily produce an interaction bow shock, standing some distance from the body, around which the shock is diverted. The main reason for this is that the surface presents a finite area to the flow, rather than the singular point seen in a conic. The greater the radius of curvature, the more the body resembles a wall or piston in the flow and the closer the bow shock comes to a classical reflected shock wave. Also, unless there has been a serious problem with the design or its execution, aircraft are not really fully compressible. At least we try to make sure they're not.

The details of the shape of a finite object don't really matter much provided the radius of curvature is large enough. Imagine a shock colliding with a compressible body, for instance, a cloud in the interstellar medium subject to a supernova blast or a cloud falling supersonically through the galactic plane. (See Fig. 4.8.) The body is hit by a blast wave (in either frame the phenomenology will be the same) with a velocity v_s.

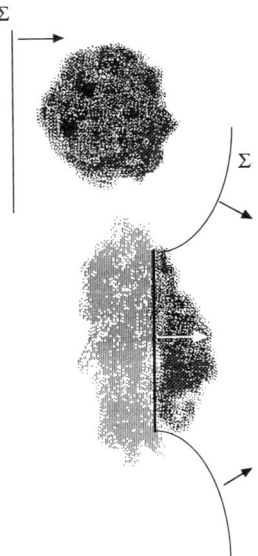

Figure 4.8 Collision of a shock with an interstellar cloud.

At a distance d from the body, there will be a reflected stagnant bow shock. The gas flow across this shock undergoes compression as for a strong shock, and the shock on the downstream side (postshocked gas) eventually comes to a halt at the surface of the cloud. One thing to note is that we are taking a medium which has a small density with respect to the cloud and slamming it into the denser medium. Even if a stagnation point is reached, there will be external pressure, higher than that within the cloud, that drives a shock into the cloud. Two physical effects follow from this. One is that the cloud is compressed. The other may be that it is accelerated by the converging shock. But we will return to this momentarily. Let's proceed with the analysis (see Fig. 4.8).

When the shock collides with a surface, two things happen. First, a shock is transmitted into the surface (if it isn't perfectly rigid) and the rest of the incident shock continues to move past the body. Second, the collision generates a reflected shock that moves into the matter heated by the incident shock. This material has a higher sound speed and higher density, so the strength of the reflected shock is not simply the multiple we would have expected from the first collision. Also, the Mach number may be much lower in this flow. Remember, this material is moving at a lower speed than the incident front. To quantify this discussion, the gas will be

4.5 A Quodlibet of Applications of Shocks to Astrophysical Problems

assumed to be $\gamma = \frac{5}{3}$. Then the compression ratio for a strong shock gives $\rho_1/\rho_0 = 4$ and the speed of the gas is then $v_1 = 3v_\Sigma/4$, where v_Σ is the incident shock speed. Now the postshock sound speed is $a_{s,1}^2 = (\gamma p_1/\rho_1)^{1/2} = 5^{1/2} v_\Sigma/4$ and therefore $M_1 = 3/5^{1/2} = 1.34$. This is usually much smaller than the incident Mach number, M_0, which can be as large as we wish. The reflected shock therefore has a Mach number given by $M_2^2 = (M_1^2 + 3)/(5M_1^2 - 1) = 2.4$ for a perfect gas. The pressure in the post-reflected shock gas is therefore $p_2 = 3p_0$. Just as an aside, you might have noticed that if $M_1 = 1$ then $M_2 = 1$ also.

Recall that the Bernoulli equation for the postshocked flow is

$$\frac{1}{2}v^2 + \frac{a_s^2}{\gamma - 1} = \frac{a_{s,0}^2}{\gamma - 1} \qquad (60)$$

for the stagnating gas near the cloud interface. Therefore, we have the condition that

$$\left(\frac{a_{s,0}}{a_s}\right)^2 = 1 + \frac{1}{2}(\gamma - 1)M^2 \qquad (61)$$

defining M as the Mach number for the initial flow (that just across the bow shock surface and away from the stagnation point). Now for a polytropic gas,

$$a_s^2 \approx \rho^{\gamma - 1} \approx \rho^{(\gamma - 1)/\gamma}$$

so that we can determine either the density *or* pressure change between the postshock region and the surface of the body. Thus

$$\frac{p}{p_0} = \left(1 + \frac{1}{2}(\gamma - 1)M^2\right)^{\gamma/(\gamma - 1)} \qquad (62)$$

Notice that this is the pressure drop with respect to the *postshock* Mach number, *not* the upstream value.

What we now need to do is find the Mach number for the downstream flow in terms of the preshock values. Recall that

$$v = \left(\frac{1}{2}\frac{(\gamma + 1)^2}{\gamma + 1}\frac{p}{\rho_{ps}}\right)^{1/2} \qquad (63)$$

for a strong shock. Here ρ_{ps} is the preshock density and in general, *ps* will refer to the gas parameters upstream of the bow shock. Therefore:

$$v = a_{s,ps}\frac{2 + (\gamma - 1)M_{ps}^2}{(\gamma + 1)M_{ps}^2} \qquad (64)$$

so that

$$p_0 = p_{\rm ps}\left(\frac{\gamma+1}{2}\right)^{(\gamma+1)/(\gamma-1)} \frac{M_{\rm ps}^2}{[\gamma-(\gamma-1)/(2M_{\rm ps}^2)]^{1/(\gamma-1)}} \quad (65)$$

which reduces to

$$p_0 = p_{\rm ps}\left(\frac{\gamma+1}{2}\right)^{(\gamma+1)/(\gamma-1)} \gamma^{-1/(\gamma-1)} M_{\rm ps}^2 \quad (66)$$

for a strong bow shock in the upstream gas. The stagnation pressure p_0 is the physical result we have been looking for, because it is this pressure that drives the shock that penetrates the cloud. It matters little whether the shock is traveling into the cloud or the cloud into the shock. The velocity of the preshock gas, $v_{\rm ps}$, is $v_{\rm ps} = 2v_{\rm s}/(\gamma-1)$ with respect to the incident blast wave or shock.

The overall result of these calculations is that the speed of the shock into the cloud scales as

$$v_c = v_{\rm s}\left(\frac{\rho_{\rm ps}}{\rho_c}\right)^{1/2} \quad (67)$$

where $\rho_{\rm us}$ is the density of the unshocked (ambient) gas. Recall that the density increase behind the blast wave is a factor of 4 for a strong, nonisothermal shock. What we have, therefore, is a simple scaling relation for the propagation of a shock into the denser cloud material and also the compression of that material (Sgro 1975; McKee and Cowie 1975; Tenorio-Tagle 1981). This means that the shock velocity is determined essentially by the density ratio between the ambient gas and the cloud so that, for large density contrast, the shock will move very slowly through the cloud. However, because the temperature of the cloud is low, the speed of the shock may still be supersonic through the cloud and the effect on its internal structure may be substantial.

Notice that we have made no assumptions about the ratio of specific heats, only that there is no dependence of γ on the strength of the shock. This may not be correct for the treatment of a shock in a molecular cloud, especially if ionization or dissociation takes place within the denser medium. The ratio for an ideal monoatomic gas, $\frac{5}{3}$, is considerably larger than the $\frac{7}{5}$ that should be more typical of a molecular cloud (which is, after all, composed of diatomic and polyatomic molecules). An ionization of this region by the precursor, if the medium is optically thin enough for this front to separate substantially from the incident shock, can also change the jump conditions (in effect, the addition of a degree of freedom to the gas reduces the velocity of the shock because of an increase in the effective

4.5 A Quodlibet of Applications of Shocks to Astrophysical Problems

enthalpy in the postshocked gas and alters the relation between the pressure and velocity of the postshocked environment).

More Exact Calculation of Shock–Sphere Interactions

The collision of a planar shock with an incompressible sphere is an interesting application of the Kelvin circulation theorem. For this reason, and to illustrate how such a computation is performed, we'll spend a little time on the problem. Since a sphere has a fixed radius of curvature, we would expect that the bow shock should as well. Also, if the sphere can't compress, we have only one shock to worry about. So let's look at the equations.

First, for an axisymmetric system we can write the velocity in terms of the radial component v_r and the angular component v_θ. For streamlines, these are both represented by the velocity potential ψ such that $\mathbf{v} = \nabla\phi$ or

$$v_r = \frac{\partial \Phi}{\partial r} = \frac{1}{r^2 \sin\theta} \frac{\partial \psi}{\partial \theta} \tag{68}$$

$$v_\theta = \frac{1}{r}\frac{\partial \phi}{\partial \theta} = -\frac{1}{r \sin\theta} \frac{\partial \psi}{\partial r} \tag{69}$$

where the second representation is called the *Stokes* form (Batchelor 1967). The vorticity is given by $\omega = \nabla \times \mathbf{v}$ so in component form becomes

$$\omega = \frac{1}{r}\frac{\partial}{\partial \theta} v_\theta - \frac{1}{r}\frac{\partial}{\partial r}(r v_r) = \text{constant} \tag{70}$$

Take $x = r\sin\theta$ to be the distance from the symmetry axis of the stream. Then at large distance, the flow should look like $\omega \sim \mathscr{C} x$ where $\mathscr{C}$ is a function only of ψ and is constant on a streamline. The equation for the potential is therefore given by

$$\nabla \cdot \nabla \phi = 0 \tag{71}$$

for incompressible steady-state flow. You are also free to represent the *mass flux* along the streamline, $\rho\mathbf{v}$, rather than the velocity by $\nabla\phi$. Lighthill (1957) solves the problem by using this technique for compressible fluids by taking the jump conditions across a surface of constant curvature and obtains

$$\omega = \frac{(2\gamma)^2}{(\gamma-1)(\gamma+1)} \frac{V_\Sigma}{R_s} \sin\theta \tag{72}$$

where V_Σ is the shock speed and R_s is the radius of the spherical obstacle. To derive this solution requires solving the continuity equation for the potential and then matching the streamlines across the shock boundary.

The basic form of the velocity law, the familiar factor of $2\gamma/(\gamma+1)$, indicates that the Bernoulli equation is again at work. Vorticity is generated by a blunt body. The postshock environment is therefore a good potential source for feeding turbulence. It also means that the body will experience an aerodynamic drag, if incompressible, from the passage of the shock.

4.5.4 H II Regions

Strömgren Spheres and Ionization Fronts as Shocks

One of the frequently encountered applications of astrophysical shock phenomena is in the formation and evolution of H II regions. (See Fig. 4.9.) These are regions of ionized hydrogen formed in the vicinity of hot, young stars which have considerable ultraviolet flux. The static radius of such regions is easily computed using what in effect is photon counting. The number of ionizing photons passing through a region of radius R is

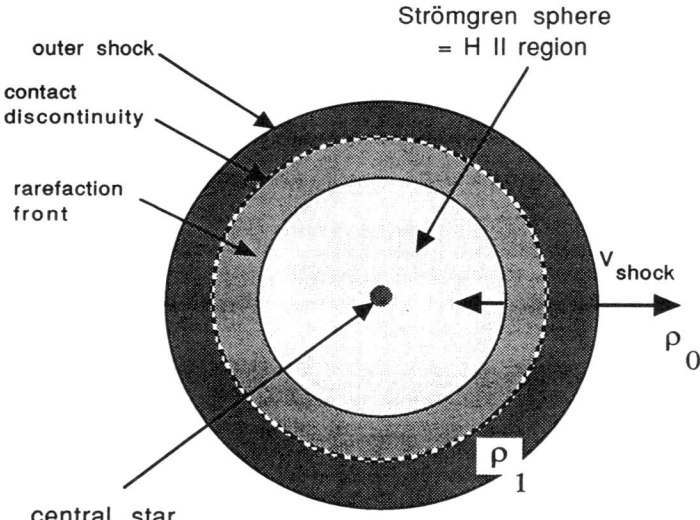

H II region expansion

Figure 4.9 An expanding H II region, late in the evolution of the nebula when the front is dynamically unstable.

4.5 A Quodlibet of Applications of Shocks to Astrophysical Problems

N_{UV} per second (where this will be examined momentarily), while the number of recombinations in a region which consists primarily of hydrogen is given by:

$$\text{Recombinations per unit volume} = \alpha(T) n_e^2 R^3 \tag{73}$$

where $\alpha(T)$ is the recombination coefficient per unit volume and n_e is the electron density. In ionization balance, therefore, we would expect that the radius of the ionized region, the so-called *Strömgren sphere*, is given by

$$R_{\text{H II}} \sim n_e^{-2/3} N_{UV}^{1/3} \tag{74}$$

and decreases with increasing density, while increasing with the number of UV photons available. No dynamical statement is contained in this equation — it merely represents the equilibrium radius which will be reached if a source is turned on and the photons diffuse outward. It is well known, however, that such a situation cannot prevail for long. The energy is continuously dumped into the medium by the stars, and since the region has low density but high pressure (hot) compared with the surroundings, it must eventually expand. It does so under the influence of gravity, but due to the fact that the heat source is still alive and active in the interior.

The ionization front expands in a fashion different from the ones we have thus far considered. The ionization behind the front means that additional free energy is available. Put another way, matter flowing through the front will suddenly find itself bathed in the UV photons from the central star and thus will ionize as well. The amount of ionized gas will therefore grow with time. Further, there will be a compressional shock at the interface between the H II region and the surrounding gas, but this will be followed by a very hot, low-density region. Within this region, the sound speed is high and therefore the matter will be moving quite subsonically in general (the sound speed across the front undergoes a jump).

Historically, Savedoff, Mathews, Vandervoort, and Lasker were among the first to consider the expansion of such regions [Spitzer (1979) and Goldsworthy (1983) give some of the "cultural" background of this work], and their dynamics have been of interest since computational fluid mechanics became an astronomical household word. We shall follow the simple approach, first outlined by Kahn (1954), for the classification and evaluation of such shocks. First take the energy equation, which now includes the ionization behind the front:

$$\frac{1}{2} v_1^2 + \frac{\gamma}{\gamma - 1} \frac{p_1}{\rho_1} = \frac{1}{2} v_2^2 + \frac{\gamma}{\gamma - 1} \frac{p_2}{\rho_2} + \chi \tag{75}$$

Here χ is the increase in the internal energy due to the ionization of matter through the front. It is assumed that the ratio of specific heats does not change across the ionization boundary. Now, the front moves at a very particular speed. As new material is ionized, it is said to have entered the region behind the front, and therefore the mass flux is fixed by the flow of ionizing photons through the medium. Therefore, there is a fixed value of J, the mass flux, which solves the mass conservation condition across the ionization front:

$$\rho_1 v_1 = \rho_2 v_2 = \mu_I J_I \tag{76}$$

where here J_I is the flux of photons through the planar front and μ_I is the mean molecular weight of the newly ionized gas. It is this factor which contributes so immediately to the equation of state behind the front and therefore to the equation of motion for the shock. Initially we have a flow of ionization rather than mass. But the addition of electrons to the post-front gas causes the medium to heat up and start expanding. We will see this shortly, but for now we have the idea that this front propagates as a sort of detonation rather than a simple explosive product.

Ionization Fronts and Detonation Waves. I.

Phase transitions in the interstellar medium take place primarily because of the external influence of stars. Nowhere is this better illustrated than in the case of the formation of H II regions, locales of ionized matter situated in the vicinity of hot luminous stars. We have already discussed Strömgren spheres in a general way. Now we have to examine the consequences of forming these regions and what their dynamical state is in the interstellar gas.

Before beginning this examination, we need to consider something of the physics of a different kind of shock wave, the *detonation wave*. Physically, this is a powered shock. That is, the input of energy from the combusting products ignited after the passage of a shock through a material adds energy to the front and keeps it moving into the undisturbed medium upstream of the front. A simple analogy is an explosion of a chemical mixture. The initial explosion is touched off by some input of energy, which may be dynamical or may be due to an instability within the material and the rapid (runaway) input of energy from reactions. This increase in energy produces a rapid rise in the temperature and pressure, initiating the expansion of the medium. If the explosion proceeds slowly enough, the expansion will cool the material and the energy input will cease. The expansion may therefore stall, depending on the conditions in the background into which the explosion products are expanding. However, if the energetics are such that the cooling by expansion is ineffective, the further propagation of the front may bring new, unburnt, material into the reaction zone and increase both the power and speed of the expansion. The difference between these two conditions is the difference between a fizzle and a bang and is therefore an important physical process to consider.

4.5 A Quodlibet of Applications of Shocks to Astrophysical Problems

To detail the behavior of a detonation under astrophysical conditions, we could imagine that a material is initially shocked. On compression and heating, nuclear processing may begin and the resultant release of energy powers the further propagation of the front. The condition, called a thermonuclear detonation, was suggested early in supernova research as a mechanism for promoting the powering of the shock wave expansion in supernovae. The specific model was called the "carbon detonation" model because the primary reaction was $^{12}C + ^{12}C$, assumed to take place under degenerate conditions in which the medium cannot respond mechanically to the increase in the energy input.

To detail physically how to treat a detonation we turn to the initial shock equations. First we note that the shock adiabat is

$$j^2 = \frac{p_2 - p_1}{(1/\rho_1) - (1/\rho_2)} \tag{77}$$

For a given compression ρ_2/ρ_1 we have a pressure differential across the front. Now if we increase the energy content behind the front by the reaction zone, the temperature will be higher in the postshock region than would have been expected. For any compression, the resulting pressure is higher and also the sound speed is higher. Therefore, the burning takes place in a completely subsonic regime and actually looks a lot like a flame. Let's follow the treatment of Landau, Lifshitz, and Zel'dovich for a moment. This will be especially useful because it leads to a postshock condition on the velocity and also because it helps clear up one of the important features of an H II region, namely that it is essentially a detonation front.

Recall that $a_s^2 = dp/d\rho$. Because the shock adiabat gives $j^2 = (p_2 - p_1)/(V_2 - V_1) = \Delta p/\Delta V$, where V is the specific volume (the inverse of the density), we see that $j^2 = v_1/V_1 = v_2/V_2$ so that $a_s^2 > j^2 V_2^2 = v_2^2$ and therefore $M_2^2 < 1$ in general. But for a detonation wave, the added energy causes the postshock sound speed to rise above that predicted simply by compression, and the decreased velocity of the gas relative to the front can yield the condition that $M_2 < 1$. In other words, in the postshocked gas the combusted material can move slower than the sound speed. In the case of an H II region this is precisely what happens. The increase in the random component of the energy, the pressure, drives the front but does not increase the bulk kinetic energy of the flow. In fact, the material inside the ionization front moves with the local sound speed. It may be preceded by a shock that runs into the background gas, and in fact this shock will separate from the "piston" (the ionization front) and move into the undisturbed gas. This happens because the front is still expanding supersonically into the surrounding medium and creates an initial compression in the still neutral medium.

The equations for the evolution of an H II region in steady state are

$$\rho_0 v_0^2 + p_0 = \rho_1 v_1^2 + 2p_1 + \frac{E}{c} W F_{\text{LyC}} \qquad (78)$$

where $E = E_0 + \chi$ is the energy of an electron liberated by ionization, F_{LyC} is the incident Lyman continuum photon flux, and W is the dilution factor for the central star(s). The factor of 2 in front of the p_1 term comes from the electrons liberated by ionization behind the front. We also have

$$\rho_0 v_0 \left(\frac{1}{2} v_0^2 + \frac{\gamma_0}{\gamma_0 - 1} \frac{p_0}{\rho_0} \right) = \rho_1 v_1 \left(\frac{1}{2} v_1^2 + 2 \frac{\gamma_1}{\gamma_1 - 1} \frac{p_1}{\rho_1} \right) + EW F_{\text{LyC}} - F_\gamma \qquad (79)$$

for the energy, where F_γ represents the radiative energy losses. The ratio of specific heats, γ, may change across the front. This won't happen for normal H II regions but might be important if the front enters a molecular region where dissociation of H_2 is followed by ionization. Just in case, we write this explicitly even though in what follows this more general case will be neglected.

Finally, the rate of flow of material across the front is due to the flux of ionizing photons from the central star, so that as we have seen

$$\rho_0 v_0 = \rho_1 v_1 = \frac{F_{\text{LyC}}}{m_p} \qquad (80)$$

This condition is unique to the H II region problem. The rate of expansion of the region is governed not only by the increased pressure behind the front but also by the rate of energy input from the star. Solving for the case where $p_1 \gg p_0$ and assuming that the radiative losses are intense enough to compensate for the increased energy from the incident shock, we obtain $v_0 = 2a_s$.

First, you may be wondering how an ionization zone is like a detonation. The increase in the pressure within an H II region is due to the input of Lyman continuum (LyC) photons that ionize the gas. The newly released electrons rapidly thermalize and add to the pressure and internal energy. As the front expands, newly ionized atoms add electrons and effectively power the expansion of the front. This continues for as long as there are incident photons from some central source, so that, ultimately, stars are the power source for H II regions. Assume that the media on the two sides of the H II region are simple gases with equations of state given by $p = \rho a_s^2$. Now take the Rankine–Hugoniot conditions to give

$$\rho_1 (v_1^2 + a_{s1}^2) = \rho_2 (v_2^2 + a_{s2}^2) \qquad (81)$$

4.5 A Quodlibet of Applications of Shocks to Astrophysical Problems

which with the mass conservation condition gives

$$\frac{\rho_1}{\rho_2} = \frac{(v_1^2 + a_{s1}^2) \pm [(v_1^2 + a_{s1}^2)^2 - 4a_{s2}^2 v_1^2]^{1/2}}{2v_1^2} \tag{82}$$

It follows that only positive real roots are allowed for this equation so that $(v_1^2 + a_{s1}^2)^2 \geq 4v_1^2 a_s^2$. This sets the condition on the velocity of propagation of the front in the neutral medium $(M_1^2 + 1)^2 \geq 4M_1^2(a_{s2}/a_{s1})^2$. Once we have found the velocity of the front, it is straightforward to compute the density ratio across the front and also to calculate the density of the postshocked gas, because we have the auxiliary condition that the flux of material through the front is the same as the photon flux responsible for forming the H II region in the first place, that is that $\rho_1 v_1 = \rho_2 v_2 = F_\gamma$. There are two critical velocities for the front:

$$v_R = a_{s2} + (a_{s2}^2 - a_{s1}^2)^{1/2} \tag{83}$$

where the R notation, meaning *rare*, is due to Kahn (1954). Since usually $a_{s2} \approx 10 a_{s1}$ under typical interstellar conditions (the temperature of the neutral medium is about 100 K and that of the H II region is about 10,000 K) v_R is approximately $2a_{s2}$ and represents an upper limit to the speed of the front. The lower limit is given by

$$v_D = a_{s2} - (a_{s2}^2 - a_{s1}^2)^{1/2} \tag{84}$$

where D stands for *dense*, and is given approximately by $v_D \approx a_{s1}^2/(2a_{s2})$.

Ionization Fronts and Detonation Waves. II.

Now let's go back to the picture of a detonation and look at the dynamical equations in more detail. One of the advantages of H II regions is that they are comparatively simple beasts. They consume photons and belch out electrons and expand because of the excess energy the electrons derive from ionization. As long as the central source remains constant, the region grows and its dynamics can be treated exactly. So let's look further at the evolution equations for a non–steady-state problem. Call the ionizing flux F_{UV} and the ionization fraction x. For simplicity, assume that the region is pure hydrogen (or at any rate only one species). Then the ionization fraction is given by

$$\frac{dx}{dt} = -\alpha\rho(1-x) + \beta x F_{UV} \tag{85}$$

Here β is the photoionization cross section and α is the recombination rate. The recombination depends on the temperature, but we can assume that the medium rapidly becomes isothermal, at least in the initial stages of the expansion. The energy equation is

$$\rho\left[\frac{d}{dt}\left(\frac{3}{2}\frac{p}{\rho}\right) - \frac{p}{\rho^2}\frac{d\rho}{dt}\right] = \beta x\rho(1-x)\chi_{\text{ion}} - \alpha\rho^2 x^2\left[\frac{3}{2}\frac{p}{\rho} + \Lambda(T)\right] \quad (86)$$

where λ_{ion} is the energy given to the electron through ionization and Λ is the line-dominated cooling rate (a function of temperature). The gas inside the H II region, indeed in the neutral ISM in general, is well approximated by a perfect gas, so the equation of state is

$$p = \rho(1+x)kT/m_p \quad (87)$$

This equation also explicitly includes the effects of ionization.

Just as an aside, the ionization equation is something like a flame equation, which is a nonlinear reaction-diffusion equation. What is different about the H II region problem is that it lacks a diffusion term. This may be present if the dynamics generate turbulence and entrainment of external material takes place. One could imagine that this could happen somewhat as follows. Across the ionization front, if the gas into which the H II is expanding has a very low temperature and high density, a Rayleigh–Taylor instability may develop. This causes fingers to form along the front, mixing the material between the media across the interface. Since this mechanism is a way of generating turbulence and looks like a good way to entrain matter, it can act like a diffusion term in the ionization equation. Such a mechanism may play a role across shocks in stellar envelopes; it appears to play a role in numerical simulations of dynamical H II regions within molecular clouds. But this is all we shall say about it; further developments would make an interesting problem.

We also require some equation for the radiative transfer. Remember that ultimately the source for the ionization is the absorption of stellar ultraviolet photons. This is done using the fact that the medium surrounding the central source is the only consumer of photons and that once a Lyman continuum photon has been absorbed it is never replaced. We have already briefly discussed this, but here it's appropriate to mention an interesting observational consequence of this fact. The essential feature of H II region photoprocesses is that they all destroy Lyman continuum radiation and redistribute the energy into other parts of the spectrum. For this reason, one can use the radio thermal continuum to measure the ionization rate. Any dust mixed into the ionized region will be heated and radiate in the infrared. This also provides a measurement of the ionization rate, because the dust is heated by Lyα photons emitted during the cascade to the ground state of neutral hydrogen following recombination. In view of this picture, we can write

$$\nabla \cdot \mathbf{F}_{\text{UV}} = \beta\rho F_{\text{UV}} \quad (88)$$

4.5 A Quodlibet of Applications of Shocks to Astrophysical Problems

where now $\mathbf{F}_{UV}$ is the radial component of the flux. Supplementing this by the equation of motion

$$\rho \frac{d}{dt}\mathbf{v} = -\nabla p \tag{89}$$

and by the continuity equation

$$\frac{\partial \rho}{\partial t} + \nabla \cdot \rho \mathbf{v} = 0 \tag{90}$$

we are ready to start our solution. First, we will see if we can form a dimensionless system of equations for the dynamics, and second, we will attempt to solve the evolution of the H II region. This second step is much harder than the first. The equations we have just written down are very nonlinear, and with good physical cause. They reproduce the complexity of the coupling between the ionization, the pressure (hence the dynamics), and the radiative transfer rate. As the ionization of the medium increases, the divergence of the flux goes down and eventually reaches zero. But the rate of heating also changes, and thus so does the rate of expansion. So we have a system that really must, unfortunately, be solved numerically in general.

The first task is *much* more straightforward. There is a characteristic time scale here, the rate of recombination. If the medium is to relax back to its neutral state, it will do so on a time scale $t_R = 1/(\alpha \rho_0)$ if the density is ρ_0. The rate of photon input from the central star, $L_\star$ (in number of Lyman continuum photons per second), also gives a characteristic time scale $L_\star^{-1}$. A length scale is provided by the penetration of continuum photons into the medium, given by $l_i = 1/(\beta \rho)$. Finally, we have a scale for the energy, χ_{ion}, so we also have a scale for the temperature of the ionized region, $T_\star \sim \chi_{\text{ion}}/k$. The density provides another length scale, $\rho_0^{-1/3}$, and we have in hand the relevant scaling parameters. Now rewrite the evolution equations in *dimensionless* form. Following the scaling of the luminosity, define a new variable $K = F_{UV} r^2$. The ionization changes as a function of time according to

$$\frac{\partial X}{\partial \tau} = \frac{K(1-X)}{\xi^2} - \mathscr{D}\alpha' X \tag{91}$$

where X, K, ξ, and τ are the scaled ionization fraction, photon luminosity, radius, and time, respectively. The flux equation now becomes

$$\frac{\partial K}{\partial \xi} = -K(1-X) \tag{92}$$

and the conservation of energy is given by

$$\frac{\partial}{\partial \tau}\left(\frac{3}{2}P - X\right) = \mathcal{D}\alpha X^2\left[1 - \frac{3}{2}\frac{T}{T_{\text{star}}} - \frac{\Lambda(T)}{\alpha' kT_\star}\right] \qquad (93)$$

Here α' is the scaled recombination rate and the scaled pressure P is given by

$$P = (1 - X)T' \qquad (94)$$

with T' being the scaled value for T. The dimensionless momentum conservation equation becomes

$$\frac{\partial V}{\partial \tau} + \frac{\partial P}{\partial \xi} = 0 \qquad (95)$$

The number $\mathcal{D}$ is called the *Damkohler* number. This number comes up frequently in flame theory (Goldsworthy 1983). Here it is the signature of the detonation-like process of H II region formation. The Damkohler number, at least when applied to flames, is the ratio of the dynamical time scale to some characteristic thermal time. For H II regions, $\mathcal{D}$ is given by $L_\star$, the dynamical time scale, to the recombination, or essentially cooling, time $(\alpha \rho_0)^{-1}$. If $\mathcal{D}$ is small, we can ignore recombination effects. Once it becomes large, and this means that we look at the development of the Strömgren sphere late in its life as it starts to expand, the cooling terms come into play.

The early stages of H II region formation, corresponding to the Strömgren sphere phase, are characterized by small values for $\mathcal{D}$. In this limit we get

$$\frac{\partial K}{\partial \xi} = -K(1-X), \qquad \frac{\partial X}{\partial \tau} = \frac{K(1-X)}{\xi^2} \qquad (96)$$

We've been looking at the development of a front, so let's look at the case where $r \approx$ constant. This gives an equation for K, hence for X, of

$$\frac{\partial}{\partial \tau}\left(\frac{1}{K}\frac{\partial K}{\partial \xi}\right) = -\frac{\partial K}{\partial \xi}, \qquad K(\xi = 0) = 1 \qquad (97)$$

the boundary condition coming from the assumed steady central source of ionizing radiation. The solution to Eq. (97) is

$$K = [1 + e^{-\tau}(e^\xi - 1)]^{-1} \qquad (98)$$

with the ionization x having the same dependence as $K = F_{\text{UV}} r^2$. Still in the small $\mathcal{D}$ limit we can look at the early dynamics of the ionization front.

Now once the H II region begins to expand, the time derivatives are replaced by scaled radial derivatives, as in

$$\frac{dx}{dt} = \frac{dR}{dt}\frac{\partial x}{\partial r} \qquad (99)$$

Here the radial size of the H II region changes so

$$-\dot{R}\frac{\partial x}{\partial r} = R^{-2}K(1-x), \qquad \frac{\partial K}{\partial r} = -K(1-x) \qquad (100)$$

so that $\dot{R} \sim R^{-2}$ and therefore $R \sim t^{1/3}$. The characteristic radial scale at which this starts to holds is the Strömgren radius. Actually, you could have guessed this because the H II region at this stage is formed by a balance between the rate of ionization and the rate of recombination and so depends only on the total number of particles. So if the density goes down, the volume goes up accordingly. Most important is the fact that we have just ended up with a similarity solution! The scaling parameters are the rate of input radiation and the density of the medium. In fact, from the recombination timescale, the rate of energy input, and the density, one can form a dimensionless number and this is the basis for the similarity approach. More on this in a separate chapter. It is just that, when such a simple, and profound, result pops out of the aether, it should be admired.

Appendix A: Details of the Spiral Shock Picture

We now follow Roberts's original derivation. Assume a cylindrically symmetric coordinate system. We write the equations of motion as

$$\frac{\partial u_r}{\partial t} + u_r \frac{\partial u_r}{\partial r} + \frac{u_\phi}{r}\frac{\partial u_r}{\partial \phi} - \frac{u_\phi^2}{r} = -\frac{a_s^2}{\rho}\frac{\partial \rho}{\partial r} - \frac{\partial \Phi}{\partial r} \qquad (101)$$

$$\frac{\partial u_\phi}{\partial t} + u_r \frac{\partial u_\phi}{\partial r} + \frac{u_\phi}{r}\frac{\partial u_\phi}{\partial \phi} + \frac{u_r u_\phi}{r} = -\frac{1}{r}\frac{\partial \Phi}{\partial \phi} \qquad (102)$$

Here $\Phi(r,\phi)$ is the gravitational potential. To this we add the continuity equation

$$\frac{\partial \rho}{\partial t} + \frac{1}{r}\frac{\partial \rho u_r}{\partial r} + \frac{1}{r}\frac{\partial \rho u_\phi}{\partial \phi} = 0 \qquad (103)$$

We assume that the unperturbed flow is Keplerian, so the gaseous disk is assumed to be in centrifugal equilibrium in the absence of perturbations:

$$\frac{\Omega^2(r)}{r} = -\frac{d\Phi_0}{dr} \qquad (104)$$

We take the flow perturbations around this local circular motion. A perturbation is now placed on the gravitational potential of the disk. This potential comes from the stellar component, assumed to carry the bulk of the mass. To do it self-consistently means solving the full Vlasov equation along with detailed solution of the Poisson equation. However, we can take a simpler approach. Rather than discuss this equation in detail, we simply note that there is a nonaxisymmetric term which results from the angular perturbation of the density field, so that there is a torque at the spiral arms. We set up a local coordinate system, so that we can *locally* treat the flow by using Cartesian coordinates. To do this, we introduce

$$\xi = -\ln\left(\frac{r}{r_0}\right)\sin i + (\phi - \Omega_p t)\cos i \tag{105}$$

the coordinate along the spiral arms, which has an inclination i and a pattern speed Ω_p; the perpendicular coordinate given by

$$\eta = \ln\left(\frac{r}{r_0}\right)\cos i + (\phi - \Omega_p t)\sin i \tag{106}$$

Both are derived assuming a logarithmic spiral pattern. The velocities can also be decomposed into parallel and perpendicular components by the rotation matrix in i:

$$w_\perp = (\Omega - \Omega_p) r \sin i + u_r \cos i + u_\phi \sin i \tag{107}$$

$$w_\parallel = (\Omega - \Omega_p) r \cos i - u_r \sin i + u_\phi \cos i \tag{108}$$

where (u_r, u_ϕ) are the velocity perturbations. We should just note that this approach to spiral shocks has also been applied to accretion disks. Even if this approach to spiral galaxies is not the main engine for powering star formation, and it doesn't appear to be, it is still important in any flow under the action of a background large-scale variable gravitational field. Not only is the basic approach illustrative of an application of hydrodynamics to understanding large-scale structure in galaxies, but also any self-gravitating disk system may display similar phenomenology. Now we return to the flow equations and look at their solution.

Now what is actually happening here? The matter is accelerated toward the arm by the locally higher density there. This results from the fact that, to first order, $\Phi' \sim \Phi'_0 + 4\pi G \rho s$, where s is a characteristic length. The gravitational acceleration changes sign across the arm. At the arm, material becomes sonic. Because the arm is oblique to the circulation, the streamlines refract outward there, diverting the flow along the inner portion of the arm until the matter relaxes back to streamlines. Then, on the other half of its orbit, the material encounters the other arm with the same results.

To first order, Roberts (1969) derives the following equations:

$$w_\perp \frac{\partial \rho}{\partial \eta} + (\rho_0 + \rho)\frac{\partial w'_\perp}{\partial \eta} + \left(\rho_0 + \rho + r\frac{d\rho_0}{dr}\right)\chi_1 = 0 \tag{109}$$

where $w'_\perp = w_\perp - (\Omega - \Omega_p) r \sin i$, and $\chi_1 = (u_r \cos i + u_\phi \sin i)\cos i - (-u_r \sin i + u_\phi \cos i)\sin i$:

$$w_\perp \frac{\partial w'_\perp}{\partial \eta} - 2\Omega r w'_\parallel + \frac{a_s^2}{\rho_0 + \rho}\frac{\partial \rho}{\partial \eta} + \frac{\partial \Phi'}{\partial \eta} + \chi_2 = 0 \tag{110}$$

and finally

$$w_\perp \frac{\partial w'_\parallel}{\partial \eta} + \left(\frac{k^2}{2\Omega}\right) r w'_\perp + \chi_3 = 0 \tag{111}$$

Appendix A

We have to define the perturbation for the gravitational field as

$$\Phi' = A \cos\left(\frac{2}{\sin i}\eta + \phi\right)$$

where A is the amplitude and ϕ is the phase of the wavelike perturbation. For compactness, we have also defined

$$\frac{\kappa^2}{2\Omega} = 2\Omega\left(1 + \frac{1}{2}\frac{d\ln\Omega}{d\ln r}\cos^2 i\right)$$

This is the epicyclic frequency in the oblique frame, the natural frequency that results from the differential rotation of the disk. The two auxiliary functions χ_2 and χ_3 are given by

$$\chi_2 = -r^2\frac{d\Omega}{dr}(w'_\perp \cos i - w'_\parallel \sin i) - w'_\perp(w'_\perp \sin i - w'_\parallel \cos i) - \frac{a_s^2 r \cos i}{\rho_0 + \rho}\left(\frac{d\rho_0}{dr}\right)$$

$$\chi_3 = -r^2\frac{d\Omega}{dr}\cos i w'_\parallel \sin i + w'_\perp(w'_\perp \sin i + w'_\parallel \cos i) - \frac{a_s^2 r \sin i}{\rho_0 + \rho}\left(\frac{d\rho_0}{dr}\right)$$

These functions, along with the forcing function $f = -(1/r)\partial\Phi'/\partial\eta$, can be used to solve for the gradients in the velocity parallel and perpendicular to the front. These are used to define the angle of the flow:

$$\tan\lambda = \frac{\partial\xi}{\partial\eta} = \frac{w_\parallel}{w_\perp} \tag{112}$$

The final flow equations are, therefore, for the perpendicular flow:

$$\frac{\partial w'_\perp}{\partial\eta} = -\frac{w_\perp(2\Omega rw'_\perp + rf - \chi_2) + a_s^2\chi_1(1 + [r/(\rho_0 + \rho)]\,d\rho_0/dr)}{a_s^2 - w_\perp^2} \tag{113}$$

and for the parallel flow:

$$\frac{\partial w'_\parallel}{\partial\eta} = -\frac{(K^2/2\Omega)rw'_\perp + \chi_3}{w_\perp} \tag{114}$$

Having these equations in hand, Roberts presented a solution which showed that the flow will in fact become supersonic at the spiral arms and that the probable site for star formation due to compression of the clouds will be there. The imposition of the spiral pattern is the result of the global oscillation of the stellar phase of the disk, and this in turn is due to the intrinsic instability of the disk to spiral wave formation.

We have gone into this example in such detail because of its general applicability to the stability of disks. Any self-gravitating disk will be unstable to the formation of such shocks, although they will not be time independent (see Toomre 1977). The idea that there are shocks at the point of the potential minimum is not a surprising one, since we have already seen from the example of traffic flow that the flow will accelerate on going into the potential well at the gravitational perturbation that the arms represent and will decelerate on leaving the arm region. The essential result from Roberts' derivation is that the flow becomes sonic at the arms even in the case of very small perturbations (it has been suggested that the arms are no more than a few percent enhancement on top of the disk potential in many galaxies).

You should also recall the discussion of the Bernoulli equation and the idea that a flow accelerates if it is constricted. What happens in the density wave picture is somewhat simpler, but it still contains a critical point. The basic physical mechanism for generating the shocks is that the gas is supersonic to begin with because of its circulation speed. If it is accelerated on the way into a spiral arm, because of the locally higher surface density, and then decelerated on the way out, the gas will plow into itself supersonically. It does not have any time to adjust to the change in the sign of the acceleration across the arm. Nor does any mechanism exist for propagation a signal upstream to warn the incoming material to slow down. The characteristics provide propagation at the local sound speed, far too slow for any readjustment of the dynamics of the large-scale flow to the presence of this gravitational obstacle.

Appendix B: Bending of Jets by a Supersonic Cross-Flow

The standard analysis proceeds as follows. Assume that a jet is placed in a cross-stream. Its momentum flux is ρv_J^2 and if the stream velocity is v_S and the density is ρ_S then the radius of curvature, R, produced by the jet being bent is responsible for the momentum balance. Therefore, the condition assumed to govern the equilibrium is

$$\frac{\rho_S v_S^2}{l_J} = \rho_J \frac{v_J^2}{R} \tag{115}$$

where l_J is the scale length for the jet, over which the background medium influences the motion. On the other hand, there has been laboratory work indicating that formation of a trailing edge vortex may also play a role in the bending of the jet. The axial flow presents a cylinder to the surrounding fluid. If the jet is normal to a cross-stream and the core is supersonic, it can be treated as almost incompressible. Then the flow of the cross-stream around the axis creates a vortex on either side (see the discussion of wake vortices in the chapter on vorticity and rotation) and the jet may even bifurcate as a result. List (1982) includes some examples of this, and it may be important for astrophysical jets, both in the promotion of instabilities and in generating some of the complex structures seen in the outer regions of wide-angle tail sources. The basic laboratory data are well illustrated by Moussa et al. (1977). Experiments show that a supersonic jet appears to be a flexible obstacle to a cross-stream flow. The compliance of the emerging jet leads to partial penetration of the cross-stream velocity field, which passes through a bow shock and is shed by the jet column as if it were a cylinder in the stream. A vortex sheet transports some momentum downstream, and a vortex sheet attaches to the leeward side of the jet, producing drag and a pressure deficit that may lead to enhanced rapid jet bending.

Appendix C: Collisions of Galaxies with an Intracluster Gas

Galaxies are comparatively fragile objects, having lower escape velocities than stars and being composed of regions of dramatically different densities. These objects, the largest self-gravitating units in the universe that we usually study, are also amazingly subject to their environments for their stability and evolution. The escape velocity is lower than that of a point source because of the large-scale distribution of mass within the system. The gas is loosely bound to the disk of a galaxy.

Consider a galaxy moving through a background gas within a cluster. We take the density of the background gas to be n_{ICM} and assume that this gas has a temperature T_{ICM}.

The temperature must be large, because this gas is assumed to be in equilibrium within a total cluster potential in which $T_{ICM} \sim \sigma$, where σ is the velocity dispersion of the galaxies. Generally, with observed dispersions of order 10^3 km s^{-1}, the temperature must be of order 10^8 K. Now consider a galaxy moving through this medium. Take the density of the gas in the galaxy to be η_0 and its temperature to be T_0. Assume that the galaxy accelerates into this gas, which is densest near the center of the cluster potential, and that this is moving through the galaxy. The analogy to a piston will allow us more easily to examine the central question, that is, how this cluster gas sweeps up matter within the galaxy. The first step is to picture what is actually happening.

The typical density of the interstellar medium within a galaxy has a temperature of about 100 K, so that the intracluster medium (ICM) propagates through the galaxy provided $P_0 < n_{ICM} v_g^2$. This forms a wall that moves through the ISM of the galaxy in the frame of the galaxy (and through which the galaxy moves in the frame of the cluster).

Since the matter from the cluster has a far higher density, it sweeps through the galaxy and drives a shock ahead of it. To compute the velocity of this snowplowed material, we have to assume that the piston moves at a velocity v_g. Then as it moves into the gas at rest, we see that

$$v_g^2 = 2(p_2 - p_1)\left(\frac{1}{\rho_1} - \frac{1}{\rho_2}\right) \quad (116)$$

which comes from the shock adiabat (assuming that momentum and mass are conserved across the shock interface). Now notice that for a compressive shock, we have ρ_2/ρ_1 in terms of p_2/p_1. This gives the velocity of the shock front lying ahead of the piston because we have

$$\rho_1 v_s = \rho_2 v_g \quad (117)$$

The thickness of the layer grows with time but is approximately

$$\Delta x = (v_s - v_g)\Delta t \approx \left(\frac{\rho_2}{\rho_1} - 1\right)v_g \Delta t \quad (118)$$

Now we see that for the case of a strong shock, which is probably going to be the situation in a massive cluster of galaxies because of the velocity dispersion of the constituent galaxies, the shock velocity is about $3v_g$ for an ideal gas. Therefore we see that the layer thickness grows and that the mass of the shell being swept up by this shock grows. The piston moves rapidly through the galaxy and the matter in the system is completely removed. Only the molecular clouds, by virtue of their surface densities, are likely to survive this process, and observations show that the deficiency of diffuse neutral gas in cluster galaxies is systematically greater than observed for the molecular clouds. What happens to this material? It is sent as a hot shock into the intracluster medium. It will sit there, behind the galaxy which is now cleared out, and eventually may collapse to form stars or disperse if it is not self-gravitating.

References

Axford, W. I. (1961). Ionization fronts in interstellar gas: The structure of ionization fronts. *Philos. Trans. R. Soc. London* **A253**, 301.
Blackmore, J. T. (1972). *Ernst Mach: His Work, Life, and Influence.* [See especially p. 105 ff.] Berkeley: University of California Press.
Bleakney, W., and Taub, A. H. (1949). Interaction of shock waves. *Rev. Mod. Phys.* **21**, 584.

Chevalier, R., and Theys, J. (1975). Optically thin radiating shock waves and the formation of density inhomogeneities. *Astrophys. J.* **195**, 53.

Draine, B. T. (1980). Interstellar shock waves with magnetic precursors. *Astrophys. J.* **241**, 1021. [See also (1981), *Astrophys. J.* **246**, 1045 (*errata*: specifically the energy equation for dust component).]

Dyson, J. E., and Williams, D. A. (1980). *The Physics of the Interstellar Medium*. Manchester, England: Manchester University Press.

Glasstone, S. (1977). *The Effects of Nuclear Weapons*. Washington, D.C.: U.S. Department of Defense. [This is a very important book, containing the best photographs of the initial stages of strong blasts ever published (mainly from the Trinity tests). See also Taylor, G. I. (1950). The formation of a blast wave by a very intense explosion. I. Theoretical discussion. *Proc. R. Soc London* **A201**, 159. (The paper that follows this is on the Trinity test of 1945; see especially plates 2 and 3 in the second paper for the expansion of the fireball.)]

Goldsworthy, F. A. (1983). A mathematical approach to the evolution of H II regions. *IMA J. Appl. Math.* **32**, 147.

Hugoniot, H. (1889). Memoire on the propagation of motion in bodies, especially in a perfect gas. *J. Ecole, Polytech.* **58**, 1.

Igumentshchev, L. V., Shustov, B. M., and Tutukov, A. V. (1990). Dynamics of supershells: Blow-out. *Astron. Astrophys.* **234**, 396.

Jones, E. M., *et al.* (1979). Interacting supernova remnants: Tunnels in the sky. *Astrophys. J.* **232**, 129. [An unusual paper and very useful for building up intuition about interacting shocks.]

Kahn. F. (1954). The acceleration of interstellar clouds. *Bull. Astron. Inst. Netherl.* **12**, 187.

Kaplan, S. (1966). *Interstellar Gas Dynamics*. Oxford: Pergamon.

Lazareff, B. (1983). Dynamics and energetics of the interstellar medium. In *Diffuse Matter in Galaxies: Cargese 1982*, J. Audouze *et al.* (eds.). Dordrecht, The Netherlands: Reidel.

Lighthill, M. J. (1957). *J. Fluid Mech.* **2**, 1.

Lin, C. C., and Roberts, W. W. (1981). Some fluid–dynamical problems in galaxies. *Annu. Rev. Fluid Mech.* **13**, 33.

List, E. (1982). Turbulent phenomena in buoyant jets and plumes. *Annu. Rev. Fluid Mech.* **14**, 189. [See also Crabb, D., Durao, D. F. G., and Whitelow, J. H. (1981). A round jet normal to a cross-flow. *Trans. ASME J. Fluid Eng.* **103**, 142.]

McKee, C. F. (1987). Astrophysical shocks in diffuse gas. In *Spectroscopy of Astrophysical Plasmas*, (p. 226), A. Dalgarno and D. Layzer (eds.). Cambridge: Cambridge University Press.

McKee, C. F., and Cowie, L. L. (1975). The interaction between the blast wave of a supernova remnant and interstellar clouds. *Astrophys. J.* **195**, 7–15.

McKee, C. F., and Hollenback, D. J. (1980). Interstellar shock waves. *Annu. Rev. Astron. Astrophys.* **18**, 219.

Merzkirk, W. K. (1970). Mach's contribution to the development of gas dynamics. In *Ernst Mach: Physicist and Philosopher*, R. S. Cohen and R. J. Seeger (eds.). [See also Seeger, R. J. (1970). On Mach's curiosity about shock waves, (p. 60). *Ibid.*]

Moussa, Z. M., Trischka, J. W., and Eskinazi, S. (1977). The near field in the mixing of a round jet in a cross stream. *J. Fluid Mech.* **80**, 49.

Narain, U., and Ulmschneider, P. (1990). Chromospheric and coronal heating mechanisms. *Space Sci. Rev.* **54**, 377.

Prigogine, I., and Herman, R. (1971). *Kinetic Theory of Vehicular Traffic*. New York: American Elsevier. [See also Herman, R., and Prigogine, I. (1977). A two-fluid approach to town traffic. *Science* **204**, 148.]

Rankine, W. J. M. (1870). *Philos. Trans. R. Soc. London* **160**, 277.
Raymond, J. C. (1979) Shock waves in the interstellar medium. *Astrophys. J. Suppl.* **39**, 1.
Raymond, J. C. (1984). Observations of supernova remnants. *Annu. Rev. Astron. Astrophys.* **22**, 75.
Roberts, W. W. (1969). Large-scale shock formation in spiral galaxies and its implications for star formation. *Astrophys. J.* **158**, 123.
Sandford, M. T., Whitaker, R. W., and Klein, R. I. (1984). Radiatively driven dust-bounded implosion: Formation and stability of dense globules. *Astrophys. J.* **282**, 178.
Sgro, A. G. (1975). The collision of a strong shock with a gas cloud: A model for Cassiopeia A. *Astrophys. J.* **197**, 621.
Skalafuris, A. J. (1968). Radiative shock structure-theory and observations. *J. Quant. Spec. Rad. Trans.* **8**, 515.
Shull, J. M., and McKee, C. F. (1979). Theoretical models of interstellar shocks. I. Radiative transfer and UV precursors. *Astrophys. J.* **227**, 131.
Spitzer, L., Jr. (1982). Acoustics waves in supernova remnants. *Astrophys. J.* **262**, 315.
Tenorio-Tagle, G. (1981). The collision of clouds with a galactic disk. *Astron. Astrophys.* **94**, 338.
Tenorio-Tagle, G., and Rozycka, M. (1983). The hydrodynamics of clouds overtaken by supernova remnants. I. Cloud crushing phenomena. *Astron. Astrophys.* **155**, 120.
Toomre, A. (1977). Theory of spiral structure. *Ann. Rev. Astron. Astrophys.* **15**, 437.
von Neumann, J. (1963). Oblique reflection of shocks. In *Collected Works*, (p. 238), A. H. Taub (ed.). Oxford: Pergamon.
Whitney, C. A., and Skalafuris, A. (1963). The structure of a shock front in atomic hydrogen. I. The effects of precursor radiation in the Lyman continuum. *Astrophys. J.* **138**, 200.
Zel'dovich, Ya. B., and Kompaneets, A. S. (1960). *Theory of Detonation.* New York: Academic Press.

CHAPTER 5

Similarity Methods

> ... and they all look just the same.
> Malvina Reynolds, *Little Boxes*

5.1 Introduction

Several times in this book we have made use of dimensional analysis to find out qualitatively how a physical situation should evolve. The reason was that by proceeding in this way, we were often able to see what the scaling of the final dimensionless solution should be and whether there are any possible critical numbers in the problem. Sometimes, as in the case of Rayleigh–Benard convection and the Rayleigh number R, it proved to be physically insightful indeed. We are now going to look at the next step of *dynamical* problems: what general relations exist among the hydrodynamical variables when general scaling is applied. For reasons we shall discuss in a moment, these are called *similarity* or self-similar solutions.

5.1.1 Blast Waves: The Sedov Problem

Let us start by taking an all-too-familiar problem. An explosion occurs at some point, which we shall call $r = 0$. It involves release of an amount of energy E due to some source, be it dynamite, ^{235}U, or a supernova. The last one is, of course, the justification for discussing this problem under the astrophysical rubric. Assume that the explosion propagates spherically into a medium with density ρ. This ensures that the shock can be treated as *one-dimensional*. Now, having only E and ρ, what can we say about this evolution? Notice that E has the dimensions ML^2T^{-2}, and ρ varies as ML^{-3}, at least if we treat the bulk density and not surface density (we will return to this later). Therefore, we see that the ratio (E/ρ) has the dimensions L^5T^{-2}; it is the only one that combines the constants of the problem

5.1 Introduction

in such a way as to remove the dimensional dependence on mass. In the absence of any characteristic scale length for the fluid, this ratio gives a quantity that is constant; therefore $R^5 t^{-2}$ is independent of time.

Another way of looking at the problem is to ask the following question: how must a spherical blast evolve when expanding into a constant-density environment *in order to maintain constant energy*? This does not mean that the energy density does not decrease. In fact, the expansion will ensure that the internal temperature of the blast drops. But the point is that the total energy input is not decreased by the propagation of the blast wave through the ambient medium, whose density is assumed to remain constant. Thus, we have a relation for $R(t)$, assuming that E and p are constant:

$$R(t) = \alpha (E/\rho)^{1/5} t^{2/5} \tag{1}$$

where α is a dimensional numerical constant to be evaluated later. It will help at this moment to note that, in general, α is of order unity. The expression in Eq. (1), even as it stands, has applications to astrophysics. Take a supernova (SN) with an energetic release of 10^{50} erg and assume that the interstellar medium has a mean density of about 10^{-24} g cm^{-3}. Then

$$R(t) \approx 10^{-0.7} (E_{50}/\rho_{24})^{1/5} t_{yr}^{2/5} \, \text{pc} \tag{2}$$

so that if we see a supernova remnant (SNR) with a radius of about 1 parsec, it implies an age of the order of a century. The *Crab Nebula* has a radius of a bit over 3 pc, and this solution says it should be of the order of 300 to 10^3 years old.[1] Using this simple formalism, we can even predict what the velocity should be at any epoch *for which our assumptions are valid* since, taking the time derivative of $R(t)$, we get

$$V = \frac{d}{dt} R(t) = \frac{2}{5} R t^{-1} \tag{3}$$

The nebula should thus be expanding at a speed of 1000 km s^{-1}. This is indeed the order of magnitude we observe. The success, and it is really quite unexpectedly successful, of this crude estimate ought to impress you.

---

[1] In fact, in 1054 the Chinese and Arabic astronomers, and perhaps even the Amerindians, recorded an astonishingly bright (by celestial standards) but relatively short-lived *Guest Star* in the part of the sky we now call Taurus. It outshone Venus and could even be seen in daylight. In the eighteenth century, the French amateur Messier recorded a fuzzy patch of light in the same part of the sky as the first object in his famous catalog of objects to *avoid* when looking for comets. Fortunately, his advice (however economically sound it might have been at a time of rich prizes for discovery of new comets) has been at least partially ignored since, and we now can say with certainty that M1 *is* the remnant of the SN of 1054.

It displays a powerful example of how even the most trivial applications of physical methods can bring results in the face of an astrophysical problem. We could have derived the energy, or the density, by turning the argument around, knowing t and R (and we shall even try this a bit later). There is, however, a hidden assumption in the detailed solution we have been discussing up to now: the solution we have discussed occurs *if and only if* the blast wave is adiabatic. We know that at some stage in the hydrodynamic expansion the assumption of adiabaticity *must* break down. Further, we have assumed that the blast wave remains spherical. This may be too restrictive to be physically allowable for very long. Therefore, that the Crab Nebula fits the solution so well shouldn't blind you to the dangers of not carefully considering the physical parameters of the problem.

For instance, suppose the blast had been two-dimensional. Such cases as an expanding ring (sometimes involved in galactic nuclei) or the more immediately realized case of a cylindrical explosive charge will serve for visualization. Then (*and you should show this yourself*):

$$R(t) = \alpha(E/\sigma)^{1/4} t^{1/2} \tag{4}$$

where now σ is the *surface density*. Although the change in the estimate of the velocity is negligible (a factor of 0.5 instead of 0.4), the *time evolution* of the blast wave is clearly qualitatively different. We shall see more of this later.

5.1.2 The Dimensionless Dynamical Equations

Now let's break all of the relevant hydrodynamical quantities down into their basic dimensions by defining a variable:

$$\eta \equiv r^\lambda t^{-\mu} \tag{5}$$

The exponents of this scaling variable, which is intrinsically dimensionless, are determined by the physical constants that are involved in the problem. In the case of an adiabatic blast wave, the energy and background density were the constants. For a stellar wind the constants are the momentum loss rate and the background density. For the momentum-conserving stage of a blast wave, the momentum and density are the constants. The particular choice of similarity variable is tailored to the problem and therefore also cannot hold forever under all circumstances during the system's evolution.

But with that aside, let's return to the scaling procedure. For example, we have the following scalings for the velocity, pressure, density, and sound speed:

$$u = rt^{-1}U(\eta), \qquad P = r^{-1}t^{-2}\Pi(\eta), \qquad \rho = r^{-3}D(\eta), \qquad \alpha_s = rt^{-1}C(\eta) \tag{6}$$

5.1 Introduction

The derivatives of all the hydrodynamic equations are transformed as follows:

$$\frac{\partial}{\partial t} \to -\mu\eta t^{-1}\frac{d}{d\eta}, \quad \frac{\partial}{\partial r} \to \lambda\eta r^{-1}\frac{d}{d\eta} \tag{7}$$

Then, for instance, the convective derivative transforms to

$$\frac{D}{Dt} \to t^{-1}\eta(-\mu + \lambda U)\frac{d}{d\eta} \tag{8}$$

In what follows, to make the notation more compact, we will write $d/d\eta$ as d_η. One important caution on this method. The physical variables contain functions of both (r, t) and η, that is, $Q(r, t) = f(r, t)g(\eta)$, after the scaling transformation has been made. When taking the derivatives, don't forget that f depends on either r or t. For instance, the velocity time derivative becomes $\partial u/\partial t = -rt^{-2}V - \mu r t^{-2}\eta d_\eta V$ and the spatial derivative becomes $\partial u/\partial t = t^{-1}V + \lambda t^{-1}\eta d_\eta V$; notice that when multiplied by $u = rt^{-1}V$, the spatial derivative has the same dimensions as the time derivative and thus can be combined with it.

We now have to assume a geometry because, as we have seen, this changes the exponential dependence. Eventually, in almost any hydrodynamic calculation, we have to take both the gradient and divergence and therefore must be concerned with the dimensionality of the problem. For example, in the continuity equation we have

$$\frac{\partial \rho}{\partial t} + \nabla \cdot \rho\mathbf{u} = \frac{\partial \rho}{\partial t} + r^{-n}\frac{\partial}{\partial r}(r^n \rho u_r) = 0 \tag{9}$$

where n is 0, 1, or 2 depending on whether the medium has 1, 2, or 3 dimensions. For this equation we can now substitute the dimensionless similarity version:

$$-\mu\eta t^{-1}r^3 d_\eta D + ((n-4)DU + \lambda\eta d_\eta(DU))r^3 t^{-1} = 0 \tag{10}$$

Finally, we can write this equation as

$$(\lambda U - \mu)\eta d_\eta D + (n-4)DU + \lambda D\eta d_\eta U = 0 \tag{11}$$

This very nonlinear equation may appear undesirable, but it can be solved, numerically if not analytically. Let's push on. At the critical point (to be derived in a moment) we have $U = \lambda/\mu$ and therefore

$$\eta\left(\frac{dU}{d\eta}\right)_{\text{crit}} = -\frac{\mu(n-4)}{\lambda^2} \tag{12}$$

(since we are sure that n cannot equal 4). For η always positive, the value of $\eta d_\eta U$ is also always positive.

The velocity gradient can be obtained from the transformed equation of motion. Using Eq. (8) for D/Dt we get

$$(\lambda U - \mu)\eta \frac{dU}{d\eta} = D^{-1}(\lambda \eta d_\eta \Pi - \Pi) - U(U-1) \qquad (13)$$

As before, you should derive this one for yourself to verify it. It has been left in a form which mimics the equation of motion, so you can see the contribution of the individual terms. Notice that there is a source term $-U(U-1) - \Pi/D$ on the right-hand side. Since Π/D will have the same behavior as U^2, this is not a surprising combination of terms. Keep in mind though that *none of these quantities have dimensions*—they are purely algebraic. We can now add to the list of equations a coupling term between the pressure P and the density ρ. For this, we can try a polytropic approximation for the pressure; hence the entropy equation becomes

$$\frac{D}{Dt}(P\rho^{-\gamma}) = 0 \qquad (14)$$

The ratio of specific heats, γ, is constant. Using the similarity variables, we convert this to

$$(\lambda U - \mu)\eta \frac{d}{d\eta}(\Pi D^{-\gamma}) + ((3\gamma - 1)U - 1)\Pi D^{-\gamma} = 0 \qquad (15)$$

We now have the required equations as a complete system. Of course, we could have saved some trouble at the step of deriving the momentum equation if we had asserted the polytropic law then instead of deriving the more general form. However, in general, we will be dealing with nonpolytropic equations of state in astrophysical environments and it is best to have the most complete presentation of the method at hand.

By now you will have noticed that terms like $(\lambda U - \mu)\eta dU/d\eta$ have been cropping up in all of the equations. This is the origin of the statement about the critical point of the solution. Notice, for example, that in Eq. (15) we were able to derive the value of the entropy at the critical point directly in terms of the equation of state. One thing worth doing at this stage is thinking about the problem of an isothermal blast wave, in which the relation between the pressure and density is simplified by $P \sim \rho$. This is one of the most direct approximations one can make and is also an interesting one in that it generalizes the relativistic limit as a limiting equation of state.

We have repeatedly emphasized the existence of *characteristic* scales. Again recalling the experience with convection (Chapter 9), the existence

5.1 Introduction

of several characteristic time scales and a unique length scale gave rise to a bifurcation behavior. The problem could be solved in dimensionless variables, *once and for all time*, and then the system behavior scaled by the value of some critical number. Were it not for the existence of these characteristic times, the bifurcation behavior would simply not happen! The most essential feature of similarity variable methods is that *there can be no characteristic times, lengths, masses, or the like in the problem. No dynamical variable is permitted to have a characteristic value.* You have no doubt noticed that there is no diffusion term in the equation of motion we have used, and now you know why.

The importance of similarity solutions was first stressed by aerodynamicists like Prandtl and von Kármán. Later it was taken up by physicists, particularly Sedov, Taylor, and von Neumann. The reason is simple. Suppose we return to the explosion problem for a moment. We are interested in looking at the effect of the blast on the environment *after* the shock has passed over the region, or in other words in a moving coordinate system. (See Fig. 5.1.) We assume that the velocity of the front itself scales with the behavior we have already derived. The structure of the region behind the front, however, is taken to be invariant with time. In effect, then, the similarity variable η serves to stretch out the space–time structure, telling us how to distort our rubber space–time structure as the shock ages. Since there are no characteristic times or lengths, we are completely free to set our clocks and shape our rods as we wish. Knowing the time, for example, we simply change the relative length scaling. I realize you are being beaten over the head with this, but it is fundamental to an

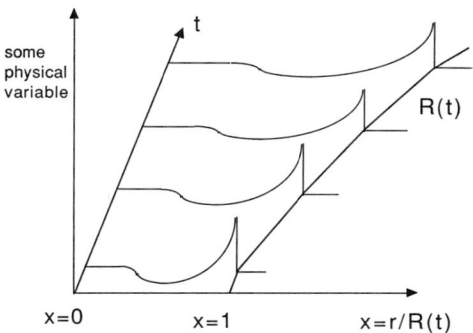

Figure 5.1 Evolution of a self-similar blast wave. The front moves along the curve $R(t)$. All curves are produced by scaling the initial blast by this radius. Otherwise, the similarity is preserved in the x variable.

understanding of why one should resort to such similarity solutions whenever possible.

Now collecting the equations of the problem, we have

$$(\lambda U - \mu)\eta \frac{d}{d\eta}(\Pi D^{-\gamma}) = -[(3\gamma - 1)U - 1]\Pi D^{-\gamma} \tag{16}$$

$$(\lambda U - \mu)\eta \frac{dU}{d\eta} = D^{-1}(\lambda \eta d_\eta \Pi - \Pi) - U(U - 1) \tag{17}$$

$$(\lambda U - \mu)\eta \frac{dD}{d\eta} = -(n - 4)DU - \lambda D\eta d_\eta U \tag{18}$$

For a choice of λ, μ, and γ these equations can be integrated numerically. Recall that n is fixed by the geometry. Now to the choice of λ and μ.

You will have noticed that there were more similarity representations defined than are actually being used here. One reason for this is that we can close the system by imposing an equation of state. As we did in the discussion of Bernoulli flow along streamlines (recall the discussion of the Rankine-Hugoniot conditions for shocks), we can invoke a polytropic representation for the pressure and so obtain a formal representation for $C(\eta)$ as a function of $\Pi(\eta)$ and $D(\eta)$. This means that we will not have the same form of the equations as, for example, in Courant and Friedrichs (1948), where the equations are expressed in terms of the sound speed, but the same basic results still apply.

Our aim has been to write the dynamical equations with η dimensionless. We obtain this by taking $\eta = r/R(t)$. Since R varies as, in the Crab nebula case we have already discussed, $t^{2/5}$, we get $\lambda = 1$ and $\mu = \frac{2}{5}$. In general, if two quantities Q_1 and Q_2 are dimensionally given by $\text{Dim}[Q_1] = ML^{n_1}T^{l_1}$ and $\text{Dim}[Q_2] = ML^{n_2}T^{l_2}$, then the radius will scale in time as $R(t) \sim t^{(l_1-l_2)/(n_1-n_2)}$. Suppose we have an equation of state of the form $T^3\rho = \text{constant}$. Since we know the functional dependence of ρ, we obtain $T \sim t^{-2/5}$, a slowly decreasing function of time. The optical depth and cooling time will also vary as the radius of the blast changes, and these can also be computed (and this is one you should do). It can be shown that the depth at which the shock becomes essentially transparent is $R_c = \tau_0^{1/2}R_0$, where τ_0 is the initial optical depth. This gives the critical time scale as $t_c \sim \tau_0^{5/4}$. If the explosion is initially optically thick (as it should be for the application of the adiabatic condition), then this is the point (scaled time) at which the approximation is expected to fail. You should also keep in mind that it depends on the initial energy and the density of the medium as well, but *you should find this result for yourself.*

5.1.3 Stagnation Pressure

The pressure in the blast wave should vary as $P \sim \rho u^2 \sim R^{-1}t^{-2}$. Taking the adiabatic as an example, since we have this case at hand, the time can be eliminated by taking

$$t = \alpha^{-5/2}(E/\rho)^{1/2}R^{5/2} \tag{19}$$

and therefore the pressure at a given blast wave radius can be obtained. This will also be a function of the initial energy and of the density of the medium. If we take the radius to be the point at which the pressure reaches a prescribed value, then the critical radius scales as

$$R_\star \sim E^{1/3} \tag{20}$$

This is perhaps the most (*in*)famous of the scaling relations to come out of the Manhattan Project research program. It states that the radius of the circle of finite overpressure (the value which is of use for military purposes, is typically several pounds per square foot) scales rather weakly with the *yield* of the explosion (assuming that all of the other conditions of the blast wave solution are met). This critical radius has astrophysically interesting properties as well. The point at which the blast wave internal pressure is greater than some multiple of the interstellar ambient pressure should scale the same way. For example, in the current work on star formation, it has been suggested that a supernova, occurring inside or near a molecular cloud, will serve to destroy the cloud and also to trigger star formation. If we assert that there must be a sufficient overpressure at some point to overcome the resistance of the cloud, then we can calculate the extent over which the SN will be effective in shaping the structure of the region. The same is true for the interstellar medium (ISM) in general—when holes are created, with tunnels resulting from their coalescence, it is possible to determine the extent to which the blast is able to remain a strong force in the medium. Again, this is assuming that all other things remain the same about the medium and the interior of the blast wave. Breakout, and the formation of the champagne phase in a molecular cloud, occurs if the internal pressure is not high enough to make the shock stall everywhere. (See Fig. 5.2.)

A stalled shock is one which expands at constant velocity. First, look at the case for an adiabatic shock. Since $R \sim E^{1/5}t^{2/5}$, the pressure of the environment scales dimensionally as $MR^{-1}t^{-2}$. Therefore P/ρ scales as $R^2 t^{-2}$ and a shock expanding at constant pressure then has a time-dependent radius:

$$R(t) = \alpha \left(\frac{P}{\rho}\right)^{1/2} t \tag{21}$$

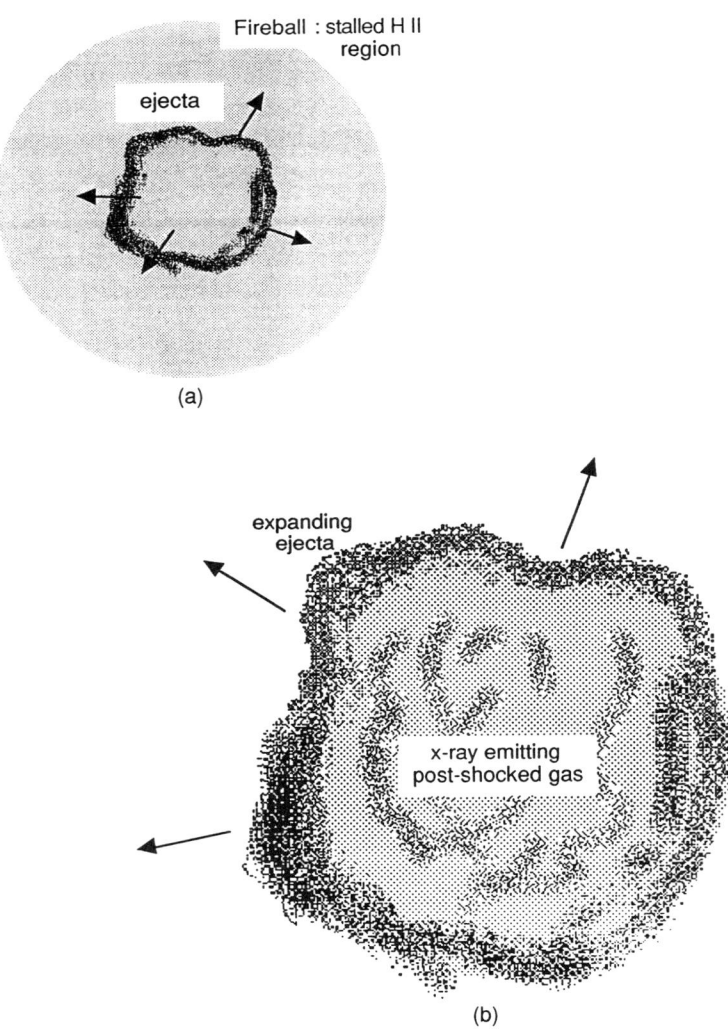

Figure 5.2 Phases of the expansion of a supernova blast in (a) the initially adiabatic (Sedov) phase and (b) the subsequent snow-plow phase. The precursor fireball is also shown.

The velocity is therefore constant. The interesting question is whether the shock stalls before the adiabatic condition breaks down. If the radiation time scale is short, the shock turns isothermal and enters the momentum-conserving stage rapidly. In order to apply the Sedov method for the interpretation of an observed structure, one needs to know whether the

5.1 Introduction

assumed conservation condition is appropriate. This can be done by looking at the expansion time, which is R/V, and compare this with the cooling times behind the shock. This time scale can be found from the observation of the spectroscopic temperature disgnostics, although these are rarely unambiguous and depend on the filling fractions of the phase of the plasma which is being used to determine the temperature of the interior.

5.1.4 Snowplow Phase

A more realistic approach is to take the following picture, using the results we have already obtained above. If the blast wave is capable of radiative cooling, there should be some stage at which the adiabatic approximation ceases to apply. The shock should begin to cool, faster than would have been expected from the adiabatic expansion, and the temperature should be determined by the conservation of momentum rather than energy. This stage in the evolution of a SNR is called the *snowplow* phase, because at this point the remnant acts like a truck being decelerated by the accumulation of material in its path. The increase in the mass swept up by the remnant causes the front to decelerate and eventually to become subsonic, at which time the *characteristic time scale* of the medium once more comes into play: the sound travel time.

It is possible, assuming that momentum and *not* energy is conserved, to derive a second approximation:

$$R(t) = \beta(MV/\rho)^{1/4} t^{1/4} \tag{22}$$

where now MV is the original momentum (or the value at the time of the solution's initial applicability). Here β is also a constant of order unity. In order to calculate the energy loss in this phase, we need merely note that the blast is decelerating to the ambient sound speed so that the difference in the energy is the difference in the squares of the velocity at the initial moment and the sound speed. In this phase of the evolution of a blast wave, it falls from the velocity at which the adiabatic solution ceases to apply to the ISM sound speed. The fractional energy loss is therefore of order

$$\delta E_{\text{SNR}} = 1 - \left(\frac{a_{s,\text{ISM}}}{V(t_{\text{isothermal}})} \right)^2 \tag{23}$$

Typically, the Mach number at which the blast wave enters the nonadiabatic stage of its evolution is of order 5, so we would expect that about 95% of the energy of the SNR is dissipated *before* it becomes sonic. The time scale for expansion can be computed from the solution for the velocity, as we did above for the adiabatic case, with a coefficient of $\frac{1}{4}$ instead $\frac{2}{5}$. The time scale is given by

$$\frac{t_{\text{sonic}}}{t_{\text{isothermal}}} = \left(\frac{V_{\text{isothermal}}}{a_s}\right)^{4/3} \quad (24)$$

The mechanical luminosity of the blast wave can then be computed, and the response of the medium to this input of energy can also be quantitatively judged.

5.1.5 Stellar Wind Bubble

Now what if the momentum input is from a continuous source? For example, if there is a stellar wind inside a bubble, which is pushing on the medium and causing it to expand, what is the solution? It is very easy (*now*) to show that it is $R \sim t^{1/2}$. The dimensionless approximation is that, for continuous (constant) mass loss, $MV = \dot{M}Vt$. Now we can specify the source as we wish. It may be a stellar wind driven by radiation pressure (in which case it is a function of the luminosity and can be so parametrized within the formalism). It may be due to wind which is driven by turbulence (as in the sun). Whatever the mechanism for generating the outflow, it can be incorporated. The same is true for the case of an adiabatic outflow, but here we have a more manageable formulation.

The essential feature of the blast wave solutions we have been discussing is that once solved numerically, they need never be solved again. Aside from the utility of scaling the motion of the front, at which the Rankine–Hugoniot conditions are applied as boundary conditions on the physical variables, the interior solution for the blast is one that we need only solve *once*! Regardless of the degree of expansion, as long as the physical approximation holds, we can treat the structure of the interior as a function only of η and forget about anything else.

5.1.6 Planetary Nebulae as a Special Case

Planetary nebulae present a special case for Sedov-type analysis, because in this case the envelope of a star is the environment into which the shock is expanding. The physical picture, still developing but reasonably secure, is that in the late stages of the evolution of a low- or intermediate-mass star, the red giant possesses a strong wind which is in steady state, at least for a while. The mass loss rate is therefore nearly constant, at least on the hydrodynamical time scale, and the envelope develops a density structure $\rho \sim r^{-2}$. We assume that the terminal velocity of the wind is constant. At some stage in the life of the now dying star, the core develops a very strong *superwind*, which has a high $\dot{M}$ and a large terminal velocity but may have a lower density than the red giant's old wind into which it plows. This wind–wind interaction is assumed to be spherically symmetric, centered on the core of the evolved star.

The mechanical luminosity of the wind is essentially constant as the core evolves, since a radiative stellar wind has rough relation between the mechanical and photon luminosity, so that $\dot{M}v_\infty^2$ is approximately constant. If the wind expands in an essentially stalled state, in which the dynamical pressure of the wind is about the same as that of the overlying envelope, we would easily be able to calculate the speed of the superwind. However, if the envelope constant is taken to be approximately ρr^2 then $R \sim (\dot{M}v_\infty^2 v_0/\dot{M}_0)^{1/3} t$ where $\dot{M}_0$ is the mass loss from the red giant wind, which has a terminal velocity v_0. Thus the expansion rate is approximately $6 \times 10^3 (\mu v_{0,2} v_{\infty,3})^{1/3} t_{yr} \, R_\odot$ where the velocities are in units of 100 and 1000 km s^{-1}. Thus for $\mu \approx 10$ and $R \approx 4 \times 10^5$ the time is of order one century. Certainly this is an underestimate, for we do not generally see this extreme youth in most planetaries, but it is in the right ballpark—most of the youngest planetaries cannot be more than an order of magnitude older than this.

This is just the beginning of the calculation, though. It shows what is possible, but we have neglected much. For instance, we have ignored the interaction between the wind and the shell and the formation of the snowplow phase in the background medium. To see what happens, consider that we have a shell formed which "absorbs" matter as it expands. More of this is discussed in Chapter 8. Here we have just been concerned with the similarity applications.

5.2 Similarity Solution for Static Configurations: Polytropes

In the paleolithic days of astrophysics, about a century ago, the problem of the interior structure and future development of the sun was of major concern. Without any knowledge of the power source for this gas bag, it was still shown to be possible to solve for its interior structure. This is one of the triumphs of the nineteenth century and was accomplished first by H. Lane in 1869, and later added to by T. J. J. See and, in a more complete exposition, by R. Emden in 1909 and Chandrasekhar (1939); see also the notes accompanying the references to earlier work). Remarkably, the treatment of the *hydrostatic* structure of a gaseous sphere can be so generalized, within the context of an equation of state, that it can be used to indicate what possible interior energy sources might be accessible. We shall treat the problem somewhat unconventionally and derive the problem in a mathematical vein somewhat like the early approaches.

To this point, we have made little resort to the gravitational field equations. In the case of a self-gravitating hydrostatic gaseous sphere, however, the Poisson equation provides one of the equations of structure:

$$\nabla^2 \Phi = -4\pi G \rho \qquad (25)$$

where now we take Φ to be the gravitational potential (we can later change this to accommodate general relativity). The equation of motion, *sans* the convective derivative and the time derivative, becomes

$$\nabla P = -\rho \nabla \Phi \qquad (26)$$

where again P is the pressure. For a spherically symmetric configuration, the hydrostatic condition is given by the two equations

$$\frac{1}{r^2}\frac{d}{dr}r^2\frac{d}{dr}\Phi = -4\pi G \rho \qquad (27)$$

$$\frac{d}{dr}P = -\rho \frac{GM(r)}{r^2} = -\rho \frac{d}{dr}\Phi \qquad (28)$$

5.2.1 The Lane–Emden Equation

Now let us take the approximation that the pressure is a function *only* of the density ρ of the form

$$P = P_0(\rho/\rho_0)^m = K\rho^m$$

which is, by definition, a *polytropic equation of state*. The constant K is called the entropy constant in some of the literature. We assume that n is constant. The two coupled equations can therefore be combined, resulting in

$$\frac{1}{r^2}\frac{d}{dr}\left(r^2 m\rho^{m-2}\frac{d}{dr}\rho\right) = -\frac{4\pi G \rho}{K} \qquad (29)$$

which is highly nonlinear and seemingly impossible to solve analytically. Let's go a bit further anyway. Notice that we can write the quantity $\rho^{m-2}d\rho/dr$ as

$$\left(\frac{m}{m-1}\right)\frac{d}{dr}\rho^{m-1}$$

Define a new auxiliary variable θ by

$$\rho = \rho_0 \theta^n \qquad (30)$$

5.2 Similarity Solution for Static Configurations: Polytropes

and *force* the derivatives to be linear in θ. Then we get

$$\frac{1}{r^2}\frac{d}{dr}r^2\frac{d}{dr}\theta = -\frac{4\pi G}{(n+1)K}\theta^n \tag{31}$$

Now take $n = 1/(m-1)$. Notice that the dimension of the left-hand side is L^{-2}. Therefore the constant must have the dimensions of L^{-2}. We can now form a dimensionless variable ξ given by

$$\xi = \left(\frac{4\pi G}{(n+1)K}\right)^{1/2} r \tag{32}$$

to obtain

$$\frac{1}{\xi^2}\frac{d}{d\xi}\left(\xi^2\frac{d}{d\xi}\theta\right) = -\theta^n \tag{33}$$

This is called the *Lane–Emden equation*. It was the first equation to succeed in describing the internal structure of a self-gravitating polytropic body. For nearly a century (from about 1867, when Lane first derived it, to the early 1950s), it was the primary model for the discussion of stellar interiors.

Now notice that this equation depends only on n, the index of the equation of state. It can be shown that the index in the case of a realistic adiabatic sphere is γ, the ratio of specific heats, and so it is obvious that $\gamma = 1 + 1/n$. The boundary conditions are also obvious. Since the value of θ is scaled to the central density, we should take ρ_0 as a free parameter. Then $\theta(0) = 1$. The central density must also be a maximum and monotonic with radius. Thus, $d\theta/d\xi = 0$ at $\xi = 0$. The range in ξ is $[0, \infty)$. However, there is still a problem: the surface has not yet been defined. Simply put, the surface of a polytrope occurs when the density vanishes, so that if $\xi_\star$ is the radius at which $\theta(\xi_\star) = 0$ we have an eigenvalue problem for $\xi_\star$. Again, we see that once we have solved for $\theta(\xi)$, we have done it for all values of the central density and the adiabatic constant. The radial variable is simply scaled—exactly as in the blast wave solution. We can derive the mass easily enough by noting that in the scaled variables, we obtain an equation for the scaled mass as a function of radius:

$$M(r) = 4\pi \int_0^r \rho(r) r^2\, dr \to M(\xi) = \int_0^\xi \theta^n \xi'^2\, d\xi' \tag{34}$$

Again, recall that this is also *dimensionless*. Noting from the Lane–Emden equation the representation of the integrand, and also knowing the boundary conditions, we get the result that

$$M(\xi) = -\left(\xi^2\frac{d\theta}{d\xi}\right)\bigg|_\xi \tag{35}$$

Since this latter quantity is always negative, the mass is always positive definite (as it should be). In all cases, the θ we employ is *specific* to a certain value of n. The total mass of the star is therefore (in the scaled variables) $M(\xi_\star)$. It too will be specific to a polytropic equation of state. The ratio of central to mean density and the run of temperature and pressure through the gas ball are now all specified, once and for all. In principle, this is another way of utilizing the power of dimensionless representations.

The polytropic approach has also proved useful for modeling any self-gravitating system which retains its finite size by virtue of a velocity distribution function. In this way, the polytropic model has been applied to stars and star clusters and to galaxies and galaxy clusters with equal alacrity. Whether this stands up to careful scrutiny depends on how seriously the assumptions of uniform structure and effective temperature are taken. To see this, let's derive the isothermal sphere from the general solution of the Lane–Emden equation we have in hand.

5.2.2 From General Polytropes to the Isothermal Sphere

We can examine the *formal* structure of the solution of the Lane–Emden equation easily. First of all, we can be sure that the only terms which appear in the solution are *even* in the radial variable. After all, that is why we chose spherical coordinates. Second, we know that the first term must be unity (*Why? This one's for you!*). Therefore, since we know that the series which solves the problem is convergent (after all, I have been asserting that this even applies to real stars and its longevity should tell you *something* about its success), we can take the following approximation.

Assume that all terms higher than 1 are small. This isn't as unorthodox an approximation as it might appear to be. You will see why momentarily. We can write θ as

$$\theta(\xi) = 1 - \sum_{n=1}^{\infty} a_n \xi^{2n} = 1 - \Sigma \qquad (36)$$

(*How can we be sure that the first term must be negative?*) Asserting that Σ is small and inserting into the Lane–Emden equation, we get

$$\frac{1}{\xi^2} \frac{d}{d\xi} \xi^2 \frac{d}{d\xi} \Sigma = (1 - \Sigma)^n \qquad (37)$$

Finally, taking the approximation that $n \to \infty$, we arrive at the proper form for the isothermal Lane–Emden equation:

$$\frac{1}{\xi^2} \frac{d}{d\xi} \xi^2 \frac{d}{d\xi} \Sigma = e^{-\Sigma}$$

5.2 Similarity Solution for Static Configurations: Polytropes

The step of taking $n \to \infty$ was motivated by the equation of state. Since the temperature varies as $T \sim \rho^{\gamma-1}$, the isothermal sphere means that n, the polytropic index, is infinite. A simple examination of even the first two terms of the Lane–Emden equation shows that as the index is increased, the radius at which the surface condition is met also increases. For the isothermal sphere, in effect, it is unbounded. It should also be clear now why we were really quite right to assert that the higher-order terms in the series are small. Since the scaling is in terms of the radius and the polytropic index, and since ξ varies as $n^{-1/2}$, we are correct in noting that all of the terms of the series are of order n^{-1} and consequently convergently small. In addition, they satisfy the requirements for the expansion of the exponential. As an alternative approach, we could have gone back to the beginning, substituted in the isothermal equation of state, and then plugged away at the solution. This way makes it, probably, more obvious that the isothermal sphere represents a true limiting configuration to the polytropic sequence.

The derivation we have just gone through will be complete only when we have specified the boundary conditions for Σ. Notice, for example, that the derivative condition is unchanged. The center must still be a maximum. However, now $\Sigma(0) = 0$. Thus, we have a complete specification. The density profile will be given completely. The usual procedure is to rewrite the equation of hydrostatic equilibrium for an isothermal equation of state and then proceed as one would do ordinarily in deriving the Lane–Emden equation. This should tie it in more immediately with the behavior of the solutions.

The case of $n = 0$ can be solved directly, since it is linear, to give $\xi_\star = \sqrt{6}$. It can easily be seen, then, that the first term in the series solution for $\theta(\xi)$ is $a_1 = -\frac{1}{6}$. The second term, which can be found from a perturbation solution by applying the same kind of reasoning that we have used to get the isothermal sphere, is $n/120$, so that to second order

$$\theta(\xi) \simeq 1 - \frac{1}{6}\xi^2 + \frac{n}{120}\xi^4 \tag{38}$$

With this in hand, you can derive all of the basic results concerning the mass, central density to mean density ratio, and the runs of the thermodynamic variables.

As in the case of the blast wave solution, however, you will not be able to do anything but stare at the dimensionless results. In order to find out how to apply these solutions to real stars, we must go back to the equations of structure themselves. The dimensional analysis of the pressure gives

$$P_c = \frac{GM^2}{R^4} \tag{39}$$

where M is the total mass and R is the radius. The central temperature is given by

$$T_c = P_c/\rho_c \qquad (40)$$

The only variable still left free in the problem is the central density. This is why, in the famous solution for the structure of white dwarf stars by Chandrasekhar (1939), the only parameter which is required to specify the internal structure, and the total mass of the configuration, is ρ_c. We could easily substitute the mean density for an approximation and so calculate more exactly the temperatures and pressures in the interiors of our *stars*, but this leaves the problem as one for you to try.

5.2.3 A Very Few Applications

How can we show that there *must* be a polytropic representation for even extreme equations of state? Let us consider the case of a pure electron sphere in the classical limit—assume that quantum mechanics has not been invented yet (something some of you may wish were true). The equation of state derived from the dimensional analysis of a purely repulsive equation of state is $P \sim r^{-4}$ (where we have assumed Coulombic repulsion only). The mass will, of course, now enter since the final form must be in terms of the mass density. We obtain $P \sim \rho^{4/3}$ with a coupling constant that depends on the charge and mass of the electrons. This is the same as a polytrope with index $n = 3$, which is also the one characterizing radiation-dominated spheres (in other words, ones which have $T^3 \rho =$ constant and $P = \frac{1}{3} a T^4$). This is the limit of a completely relativistic gas and is the limit of white dwarf stars with the proper treatment of the equation of state using Fermi–Dirac statistics. A limiting equation of state in the ultrarelativistic limit is $P = \frac{1}{3}\rho c^2$, where c is the speed of light. Now we can also, as a final step, show that there must exist a maximum mass of such configurations, again by dimensional analysis. Since the radius, as we have already shown, for all polytropes of constant index is the same in the scaled variables, we need merely to assert that the radius *in dimensional form* must be greater than the Schwarzschild radius for stability (or some multiple thereof). Thus, there must be a maximum mass for the star, since the radius *decreases with increasing mass to the limit of the Schwarzschild radius*. The details are trivial to show for yourself. It is also, by the way, easy to show that there is a dependence on the mass of the particle and that the limiting mass for electron configurations will be different from that for neutrons (where we must invoke the proper quantum statistics or assert some quasi-classical mock-up for the equation of state) and protons. Once again, it is necessary to emphasize that the configurations all have the same

5.2 Similarity Solution for Static Configurations: Polytropes

relative internal structure—that one need only know the global properties to scale one of the stars to the other internally.

It was this property which first spurred work on the interior structure of stars, in the realization that it was indeed possible to get a beginning model for, as a case in point, the main sequence. It was even true at the beginning of the more modern era of interior model calculations that the first guess at the interior structure was of polytropes. Nowadays, with the discovery of diffuse x-ray emission from gas in clusters of galaxies, the interest in polytropes has been born again, so to speak.

It has been realized that the physical conditions in such clusters can be looked at in terms of gas dynamics in the following way. We consider the collection of galaxies we call a cluster to consist of particles moving within a potential well of their own creation, with a global velocity dispersion. Since the system is, in the picture we are imposing, collisionless, it is not really correct to say that they have an equation of state, a pressure; instead, the velocity dispersion is responsible for the finite size of the cluster. It is the Red Queen's comment all over again, only this time to the galaxian members and not to Alice: "Nowhere, you see, it takes all the running *you* can do, to keep in the same place." The galaxies have a global velocity distribution function and consequently a global velocity dispersion. Therefore they *behave* like an isothermal gas and will *appear* to construct an isothermal sphere. The properties of the cluster will then be completely specified. The virial theorem will hold, the mass can be determined, the profile of the surface density of galaxies is known. The model can then be used to describe the potential well in which the gas resides, and all of the properties can be given for that gas. If the temperature, well depth, and density at any point have been given, it is then possible to calculate the abundances in the gas and to discuss heating and cooling mechanisms. This is a gross oversimplification of what is really going on, but it makes nice bedtime reading and is a very useful and robust first approximation. Another way to look at this harkens way back to the introductory chapters, where the Vlasov equation was discussed. The distribution function in the presence of a background gravitational field must include in the energy E the gravitational potential (this is the origin of the isothermal sphere, after all). The system is bound by the escape velocity, which is determined by the total mass of the system. If we have identical particles, then we can write the Poisson equation as an integrodifferential equation for the potential and from this calculate the expected density as a function of distance. This comes from noting that the distribution function $f(v)$ yields a density $\rho \sim \int_0^{v_x} \exp-(\beta v^2 + \Phi)v^2 dv$ so that $\nabla^2\Phi = -4\pi G\rho$ is an equation for Φ, the potential, in terms of β, the inverse "temperature" of the stellar velocity distribution.

Interest has even been spurred in using isothermal spheres to model interstellar molecular clouds, which are turbulently supported against collapse and pressure bounded by the diffuse ISM, but a detailed examination of this problem is beyond the scope of this book (although one which would certainly repay your efforts handsomely). Pressure-bounded polytropes provide a good approximate description of molecular clouds and also yield insight into the relation between size and internal velocity dispersion. Pressure-bounded spheres have been computed for negative-index polytropes, and the literature on this subject has been growing in the past few years because of its application to interstellar problems.

Before leaving this discussion, a mention, however brief, should be made of the idea of galactic winds. This lovely idea, which has yet to be completely exhausted as a physical model, derives from the same kind of thinking that gave rise to the idea of magnetic heating of the solar corona and nonthermal or radiative driving of stellar outflows. The argument goes like this. In the case of a galaxy, we have a gravitational potential which has a small amount of gas sitting in it. This gas, being constantly subjected to the blast wave and radiative input of the stellar population that makes up the bulk of the galactic mass, will be heated. *If* this heating takes place on time scales shorter than either the conductive or radiative time scales for the medium, there will be no other solution for the gas but to begin to expand away from the disk. Although more complicated by the detailed geometry, the problem is nonetheless similar to the stellar outflow case — the galaxy may be able to set up a wind. Indeed, if nothing else, it may end up possessed of a hot corona. The additional heating by high-energy particles, cosmic rays, and their interaction with the material in the disk also complicates this a bit. The thinking, however, in the stellar case is also moving more toward the presence of high-energy particles, magnetic fields, and nonthermal heating as the source of at least the beginning of the mass outflow, so the cases may not be terribly different. At any rate, we have now come full circle. Blast waves giving rise to galactic winds, which move within a polytropic galaxian and cluster potential — hydrodynamics and astrophysics are indeed a match made, if not in heaven, then at least in the laboratory.

5.3 Classical Gravitational Collapse

5.3.1 Pressureless Collapse

The problem of free fall is one that relates well to the idea of a similarity analysis. Here we treat only the classical problem; any relativistic effects you'll have to seek elsewhere. If the collapse is spherical and the

5.3 Classical Gravitational Collapse

mass finite, then it is possible to form a characteristic time scale from $G\rho$, but this will continually change in the course of the collapse. If the mass is a constant, then assuming the initial density is ρ_0 and the initial radius is r_0,

$$\ddot{r} = -\frac{GM}{r^2} = -\frac{4\pi G\rho_0 r_0^3}{3}\frac{1}{r^2} \tag{41}$$

As a reminder, the gravitational self-energy of the sphere is $3GM^2/5R$. We define a time $t_\star^{-1} = 4\pi G\rho_0/3$. Then the radius scales as $\xi = r/r_0$ and the time scales as $\tau = t/t_\star$. The dimensionless equation is then $\xi'' = -\xi^{-2}$. This gives

$$\dot{\xi}^2 = \frac{2}{\xi} - c \tag{42}$$

where c is a constant of integration. Defining $\xi = \cos^2\beta$ gives as an equation for collapse

$$\cos^2\beta\dot{\beta} = \frac{1}{\sqrt{2}} \tag{43}$$

The solution to this equation is a cycloid,

$$\frac{1}{2}\beta + \frac{1}{4}\sin 2\beta = \frac{1}{\sqrt{2}}(\tau - c_1) \tag{44}$$

where c_1 is another integration constant. Since $\xi = 1$ at $\tau = 0$, $c_1 = 0$. The collapse time is then found by taking $\xi = 0$, giving $\beta = \pi/2$. This yields the free-fall time more precisely than we have previously defined it, $t_{\rm ff} = (32G\rho_0/3\pi)^{-1/2}$. Notice that this is independent of r_0; it depends only on the initial density (in that sense, it depends on the initial radius and the mass). But all pressureless spheres undergo identical collapse. When this time scale is approximately the same as the thermal time, the body is unstable to gravitational collapse. You will meet this again soon, in the guise of the Jeans instability. Now we can look at the self-similar aspects of this collapse because we suspect that the presence of a scale-free time scale yields some sort of similarity solution.

5.3.2 Isothermal Collapse: Effects of Pressure

For an isothermal sphere, there is another possibility. We have several dimensioned quantities in this case, the sound speed itself, a_s, the total mass of the body, M_0, and the gravitational constant G. Choose a similarity variable $\xi = a_s^{-1}rt^{-1}$. The dimensions of G are $L^3M^{-1}T^{-2}$. Thus density scales as $\rho(r,t) = G^{-1}t^{-2}D(\xi)$, velocity scales as $v(r,t) = aV(\xi)$, and mass scales as $M(r, t) = a^3 G^{-1} t \mathcal{M}(\xi)$. Then the derivatives become $\partial/\partial r = (\xi/r) d/d\xi$ and $\partial/\partial t = -(\xi/t) d/d\xi$.

The equation of mass conservation is transformed as follows:

$$\frac{\partial \rho}{\partial t} + \frac{1}{r^2} \frac{\partial}{\partial r}(r^2 \rho u) = 0 = -2D + \xi D' + \frac{2}{\xi} DV + (DV)' \tag{45}$$

and the momentum conservation equation becomes

$$\frac{\partial u}{\partial t} + u \frac{\partial u}{\partial r} + \frac{a_s^2}{\rho} \frac{\partial \rho}{\partial r} + \frac{GM}{r^2} = 0 = -\xi DV' + DVV' + D' + \xi^{-2} \mathcal{M} \tag{46}$$

Then, using the mass conservation condition, we obtain

$$\left[\frac{\partial}{\partial t} + u \frac{\partial}{\partial r}\right] M = 0 = \mathcal{M} + (V - \xi) \mathcal{M}' \tag{47}$$

Transforming the mass equation, we get $\mathcal{M}' = 4\pi D \xi^2$, and this equation is used to eliminate $\mathcal{M}$ from the dynamical equation. Now the mass is given by $\mathcal{M} = 4\pi D \xi^2 (\xi - V)$. We have completed the system of equations. Notice that the difference between the pressured collapse problem and that of simple free fall is the presence of a second dimensioned variable, the sound speed. We still cannot create dimensionless combinations solely from the physical constants of the problem because we cannot eliminate the mass. We now write the equations for the velocity and density, obtained by substituting the transformed mass just derived back into the evolution equations. We get for the velocity

$$[(\xi - V)^2 - 1] V' = -[(\xi - V) D - 2\xi^{-1}](\xi - V) \tag{48}$$

and, for the density, we find that

$$[(\xi - V)^2 - 1](\ln D)' = [D - 2\xi^{-1}(\xi - V)] D(\xi - V) \tag{49}$$

This system of equations has been discussed by Shu (1977). He pointed out that the singular point is just a constraint condition on the problem of the flow—that the flow does not experience shocks at this point and that it is analogous to the condition for the de Laval nozzle or for the Parker solution for the solar wind. In the dynamical collapse, this is the matching condition that ensures that no shocks occur. For the static sphere, $V = 0$, the only allowable self-similar solution found by Shu for the envelope density is $D \sim \xi^{-2}$ or $\rho \sim r^{-2}$. This is the same one found in 1969 by R. Larson and M. Penston in the context of collapse at the start of star formation. The core, inside the critical point, is the interior onto which this self-similar envelope has to be grafted.

Appendix: Diffusion Equations

The simplest parabolic equation, the heat conduction equation, presents an interesting example of how a similarity approach can be used. As we said at the beginning of the chapter, the essential feature of similarity transformations is that there are no fixed scale lengths in the system. In the case of heat conduction, we know that there is a typical time scale over which the solution decays, dependent on the diffusion coefficient and the size of the system. But if the size of the system is not specified, the coefficient of the Laplacian is only a dimensioned parameter. As in the case of pressurized free fall, where we have only G and a_s, here we have only ν.

Transformation of the Equation

Take an equation of the form

$$\frac{\partial y}{\partial t} = \nu \nabla^2 y \tag{50}$$

where ν is a diffusion coefficient. Dimensional analysis leads us to the transformed dimensionless variable

$$\xi = \nu^{-1/2} r t^{-1/2}$$

Here, as in the blast wave and the collapse problem, we have $\xi = \xi(r, t)$. The transformation of the spherical equation gives, as before for the Sedov solution,

$$\frac{\partial}{\partial t} = -\frac{1}{2}\frac{\xi}{t}\frac{d}{d\xi}, \qquad \frac{\partial}{\partial r} = \frac{\xi}{r}\frac{d}{d\xi} \tag{51}$$

Substituting this into the equation for spherical heat conduction, using $n = 2$ (see the earlier discussion of the Lane–Emden equation), we get

$$-\frac{1}{2}\xi^2 y' = \nu\left(y' + \frac{d}{d\xi}[\xi y']\right) \tag{52}$$

or, calling $\xi y' = w$, we obtain

$$\frac{dw}{d\xi} = -\left(\xi + \frac{1}{2\xi}\right)w \tag{53}$$

The integral, which will depend on the details of the boundary conditions, is given by

$$y(r, t) = \int_{\xi_0}^{\xi} w(\xi') e^{-\xi'^2/4} \frac{d\xi'}{\xi'^2} \tag{54}$$

Notice that we have a solution in which the dependence is given implicitly on (r, t). The initial conditions for w must be specified in terms of ξ, not in terms of (r, t), however. Another reason for noting this method for the diffusion equation is that, in the case of nonlinear systems, such transformations often make the problem more tractable—that is, they can make the equation a less imposing threat. We could have simply taken the Fourier transform of the solution with respect to the spatial coordinates, integrated the time-dependent equation, and then taken the inverse transform (as is the usual way of solving the problem). However, you will notice that, as in the Sedov approach earlier, we can tailor our transformation to the coordinate system.

References

Barenblatt, G. I., and Zel'dovich, Ya. B. (1972). Self-similar solutions as intermediate asymptotics. *Annu. Rev. Fluid Mech.* **4**, 285.

Chandrasekhar, S. (1939). *Introduction to the Study of Stellar Structure*. Chicago: The University of Chicago Press.

Chevalier, R. A. (1977). The interaction of supernova remnants with the interstellar medium. *Annu. Rev. Astron. Astrophys.* **15**, 175.

Chevalier, R. A., and Liang, E. P. (1989). The interaction of supernovae with circumstellar bubbles. *Astrophys. J.* **344**, 332. [See also Chevalier, R. A. (1982). *Astrophys. J.* **258**, 790.]

Gratton, J. (1991). Similarity and self-similarity in fluid dynamics. *Fund. Cosm. Phys.* **15**, 1.

Hamilton, A. J. S. (1985). Similarity solutions for the structure of supernova blast waves driven by clumped ejecta. I. Undecelerated clumps. *Astrophys. J.* **291**, 523.

Kesteven, M. J., and Carswell, J. L. (1987). Barrel-shaped supernova remnants. *Astron. Astrophys.* **183**, 118.

McRay, R., and Snow, T. P. (1980). The violent interstellar medium. *Annu. Rev. Astron. Astrophys.* **17**, 213.

Ostriker, J. P., and McKee, C. (1988). Blast waves and similarity solutions. *Rev. Mod. Phys.* **60**, 1.

Roger, R. S., and Landecker, T. L., (eds.). (1988). *Supernova Remnants and the Interstellar Medium*. Cambridge: Cambridge University Press.

Sedov, L. (1957). *Similarity Methods in Mechanics*. New York: Academic Press.

Shklovskii, I. (1966). *Supernovae*. New York: Interscience.

Shu, F. H. (1977). Self-similar collapse and isothermal spheres and star formation. *Astrophys. J.* **214**, 488.

Tenorio-Tagle, G., and Bodenheimer, P. (1988). Large-scale expanding superstructures in galaxies. *Annu. Rev. Astron. Astrophys.* **26**, 145.

Trimble, V. (1983). Supernovae. II. Aftermath. *Rev. Mod. Phys.* **55**, 511.

CHAPTER 6

Magnetic Fields in Astrophysics

> *Magnetic fields are to astrophysics as sex is to psychology.*
> H. C. van der Hulst

6.1 Historical Introduction

Cosmic magnetic phenomena were discovered with the observation, reported by Gilbert in 1600, that the magnetic field of the Earth has a similarity to the magnetic field produced by a terrella, a spherical lodestone. The fact that the Earth possesses a dipolar magnetic field, first announced in Gilbert's *De Magnete*, indicated that there is a simple explanation for the behavior of the compass on the planet and that the field forms a fixed coordinate system on the planet. It was only some 35 years later that Henry Gellibrand determined the secular variation in the magnetic declination and variation at London. The determination of the rate remained controversial for some time, but the idea that the magnetic field of the Earth is not constant dates back to the founding years of the subject. Halley in 1683 and 1692, in addition to producing a nautical magnetic chart, also introduced the idea of a multipole field as the best representation of the global magnetic properties of the planet. He further was the first to describe a dynamical model for the interaction of the core with the crust of the Earth such as to produce a temporal variation in the local magnetic properties of a site on the surface—that is, he was the originator of the argument that leads to dynamo theory. Gauss succeeded in describing the global magnetic field for the Earth in the opening years of the nineteenth century. This led to the founding of the worldwide magnetic observatory network. This network not only served to monitor the changes in the field strength and its detailed structure, it also ushered in a new phase in the explanation of the origin of the field, dynamo theory.

Three lines of development led to the discovery of nonterrestrial magnetism. First, Faraday's discovery of the dynamo phenomenon, that a current traveling in a closed loop is able to generate a magnetic field in the form usually associated with a dipole, indicated for the first time that the field of the Earth can have a dynamical origin. Second, at the close of the nineteenth century, Zeeman discovered the remarkable fact that a magnetic field, when placed across a flame containing ions of sodium and other elements, produced a wavelength splitting of the normally single lines of the elements in an amount that is linearly proportional to the strength of the imposed field. Finally, in the decade immediately following the announcement of this discovery and its explanation in the context of what was then called electron theory by Lorentz, Hale observed for the first time the magnetic fields associated with sunspots. Even more dramatic was the discovery by Hale that there was a cycle to the magnetic structure which was linked to the sunspot cycle. In particular, Hale discovered the two-period phenomenology in the cycle—the sunspot polarities reverse on essentially twice the sunspot cycle time scale.

Hale also discovered that the typical strength of the observed fields was large in comparison with the Earth—about a factor of 100 to 10^3 times greater. However, attempts in the decades that followed to discover an ordered field for the Sun—that is, a global dipolar field—proved fruitless. There were several announcements for this field, but they all proved spurious on closer examination. This is a very basic result and one that was largely unappreciated until recently. Specifically, in the 1940s, Babcock discovered the existence of ordered dipolar magnetic fields in a class of A stars, the Ap stars, that have peculiar abundances. These fields were of the same order as those observed in the sunspots by Hale but were globally dipolar and periodically variable. Stibbs and Schwarzschild in 1950 came up with the idea of an oblique rotator; that is, the fields in these stars vary because of the rotation of a nonaxisymmetric field relative to the line of sight. The fields of the magnetic A stars do not show the kind of behavior seen for the solar field. Their polar strengths are constant to very high accuracy and their projected field variations are periodic.

The discovery of the fields in the interstellar medium was an accident. Attempting to observe the polarization due to electron scattering in rotationally distorted stars, Hall and Hiltner in 1947 showed that the polarization was the result of the orientation of particles in the interstellar

6.1 Historical Introduction

medium. Davis and Greenstein proposed the model that this is due to the orientation of ferrite-bearing graphite grains in the magnetic field of the galactic disk and that collisions produced precession of the grains about the field lines. The magnetic fields inferred for this orientation mechanism are of order 10^{-6} G, extremely small by stellar standards but having a considerable energetic effect on the dilute medium of the galactic disk. Later direct measurements of the Zeeman splitting of the 21-cm line of hydrogen, by Verschuur and Heiles, have confirmed this order of field strength.

The discovery of radio emission from Jupiter at 10-m wavelengths by Burke and Franklin, a chance observation, indicated the presence of magnetic fields in the planets other than Earth. Later observations of the planet, especially by satellite measurements from Voyager and Pioneer, confirmed and extended our knowledge of the field. Additional observations have shown that Saturn, Uranus, and Neptune also possess strong fields. Neptune is perhaps the most interesting case, because its field bears a striking resemblance to the oblique rotator configuration observed in the magnetic upper main sequence stars.

The most important tool for the discovery of cosmic magnetic fields has been that of synchrotron radiation from a variety of galactic and extragalactic objects, which permits the measurement of extremely weak fields in large-scale structures. The first discrete radio sources discovered, the Crab Nebula and Cygnus A among them, showed spectra that indicated extreme brightness temperatures, well in excess of values expected for celestial sources. These temperatures were also accompanied by spectra that did not fit well the form of a thermal distribution. Instead, they appeared to be more consistent with power laws of the sort which had been observed in the Sun during maximum activity phases. It was Shklovskii, following the original derivation by Schwinger of the radiation of a relativistic electron, who explained the emission by synchrotron processes of extremely energetic particles in the presence of weak but ordered magnetic fields. The discovery of polarization in the Crab and several other sources at levels far higher than that observed in the interstellar medium, and the later observation of truly thermal radio sources, confirmed the explanation and added a major tool to the already full box available for the understanding of plasmas in astrophysical environments.

Stellar activity cycles have been determined for main sequence G- and K-type stars by Wilson and collaborators, indicating the presence of dynamos like the Sun's. Periodic light variations have been discovered for many T Tauri stars, with periods consistent with the rotational time scale. Radio, x-ray, and ultraviolet observations have revealed coronal structures and flaring on the RS CVn stars and Algol binary systems. It even appears that magnetic fields play an important role in generating and maintaining disks in the protostellar phase of stellar evolution and in the origin of bipolar mass outflows in pre–main sequence stars. Pulsars and magnetic white dwarfs show that extremely strong fields, upward of 10^{12} G, can exist in collapsed objects. All of these bespeak the importance of the physics of plasmas in magnetic fields and point to the need to examine the phenomena more closely. We will not attempt an exhaustive treatment here.[1] Rather, this chapter is intended to serve as a guidebook through the maze of basic physical processes associated with magnetic fields in an astrophysical fluids context.

6.2 The Basic Equations

Let us begin with Maxwell's equations for the electromagnetic field. It is important to note that the original derivation of these equations was based on a hydrodynamic analogy, and as we shall see shortly these equations lend themselves quite readily to a fluid dynamical treatment. The first equations we require are the conservation equations, both of which are divergence conditions for the fields. Conservation of magnetic flux, **B**, is given by

$$\nabla \cdot \mathbf{B} = 0 \tag{1}$$

As you know, this arises because there are no chargelike sources for the field (or, put otherwise, there are no magnetic monopoles). The electric field, on the other hand, has a source term. Its strength, **E**, obeys the conservation law

$$\nabla \cdot \mathbf{E} = 4\pi\rho \tag{2}$$

where ρ is the charge density. The time-dependent equations describe the evolution of the field components, first for the induction field:

$$\nabla \times \mathbf{E} = -\frac{1}{c}\frac{\partial \mathbf{B}}{\partial t} \tag{3}$$

and then Ampère's law:

$$\nabla \times \mathbf{B} = \frac{\partial \mathbf{D}}{\partial t} - \frac{4\pi}{c}\mathbf{J} \tag{4}$$

Here **D** is the displacement current and **J** is the charge current.

---

[1] You are urged to consult the masterpiece by Parker (1979) for the most comprehensive overview of the physics and mathematics of astrophysical magnetic fields.

6.2 The Basic Equations

This last equation is the key to the connection between hydrodynamical and magnetic phenomena—the current behaves like the fluid mass current of the normal mechanical equations with which we have been dealing. This equation is the one that contains all of the interesting magnetohydrodynamic (MHD) phenomena and that serves as the focus of the discussions that follow. It should not come as a great shock that this fluid analogy comes about. One way of looking at the structure of the field equations, in the context of MHD, is that two currents are possible. One is the displacement due to the static charges in the medium or due to the changes in the spatial distribution of these otherwise fixed charges, the analogy to shear in an elastic or fluid medium. The other is the free current, due to the charges that can respond dynamically to imposed fields. This free current is responsible for the generation or cancellation of electrical potential differences within the medium. The original derivation by Maxwell used this sort of argument. Here we see that it can be put in a statistical context, in saying that there are two components to a plasma-generated magnetic field—one due to the component of the particle spectrum with vanishing mean velocity and the other due to the dynamic component of the charge distribution.

What is the justification for the fluid approximation? First of all, the current is carried by one of the components of the medium, specifically the electrons, and they are strongly interacting with the protons by collisions. We assume that the medium is completely neutral (that is, in strict charge balance) in order both to simplify matters and (more critically) to satisfy the observational constraint that there cannot be a charge excess in stable astronomical bodies of more than 1 part in about 10^{39}. Finally, we assume that the ions and protons are in statistical equilibrium with the electrons, and that ionization balance occurs in the plasma.

We assume that the collision time between electrons and ions is very short. Thus if a high-frequency field starts to accelerate the electrons, they collide with the protons before they have much gain in velocity. Local electric fields generated by the attempted charge separation are sufficient to couple the two particles (after all, that is what we mean by a collision in the impulse approximation). Another way of saying this is that if a single electron is accelerated by an incident electric field with frequency ω, the collision frequency τ^{-1}, where τ, the mean collision time, acts like a damping to the motion of order $m_e v_e \tau^{-1}$. The presence of an electric field across the medium causes a current to flow, but the collisions thus act to produce a bulk motion of the ions and electrons. The displacement current can be neglected, and the bulk variables of density, pressure, and mean velocity can be used to describe the plasma. This is the *magnetohydrodynamic limit*, the dense limit of a plasma where the hydrodynamic equations apply to the description of the motion of the fluid. This is why we will be safe in assuming that Ampère's law connects the charge and bulk densities and thus provides the current from the fluid equations rather than the Boltzmann equation.

In other words, we can look at whether charge separation effects are likely within the plasma. If the medium is dense enough that there are no regions where we would be able to effect such separation on a time scale equal to ω_p^{-1}, the plasma oscillation time scale, then we are in the fluid mode. Put another way, we can scale the behavior of the medium by using the collision time τ_c through a dimensionless number $\omega_p \tau_c \ll 1$. A final way to state this condition is that the mean free path for a particle is small compared with the Debye length. This last statement is more physically significant, however, because it states something about the possible electric fields within the gas. If the medium is low enough in density that the collective microfield dominates the motion of a gas, and that the gas can be considered collisionless, then we are in the regime of kinetic theory. Provided that the plasma cannot support any internal electric fields, the current is determined only by the magnetic field, and the changes do not separate, then we are justified in choosing a fluid description of the plasma.

With this justification, we see that we can write out the equation that connects the magnetic and electric fields. It is simply the continuity equation:

$$\frac{\partial \rho}{\partial t} + \mathbf{\nabla} \cdot \mathbf{J} = 0 \tag{5}$$

We have connected the Ampère equation, which provides the evolution equation for the current, with the conservation of charge represented by Gauss's equation. The equation for the electric field in a moving medium gives

$$\mathbf{E}' = \mathbf{E} + \frac{1}{c} \mathbf{v} \times \mathbf{B} \tag{6}$$

so that we obtain from Ohm's law

$$\mathbf{J} = \sigma \mathbf{E}' = \sigma \left(\mathbf{E} + \frac{1}{c} \mathbf{v} \times \mathbf{B} \right) \tag{7}$$

The conductivity, σ (which is the inverse of the resistance), is the main object of concern at this stage. This parameter determines the characteristic time scale for all magnetic phenomena that are connected with the dissipation of momentum via collisions. Ultimately, since there is a collisional redistribution of the momentum of the charges gained by the imposition of external fields in the medium, the emergent magnetic field must decay. The currents cannot continue forever and must suffer a slow decay. The magnetic field cannot therefore be supported forever in the medium, and there will be a slow secular change in the dipole moment. We shall return to this presently under the guise of Cowling's theorem for dynamos, but here state that it is important to keep in mind that the magnetic fields generated by the application of a naive dynamo model cannot be time independent.

The resistivity (the inverse conductivity) for a plasma was derived by Spitzer (1962):

$$\eta = 3.8 \times 10^{12} \frac{Z \ln \Lambda}{T^{3/2}} \text{ esu} \tag{8}$$

where Z is the mean charge of the plasma. Here Λ is the so-called Coulomb integral, which is given by

$$\Lambda = \Lambda(T, n_e) = \frac{3}{2Ze^3} \frac{(kT)^{3/2}}{(\pi n_e)^{1/2}} \tag{9}$$

A typical value for Λ is around 15 ± 5 for temperatures between $10^3 \leq T \leq 10^6$ K and $1 \leq n_e \leq 10^{12}$ cm^{-3}. Note that turbulence and wave interactions tend to increase this value (hyper- and anomalous resistivity, respectively).

6.2.1 Diffusion

In order to remove the electric field, we take the curl of both sides of the equation. If we had only the displacement current and not the charge current, we would at this point emerge from the exercise with the wave equations for the electromagnetic field. However, here we have one less time derivative, since we have argued previously that there is no displacement current, and therefore we will come out with a diffusion equation. To see this, let us go ahead and perform the operation required:

$$\nabla \times (\nabla \times \mathbf{B}) = \frac{4\pi\sigma}{c^2} \nabla \times (\mathbf{v} \times \mathbf{B}) - \frac{4\pi\sigma}{c^2} \frac{\partial \mathbf{B}}{\partial t} \tag{10}$$

under the assumption that there is no spatial dependence to the conductivity (that the fluid is at least homogeneous). Since we know that the magnetic field is divergenceless, we obtain

$$\frac{\partial}{\partial t}\mathbf{B} = \eta \nabla^2 \mathbf{B} + \nabla \times (\mathbf{v} \times \mathbf{B}) \tag{11}$$

This is the basic equation of MHD and one that has the characteristic "diffusion coefficient" for ohmic dissipation given by

$$\eta = \frac{c^2}{4\pi\sigma} \tag{12}$$

This simple equation is of enormous astrophysical importance, for it provides a useful estimate of the time scale over which the diffusion of the magnetic field through *ohmic dissipation* will produce a decay of the magnetic field strength in the plasma. Assume that there is a characteristic scale length, l, over which the medium is uniform. Further assume that there is no turbulence in the medium and that only a simple diffusion equation is needed to determine the global evolution of the magnetic field. We then obtain the time scale for secular change of the field:

$$\tau_{\text{decay}} = \frac{4\pi\sigma l^2}{c^2} \qquad (13)$$

In other words, in the absence of the "source term" that is connected with the fluid motion we would expect the magnetic field to decay on this time scale. The details of this decay were first discussed by Wrubel and Cowling for stars, and we shall return to the description of this phenomenon shortly. However, it is important to note that the conductivity is simply of the same order as the mean collisional frequency. This is related to the kinetic equations via the Boltzmann equation. For the case of a uniform gas, this will simply be the same as the mean free path λ times the velocity dispersion, $\nu = \lambda \langle u^2 \rangle^{1/2}$. You can see that the time scale is then related to the number of mean free paths contained in the medium: the more there are, the longer the field will take to decay because there is more surface through which the flux is passing. For the typical densities of main sequence stellar interiors, this provides a decay time of the order of 10^9 years, while for the interiors of planets, notably the Earth (where the conductivity is determined by the details of the crystalline phases of the core and mantle), we find a far shorter time—of order 10^4 years. Clearly, the fact that we observe any magnetic fields in such bodies at all argues strongly for the continued generation of such fields by internal processes related to dynamo actions.

6.2.2 Flux Freezing

Any attempt to maintain an internal electric field in a perfectly conducting medium will be thwarted by the mobility of the charges, which immediately move to cancel any potential difference. The time scale for this cancellation is very short in comparison with the time scale for the field to begin building in the medium—that is, they take place on times short in comparison with the actual fluid motions—so there will be no net electric field. The meaning of the diffusive term for the evolution of a magnetic field can thus be explained more clearly by thinking about the effect of mass motions in a magnetic field. For a net current to result from the fluid

6.2 The Basic Equations

motion in some medium, there must be uncanceled potential differences that manage to survive within the fluid. In a highly conducting medium, there is a simpler amplification mechanism for the field, which acts even without recourse to a dynamo. The magnetic field appears to move as if "frozen" into the medium, the spatial energy density of the magnetic field precisely following the fluid density. So if the density increases locally, so does the magnetic field.

To see this, assume that σ is the value usually quoted for the conductivity of stellar plasma, $>10^{15}$ s^{-1}. This conductivity is large enough to ensure that the magnetic diffusion term effectively vanishes. Then

$$\frac{d\mathbf{B}}{dt} = -\mathbf{B}\nabla\cdot\mathbf{v}, \qquad \frac{d\rho}{dt} = -\rho\nabla\cdot\mathbf{v} \qquad (14)$$

The magnetic flux is therefore simply a scalar functional multiple, f, of the mass density, or $\mathbf{B} \approx f\rho\hat{\mathbf{b}}$. Thus, if the density is locally increased, so is the magnetic field strength. The magnetic field seems to move with the fluid, hence the appellation "frozen." Now imagine that there is a small deviation from perfect conductivity. As the fluid moves, there will be some slippage of the mass through the field. This appears to change the magnetic field strength in the comoving fluid. At the same time, the fact that the field is changing induces the formation of a potential difference, which, in a finite-conductivity environment, induces the generation of a current. All of this is at the expense of the magnetic field. The field consequently decreases in local strength and will do so everywhere throughout the fluid in time. In fact, the form of the magnetic field decay is the same as that of the heat equation, so the field can be said to be diffusively lost. The energy simply goes over into heat, since the field generates dissipative currents that lose their energy through collisions throughout the fluid, and the result is the gradual fading away of the field with time.

6.2.3 Ambipolar Diffusion

The relation between the strength of a magnetic field and the density of the ambient medium strictly holds only if the medium is highly ionized. In stellar plasmas, this is almost certainly a good assumption, as it appears also to be for large-scale extragalactic radio jets and lobes. But in interstellar clouds, where the densities may become large enough and the opacity high enough, ultraviolet radiation is effectively screened out of the cloud cores. Thus, in the densest part of a molecular cloud, the ionization is expected to be due primarily to cosmic ray penetration of the cloud, leaving ionization fractions $<10^{-5}$ in the regions where the density is

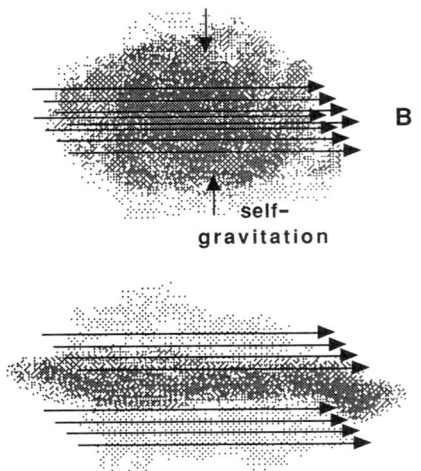

Figure 6.1 Ambipolar diffusion. Top: Ions (light) and neutrals (dark) are mixed with supporting field lines. Bottom: Neutrals have separated under self-gravity into the midplane, leaving the ions supported by the field lines.

$>10^5$ cm^{-3}. When the temperature in this region falls below about 100 K, it becomes locally unstable to gravitational collapse. A trapped magnetic field will be transported inward with the collapsing gas but, because of the low conductivity, will not be amplified as rapidly as one would expect from flux freezing. Instead of increasing in strength as $\rho^{2/3}$ or ρ, the matter separates from the field. Models indicate that $B_{\text{collapse}} \sim \rho^{1/2}$, a weaker dependence than for strictly flux-frozen fields (see Mouschovias 1978; Mestel 1985). The separation of the field and the material is due to ambipolar diffusion, the tendency of ions to separate in an external magnetic field, and also the field line slippage relative to the gravitationally contracting medium. (See Fig. 6.1.) Direct measurements support this approximate scaling relation although models constrain the exponent n in the relation $B_{\text{collapse}} \sim \rho^n$ only to lie between about 1/3 and 2/3.

Although not strictly a hydrodynamic process, the diffusion of gas in the presence of a gravitational field does pose some problems similar to those of flows. In particular, one essential feature of ambipolar diffusion is quite clear. The fact that the gas is multiphase, that it can make the transition between neutral and ionized phases, is vital to the slow diffusion of the matter through an ambient magnetic field. The gravitational acceleration does not know about charge, of course, so that ions and neutrals should diffuse in like manner. But if a magnetic field is present, the diffusion is slowed by the effective change on the ions. If there are several stages present, and if the medium has a dominantly neutral phase, the gas will not be able to remain supported by the field and will slowly slip out of the field-dominated regions.

6.3 Alfvén Waves

We know that wave solutions exist for the electromagnetic equations in a vacuum and also in a dielectric medium. In these cases, the role played by the free charge is not important. The collision time becomes the characteristic time for the damping of the wave, and the existence of the stationary charges that give rise to $\mathbf{D}$ provides a characteristic response time for the dielectric. Now in the fluid approximation we can see that if the field were to be jostled, the motion induced in the fluid would be through the existence of wave motions that would be able to transmit momentum to the medium. These are the Alfvén waves.

Let us take a medium that is supported by pressure of both gas and the magnetic field. We will treat this in terms of perturbation theory (which is discussed in more detail in the chapter on instabilities) but it is most useful to introduce this phenomenon here. Begin with the perturbation of the equation of hydrostatic support:

$$\rho \frac{\partial \mathbf{v}}{\partial t} + \nabla \delta p - \frac{1}{4\pi}[(\nabla \times \mathbf{b}) \times \mathbf{B} + (\nabla \times \mathbf{B}) \times \mathbf{b}] = 0 \tag{15}$$

We have already assumed that the medium is initially at rest, to remove the advection terms, and we have retained only the first-order terms in the expansions. The equation of state for an isothermal medium, $p = \rho a_s^2$, will be needed in a moment as well, but we could also have chosen a polytropic form. The equation for the magnetic field perturbation becomes

$$\frac{\partial \mathbf{b}}{\partial t} - \nabla \times (\mathbf{v} \times \mathbf{B}) = 0 \tag{16}$$

again retaining only the first-order terms. Finally, the continuity equation is

$$\frac{\partial \delta \rho}{\partial t} + \rho \nabla \cdot \mathbf{v} = 0 \tag{17}$$

The plane wave assumption gives rise to a dispersion relation, which is seen from the substitution of $\partial/\partial t = -i\omega$ and $\nabla = i\mathbf{k}$, so $-i\omega\delta\rho + i(\mathbf{k} \cdot \mathbf{v})\rho = 0$ for the continuity equation. This can also be written as

$$\delta \rho = -\frac{1}{\omega} \rho \mathbf{k} \cdot \mathbf{v} \tag{18}$$

if we choose instead to retain the option of the choice of spatial dependence. But we already have an important piece of physics here. Longitudinal waves are those with $\mathbf{k} \cdot \mathbf{v} = kv$. These, by Eq. (17), are seen to be compressive. Transverse waves are not. So the existence of Alfvén modes does not necessarily require that the medium be compressible. Unlike sound waves, only the magnetic field is needed for Alfvén wave propagation. The plasma can even be collisionless, so pressure need not play any role in the medium. The assumption of a uniform initial state of the fluid is very important. It means that the direction of the resultant propagation will be anisotropic (there is a direction imposed by the initial magnetic field) but the amplitude will not change. This is a simple wave and consequently we anticipate the existence of a wave speed in the solution. For completeness, the equations for momentum and induction transform under the plane wave *ansatz* to

$$-i\omega\rho\mathbf{v} + ia_s^2\delta\rho\mathbf{k} - \frac{1}{4\pi}[(i\mathbf{k} \times \mathbf{b}) \times \mathbf{B} + (\boldsymbol{\nabla} \times \mathbf{B}) \times \mathbf{b}] = 0$$

$$-i\omega\mathbf{b} + i\mathbf{k} \cdot \mathbf{v} - \mathbf{v} \cdot \boldsymbol{\nabla}\mathbf{B} = 0 \qquad (19)$$

respectively. Again, assume an isothermal equation of state, so that the pressure perturbation is given by $\delta p = a_s^2 \delta\rho$.

The equation for the velocity perturbation reduces to[2]

$$-\omega\mathbf{v} - \frac{a_s^2}{\omega}\boldsymbol{\nabla}\boldsymbol{\nabla} \cdot \mathbf{v} = \frac{B^2}{4\pi\rho\omega}\{-\boldsymbol{\nabla}[\mathbf{n} \cdot (\mathbf{n}\boldsymbol{\nabla} \cdot \mathbf{v} + \mathbf{n} \cdot \boldsymbol{\nabla}\mathbf{v})]$$
$$+ \mathbf{n} \cdot \boldsymbol{\nabla}[\mathbf{n}\boldsymbol{\nabla} \cdot \mathbf{v} + \mathbf{n} \cdot \boldsymbol{\nabla}\mathbf{v}]\} \qquad (20)$$

where we have removed the constant background magnetic field by replacing it with $\mathbf{B} = B\mathbf{n}$. Also, we have removed the time derivatives everywhere using $\partial/\partial t \to -i\omega$. Now since B^2 has the magnitude of a pressure and the sound speed is dimensionally $(P\rho^{-1})^{1/2}$, we define the quantity

$$v_A^2 \equiv \frac{B^2}{4\pi\rho} \qquad (21)$$

[2] We use the following expansion:

$$(\boldsymbol{\nabla} \times \mathbf{B}) \times \mathbf{B} = -\tfrac{1}{2}\boldsymbol{\nabla}B^2 + \mathbf{B} \cdot \boldsymbol{\nabla}\mathbf{B}$$

where the first term on the right is the magnetic pressure. The perturbation gives a term of the form $(\boldsymbol{\nabla} \times \mathbf{b}) \times \mathbf{B} + \boldsymbol{\nabla} \times \mathbf{B}) \times \mathbf{b}$. The pressure perturbation for the system is thus $\delta p + 2\mathbf{b} \cdot \mathbf{B}$, including the magnetic field. For the Lorentz force, we use

$$\boldsymbol{\nabla} \times (\mathbf{v} \times \mathbf{B}) = -\mathbf{v} \cdot \boldsymbol{\nabla}\mathbf{B} - \mathbf{B}\boldsymbol{\nabla} \cdot \mathbf{v} + \mathbf{v} \cdot \boldsymbol{\nabla}\mathbf{B}$$

and use the continuity equation to remove $\boldsymbol{\nabla} \cdot \mathbf{v}$ for the final form of Eq. (20). Notice that the incompressible gas would have removed the divergence here.

6.3 Alfvén Waves

to be a magnetohydrodynamic analog of the sound speed. This is the *Alfvén speed*. To see more clearly how it enters the propagation equations, use the substitution for a simple plane wave that $\nabla \to i\mathbf{k}$ to obtain the final form of the dispersion relation for the waves:

$$\omega^2 \mathbf{v} - a_s^2 \mathbf{k}(\mathbf{k} \cdot \mathbf{v}) = -v_A^2[\mathbf{k}(\mathbf{k} \cdot \mathbf{v}) + \mathbf{k}(\mathbf{n} \cdot \mathbf{v})(\mathbf{n} \cdot \mathbf{k})]$$
$$- v_A^2[-\mathbf{n} \cdot \mathbf{k}(\mathbf{k} \cdot \mathbf{v})\mathbf{n} - (\mathbf{n} \cdot \mathbf{k})^2 \mathbf{v}] \qquad (22)$$

This doesn't look much better than the full form of the equations, but it is more tractable. For one thing, we see that both longitudinal and transverse waves exist and that, if we take $\mathbf{k} \cdot \mathbf{v} = 0$, the phase speed of a transverse wave is v_A.

The dispersion relation in Eq. (22) still depends on the velocity perturbation but is linear in $\mathbf{v}$. It is therefore independent of the amplitude of this perturbation and its phase speed depends on the *direction* of propagation with respect to the imposed magnetic field, which has the unit vector $\mathbf{n}$ in this treatment. (See Fig. 6.2.) This is actually what you would expect, because the magnetic field imposes a direction on the medium and serves as an additional tension in the fluid.

Alfvén waves exist as both transversely and longitudinally propagating waves. There are different indices of refraction for the medium in these two modes, but essentially their velocity is v_A. It is instructive to consider the following astrophysically interesting case. We begin with a magnetic field of uniform strength. A wave is started in the medium (the source is, for the moment, unimportant). The wave may begin at subsonic speed, but if the density decreases the wave can eventually accelerate and become supersonic. They also reflect and refract, like sound waves and seismic waves, in the presence of density gradients. Especially in the case of an isothermal atmosphere, for which the sound speed is constant and the density is exponentially decreasing, there will be a point at which the wave

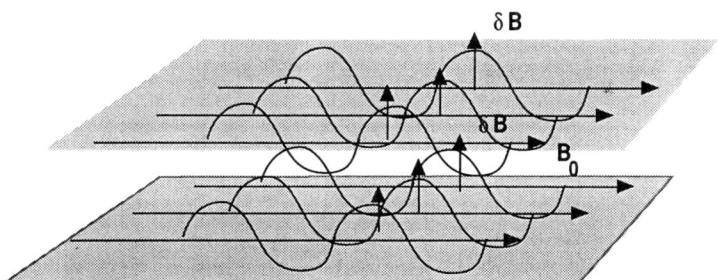

Figure 6.2 Alfvén wave propagation in a magnetic field.

will have to steepen and become a shock wave. At that point, there is a rapid dissipation of energy with resultant heating of the medium. This nonthermal source of energy will show up as an excess temperature of the medium, compared with what we would have expected from radiative transfer models alone. In the Sun, this may be part of the source for the heating of the base regions of the corona. It is important to note here that the magnetic energy that is available from the generation of these waves represents a significant and largely unknown source of heating for stellar atmospheres and the interstellar medium. It is one of the most interesting and important areas of research to find what the most reasonable sites for this nonthermal heating are and what the generation mechanisms are for the fields.

It is important as well to look at the interstellar medium, since here in the most dilute of astrophysical plasmas the magnetic field plays a central role in support and energetics of the medium. The ratio of the thermal to magnetic energy density is

$$\Gamma = \frac{4\pi\rho a_s^2}{B^2} = \frac{a_s^2}{v_A^2} \tag{23}$$

This is just analogous to the usual plasma parameter, $\beta = p_{\text{thermal}}/p_{\text{magnetic}}$. Equation (23) is the inverse magnetic Mach number and provides a more astrophysically useful way of thinking about the fields. Now let's consider the diffuse interstellar medium. We take the standard scaling values of 1 μG for the field and of the order of 10 km s^{-1} for the sound speed. We further assume that the mean density of the medium is of order 1 cm^{-3}. Therefore the value of Γ is of order unity—the magnetic field provides roughly the same contribution to the energetics of the medium as does the thermal pool. We should add that the energetic particles (cosmic rays) also provide a similar contribution. This *equipartition* is one of the basic arguments for the connection between interstellar magnetic fields and the generation of cosmic rays, and it provided part of the arsenal of arguments first used by Fermi for the identification of interstellar turbulence as the source of cosmic ray acceleration. Broad molecular lines are observed routinely from nearly all molecular clouds. The velocities are always well in excess of the sound speed and usually are a few kilometers per second. Turbulent Alfvén waves, perhaps generated in the intercloud diffuse interstellar medium, may be responsible for this broadening. At any rate, the densities of the clouds are about 10^4 to 10^5 cm^{-3}, so the field strengths derived from the Alfvén wave assumption are a few hundred microgauss. Such a magnetic field strength is consistent with other direct measurements for molecular clouds. If the broadening is due to Alfvén waves, it also indicates that the ionization must be high enough for collisions to couple the ions with the neutral gas.

6.4 Magnetic Equilibrium: A Basic Example—Pinch Equilibrium

One of the standard textbook examples of fluid equilibrium in a magnetic field is actually of astrophysical interest. Suppose we want to build a monodirectional current J_z along some axis and have this current confine itself by the generation of a circumferential magnetic field. That is, from basic physical considerations, one would think that it is possible to take a column of plasma, or fluid in the sense we've been arguing for in the MHD picture, and by passing a current through it both heat it and form the magnetic field that confines it. This was the dream of plasma physicists at the start of the herculean effort to produce controlled fusion several decades ago. We see that magnetic field structures of this sort, called prominences, exist in the solar atmosphere, and are apparently stable for long times. It seems likely that such a configuration would be possible in the laboratory.

The basic condition for magnetic equilibrium is that $\nabla p \sim \mathbf{J} \times \mathbf{B}$, neglecting the constants for a moment. If the only structural support for the fluid derives from the magnetic field, then it follows that

$$\mathbf{J} \cdot \nabla p = 0, \quad \mathbf{B} \cdot \nabla p = 0 \tag{24}$$

and the isobars are also surfaces of constant field strength and current density. You see here something analogous to a Taylor–Proudman theorem. We already know that the potential vorticity is conserved within a fluid. Here we have something very much like it, in that p is also a scalar. We would have to identify $c_1 \mathbf{J} + c_2 \mathbf{B}$ with ϖ/ρ to complete the analogy [this is similar to the Elässer variables; see Montgomery (1989) for further discussion of these coordinates]. Consider the equations for a monodirectional current. The conductivity is assumed to be infinite, so that

$$(\nabla \times \mathbf{B}) \cdot \hat{\mathbf{z}} = \frac{4\pi}{c} J_z \tag{25}$$

which gives

$$\frac{1}{r} \frac{dr B_\phi}{dr} = \frac{4\pi}{c} J_z \tag{26}$$

Now let's complete writing down the equations describing this axisymmetric configuration:

$$\frac{dp}{dr} = -\frac{J_z B_\phi}{c} \tag{27}$$

$$\frac{1}{r}\frac{d}{dr} rJ_z = 0 \tag{28}$$

$$\eta J_z = E_r - \frac{1}{c} V_z B_\phi \tag{29}$$

$$E_z = \eta J_z \tag{30}$$

Now take the current to be a constant and the density to be constant, along with the conductivity. The equation for the induced field can be immediately integrated to give $B_\phi \sim r$ so that the azimuthal field must increase with increasing distance from the axis. The pressure is given by

$$p(r) \approx (a^2 - r^2) \tag{31}$$

which states that the pressure is largest in the interior of the plasma and decreases out to the edge. This in turn produces an equilibrium condition that the total pressure is given by

$$p + \frac{B_\phi}{4\pi} = \text{constant} \tag{32}$$

All of this assumes that the current is kept constant. But even here, it is probable that we will not be able to maintain stability. If the plasma column is perturbed, and this means that we will have to consider a more detailed solution eventually, the changes in the current density at any point along the column feed back into the induced magnetic field, which causes the column to buckle in a sausage. This causes the column ultimately to disperse by the formation of blobs, the so-called *sausage instability*. We defer the analysis for the moment but will return to it later. The point of this exercise is that the MHD approximation permits a simple fluid solution, with the additional action of the background fields, of a wide variety of problems.

6.4.1 Flux Conservation and Stellar Magnetism

The fact that we observe strong, global fields in stars can be attributed to a quirk of stellar evolution. Massive stars burn their fuels on short time scales, and so their lifetimes on the main sequence are shorter than those of low-mass stars. The break-even point for the competition between ohmic

6.4 Magnetic Equilibrium: A Basic Example—Pinch Equilibrium

decay of the fields and the lifetime of the star is at about 2 $M_\odot$, in the A stars. The fact that the strongest observed magnetic field among main sequence stars is observed in this range, among the Ap and Bp stars, is a consequence of this. The lower-mass stars simply lose their fields through dissipation. This occurs even in the absence of turbulent destruction of the fields. To date, no star of the order 1 $M_\odot$ has been observed to display anything but transient, sunspot-like, fields. On the other hand, the fields observed in the Ap stars are strictly periodically variable and stable (as far as we can tell from about 60 years of measurement of the abundance patches on the surfaces or by direct detection of the fields).

Strong fields have also been detected among the white dwarfs. These are of order 10^6 G, considerably stronger than *anything* seen in normal stars. The assumption is that this is a fossil field, relic from the post-main sequence phase of stellar evolution and retained and amplified due to the contraction of the star. To see this, consider the spherical form of the flux conservation equation:

$$\nabla \cdot \mathbf{B} = \frac{1}{r^2} \frac{\partial}{\partial r} (r^2 B_r) = 0 \qquad (33)$$

so that $B_r \sim r^{-2}$. As the core contracts the magnetic field will increase, so if we take the radius of the star on the main sequence, for a 1 $M_\odot$ star, to be of order 1 $R_\odot$, then the radius of a typical white dwarf is of order 0.01 $R_\odot$, giving rise to an amplification factor of the order of 10^4. The origin of the field in the red giant remains a major research question which has yet to be resolved. Many magnetic white dwarfs have been directly observed using Zeeman polarimetry. This method is an extension of the procedure first used to detect fields on the Ap stars. A circular polarimeter is connected with a tilted Hβ filter. The opposite wings of the line are observed through alternating senses of circular polarization. In the presence of a global magnetic field, the effective field—the longitudinal field—produces a small wavelength shift in the position of the wing due to the Zeeman effect. This causes a variation in the brightness of the star measured at a fixed wavelength in the wing, the signal being detected by a photometer and the brightness fluctuation being correlated with the slope of the line and the strength of the longitudinal magnetic field. This technique has been exploited extensively by Angel, Landstreet, and their collaborators. The normal field is dipolar. Fields are also inferred for a class of magnetic white dwarfs which are the accreting objects in very close binary systems—the AM Her stars. The magnetic fields measured in the emission lines trapped in the polar accretion columns of these stars bespeak of strong surface fields, but the periodicity of the accretion-linked emission lines is the primary tool for inferring the presence of fields in these stars.

Neutron star magnetic fields appear to have a different origin. Fields cannot be directly measured for the majority of these objects, although pulsar emission properties allow the inference of fields as high as 10^{13} G. While the collapse of a star from the size of a main sequence radius to that of a typical neutron star (of order 10 km) will produce fields of the order of 10^{12} G, it appears that there are processes in the crusts of these stars which are capable of *in situ* generation of the fields. These processes have been discussed by Blandford and his collaborators (1983)[3] and are not related to dynamo processes responsible for the fields observed in less pathological objects.

6.5 Force-Free Fields

One of the most frequently encountered fields in astrophysical plasmas is the force-free field. The solar corona displays perhaps the best available example of such fields in action in a plasma. Since the Lorentz force is $\mathbf{J} \times \mathbf{B}$, which in the case of the MHD equations we have been using is $(\nabla \times \mathbf{B}) \times \mathbf{B}$, the absence of this force yields a special solution for the details of the field structure. This is one in which the field is given by the solution of

$$\nabla \times \mathbf{B} = \lambda \mathbf{B} \tag{34}$$

where λ is a constant. The solution is clearly dependent upon the geometry of the configuration.

A very basic result, *Woltjer's* theorem, states that the force-free field is a minimum energy configuration of the field. This is of central importance to astrophysical arguments, because we often do not know what the source or driving current may be for a field but can make some arguments on the minimum energy required for its maintenance. The magnetic field is obtained from the vectorial potential $\mathbf{A}$ by $\mathbf{B} = \nabla \times \mathbf{A}$ and the fact that no external scalar potentials are present. We know, of course, the additional constraint that $\nabla \cdot \mathbf{A} = 0$. Therefore we have as an energy condition to the field the equation

$$E_M = \frac{1}{2} \int (\nabla \times \mathbf{A})^2 \, dV \tag{35}$$

Minimization of this energy is accomplished by taking $\delta E_M = 0$, where we have taken the Lagrangian variational. Notice that the medium is assumed

---

[3] Blandford, R. D., Applegate, J. H., and Hernquist, L. (1983). *Mon. Not. R. Astron. Soc.* **204**, 1025.

6.5 Force-Free Fields

to be static and that there are no electric fields driving the currents other than those consistent with the magnetic field alone. This variation gives

$$\int \nabla \times \mathbf{A} \cdot \delta(\nabla \times \mathbf{A}) \, dV = 0 \tag{36}$$

Actually, as we shall see in a moment, this leads to a very interesting quantity, the magnetic helicity, $H_M \equiv \int \mathbf{B} \cdot \mathbf{A} \, dV$. The last statement is that the magnetic helicity of a force-free field is minimal. This looks quite a bit like the result for Kelvin's theorem for an inviscid fluid, serving as yet another example of the importance of the vorticity analogies as we will discuss below.

We now take a sidestep, to look at the product of $\mathbf{B} \cdot \mathbf{A}$. Having called attention to its importance, it turns out that for magnetic fields, it has some very interesting properties. These are useful for determining the force-free energy configuration. Another way of writing the equation for the vector potential is

$$\frac{\partial}{\partial t} \mathbf{A} = \mathbf{v} \times \mathbf{B} \tag{37}$$

This is obtained by removing the *curl* from the Maxwell equation for the Lorentz force. We take the integral

$$\frac{\partial}{\partial t} \int \mathbf{B} \cdot \mathbf{A} \, dV = \int \left(\frac{\partial \mathbf{A}}{\partial t} \cdot \mathbf{B} + \mathbf{A} \cdot \frac{\partial \mathbf{B}}{\partial t} \right) dV \tag{38}$$

and substitute the equations for the time derivatives:

$$\int [(\mathbf{v} \times \mathbf{A}) \cdot \mathbf{B} + (\nabla \times (\mathbf{v} \times \mathbf{B})) \cdot \mathbf{A}] \, dV = \int \left[\left(\frac{\partial}{\partial t} \mathbf{A} \right) \times \mathbf{A} \right] \cdot d\mathbf{S} = 0 \tag{39}$$

and therefore the volume integral of the product is a constant. This is also the same as the statement that the current has an energy of

$$E = \int \mathbf{J} \cdot \mathbf{A} \, dV \tag{40}$$

which is a nonpotential field contribution to be added to the total energy of the potential field (the volume integral of the square of the field strength that we had previously written down). Most of these equations are analogous to the ones you worked with for vorticity. The difference is that there is no separate field corresponding to $\mathbf{v} \times \mathbf{B}$ to make the notation precisely the same. But since historically the two fields developed in parallel, it isn't a surprise to see how deep these analogies go.

Now we can take the variational of the combined integrals:

$$\delta \int \mathbf{B}^2 \, dV - \lambda \delta \int \mathbf{B} \cdot \mathbf{A} \, dV = 0 \qquad (41)$$

where the (undetermined) Lagrange multiplier λ has been added to the formula. The variation is given by

$$\int [\mathbf{B} \cdot \delta \mathbf{B} - \lambda (\mathbf{B} \cdot \delta \mathbf{A} + \mathbf{A} \cdot \delta \mathbf{B})] \, dV = 0 \qquad (42)$$

keeping the magnetic field explicit. We use the representation of the magnetic field:

$$\delta \mathbf{B} = \nabla \times \delta \mathbf{A} \qquad (43)$$

and substitute this into the equation for the variational. This reduces the integral to

$$\int [(2\mathbf{B} - \lambda \mathbf{A}) \times \delta \mathbf{A}] \cdot d\mathbf{S} + 2 \int (\nabla \times \mathbf{B} - \lambda \mathbf{B}) \cdot \delta \mathbf{A} \, dV = 0 \qquad (44)$$

The first term, by Gauss's theorem, becomes a surface integral. Since the variation vanishes at the surface, this integral vanishes *independently* of the second. This is a familiar assumption in cosmic arguments: that the surfaces can be extended to infinity and therefore the fields passing through the surface must vanish asymptotically. Therefore, the force-free field equation is obtained and this is the minimum-energy configuration of the magnetic field, as we sought to show. The value of λ is left undefined, short of the specification that it is a constant. This is determined by the boundary conditions of the problem and specifically by the geometry of the problem.

One of the traditional solutions for such a field is to look at a cylindrical flux tube. This can be used as well to introduce some of the later discussion of such configurations, which are a product of dynamo activity in stars and accretion disks. We assume that the field solves the equations

$$\frac{\partial B_z}{\partial r} = -\lambda B_\phi, \qquad \frac{1}{r} \frac{\partial}{\partial r}(r B_\phi) = \lambda B_z \qquad (45)$$

Thus we have the solution for the combined equation for the axial field:

$$B_z = C J_0(\lambda r) \qquad (46)$$

The values of λ are determined by the boundary conditions for the specific problem. The azimuthal field is determined by the radial derivative of B_z and so varies as $J_1(\lambda r)$. There is no radial component to the field other than a constant factor, since the derivative vanishes.

Why is this field structure special? First of all, it permits us to obtain both the magnetic field configuration and the properties of the current that supports it. We don't have to specify the current configuration in advance—it is provided by the condition of the equation itself. Second, it *is* the minimum-energy configuration for the field and therefore the one to which the fields should be attempting to decay if they are created by some dynamo-related process. The derivation of the field in this form also provides us with the energy of the field itself. Therefore, if by some chance we have been given an initial field which is not force-free, we can calculate the energy that a transformation of the state of the field is expected to release. This is one of the most immediately useful results. Having been able to solve for the structure of a force-free field, the concept of a flux tube may seem more compelling. It is a structure which consists of wrapped fields, oriented in a cylindrical geometry as specified by the Laplace equation and, because of the divergence condition, closed on its ends. The fields are strictly circumferential and helical, lacking a net divergence. Flux tubes have been directly observed in the tails of comets, in the magnetospheres of several planets (especially the Earth), in the ionosphere of Venus, and possibly in the solar wind. They are usually taken as the basic model for the dynamo by-product from which large-scale field structures coalesce.

6.6 Vorticity Analogy and Magnetic Helicity

As we discussed in the chapter on rotating fluids, we obtained an evolution equation for the vorticity of the form

$$\frac{\partial \omega}{\partial t} = \nu \nabla^2 \omega + \nabla \times (\mathbf{v} \times \boldsymbol{\omega}) \tag{47}$$

where $\boldsymbol{\omega} = \nabla \times \mathbf{v}$ and ν is the kinematic viscosity. This was derived from the Navier–Stokes equation by taking the curl in a similar fashion to the operation we performed with the Ampère equation. Now we also note that $\nabla \cdot \boldsymbol{\omega} = 0$, analogous to the flux condition. It has thus been argued, especially by Batchelor, that magnetic fields can be generated by vortex-dominated turbulence. This analogy also leads to the term "magnetic viscosity" for the coefficient η that we introduced earlier. We also have a new dimensionless number, the *magnetic Reynolds number*, which is defined by

$$\mathrm{Re}_M \equiv \frac{Ul}{\eta}$$

Since we were able to see in the chapter on rotation that vorticity leads to dissipative processes, in the case of turbulence we expect that magnetic dissipation will be enhanced over the ohmic rate. This was first discussed by Spitzer (1957) in connection with the magnetic fields of stellar interiors.

Now we come to a truly beautiful and elegant result, that the analogy between the fluid and magnetic equations leads to a precise definition of the force-free field configuration. To see this, we first note again that $\mathbf{B} = \mathbf{\nabla} \times \mathbf{A}$ in the same way that $\boldsymbol{\omega} = \mathbf{\nabla} \times \mathbf{v}$. Now the equation for the field evolution with magnetic viscosity is given by

$$\frac{\partial \mathbf{B}}{\partial t} = \mathbf{\nabla} \times (\mathbf{v} \times \mathbf{B}) + \eta \nabla^2 \mathbf{B} \tag{48}$$

For magnetic fields, the last term is the same as $\eta \nabla^2 \mathbf{B} = - \eta \mathbf{\nabla} \times (\mathbf{\nabla} \times \mathbf{B})$. Removing the curl from this equation gives an equation for the vector potential $\mathbf{A}$:

$$\frac{\partial \mathbf{A}}{\partial t} = \mathbf{v} \times \mathbf{B} - \mathbf{\nabla}\Phi - \eta \mathbf{\nabla} \times \mathbf{B} \tag{49}$$

where we have now had to include the possibility of a scalar potential Φ for any static field. So far, this is playing with symbols, it seems; the equations are really saying the same thing. Yet if we take the scalar product of Eq. (48) with $\mathbf{A}$ and of Eq. (49) with $\mathbf{B}$ and add them, we get

$$\frac{\partial}{\partial t}\mathbf{A} \cdot \mathbf{B} = \mathbf{\nabla} \cdot [(\mathbf{v} \times \mathbf{B}) \times \mathbf{A}] + (\mathbf{\nabla} \times \mathbf{A}) \cdot (\mathbf{v} \times \mathbf{B}) + \mathbf{B} \cdot (\mathbf{v} \times \mathbf{B})$$
$$- \mathbf{\nabla} \cdot \Phi \mathbf{B} - \eta \mathbf{\nabla} \cdot [(\mathbf{\nabla} \times \mathbf{B}) \times \mathbf{A}] - \eta(\mathbf{\nabla} \times \mathbf{A}) \cdot (\mathbf{\nabla} \times \mathbf{B})$$
$$- \eta \mathbf{\nabla} \cdot (\mathbf{B} \times \mathbf{B}) - \eta(\mathbf{\nabla} \times \mathbf{B}) \cdot \mathbf{B} \tag{50}$$

The second, third, and seventh terms vanish immediately (they could have been dropped, but that would have been too confusing). The sixth and eighth terms are *identical* (more on this in just a moment). Finally, the rest are divergences of their respective quantities. Now we take the integral over volume and assume the normal boundary condition that there are no surface terms in the equations, so that all of the divergences vanish on integration over volume. Therefore, we arrive at the evolution equation:

$$\frac{\partial}{\partial t}\int \mathbf{A} \cdot \mathbf{B}\, d\mathbf{x} = -\eta \int (\mathbf{\nabla} \times \mathbf{A}) \cdot (\mathbf{\nabla} \times \mathbf{B})\, d\mathbf{x} - \eta \int (\mathbf{\nabla} \times \mathbf{B}) \cdot \mathbf{B}\, d\mathbf{x}$$
$$= -2\eta \frac{4\pi}{c}\int \mathbf{J} \cdot \mathbf{B}\, d\mathbf{x} \tag{51}$$

6.6 Vorticity Analogy and Magnetic Helicity

The last integral resulted from substituting $\nabla \times \mathbf{B} = (4\pi/c)\mathbf{J}$ and $\mathbf{B} = \nabla \times \mathbf{A}$.

Look on this result with a true sense of wonder! It is the evolution equation for the magnetic helicity. It states that the evolution of magnetic helicity in a fluid is driven by dissipation and that if $\mathbf{B}$ is parallel to $\mathbf{J}$, the magnetic helicity decays. Otherwise, it increases. The coupling constant for the interaction between the field and current is the magnetic viscosity. Moreover, in a medium with $\eta \to 0$, the magnetic helicity is a conserved quantity. Now the connection between vorticity and magnetic fluid phenomena is clearer. Just as in the case of a fluid, the magnetic helicity serves as a topological tool for understanding the dissipative mechanisms and the instabilities of the medium. Tangles of magnetic field interact with one another, merge, split, and ultimately decay. (See Fig. 6.3.) The medium has the same kind of interactions at the level of the field that a fluid does in the presence of shear. In fact, this is because the current can be thought of as being related to the shear of the magnetic field, again a concept from the nineteenth century.[4]

But this analogy goes even deeper, and this is an argument originally due to Taylor (1986). Since in a "magnetically inviscid" fluid the magnetic helicity is strictly conserved, and the energy is conserved, the variation of the two must vanish. We have used the energy alone for Woltjer's theorem. Now if we put H_M and E together, what do we get? Assume that the medium is at rest. Then $\mathbf{v} = 0$ and the total energy of the system is $E = E_M = (1/8\pi) \int B^2 \, d\mathbf{x}$. Then

$$\delta(\lambda H_M + E_M) = 0$$

where λ is again a Lagrange multiplier. Also, λ is a measure of the partitioning between the magnetic energy and the helicity. This is *exactly* the statement of the equations that led us to the force-free field! You see that had we started out with the topological arguments in mind, we would have arrived at the same point but more physically.

---

[4] Doing violence to units, as is often done in MHD work, we can write the total energy of the fluid as

$$E = \int \frac{1}{2}(v^2 + B^2)\, d\mathbf{x}$$

Then it can be shown using the fluid equations that for an incompressible fluid:

$$\frac{\partial E}{\partial t} = -2\nu \int \omega^2 \, d\mathbf{x} - 2\eta \int J^2 \, d\mathbf{x}$$

This connects the idea of enstrophy with the magnetic field decay. The viscous dissipation of enstrophy and the interaction between currents dissipate energy, and at the same time the interaction between the currents and the fields they create alters the magnetic helicity.

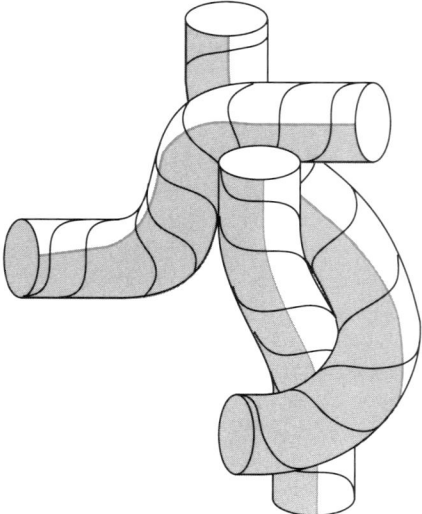

Figure 6.3 Schematic of flux ropes, illustrating the topological approach to their interaction and merging.

6.7 Magnetic Dynamos

There is a voluminous literature on the general problem of magnetic field generation by dynamos. Most of the work now centers on the calculation of numerical models for various types of dynamo configurations, mostly those linked to differential rotation and convection in spherical shells. However, some of the basic analysis can yield an insight into the behavior of these models and also provide the underlying equations required for the models. We shall discuss the schematic dynamo model, first formulated by Babcock and Leighton, and then discuss the more detailed model by Parker, Roberts, Soward, Malkus, and Krause for the connection between differential rotation, convection, and the generation of the poloidal global fields we observe in stars.

6.7.1 Cowling's "Antidynamo" Theorem

Since the fields observed in the Sun and Earth are obviously not steady state and must be generated by some kind of dynamo action, it is natural to hope that the symmetric motion of fluids in a rotationally driven loop will be capable of creating the surface fields. While this smacks of perpetual motion, it would not be outlandish to expect that something like this

6.7 Magnetic Dynamos

can happen. The rotation of the star and radiation are, after all, sources of energy which seem inexhaustible. However, this hope was dashed with the theorem by Cowling (1957) for the generation of magnetic fields by dynamos—that it is impossible for an axisymmetric configuration to maintain a steady-state magnetic field.

To see qualitatively how this comes about, consider the motion of the rotating fluid. The field lines are generated by the current, which has a curl about the rotational axis, and consequently has a representation J_ϕ. The field that is generated has what is called an O-type neutral point about the current lines. It is therefore conceivable that the field lines can be shrunk to points about the current lines without loss of topological symmetry. This is equivalent to the decrease in the magnetic flux. Consequently, there is no way to prevent the decay of the field by these motions and the field will eventually disappear on the ohmic time scale. To make this quantitative, consider the equations for the generation of the field by the current:

$$(\nabla \times \mathbf{B})_\phi = \frac{4\pi}{c} J_\phi \tag{52}$$

$$(\nabla \times J_\phi \hat{\phi}) = -\frac{1}{c}\frac{\partial \mathbf{B}}{\partial t} \tag{53}$$

The combination of these leads to the diffusion equation, which must show the decay that we have been discussing. In other words, as a strictly formal result, the field must decay on the ohmic time scale. Now it is even easier to see the result by following the line of argument that Cowling introduced. (See Fig. 6.4.) We take the integral, around a streamline, of the current:

$$\oint \mathbf{J} \cdot d\mathbf{s} \neq 0 \tag{54}$$

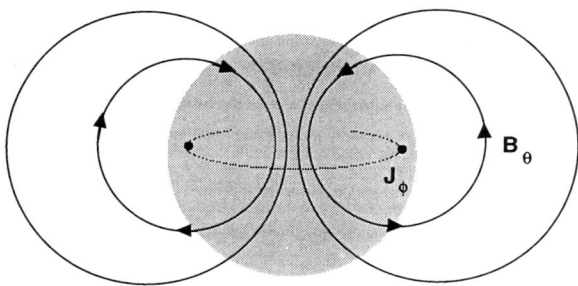

Figure 6.4 Illustration of the conditions for Cowling's theorem for an axisymmetric dynamo.

while we see that, from Ohm's law and Stokes' theorem, this is the same as

$$\oint \mathbf{J} \cdot d\mathbf{s} = \sigma \left(\int (\nabla \times \mathbf{E}) \cdot d\mathbf{S} + \int (\nabla \times (\mathbf{v} \times \mathbf{B})) \cdot d\mathbf{S} \right) \tag{55}$$

The second term vanishes at the O-type neutral point. The first term gives, from Maxwell's equations,

$$\oint \mathbf{J} \cdot d\mathbf{s} = -\frac{\sigma}{c} \frac{\partial}{\partial t} \int \mathbf{B} \cdot d\mathbf{S} \tag{56}$$

The integral over the surface is the flux. Therefore, since the line integral of the current does not vanish, there must also be a nonvanishing value of the time derivative of the magnetic flux—it must decay since the current integral is positive definite.

The discovery of this theorem has had a profound effect on thinking about magnetic field generation. It ensures that the motions responsible for the fields cannot be axisymmetric. Therefore some mechanism must be asserted for the breaking of this symmetry. It was Elasser who first suggested that the motions of convection and meridional circulation would be the best source of this motion. A feedback is then possible between these motions and the global magnetic field. Perhaps the easiest way of seeing this is that in the presence of nonaxisymmetric fluid motion, there is no longer an O-type neutral point. In fact, there is a very nice pictorial view of this process, championed by Parker, that allows an understanding of this process. Imagine that we have a field which is wrapped up by the rotational motion of the medium—specifically differential rotation. This is a way of converting the initially dipolar field from a symmetric configuration along meridional lines to one with a B_ϕ component. We now switch our view to this horizontal field. Convection cells rising through this set of field lines have a helicity due to the differential rotation. They therefore introduce a complex set of currents into the flow. By rising, they create a new field that has both B_r and B_θ components. These can then be pinched off, reconstructing after annihilation of opposite polarities into a global dipolar field. Depending on which of the individual cell helicities predominates in this stochastic process, the global field may have a different polarity than the one that existed at the start. This is then an attractive way of generating flips in the polarity as well. This is one of the essential features of celestial fields—especially those of the Earth and Sun. In the case of the terrestrial field the flip occurs quickly (of order one ohmic time scale), as revealed by the fossil fields preserved in the thermal remnant magnetism in terrestrial rocks. The field flips are apparently chaotic, in that it has thus far proved impossible to find a unique period to the phenomenon. In the solar case, however, the period is a bit more clearly defined. It is about twice the sunspot time scale.

It is important to note that the magnetic fields that have been observed in stars, that are not dipolar fields, are the ones which have been seen in convective stars. The RS CVn, BY Dra, and T Tau stars have deep convective envelopes and also show the best evidence yet observed for strong and complex magnetic structures (starspots and prominences). Solar analog stars on the main sequence are also seen only among stars with surface and envelope convection. It would seem, then, that Cowling's theorem points the way to an understanding of these observations—bodies that are differentially rotating and have convection are those that show strong, and variable, magnetic fields. Stable, radiative atmosphere and envelope stars do not show such phenomena and either are not magnetic or possess the relics of the pre–main sequence or even primordial magnetism with which they arrived on the main sequence.

6.7.2 A Simple Phenomenological Dynamo Model

We begin with the basic scenario. The picture is that initially we have a global dipolar magnetic field which is frozen into a differentially rotating, self-gravitating sphere—for want of a better word called a star. The first step is to use the *Ferraro* isorotation theorem, which states that

$$\mathbf{B} \cdot \nabla \Omega = 0 \tag{57}$$

or that the field is constant along lines of constant rotation. Notice that this is a magnetic version of the Taylor–Proudman theorem. It comes directly from assuming that $\mathbf{v} = r\Omega\hat{\phi}$ and Ω = constant; substituting this into $\nabla \times (\mathbf{v} \times \mathbf{B}) = 0$ (steady-state field) gives the result. This is another way of stating that the differential rotation wraps the poloidal field lines up, generating a *toroidal* field. (See Fig. 6.5.) The initial field has components B_r and B_θ, so that as the field is wrapped, there is a conversion of the field lines to a weak B_ϕ. We see that

$$\frac{B_\phi}{B_\theta} = \cos\phi \frac{d\phi}{d\theta} \tag{58}$$

The differential rotation is given by some law which relates the rotational frequency to the latitude. Following the observed relation for the Sun, Babcock suggested that one use

$$\Delta\omega = -\alpha \sin^2\theta \tag{59}$$

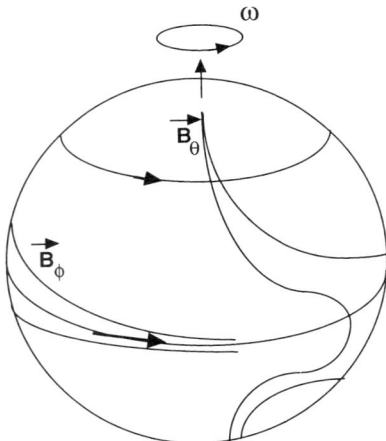

Figure 6.5 Schematic of the Babcock mechanism. A poloidal field is concerted into a toroidal one through differential rotation.

so that the time that it takes for the field line to wrap up on itself is

$$\left(\frac{B_\phi}{B_\theta}\right)_c = \alpha \Delta t \sin 2\theta \cos \theta \tag{60}$$

This model then provides a simple phenomenological machine for the calculation of the time scale for buildup of the field at any magnetic latitude. It is most important to recognize that this model is completely without *any* dynamical input. We have already argued that the generation of the magnetic field must be related to the occurrence of convection in the stellar envelope and that some turbulence is required to break the symmetry of the dynamo. Therefore, despite the apparent success of this simplified model, something more is clearly required.

6.7.3 Building Dynamos: Differential Rotation and Turbulence

The simplest way to imagine that a cosmic body will generate a magnetic field is if it is rotating. Since most plasmas have high conductivities, the amplification of an external field by collapse or compression (like shocks), coupled with rotation, should generate a strong Lorentz force. This field is not, however, supported for indefinite lengths of time. Instead, Cowling's theorem—as we have seen—states that no stationary (time-independent) axisymmetric dynamo is possible. But shear is essential to breaking the spherical symmetry.

6.7 Magnetic Dynamos

The symmetry is further broken, and the effect of the rotation translated into a poloidal field, through the combined action of circulation and turbulence. An initially axisymmetric field is sheared by differential rotation, and if it is initially cylindrical (B_z) or poloidal (B_r, B_θ), then an azimuthal field (B_ϕ) results. Here r and θ are the radius and latitude, respectively. A poloidal field results from a toroidal potential field, $\mathbf{B}_p = \nabla \times \mathbf{A}_\phi$, so that the toroidal magnetic field results from a distortion of the poloidal field. Finally, in order to convert the toroidal field back into a toroidal potential, some additional symmetry breaking is required. Turbulence in a rotating medium has vorticity, or handedness, which is parallel to the local angular velocity vector and neither radial nor even hemispherically symmetric.

In an electrically conducting fluid, buoyant turbulent cells produce a helical twist to the toroidal field and induce a poloidal conversion. This is the basis of the α–ω dynamo model. The electromotive force is schematically given by $\mathscr{E} \sim \alpha \mathbf{B}$, where α is related to the velocity correlation function and essentially measures the amplitude of velocity fluctuations in the fluid. We will discuss this in some detail shortly, but for now just notice that the assumption is that the turbulent fluctuations in the fluid velocity field drive buoyant motions (through the density fluctuations or temperature) and that these translate locally into some kind of electromotive force. In a fluctuating medium, the velocity breaks into a mean component, $\mathbf{V}$, plus a fluctuating part, $\mathbf{u}$ (which has a vanishing mean value but for which $\langle u^2 \rangle$ does not vanish). Here the brackets $\langle \ \rangle$ represent ensemble averages over the turbulent spectrum of the eddies in the fluid. The magnetic field evolution depends on both the mean field $\mathbf{B}_0$ and the fluctuating part $\mathbf{b}$ and the dynamo equation becomes

$$\frac{\partial \mathbf{B}_0}{\partial t} = \nabla \times (\mathbf{V} \times \mathbf{B}_0) + \nabla \times \langle \mathbf{u} \times \mathbf{b} \rangle + \eta \nabla^2 \mathbf{B}_0 \qquad (61)$$

where $\mathscr{E} = \langle \mathbf{u} \times \mathbf{b} \rangle = \alpha \mathbf{B}$. Thus α represents the fluctuations in the fluid and describes schematically the way that this feeds back into the magnetic field strength. The evolution of the fluctuating part of the field is given by

$$\frac{\partial \mathbf{b}}{\partial t} = \nabla \times (\mathbf{u} \times \mathbf{B} + \mathbf{V} \times \mathbf{b}) + \nabla \times (\mathbf{u} \times \mathbf{b} - \langle \mathbf{u} \times \mathbf{b} \rangle) + \eta \nabla^2 \mathbf{b} \qquad (62)$$

Then $\alpha \sim l_0 \langle u^2 \rangle / \eta$ and therefore depends on the velocity fluctuation spectrum. The turbulence therefore controls the small-scale structure, and the differential rotation (shear) controls the large-scale, ordered field and provides the symmetry breaking necessary to generating the dipole.

6.7.4 The Dynamo Number and Scaling Relations

A schematic estimate of the strength of the dynamo components, and an approximate scaling law, results from the quantitative side of this picture (see Fig. 6.6.). Differential field stretching causes poloidal-to-toroidal conversion, which takes place at a rate $\delta v_\phi/L$. Vortical motion of rising convective eddies transforms toroidal to locally poloidal field at a rate Γ, which is a pseudoscalar quantity whose sign depends on the hemisphere. The dynamo equations simplify by dimensional analysis. For the poloidal field, which is given by a vector potential field

$$\mathbf{B}_p = \nabla \times \mathbf{A}_\phi \rightarrow \frac{\Delta A_\phi}{L} \tag{63}$$

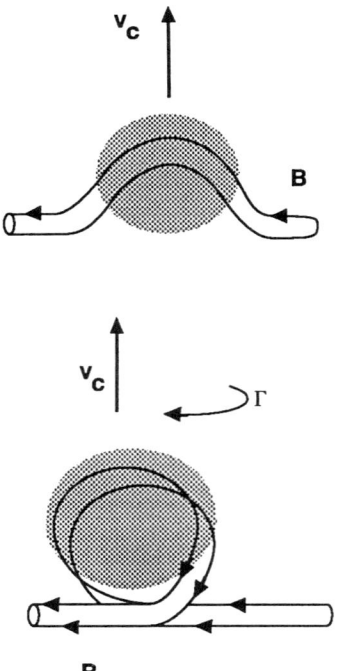

Figure 6.6 "Microphysics" of the α–ω dynamo mechanism, showing the wrapping of toroidal (horizontal) field lines to form poloidal component. Flux tubes are entrained by rising convective elements and helicity is introduced through the Coriolis effect through Γ.

6.7 Magnetic Dynamos

The rate at which poloidal field dissipates by turbulence and diffusion is, in the steady state, balanced by the conversion of toroidal to poloidal flux:

$$-\eta \nabla^2 \mathbf{A}_\phi \approx \Gamma \mathbf{B}_\phi \rightarrow -\eta \frac{\Delta A_\phi}{L^2} \sim -\Gamma B_\phi \qquad (64)$$

Finally, the poloidal field is wrapped to form the azimuthal field at a rate that depends on the shear $\Delta\Omega$

$$-\eta \nabla^2 \mathbf{B}_\phi \approx \nabla \times (v_\phi \times \mathbf{B}_p) \rightarrow -\eta \frac{B_p}{L^2} \sim \frac{\Delta v_\phi}{L} B_\phi \sim \Delta\Omega \, B_p \qquad (65)$$

The dimensionless number

$$N_D = \frac{\Gamma \Delta\Omega \, L^2}{\eta^2} \qquad (66)$$

called the *dynamo number*, serves as the scaling parameter for the generation by the α–Ω process. Notice that N_D is independent of the magnetic field and that in the steady-state case it is of order unity. For large values, the field will not be steady (what is usually meant by an active dynamo). Now notice what happens when we look at the dimensions of the quantities involved. The rate Γ has the dimension T^{-1}, the magnetic viscosity varies as $L^2 T^{-1}$, and if we assume that turbulence and buoyancy (convection, for example) provide a characteristic time scale τ_c, then $N_D \sim \Delta\Omega \, \tau_c$. So we have recovered the astronomer's version of the Rossby number! This is why there is so much interest currently in the correlation between dynamo activity and Ro.

Empirically, UV emission line strengths and L_{XR}, the luminosity of the x-ray continuum, correlate with Ro, although the precise law is still debated. However, the determination of an empirical scaling relation, drawn as it is from very indirect measures of the magnetic field strength and its rate of generation and structure, has not yet provided a serious constraint on the stellar dynamo theories. In general, the most that can be said from the measurement of activity in late-type stars is that a dynamo must be acting, that the field must be constantly emerging anew at the photosphere, and that the mechanism must depend on surface gravity, effective temperature, and rotational frequency.

6.7.5 The Basic α–ω Dynamo

The Babcock model is, of course, only schematic and phenomenological. Its chief purpose is to provide a framework within which certain physical processes can be recognized. Similarly, the idea that there is a relation between turbulence (vorticity) and magnetic field generation—the Batchelor analogy discussed earlier—is a very useful one. It forces one to recognize that the role of turbulence is central to the magnetic field generation problem. The combination of these two different mechanisms, rotation and turbulence, is known as the $\alpha-\omega$ dynamo. The α is the turbulence parameter (as we shall soon see) and ω is the rotation.

Rotation must play a role in magnetic field generation. This can be asserted on several counts. First, only the rapidly rotating main sequence stars of late type exhibit the effects of coronal heating and chromospheric activity that one expects to be associated with active dynamos. Specifically, the x-ray flux from late-type main sequence stars with high rotational velocities (and therefore high frequencies) is systematically higher than for the ones which rotate at about the rate of the Sun. The second point is that the differential rotation seen in sunspots appears to be associated with the rate of shear of active regions and that the largest spots occur on the systems which are most nearly rigidly rotating.

The stars for which turbulence plays any role at all are precisely those in which we observe stellar activity. Late-type stars possess strong convection, which dominates the structure of the envelope. The turbulence thus generated helps to break the symmetry of the rotation and thereby gets around Cowling's theorem. The stars which have deep convective zones, which nonetheless extend to the surface, are precisely the ones for which the dynamo is expected to have the greatest effect (spectral types late F to mid K). However, the gravity must come into play only weakly because there are several classes of stars that have strong x-ray emission and indications of activity which are members of binary systems. They otherwise have the same properties of turbulence and rotation as their main sequence counterparts. These include the RS CVn stars and the main sequence analogs, the BY Dra stars.

The model is quite similar conceptually to that discussed by Babcock but was outlined in some detail first by Parker in 1955. Beginning with a global field of some dipolar-like character, the field is wrapped up by differential rotation. The radial and meridional field is consequently converted to a toroidal field. So far, this is essentially the same as the phenomenological model. However, here comes the departure. As a blob rises, it moves radially because of gravity. However, the rotation produces a Coriolis force which acts at an angle to the direction of rise, thereby introducing a helicity to the field. Put another way, there is a net twist applied to any rising blob which contains the field trapped within it. The field will reconnect and thereby convert again from a toroidal configuration

6.7 Magnetic Dynamos

back to a poloidal one. This is the new kinematic feature of the model. We can therefore see that, unlike the simple phenomenological models, here we have to include the effects of diffusion and of the rotation for the toroidal and poloidal fields. The model will thus require two coupled differential equations. We assume that we can represent the poloidal field by a potential A_ϕ. The equation which describes the evolution of this field is one that contains the creation terms by currents and the destruction terms which are due to the generation of the toroidal field.

Now we can add some quantitative details. The fluid motion is assumed to consist of two parts, a deterministic or global flow field **u** and a random component **v**. The latter is assumed to be driven by some mechanism, presumably thermal convection, but may also be due to shear instabilities. The precise details do not really matter for this discussion. A random fluctuation of the velocity having a mean of zero does not necessarily have a vanishing dispersion [that would require that the velocity distribution function $g(v)$ be a delta function, which would be a very special case indeed]. It is too difficult to write down the equations of motion *and* the electrodynamic equations and obtain a closed-form solution for anything but the simplest case—that of a purely diffusive dissipation to the field—so it becomes necessary to resort to approximations in order to obtain a result. The velocity field is assumed to consist of two parts, a mean motion (which may be due to rotation, for instance, as in the Babcock dynamo) and a stochastic component whose origin is magical. That is to say, at the moment we simply assume that the random motions have some source whose detailed driving conditions may be unknown or may be imposed *ad hoc*. Then $\mathbf{u} = \mathbf{V} + \mathbf{v}$, where $\langle \mathbf{u} \rangle = \mathbf{V}$ and $\langle \mathbf{v} \rangle = 0$. Similarly, we can assume that there is a separation between the mean and random components of the magnetic field, which we shall label $\mathbf{B}_0$ and **b**, with $\langle \mathbf{b} \rangle = 0$. The dynamo equation can be written as

$$\frac{\partial \mathbf{B}_0}{\partial t} = \mathbf{\nabla} \times \mathbf{V} \times \mathbf{B}_0 + \mathbf{\nabla} \times \langle \mathbf{u} \times \mathbf{b} \rangle + \eta \nabla^2 \mathbf{B}_0 \tag{67}$$

The first term is the electromotive force, the EMF so familiar from elementary physics classes. It couples the velocity field, hence the equation of motion, with the Maxwell equations. Now define $\mathscr{E} = \langle \mathbf{u} \times \mathbf{B} \rangle$, and define a secondary fluctuating quantity which is the difference between the instantaneous EMF and the mean field electromotive force:

$$\mathscr{G} = \mathbf{u} \times \mathbf{b} - \langle \mathbf{u} \times \mathbf{b} \rangle \tag{68}$$

which is assumed to be a second-order quantity. We shall return to this point in a moment. The mean field, $\mathbf{B}_0$, can be separated out for the moment as being of zeroth order so that:

$$\frac{\partial}{\partial t}\mathbf{b} = \nabla \times (\mathbf{V} \times \mathbf{b} - \mathbf{u} \times \mathbf{B}_0) + \nabla \times \mathcal{E} + \eta \nabla^2 \mathbf{b} \tag{69}$$

Consider that the electromotive force, $\mathcal{E}$, is connected with the magnetic field and its derivatives. This is the basis of the approximation which is usually referred to as the α-dynamo. Here we say that

$$\mathcal{E}_i = \alpha_{ij} B_j + \beta_{ijk} \frac{\partial B_k}{\partial x_j} + \cdots \tag{70}$$

where we will assume that α and β are constants. To give you an idea of what these represent, imagine the twisting of a preexisting magnetic field via turbulence. The individual flux ropes can be distorted in three dimensions, and if they twist appropriately then amplification will occur. If oppositely directed fields combine because of viscous coupling, the mean field is decreased. Again, the analogy with vorticity is central to the process. Now we can continue with the step for finding the mean field created by this process.

Let's take the first-order term, which says that the velocity fluctuations directly generate the EMF, substitute the α term, and neglect the fluctuating terms. We take the Fourier components of this equation to get

$$(-i\omega + \eta k^2)\mathbf{b} + i\alpha \mathbf{k} \times \mathbf{b} = i\mathbf{k} \times (\mathbf{V} \times \mathbf{b} + \mathbf{u} \times \mathbf{B}_0) \tag{71}$$

Instead of solving the real problem, we pull the usual astrophysical ruse and solve a new one. Take the magnetic field evolution to be given by a fluid which is simply turbulent, where there are no mean motions. Also, assume that the EMF will be given by $\mathbf{v} \times \mathbf{B}_0$. Then, from the condition that $\nabla \cdot \mathbf{B} = 0 \rightarrow \mathbf{k} \cdot \mathbf{b} = 0$, the dynamo equation becomes

$$\frac{\partial \mathbf{b}}{\partial t} = \mathbf{B}_0 \cdot \nabla \mathbf{u} + \eta \nabla^2 \mathbf{b} \tag{72}$$

and taking the Fourier components, assuming that there is no mxing between the components in wave number, we obtain

$$(-i\omega + k^2\eta)\tilde{\mathbf{b}} = -i\mathbf{k} \cdot \mathbf{B}_0 \tilde{\mathbf{u}} \tag{73}$$

where $\tilde{\mathbf{b}}$ and $\tilde{\mathbf{u}}$ are the Fourier components of the magnetic field and velocity perturbations. Again, these are *random variables*, and they may be discontinuous (that is, not strictly differentiable), so that this is a tricky procedure and must be justified later (see the discussion in the turbulence chapter). We therefore have a unique representation for the magnetic

6.7 Magnetic Dynamos

field:

$$\mathbf{b} = \frac{1}{2\pi^4} \int_{\infty}^{\infty} \int_{-\infty}^{\infty} \tilde{\mathbf{b}}(\mathbf{k}, \omega) e^{i(-\omega t + \mathbf{k} \cdot \mathbf{x})} \, d\mathbf{k} \, d\omega \equiv \mathcal{F}(\tilde{\mathbf{b}}) \tag{74}$$

where $\mathcal{F}$ denotes the Fourier transform for compactness. This becomes

$$\mathbf{b} = \mathcal{F}\left(-\frac{i(\mathbf{k} \cdot \mathbf{B}_0)}{-i\omega + k^2 \eta} \tilde{\mathbf{u}}(t, \omega)\right) \tag{75}$$

You see that the power spectrum for the magnetic fluctuations is connected with the power spectrum of the velocity fluctuations, so that we can schematically assert that $\langle b_i b_i \rangle \sim \int k^2 \Phi_{ii}(k) \, dk$. You will notice here that the integral is actually the helicity energy spectrum, which is the heart of the matter. The fluctuations in the local helicity drive the amplification process in the α-dynamo. This means that locally we can amplify the field by twisting the magnetic filaments, ignoring the dissipative effects. The role played by the magnetic viscosity is to damp out the growth of the small-scale structures, thereby increasing the ordered larger-scale component of the magnetic field.

It is important here to note that the problem has been rendered linear, so that each component of the magnetic field and the velocity can be followed separately and the assumption of a single component in wave number does not critically enter into the discussion. The same cannot, however, be said for the velocity field. You will notice that we have not written down an equation for $\mathbf{v}$ and for good reason: there is no reason for believing that this is a linear problem. Remember that the fluctuations in the momentum are the result of both time variations and spatial variations in the velocity. The gradient terms in the equations of motion are the culprits responsible for the intrinsic nonlinear evolution of the turbulence spectrum. More on this point later.

The EMF can be calculated from the Fourier amplitudes of the velocity and magnetic field separately and then forming the cross product of the transforms:

$$u_j = \int e^{-i(\omega' t - \mathbf{k}' \cdot \mathbf{x})} \tilde{u}_j \, d\mathbf{k}' \, d\omega' \tag{76}$$

so that

$$\epsilon_{ijk} \langle u_j b_k \rangle = \iint d^3k \, d^3k' \, d\omega \, d\omega' \, \epsilon_{ijk} \langle \tilde{u}_j \tilde{b}_k \rangle$$
$$\times \exp[-i(\omega + \omega')t + i(\mathbf{k} + \mathbf{k}') \cdot \mathbf{x}] \tag{77}$$

Substituting the $\tilde{b}_i$ from the transform of the dynamo equation, we have in schematic form

$$\mathcal{E} = \langle \mathbf{u} \times \mathbf{b} \rangle$$

$$= \int \frac{i \langle \mathbf{u}(\mathbf{k}, \omega) \times \mathbf{u}^\star(\mathbf{k}', \omega') \rangle \mathbf{B}_0 \cdot \mathbf{k}}{-i\omega + \eta k^2} e^{i((\mathbf{k}-\mathbf{k}')\cdot\mathbf{x} - (\omega-\omega')t)} \, d\mathbf{k} \, d\mathbf{k}' \, d\omega \, d\omega' \tag{78}$$

The correlations in the kinematic helicity fluctuations are directly related to the correlation tensor [Φ_{ij} is the Fourier transform of the velocity correlation function, $R_{ij}(\mathbf{r})$, as discussed in the chapter on turbulence]. Thus the velocity correlation, which may be either positive or negative, controls the generated magnetic field. Put dimensionally, the small-scale fluctuations in velocity which have relatively large vorticity are the ones responsible for the generation of new magnetic flux. The voriticity therefore enters critically into the model for the dynamo and in the absence of the large-scale mean motions serves as the sole means for generating new magnetic flux. This is the basis of the α-dynamo. Finally, noting that $\epsilon_i = \alpha_{ij} B_{0,j}$ we obtain

$$\alpha_{ij} = i\eta \epsilon_{imn} \iint \frac{k^2 k_m k_j \Phi_{mn}(\mathbf{k}, \omega)}{\omega^2 + k^4 \eta^2} \, d\mathbf{k} \, d\omega \tag{79}$$

By virtue of the fact that the EMF is real, the correlation function must be complex (there must be phase shifts among the velocity fluctuations). We can therefore estimate that the α parameter is of order $v_t^2 L$. Notice that this is independent of any large-scale ordering in the medium. The helicity we have in this formulation is purely local, derived from the shear within the turbulent velocity field. The parameter α is actually a pseudotensor, in that it changes sign on rotation of the coordinate system. This is what we have imposed on it from the start—not something that comes out of the formalism—because one needs a way of generating the magnetic field in a monopolar model without the imposition of a large-scale rotation (which serves to break the symmetry). Put another way, one needs a way of generating oppositely directed magnetic fields in the two hemispheres by some otherwise homogeneous turbulent mechanism without having rotation provide a hemispherical perturbation. The velocity perturbation is related to the cross-correlation function R_{ij} through the Fourier components of the equation of motion.

Now consider what happens in a rotating system. The effect of rotation, aside from the symmetry breaking, is to introduce a new degree of freedom into the dynamic equations and also to produce a mechanism for ordering small-scale structure generated by the α process. Any turbulent region now develops a helicity which depends on the hemisphere in which it arises, and there is also a net current generated by Ω, the rotational frequency. Consider the generation term $\nabla \times (\mathbf{v} \times \mathbf{B})$. For a steady-state system, we see that this term is equivalent to $\mathbf{B} \cdot \nabla\Omega = 0$; Ferraro's isorotation theorem shows up again. The strong coupling between the magnetic field and the rotation enforces the condition that rotation is constant along field lines. For differential rotation to be maintained, this implies that the magnetic field cannot remain in steady state.

6.7.6 Some Examples of Dynamo Fields

Planetary Magnetic Fields as Tests for Dynamo Mechanisms

Planetary magnetic fields are the only ones for which it is possible to obtain direct *in situ* measurements of the detailed field configuration and time variability. The terrestrial field is probably the best studied. It displays complex time-variable structure, most importantly the long-term global polarity reversals on time scales between 10^4 and 10^6 years. These appear to be random and have been modeled as chaotic. The Earth's dipole field maintains its same orientation, inclined to the rotational axis by about $11°$, but its higher-order moments vary with different time scales and the geometry of the field changes slowly as a result. For scaling purposes, the Earth's magnetic dipole moment is 7.9×10^{30} G cm^3 and a polar field strength of about 0.6 G.

With the exception of Mercury, Mars, and Venus, all of the planets which have been studied by spacecraft have strong magnetic fields. Neptune and Uranus have highly inclined fields, similar to those seen in the chemically peculiar stars, and also have strong nondipole components. The Jovian field is about 4 G, it has a dipole moment about 500 times that of the Earth, and it is inclined similarly (about $9\frac{1}{2}°$). Saturn is the anomaly. Although its field is about the same strength as Earth's (0.2 G) and its magnetic dipole moment is 4.6×10^{28} G cm^3, it is almost perfectly aligned (about a $1°$ inclination to the rotation axis). It is also interesting to note that no changes in the field configuration or strength were detected for either Jupiter or Saturn between the Pioneer and Voyager flybys of these two planets. The ground-based radio measurements of Jupiter, while not very sensitive to the precise field strength, are also consistent with little change in the dominantly dipole geometry over the past decade. It will be interesting to compare the previous *in situ* measurements of the Jovian field with those from the Galileo probe, since these will sample a time scale of nearly 20 years. Uranus and Neptune have similar magnetic fields, both being highly inclined ($>40°$) and with polar field strengths of order 0.2 G. The extreme offset required for the Neptune field, discovered by Voyager 2 in the 1989 flyby, makes this planet an especially interesting test case for dynamo models and seems to indicate the presence of a complicated field geometry, more complex than observed for any of the other planets.

The absence of a strong field on Venus, despite its otherwise terrestrial bulk properties, is probably consistent with the dynamo mechanism. The planet rotates about 250 times more slowly than the Earth. Mercury rotates slowly and is too small to support a strong convective core, but it does have a very detectable dipole moment of 2.4×10^{22} G cm^3. Its field is very small, about 0.002 G, and is nearly aligned and probably a relic from the earlier stages of planetary evolution. Mars rotates with nearly the same period as the Earth but it is smaller and may support only a very small convective core. Mars has displayed volcanic activity in the past, evidence for core or mantle convection, so the planet may possess a very weak intrinsic magnetic field. The Phobos and Viking mission results are, however, still equivocal on this point.

Stellar Dynamos

Polarity reversals were observed in the Sun in the first decades of this century, long before they were recognized in the Earth. The explanation for the reversal of the terrestrial field, therefore, lends support to the dynamo explanation for the magnetic fields in late-type stars and serves to link the study of dynamo processes across a large scale of mass and size of cosmic objects. There are several classes of stars that appear to be especially magnetically active. T Tauri stars, pre–main sequence objects, have been detected as radio emitters and also periodic light variables. The prevailing model explains the periodicity in terms of large-scale spot regions, and UV observations indicate that many of these stars show enhanced chromospheric and coronal activity. The periods are characteristically short, of order a few days, and this indicates that the most rapidly rotating stars in this class show the dynamo at maximum. Similar behavior is noted in the later stages of evolution of some close binaries, the RS CVn and related stars. These stars have deep convective envelopes that are usually found in binary systems with periods shorter than a month, often shorter than one week. The rotation is assumed to be nearly in synchronism with the binary period, and the enhanced activity derives from the comparatively rapid rotation of these stars compared with evolved single field stars. They are, in effect, T Tau stars evolving in the reverse direction across the Hertzsprung–Russell diagram. The dMe stars are low-mass, nearly fully convective, main sequence stars that show enhanced radio and optical flaring activity and also rotationally modulated light output. The W UMa class are contact binary systems, existing within a common envelope of nearly uniform temperature. These stars are on or near the main sequence and show some evidence for spot cycles and activity and also strongly enhanced UV signatures of chromospheres. Finally, a more enigmatic group, the FK Com stars, also show evidence for strong dynamo activity, but these appear to be single evolved stars. The magnetic dynamo phenomenology appears to conform with expectations that only stars with deep convective envelopes and unusually rapid rotation display the evidence of magnetic field generation. But the evidence is still indirect. Most magnetic fields have been measured by using differential line broadening depending on the intrinsic Zeeman pattern. The technique, pioneered by Durney and Robinson (1982), permits estimates of magnetic field strengths and fractional hemispherical coverage. But to date, polarization measurements analogous to those for the Ap and Bp stars (which show only steady-state surface fields) have not been achieved.

Galactic-Scale Dynamos

The formation of a large-scale field by dynamo action requires the conversion of toroidal magnetic fields to poloidal configurations. The presence of turbulence and differential rotation on the galactic scale and the requirement that these act in concert in stars to produce the observed fields thus invite a comparison between the two environments. The length scales are much larger for galactic-size field structures, and the time scales over which they are generated are also longer. Thus the scaling of the dynamo process to the interstellar medium as a whole does not seem an outlandish idea and has been attempted by a number of groups (see Sofue et al. 1986; Zel'dovich et al. 1983; Zweibel 1988). Most of the details of the problem are still to be understood, but in broad outline the galactic dynamo is largely indistinguishable from a stellar one; the notable differences are the planarity of the differentially rotating disk and the nonthermal nature of the turbulence and buoyancy of the medium. These models have the advantage that α can be determined from the turbulence of the diffuse interstellar medium and that Ω is specified by differential galactic rotation.

6.8 Magnetic Reconnection

Magnetic reconnection is presumed to occur when regions of oppositely directed polarity are brought into contact by turbulent motions (Fig. 6.7). This mixing results in the annihilation of the field within a small volume and the transformation of magnetic energy into kinetic energy through the generation of extremely strong local electric fields within very small (of order 100–1000 km) length scales. The rate at which the field destruction

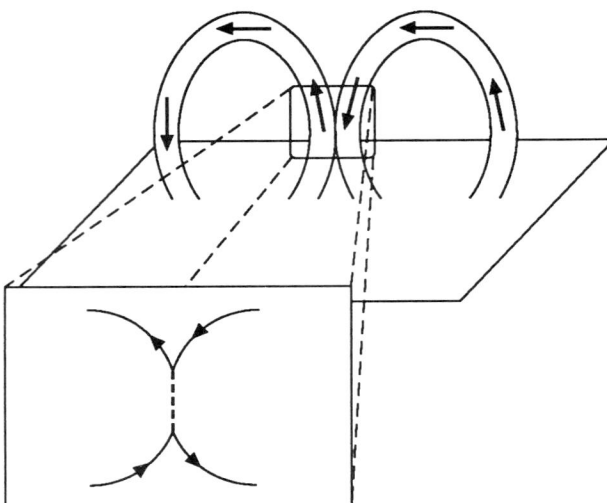

Figure 6.7 Example of the basis of magnetic-field reconnection in a solar environment. The merging of magnetic loops produces an x-type neutral point with the consequent collapse of the current sheet and dissipation of local magnetic field.

takes place is a matter of debate and significantly affects estimates of the heating of the confining plasma. The fastest time scale, the Alfvén wave crossing the time, appears to be too short (microseconds) to explain the acceleration observed in solar flares. These require an extended period of electric field amplification during the annihilation phase. Theoretical particle simulations suggest that strong MHD turbulence is probably present during the reconnection phase of magnetic field line merging and that the electrons are energized through being scattered off of these waves. Reconnection processes have been implicated in the acceleration of particles in the terrestrial magnetotail and in the generation of disconnection events in comets, when the plasma tail of a comet separates from the coma when it crosses regions in the solar wind of oppositely directed magnetic polarity (sector boundaries). There is strong evidence, from fast observations of stellar flares, that the frequency dependence and energetics of large stellar active regions mimic those observed on the Sun, where reconnection is more secure. Finally, it may be possible to see field line merging in radio lobes and jets of external galaxies. Large-scale filamentation may be associated with this same instability.

Appendix: Rikitake's Toy Phenomenological Model

Rikitake (1958) has created one of the most successful models for a stochastically stable but oscillating dynamo. The model envisions the interaction, in opposite hemispheres, between two current loops which are maintained through the effects of rotation. The currents are strictly toroidal. The schema is to look at two coupled rotating conducting disks, rotating with frequencies ω_1 and ω_2. Imagine that the coupling occurs using two loops, which are connected as shown in Fig. 6.8. The capacitance and inductance, C and L, and mutual inductance $M_{12} = M_{21}$, give the coupled system

$$L\frac{dI_1}{dt} = -RI_1 + M\omega_1 I_2 \tag{80}$$

$$L\frac{dI_2}{dt} = -RI_2 + M\omega_2 I_1 \tag{81}$$

$$C\frac{d\omega_1}{dt} = G - MI_1 I_2 \tag{82}$$

$$C\frac{d\omega_2}{dt} = G - MI_1 I_2 \tag{83}$$

The term G is schematically included as the external forcing for the system, because we have in the capacitance term the force required to maintain the system against dissipation.

We will take the time for the analysis of this system because it can be easily explored numerically, because it illustrates most of the phenomenology required for any *realistic*

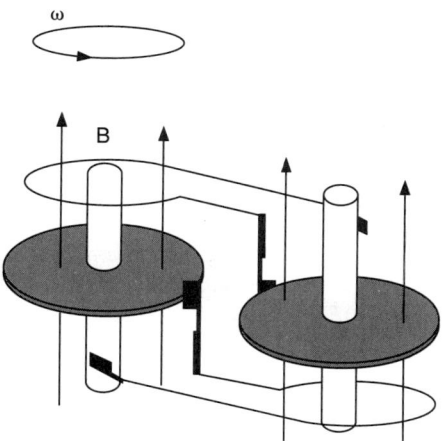

Figure 6.8 A schematic self-induction dynamo imbedded in an external magnetic field, $\mathbf{B}_0$.

dynamo model, and because it can be studied analytically. There are not many such models and we should grab them whenever they cross our path. Define the following dimensionless variables: $\tau = (MG/CL)^{1/2} t$ for the time, $\mu = (C/GML)^{1/2} R$ for the resistance, $X_j = (M/G)^{1/2} I_j$ for the current, and $Z_j = (MC/GL)^{1/2} \omega_j$ for the rotation rates. Then

$$\dot{X}_1 = -\mu X_1 + Z_1 X_2 \tag{84}$$

$$\dot{X}_2 = -\mu X_2 + Z_2 X_1 \tag{85}$$

$$\dot{Z}_1 = 1 - X_1 X_2 \tag{86}$$

where $\dot{Z}_1 = \dot{Z}_2$. This is a third-order system, and you may be aware of the probability that such systems show both oscillatory phenomenology *and* chaos. For the steady-state solutions in this much reduced system, we find that $X_1 X_2 = 1$ and $Z_1 + Z_2 = c$, where c is a constant. Also, $X_1^2/X_2^2 = Z_1/Z_2 \rightarrow X_2 = (Z_1/Z_2)^{-1/4}$. Now expand the solution about these fixed points by assuming that we have small departures from this equilibrium state. Take for each dynamical variable $\zeta = \zeta_0 + \eta$ where η is a small change from the stationary state ζ_0. Assuming that each $\eta \sim \exp \lambda t$, we have a cubic for λ. This means that multiple solutions are possible, and numerical integrations of this equation display the same behavior as observed in chaotic systems—the dynamo reverses sign at unpredictable (random) intervals in spite of the fact that the system is deterministic.

References

Asseo, E., and Sol, H. (1987). Extragalactic magnetic fields. *Phys. Rep.* **148**, 307.
Borra, E. F., Landstreet, J. D., and Mestel, L. (1982). Magnetic fields in stars. *Annu. Rev. Astron.* **20**, 191.
Cowling, T. G. (1957). *Magnetohydrodynamics*. New York: Interscience.
Cowling, T. G. (1981). The present status of dynamo theory. *Astrophys.* **19**, 115.
Cram, L., (ed.). (1990). *The F, G, and K Stars and T Tauri Stars* (NASA SP-502).
Dressler, M., (ed.). (1983). *The Jovian Magnetosphere*. Cambridge: Cambridge University Press. [See also Belton, M., West, R. A., and Rahe, J., (eds.) (1989). *Time-Variable Phenomena in the Jovian System*. (NASA SP-494).]
Durney, B. R., and Robinson, R. D. (1982). On an estimate of dynamo-generated magnetic fields in late-type stars. *Astrophys. J.* **253**, 290.
Hartmann, L., and Noyes, R. (1987). Rotation and magnetic activity in main sequence stars. *Annu. Rev. Astron. Astrophys.* **25**, 271.
Heiles, C. (1987). Interstellar magnetic fields. In *Interstellar Processes* (p. 171). D. J. Hollenbach and H. A. Thronson Jr. (eds.). Dordrecht, The Netherlands: Reidel.
Hones, E. W. (ed.). (1984). *Magnetic Reconnection in Space and Laboratory Plasmas*, AGU Monograph. Washington, D.C.: American Geophysical Union.
Jordan, S. (ed.). (1981). *The Sun as a Star*. NASA SP-450.
Krause, F., and Raedler, K. H. (1980). *Mean Field Magnetohydrodynamics and Dynamo Theory*. Oxford: Pergamon Press.
Mestel, L. (1985). Magnetic fields. In *Protostars and Planets*, Vol. II (p. 320). D. Black and M. S. Shapley (eds.). Tucson, Ariz.: University of Arizona Press.
Moffatt, H. K. (1978). *Magnetic Field Generation in Electrically Conducting Fluids*. Cambridge: Cambridge University Press.
Montgomery, D. (1989). Magnetohydrodynamic turbulence. In Lecture Notes on Turbulence: NCAR-GTP Summer School. J. R. Herring and J. C. McWilliams, eds. Singapore: World Scientific Publishers.

Mouschovias, Th. (1978). Formation of stars in planetary systems in magnetic interstellar clouds. In *Protostars and Planets* (p. 209). T. Gehrels (ed.). Tucson, Ariz.: University of Arizona Press.

Parker, E. N. (1979). *Cosmical Magnetic Fields*. London: Oxford University Press.

Priest, E. R. (1984). *Solar Magnetohydrodynamics*. Dordrecht, The Netherlands: Reidel.

Rikitake, T. (1958). Oscillations of a system of disk dynamos. *Proc. Cambr. Phil. Soc.* **54**, 89.

Schussler, M. (1979). Magnetic buoyancy revisited: Analytical and numerical results for rising flux tubes. *Astron. Astrophys.* **71**, 79.

Sofue, Y., Fuzimoto, M., and Wielibinski, R. (1986). Global structure of magnetic fields in galaxies. *Annu. Rev. Astron. Astrophys.* **24**, 459.

Spitzer, L., Jr. (1957). Influence of fluid motions in the decay of an external magnetic field. *Astrophys. J.* **125**, 525.

Taylor, J. B. (1986). Relaxation and magnetic reconnection in plasmas. *Rev. Mod. Phys.* **58**, 741.

Vasyliunas, V. M. (1975). Theoretical models of field line merging. I. *Rev. Geophys. Sci.* **13**, 303.

Zel'dovich, Ya. B., Ruzmaikin, A. A., and Sokoloff, D. D. (1983). *Magnetic Fields in Astrophysics*. New York: Gordon & Breach.

Zweibel, E. (1988). The growth of magnetic fields prior to galaxy formation. *Astrophys. J. (Lett.)* **329**, L1.

CHAPTER 7

Turbulence

> ... the steady motion was unstable for large disturbances long before the critical velocity was reached, a fact which agreed with the full-blown manner in which the eddies appeared.
> O. Reynolds, 1883

7.1 Introduction

Few problems have so challenged the ingenuity of physicist and engineer alike as that of turbulence. Its origin, development, and detailed analytic description have defied all but the most qualitative of treatments and there are reasons for believing that this will be the case for some time in the future. It is for this reason that all but the simplest treatment of the theory will be eschewed in this book and that all but the most general treatment of the problem has been overlooked by the astrophysical community in general. Specifically, we will treat the problem of the general description of the spectrum of homogeneous turbulence, the so-called Kolmogorov and Heisenberg spectra, and then treat some of the applications of turbulence to the structure of astrophysically interesting flows. In the case of the latter, there will be some overlap with the discussion in the vorticity chapter, but this should be useful in serving as a guide to the treatment of a specific problem using the more general methods of this chapter.

It is assumed that you have at some time experienced a bumpy ride in an airplane. This is perhaps the best (and for that matter *worst*) introduction to the problem of boundary-free or *clear air* turbulence. The treatment of such problems has been central both to theoretical understanding and to the problem of air traffic safety. The turbulence is usually encountered in the situation of temperature inversions, or of fronts which are crossing a region quickly, without the possibility of an equilibrium being reached. There is usually considerable energy input, of a kind which generally involves thermally generated waves, which will transport momentum across large distances. This in turn generates a velocity field which has large fluctuations from one point to another in the flow and which therefore shows some of the characteristics of chaos. This is the essential feature of turbulent flows, that there is little or no (in the case of *fully developed* turbulence) correlation between the various components in the density and velocity fields in the flow.

The best descriptions of the theoretical problem can be found in the monographs by Batchelor (1953), Tennekes and Lumley (1974), and Hinze (1975). Also, for the case of boundary layers see Schlichting (1966); these you will find in the Bibliography. The discussion in this chapter will draw heavily on the basic properties of stochastic functions, for which there is a very schematic outline in the appendix to the chapter, but it is assumed that you have some idea of probability theory, probably what has been thrown at you at some stage in a statistics course. If not, Bartlett (1965) and Gardiner (1983) are good books to consult.

7.2 Astrophysical Environments

The scales on which we find turbulence appearing in the cosmos are manifold and complex. On the size of the smallest bodies, planets, it plays a role second to none in the structuring of the atmosphere, being responsible for the transport of heat and momentum over most of the surface. In the case of the terrestrial planets, it is central to an understanding of small-scale structure, while in the Jovian planets, it is of so large a scale that it can be studied even by remote sensing. On the very smallest scale, it is responsible for the twinkling of starlight and the distortion of images of celestial objects due to the phase incoherence it introduces. On the scale of the interplanetary and interstellar medium, it causes the flickering of radio sources through basically the same mechanism. On the scale of a stellar atmosphere, it is a component in any proper theory of convective energy transport, a problem which also has implications for the internal structure of stars. Convection in high-viscosity environments has recently been discussed for the mantle of the terrestrial planets, and turbulence may also play a role in the structuring of the continents and in plate tectonics. On the scale of the interstellar medium, turbulence appears to occur in molecular clouds and plays an important role in star formation. On the metagalactic scale, it is the chaotic environment following the Big Bang in which galaxies formed, and thus turbulence plays a role in the formation of the largest self-gravitating objects that we observe—clusters of galaxies.

From the scale of the laboratory to that of clusters of galaxies is a span of scale of order 10^{25}, making turbulence one of the most important physical processes known in physics and ubiquitous in any continuous, fluidlike medium. Understanding its origin and development is, therefore, a major task of both the laboratory engineer and astrophysicist, since any theory can be tested over an enormous range of physical conditions. This chapter will merely outline the skeleton of the theory, which is still after nearly a century of concerted effort only in its infancy. It is fair to say that, like the weather it is so important in determining, everyone talks about turbulence but few do anything (or know anything) about it.

7.3 Incompressible Turbulence

7.3.1 Introduction

A word of warning to start off our discussions: we will concentrate on incompressible turbulence and only later generalize to compressible. Although it would appear unrealistic for astrophysical applications, there are very cogent reasons for this step. First, most laboratory studies have been with fluids like water, normal terrestrial liquids. Second, much of your experience is built up with such fluids. Third, it is actually possible to get some results if we make the assumption that the flow is incompressible. You see, in this case we only have to worry about moving vorticity around within the fluid, shifting the velocity field, and dissipating random kinetic energy. We don't have to worry about effects arising from the equation of state, radiation, or internal collapses and filamentation flows. The world is just too complicated for us to consider all of its properties, at least not at first.

So let's start with the the Navier–Stokes equation to describe the dynamics of the fluid. Taking the usual representation for the flow to be that of an incompressible fluid, the continuity equation simplifies to an algebraic constraint on the velocity field, namely that

$$\nabla \cdot \mathbf{u} = \frac{\partial u_i(\mathbf{x})}{\partial x_i} = 0 \tag{1}$$

The flow is assumed to be in steady state and the velocity field to be a function only of $\mathbf{x}$. The equation of motion for the fluid is

$$\frac{\partial u_i}{\partial t} + u_j \frac{\partial u_i}{\partial x_j} = -\frac{1}{\rho} \frac{\partial p}{\partial x_i} + \nu \nabla^2 u_i \tag{2}$$

7.3 Incompressible Turbulence

which, on taking the divergence of the two sides, becomes

$$\nabla^2 p = -\frac{\partial^2 \tau_{ij}}{\partial x_i \partial x_j} \tag{3}$$

Note the minus sign here. For incompressible turbulence, changes in the pressure do not produce density variations; they push fluid around and only alter the velocity field. We have made use of the continuity equation and the constancy of density, ρ, and written $\tau_{ij} = \rho u_i u_j$ as the kinetic portion of the stress tensor, T_{ij}. We shall return to this equation later; it is enough to note for the moment that it is an inhomogeneous equation of second order, which can be solved by using a Green's function approach.

7.3.2 The Correlation Tensors and Representation of the Flows

We imagine that the flow in the fluid has a field which is a touch chaotic. This is only because the material has been assumed to be in a flow which has a fixed direction, but in which there may be eddies and waves which are not necessarily parallel to the mean velocity of the material. In other words, one of the signatures on which we shall be relying for the existence of turbulence is that there is a vorticity present in the fluid, although the *mean* value of the vorticity may vanish. In fact, it is the fact that the rms value of ω does not vanish that gives rise to the basic signature of what is usually called fully developed turbulent flow. Now take the velocity field to have two basic components.

We assume that the velocity field can be separated into two parts, a mean and a random component, and that this separation can be effected for *any* component. Just as in the discussion of the Vlasov equation, we can write

$$v_i(\mathbf{x}) = U_i(\mathbf{x}) + u_i(\mathbf{x}) \tag{4}$$

where

$$\langle v_i \rangle = U_i, \quad \langle u_i \rangle = 0, \quad \langle u^2 \rangle \neq 0 \tag{5}$$

Here we have assumed that we have some quantity that controls the fluctuations. It may be that we are taking the mean value for the quantity over some volume, for some interval of time, or averaging over some velocity range. The point is that the averages are defined by the statistics and we generally need to choose some scale over which to average. Here we will generally mean spatial averages or time averages, assuming that we have taken either a large enough spatial sample or a long enough time sample to see what the mean condition really is. This is an important point to note and is embodied in the Taylor hypothesis, which states that for fully developed turbulence, the two averages are equivalent. It means that if we sit at a single point in the flow and watch it for a long enough time, its average properties are the same as those derived from a large spatial-scale snapshot at an instant in time. Now take the statistical average over two components i and j:

$$\langle v_i(\mathbf{x})v_j(\mathbf{x}')\rangle = U_i(\mathbf{x})U_j(\mathbf{x}') + R_{ij}(\mathbf{r}) \qquad (6)$$

We have used this to introduce the correlation tensor for a displacement $\mathbf{x}' = \mathbf{x} \pm \mathbf{r}$. This is the two-point correlation tensor for the velocity. It remains to show that this is related to the energy density of the turbulent velocity field and that the information contained in the two-point function is the important measure of the transport properties of a turbulent flow.

We define a point in space, call it $\mathbf{x}$, and around this point we draw a sphere of radius r. Now, take the simple picture of two points separated by a distance $\mathbf{r}$ and consider two components of the velocity field: one which is parallel to the line of centers, $u_l(\mathbf{x})$, and one which is transverse to that line, $u_t(\mathbf{x})$. Here l and t stand for longitudinal and transverse, respectively. We now *define* a quantity which is the correlation function averaged over the body of the fluid (so for all points $\mathbf{x}$):

$$R_{ij}(\mathbf{r}) \equiv \langle u_i(\mathbf{x})u_j(\mathbf{x}+\mathbf{r})\rangle \qquad (7)$$

where we have taken the displacement $\mathbf{r}$ to be a continuous function of distance. In addition, we assume that the function R_{ij} is a continuous function (since we have to assume, for the purposes of this discussion, that all inhomogeneities and discontinuities will be smoothed out in time within the medium. At infinite separation, the motion becomes uncorrelated so the function is normalizable. We define

$$\Phi_{ij}(\mathbf{k}) = \frac{1}{2\pi^3}\int R_{ij}(\mathbf{r})e^{-i\mathbf{k}\cdot\mathbf{r}}\,d\mathbf{r} \qquad (8)$$

to be the Fourier transform of the correlation tensor. Notice that if the correlation tensor can be normalized (and this is at this stage by no means certain) we have a way of determining the mean value of the separation between the correlated quanities, a characteristic length scale of the turbulence. This is just the integral of the trace of the correlation tensor:

$$R(r) \equiv \sum_i R_{ii}(r) \rightarrow \Lambda \equiv \int R(r)\,dr \qquad (9)$$

The time scale over which one would expect the smoothing to occur is a function of the Reynolds number. By dimensional analysis, you see that $t_{\text{decay}} \sim \text{Re}^{-1/2}$; also, since the Reynolds number measures the enstrophy, ω^2, this can be seen as a dissipation and redistribution time scale for that

7.3 Incompressible Turbulence

quantity. In the case of turbulent flows the Reynolds number is sufficiently high that this is short compared with the crossing time for the medium. Therefore, we can take the approximation that the function $R_{ij}(\mathbf{r})$ is continuous to at least second order.

If the turbulence is fully developed, there are several additional things we can say, even without a detailed knowledge of the physics. If the medium is also *isotropic*, the function R_{ij} will have to be quadratic in that only even powers of r can enter. This is easy to see from the fact that

$$R_{ij}(\mathbf{r}) = R_{ij}(-\mathbf{r}) \qquad (10)$$

and that therefore we can represent the correlation tensor as a function of only the unit vectors in the three orthonormal directions of the flow and of the radius r about any point in the flow:

$$R_{ij}(r) = F(r)r_i r_j + G(r)\delta_{ij} \qquad (11)$$

where we have taken the directions r_i and r_j to have unit vectors of the magnitude n_i and n_j. We further know that in the case of the correlation tensor, we can use the continuity equation to obtain

$$\frac{\partial R_{ij}}{\partial r_j} = \frac{\partial R_{ij}}{\partial r_i} = 0 \qquad (12)$$

The correlation tensor for zero lag is

$$R_{ij}(0) = \langle u_i u_j \rangle = \int \Phi_{ij}(\mathbf{k})\, d\mathbf{k} \qquad (13)$$

Thus, since this is a quantity which has the dimensions of an energy, we see that the energy is given by

$$\int E(k)k^2\, dk = \int \Phi_{ii}(\mathbf{k})\, d\mathbf{k} \qquad (14)$$

As a consequence of isotropy we have $\langle u_i u_j \rangle = 3\langle u^2 \rangle$. Here u is some velocity dispersion (we have not yet specified the nature of the statistical distribution of the velocity fluctuations). This is likely to be true at some scale of the fluid, although it will probably not be true at *every* scale. For example, the assumption should break down for a finite flow on the largest scales. But for the moment, let's press on. We can move the derivative under the ensemble average since we assume that the coordinate i is independent of the coordinate j. Thus, we can use the continuity equation in the form we see here as a measure of the divergence of the mean flow and also of the rate of energy dissipation (there we go again with that promise!).

Now, it is not enough to have defined these functions. We have to examine their qualitatively anticipated behavior. The transverse function is easy to consider as a function of lag. At every point in the system, if we go far enough from the origin, there will be a deficiency of "velocity"—that is, there will be a smaller population of fluid particles moving in the same direction as the point about which the correlation is being taken. The reason is directly connected with the reason that we have been calling turbulence a phenomenon connected intrinsically with vorticity in general. There can be no net transfer of momentum in completely developed homogeneous turbulence. Therefore, in the case of these flows it will be impossible for the correlation tensor of the transverse component to be everywhere positive.

On the other hand, as has been emphasized by every author on this subject, there is *no reason* to believe that the longitudinal component of the correlation *has to be positive definite everywhere*. In fact, it is entirely possible that it too has a negative part and that in fully developed homogeneous turbulence there would also be a deficit portion of the tensor. Since the solution for the radius (the scale length) must be finite at the origin and vanish at infinity, and also have (as we shall see in a moment) a finite second derivative, we make the approximation that the function is concave downward and one which probably behaves much like a Gaussian in its basic properties.[1]

7.3.3 Correlation Functions and What They Measure

The correlations within the flow are due to the comparisons of the transverse and longitudinal components of the velocity field. Specifically, we are interested in the *autocorrelation functions*. Therefore, we define two auxiliary functions:

$$f(r) \equiv \frac{\langle u_1(\mathbf{x})u_1(\mathbf{x}+\mathbf{r})\rangle}{\langle u_1^2(\mathbf{x})\rangle}, \qquad g(r) \equiv \frac{\langle u_t(\mathbf{x})u_t(\mathbf{x}+\mathbf{r})\rangle}{\langle u_t(\mathbf{x})^2\rangle} \qquad (15)$$

for the longitudinal component and for the transverse flow, respectively. (See Fig. 7.1.) The correlation tensor at zero lag is then defined by

$$R_{ii}(0) = \langle u_i(\mathbf{x})u_i(\mathbf{x})\rangle = \int \Phi_{ii}(\mathbf{k})\,d\mathbf{k} \qquad (16)$$

---

[1] The Cauchy function, $f(x) = 1/[1+(x/a)^2]$, which astronomers think of as a Lorentzian function, is one which is an approximation to the Gaussian and also one which has the same properties as the ones we are envisioning here.

7.3 Incompressible Turbulence

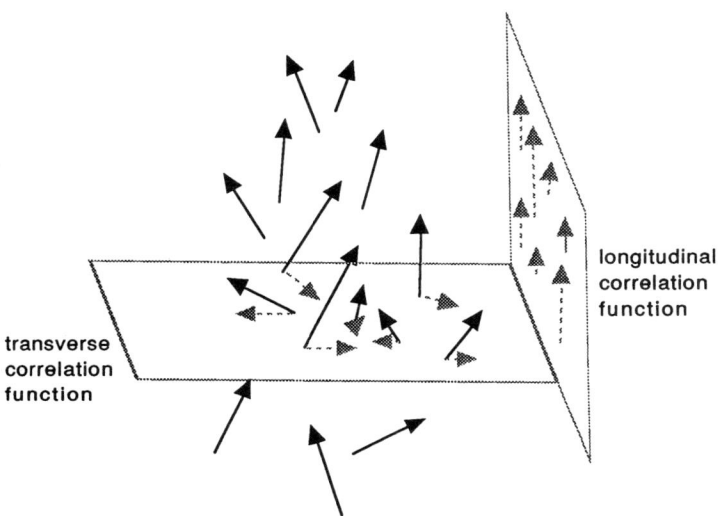

Figure 7.1 Illustration of the meaning of the correlation function for transverse and longitudinal flows. The mean flow is easily seen and indicated by the longitudinal correlation.

This measures the energy in the turbulent flow, integrated over all length scales. There is no sum over the index here, because we are taking the autocorrelation one component at a time. Note that, in light of the definitions in Eq. (11), we have

$$R_{ll} = Fr^2 + G, \qquad R_{tt} = G, \qquad R_{lt} = 0 \qquad (17)$$

as connections between observables, namely (f, g), and functional representations for use in the algebra of the theory, namely (F, G).

We now make use of the continuity equation to derive the relation between F and G and between f and g. This is simply done by taking the derivative

$$\frac{\partial R_{ij}(r)}{\partial r_j} = F(r) r_j \delta_{ij} + r_i r_j \frac{\partial F(r)}{\partial r_j} + \frac{\partial G(r)}{\partial r_j} \delta_{ij} \qquad (18)$$

which becomes, for the relationship between the quantities

$$4F + r \frac{\partial F}{\partial r} + \frac{1}{r} \frac{\partial G}{\partial r} = 0 \qquad (19)$$

We now substitute the autocorrelation functions f and g to obtain

$$R_{ij}(r) = (f - g) n_i n_j + g \delta_{ij} \qquad (20)$$

The functional forms for F and G are substituted in order to derive the relation between f and g, which are directly related to the correlation function in the transverse and longitudinal directions. We can make use of the fact that the autocorrelation should have a maximum at the origin, that is, that it should (for a zero lag) have a local maximum and also a negative curvature. This must be true in order to satisfy the property that the function be normalized (or that the integral over the infinite domain which is available for integration be finite) and that the zero-lag autocorrelation function be an effective measure of the total power in the system.

Now we can define a new scale of length. This is really what we have been leading up to in this exercise in deriving the vorticity spectrum. Recall now that the integrated correlation function, taken over all possible scalar lags r, was defined by $\Lambda \equiv \int_0^\infty R(r)\,dr$. This Λ is a global measure. We see that there is a local scale length that can be defined from the normalized R:

$$\lambda^{-2} \equiv -\frac{d^2R(r)}{dr^2}\bigg|_0 \tag{21}$$

This is called the *Taylor microscale* and represents the curvature of the correlation function at zero lag. This is the scale on which most of the energy is concentrated and measures how developed the turbulence is throughout the fluid. There is another way of stating this that relates the enstrophy and energy. Notice that the enstrophy varies as ω^2, so if we take v_t to be the rms turbulent velocity, then $\omega^2 \sim v_t^2/\lambda^2$. Since the energy in the turbulence is of order v_t^2, the Taylor microscale is a measure of the ratio of the energy to the enstrophy in the turbulence. Unlike the Kolmogorov scale, the Taylor microscale is not a fixed length in the turbulence, although it is related to the Kolmogorov scale (see below). You will notice that it does not directly depend on the energy dissipation rate or the viscosity.

In the calculation of the correlation it is useful to note that the fluid *must* be correlated at zero lag and should be (if the velocity field is to be normalizable and isotropic) completely uncorrelated at infinite separation. The two-point function must thus have a maximum for zero lag. The integral over all scales must have a finite value, and therefore the two-point function satisfies the conditions

$$R'_{ij}(r)|_{(0)} = 0, \qquad R''_{ij}(r)|_{(0)} < 0 \tag{22}$$

from which the functional form of f and g can be obtained. We expand $f(r)$ and $g(r)$ to second order about the origin in a Taylor series to obtain

7.3 Incompressible Turbulence

(recalling that both f' and g' vanish at $r = 0$):

$$f(r) \approx 1 - \frac{r^2}{2\lambda^2}, \qquad g(r) \approx 1 - \frac{r^2}{\lambda^2} \tag{23}$$

Keep in mind that this is true only in the vicinity of the origin (zero lag) even though one often finds this form used to represent the function even for large separations. While this is an excellent approximation of a Gaussian at the origin, at larger lags progressively higher-order terms play more of a role. Putting this into the form of the correlation function, we then see that the proper value for the power spectrum in the vicinity of the origin is

$$\langle \omega_i \omega_i \rangle = \nabla^2 R(r)|_0 = -15 \frac{u^2}{\lambda^2} \tag{24}$$

where ω_i is the component of the vorticity and the Laplacian results from taking the trace of the correlation tensor.[2]

Notice that the coefficient is large in comparison with unity and that this is an indication of the fact that we would have made a mistake if we had estimated the correlation function of the vorticity, which is what this value represents, by simply taking the mean velocity and scale length for the system as the estimators. This is important, as we shall see later in this chapter, in the theory of heating provided by acoustic noise generated by presence of a turbulent flow. In fact, it is essentially an amplification factor that represents the greatest possible efficiency we would expect from

---

[2] To see this, assume that the flow is dependent only on r. Now take

$$\nabla^2 = \frac{1}{r^2} \frac{\partial}{\partial r} r^2 \frac{\partial}{\partial r}$$

so that

$$\nabla \frac{r^2}{\lambda^2} = -\frac{6}{\lambda^2}$$

Then using $(f - g)/r^2 = (2\lambda)^{-2}$ and

$$\frac{\partial}{\partial r_i} \frac{r_i r_j}{2\lambda^2} = \frac{6 r_i}{2\lambda^2}$$

and differentiating this again with respect to r, we obtain

$$\nabla^2 R_{ii}|_{r=0} = -\frac{15 u^2}{\lambda^2} = -k^2 \Phi_{ii}$$

which was what we were looking for. Notice that this is valid only in the vicinity of zero lag (high correlation), in other words near the Taylor microscale length.

a source of noise which is isotropically turbulent and radiating waves which will in turn heat the surrounding medium, the *Lighthill* process (Section 7.7). If the correlation tensor is a quadratic in position, it must also be quadratic in the wave number. That is, we can represent the transform of the correlation tensor as $\Phi_{ij}(\mathbf{k}) = C_{ijlm}k_l k_m + \mathcal{O}(k^4)$, maintaining only even powers by the symmetry condition on the correlation tensor. The C_{ijkl} terms are constants. This function will be employed shortly, when we attempt to derive the evolution of the power spectrum with both time and wave number.

7.4 Kolmogorov Theory: The Role of Dissipation

To this point, if you have been noticing all of the assumptions, we have not introduced anything concerning the nature of the energy spectrum. In fact, it has been true to this point that the calculation has been implicitly adiabatic as well as incompressible. You are likely wondering how long we can continue to dwell in this fantasy land. In fact, this was the state of affairs until the end of the 1940s. Researchers dealing in turbulence theory remained content to describe, rather than derive, the nature of the turbulence spectrum. The breakthrough came with the introduction, by Kolmogorov, Heisenberg, and von Weissacker, of several extremely simple and genuinely beautiful assumptions from which a heuristic theory of the origin of the turbulence spectrum could be derived.

The picture we have to imagine is this. The turbulence of a medium is driven—that is, the eddies are made, not merely born. The idea which is most natural from observation is that of a cascade, in which the largest eddies are the ones which give rise to the smaller, which in turn divide and so on. The phenomenon is best described by a short poem from Richardson, based on a poem by Jonathan Swift:

> Big whorls have smaller whorls
> That feed on their velocity,
> And little whorls have lesser whorls
> And so on to viscosity (in a molecular sense).[3]

Encapsulated in this verse is the basic assumption that was used by Kolmogorov: the dissipation of the energy, which is derived from the largest scales of the turbulent flow, takes place at the level of the small eddies near

[3] Richardson, L. S. (1922). *Numerical Methods for Weather Prediction* (New York: Dover).

7.4 Kolmogorov Theory: The Role of Dissipation

the Taylor microscale. The energy cascade in homogeneous isotropic turbulence is in a steady state and determined by the rate at which the energy is being fed into the largest scale length. Therefore, we have a simple way of looking at the problem in terms of similarity variables.

There are two quantities which characterize the medium, the rate of energy dissipation ϵ and the coefficient of molecular viscosity ν. The density is constant, and therefore we can look at the combinations of these coefficients to obtain the length and velocity scales of the turbulence. We note first that the dimensions of ϵ are energy per unit time per unit mass $L^2 T^{-3}$ and those of ν are $L^2 T^{-1}$. Therefore, the combination

$$l_K = \left(\frac{\nu^3}{\epsilon}\right)^{1/4} \tag{25}$$

provides a length scale, which we will refer to as the *Kolmogorov* or *dissipation* length scale. This length scale is different from the Taylor scale and related to the very smallest scale of the process of dissipation, the viscosity level. By taking

$$\tau = \left(\frac{\nu}{\epsilon}\right)^{1/2} \tag{26}$$

we obtain a characteristic time scale for the dissipation of the energy by eddies of the length scale l_K. A final pass through this dimensional analysis gives the characteristic velocity:

$$u_K = (\nu \epsilon)^{1/4} \tag{27}$$

Now we are ready to make the crucial assumption of the cascade theory. This is the one first introduced in 1941 by Kolmogorov. Turbulence spectra appear to be universal. That is, they do not depend on the properties of the medium with which one is dealing. Therefore, the assumption that the spectrum is independent of the viscosity is natural and the one which drove the development of the theory. The viscosity should be a hidden parameter of the velocity field; it should simply be part of the scaling relation and nothing more. To remove the dependence of the energy spectrum on viscosity, notice that $\nu = l^{4/3} \epsilon^{1/3}$. We assume that in the inertial portion of the cascade there exists a universal spectrum of the form

$$E(k, t) = u_K^2 E_\star(l_K k) \tag{28}$$

where $E_\star$ is a universal function of the wave number, but one that is *dimensionless*. The form of the spectrum, which was merely the result of the dimensional analysis, agrees with the intuitive picture—one of the triumphs of this extremely simple argument.[4] The rate of energy dissipation is given by

$$\epsilon = -\frac{3}{2}\frac{du^2}{dt} = 2\nu \int_0^\infty E(k,t)k^2\,dk \tag{29}$$

as we have discussed before, and this is the basis of the calculation of the rate of dissipation. Again, it is worthwhile to recall where this comes from, because for compressible turbulence it is a very different result. The rate of energy dissipation we have discussed in Chapter 2. It is given by the shear σ_{ij} times the stress T_{ij}. We replace ν in Eq. (29) with the ϵ-scaling. This energy spectrum leads immediately to the form, substituting our di-

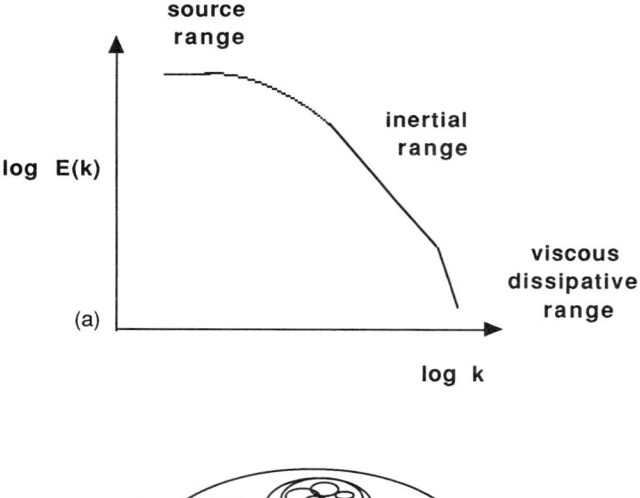

Figure 7.2 (a) The Kolmogorov spectrum, $E(k)$, for self-similar turbulence. (b) Hierarchical eddy cascade responsible for producing the spectrum.

---

[4] As an aside, it is worth saying that just because something is both simple and heuristic doesn't mean either that it is simplistic or wrong—it may just be elegant.

mensional analysis from a moment ago,

$$E(k, t) \sim \epsilon^{2/3} k^{-5/3} \tag{30}$$

This is the *Kolmogorov spectrum*. If this has appeared to be merely playing with dimensioned quantities, in reality it is much more. The underlying assumption is that there is a scale at which this dissipation of energy begins to dominate the spectrum, but that above that scale there is an "inertial" regime in which a steady state has been reached. (See Fig. 7.2.) Therefore, there is no need to keep track of the details of the spectrum at the larger scales, because the flow will merely preserve similarity. As each eddy dissipates its energy to the smaller scale, for which one can read as each eddy breaks up into smaller eddies, eddies of the same scale are recreated by the breakup of the larger ones. The result is that at every time there will be a population of the portion of the spectrum which is responsible for the creation of the eddies above the Kolmogorov length. You have seen this phenomenon many times, especially in water. The most obvious one is the feeding of the turbulence by the two-stream instability, which can often be seen in rivers at the point of the creation of rapids, in flow behind boats and bridges, and in the transition to turbulence in waterfalls. The largest scale is fixed by the shearing of the fluid and this in turn, especially in wakes, can be seen to break down into progressively smaller whorls (the inspiration for Richardson's parody).

7.5 Time Dependence of the Turbulence Spectrum

We have been assuming that the turbulence is in the inertial subrange and that there is a steady state. However, what happens if we stop paying the bills and the energy source is switched off? To put this more specifically as a physical problem, picture the wake behind a barge in a river, say the Seine. The decay of the turbulence of the wake is the result of the motion of the boat. Eventually the eddies which are created by the boat and the engine decay away, and the river will be more or less placid (unless it is in the height of the tourist season and there are frequent boats coming along). Or suppose we have a propeller spinning in water and then turn off the rotation. What is the subsequent development of the spectrum?

7.5.1 The Use of Fourier Space Methods

As an aside, it should be noted that there is another way of solving the equations for the turbulence evolution. We can immediately go from the space to wave number representation by a Fourier transform of the equations of motion. In this case, because of the convolution theorem which states that $\mathcal{F}(a \cdot b) = \mathcal{F}a * \mathcal{F}b$, where $\mathcal{F}$ denotes the Fourier transform, the nonlinear terms will turn into a convolution and the equations can at least be dealt with as finite difference rather than integral equations. To see this, we begin with

$$u(\mathbf{x}) = \sum_{\mathbf{k}} \mu(\mathbf{k}) e^{i\mathbf{k} \cdot \mathbf{x}} \tag{31}$$

where summation is assumed to occur over all possible wave numbers. Thus:

$$\frac{\partial}{\partial x_l} u_l(\mathbf{x}) = i \sum_{\mathbf{k}} k_j \delta_{jm} \mu_m(\mathbf{k}) e^{i\mathbf{k} \cdot \mathbf{x}} \tag{32}$$

$$u_n \frac{\partial u_l}{\partial x_n} = -\sum_{\mathbf{k}'} \mu_k(k') k_j' \delta_{jk} \sum_{\mathbf{k}''} \mu_l(k'') e^{i(\mathbf{k}' + \mathbf{k}'') \cdot \mathbf{x}} \tag{33}$$

and the latter can be reduced by noting that $k = k' + k''$ to

$$u_n \frac{\partial u_l}{\partial x_n} = \sum_{\mathbf{k}'} \mu_k(k') \mu_l(k - k') e^{i\mathbf{k} \cdot \mathbf{x}} k_j' \delta_{kj} \tag{34}$$

Here we have made use of the convolution theorem, that the Fourier transform of a product is the convolution of the Fourier transforms. This means that we are also able to define a projection operator that selects out only one mode of the wave number space:

$$P_{ij} k_j = \left(\delta_{ij} - \frac{k_i k_j}{k^2} \right) k_m \tag{35}$$

an expression due to Kraichen (1967). We have effectively applied this in the convolution by saying that we look at the modes that result from the interaction of the different velocity components, $P_{ijm} u_j u_m$, as for a scattering process. Now you can write the Navier–Stokes equation in the form

$$\frac{d}{dt} \mathbf{v}(\mathbf{k}) + i\mathbf{k} \sum_{\mathbf{k}'} \mathbf{v}(\mathbf{k} - \mathbf{k}') \mathbf{v}(\mathbf{k}') = -i\mathbf{k} p(k) - \nu k^2 \mathbf{v}(\mathbf{k}) \tag{36}$$

Then using the projector and the fact that $\nabla \cdot \mathbf{v}(\mathbf{x}) = 0 \rightarrow \mathbf{k} \cdot \mathbf{v}(\mathbf{k}) = 0$, this equation is seen to be equivalent to

$$\frac{d}{dt} v_l(\mathbf{k}) = -\nu k^2 v_l(\mathbf{k}) + \sum_{mn} \mathcal{M}_{lmn} v_m(\mathbf{k} - \mathbf{k}') v_n(\mathbf{k}') \tag{37}$$

where the projection operators have now been combined in the function $\mathcal{M}_{imn} = -(i/2)(k_n P_{lm} + k_m P_{ln})$. Here the advantage of working in wave

7.5 Time Dependence of the Turbulence Spectrum

number rather than in space becomes more obvious. Turbulence is the result of interactions between different velocity components in the fluid, the exchange of vorticity between different portions of the mean flow. This takes place on many spatial scales at once and throughout the medium. The Fourier approach makes it clearer that this is the same as collisions of waves exchanging momentum and that this happens in any elementary volume of the medium. This approach to the turbulence problem, aside from its use in our discussions, is also the basis for much of the current diagrammatic calculations in this field (see Forster et al. 1977).

7.5.2 Development of the Equations in Wave Number

To answer the question of the form of the spectrum for the energy, resulting from the combined effects of generation and dissipation of energy, we are asking for a function in wave number. That is, we want something equivalent to a scaling law, like the one from the Kolmogorov analysis. We'll work now only with the Navier–Stokes equation and examine the linearized incompressible problem. That should be hard enough and provide enough physical insight for the moment. We thus begin with

$$\frac{\partial u_i}{\partial t} = \nu \frac{\partial^2 u_i}{\partial x_j \, \partial x_j} \tag{38}$$

Taking the Fourier transform of both sides with respect to the wave number k gives

$$\frac{\partial}{\partial t} \mu_i = -\nu k^2 \mu_i \tag{39}$$

where μ_i is the transform of the velocity. Then we have

$$\mu_i \sim e^{-\nu k^2 t} \tag{40}$$

which yields the turbulence correlation function

$$\Phi_{ij} = C_{ijmn} k_m k_n e^{-2\nu k^2 t} \tag{41}$$

where the coefficients C_{ijmn} are constants. By taking the inverse transform, we arrive at the two-point correlation function:

$$R_{ij}(r) = \int C_{ijmn} k_m k_n e^{-2\nu k^2 t} \, d\mathbf{k}$$

$$= \frac{(2\pi^3)^{1/2}}{16[\nu(t-t_0)]^{5/2}} \left[C_{ijll} - C_{ijlm} \frac{r_l r_m}{4\nu(t-t_0)} \right] \exp\left(-\frac{r^2}{8\nu(t-t_0)} \right) \tag{42}$$

The second term decays faster in time than the first, so let's throw it away. The turbulence correlation function, and as a result the energy, therefore decays according to

$$R_{ij}(r) \sim (\nu(t-t_0))^{-5/2} \exp\left(-\frac{r^2}{8\nu(t-t_0)}\right) \tag{43}$$

We have used the fact that $k^2 = k_i k_j \delta_{ij}$ in order to obtain our result. The energy dissipation at large times after the energy source for the turbulence has been turned off varies as

$$\int E(k,t) k^2 \, dk \sim t^{-5/2} \tag{44}$$

We could have obtained this more easily by noting that the integral is dimensionally $t^{-5/2}$ and then assumed that the exact coefficient is a matter of empirical choice. But while this works for laboratory studies, we don't have the option of performing the experiments.

Now, let's go further into this problem of energy dissipation in turbulent flows. We write down the complete Navier–Stokes equation, making the assumption that v is the total velocity and that U and u are the mean and fluctuating components, respectively:

$$\frac{\partial U_i}{\partial t} + U_l \frac{\partial U_i}{\partial x_l} = -\frac{1}{\rho}\frac{\partial \langle p \rangle}{\partial x_i} + \nu \frac{\partial^2 U_i}{\partial x_l \partial x_l} - \frac{\partial}{\partial x_l}\langle u_i u_l \rangle + F_i \tag{45}$$

Now define **x** and **x'** such that

$$\frac{\partial}{\partial x_i} = -\frac{\partial}{\partial r_i}, \quad \frac{\partial}{\partial x'_i} = \frac{\partial}{\partial r_i} \tag{46}$$

where we have introduced the lag, **r**, into the calculations. Thus

$$u'_j \frac{\partial u_i}{\partial t} + u_k u'_j \frac{\partial u_k}{\partial x_k} = -\frac{1}{\rho} u'_j \frac{\partial p}{\partial x_i} + \nu u'_j \nabla^2 u_i \tag{47}$$

and

$$u_i \frac{\partial u'_j}{\partial t} + u_i u'_k \frac{\partial u'_j}{\partial x'_k} = -\frac{1}{\rho} u_i \frac{\partial p'}{\partial x'_j} + \nu u_i \nabla^2 u'_j \tag{48}$$

Combining these two equations gives

$$\frac{\partial}{\partial t}\langle u_i u'_j\rangle + \frac{\partial}{\partial r_l}(\langle u_i u'_j u_l - u_i u'_j u'_l\rangle)$$

$$= +\frac{1}{\rho}\left(\frac{\partial}{\partial r_i}\langle p u'_j\rangle - \frac{\partial}{\partial r_j}\langle p' u_i\rangle\right) + 2\nu \nabla_r^2 \langle u_i u'_j\rangle \tag{49}$$

7.5 Time Dependence of the Turbulence Spectrum

We define

$$\langle u_i u_j' \rangle (\mathbf{r}) = \int \Phi_{ij}(\mathbf{k}) e^{i\mathbf{k}\cdot\mathbf{r}} d\mathbf{k} \tag{50}$$

Now we obtain an equation for the evolution of the energy spectrum with the appropriate source and sink terms:

$$\frac{\partial}{\partial t} \Phi_{ij} = \Gamma_{ij} + \Pi_{ij} - 2\nu k^2 \Phi_{ij} \tag{51}$$

If we deal only with the real quantities, the Fourier transforms can be simplified. First, define two new quantities:

$$T(k) = -\frac{2}{\pi} \int_0^\infty \frac{\partial}{\partial x_l} [\langle u_i u_l' u_i' \rangle - \langle u_i u_l u_i' \rangle] kr \sin kr \, dr = 2\nu k^2 E(k)$$

$$S(k) = \int_k^\infty T(k) \, dk \tag{52}$$

The quantity T is related to the energy flux, since you see it measures the divergence of the triple correlation in the velocity. You may suspect that its evolution will yield an energy conservation equation, and this will emerge shortly. The energy is defined in the usual way by

$$E(k) = \frac{2}{\pi} \int_0^\infty R(r) \sin kr \, dr \tag{53}$$

Keep in mind that this development is for incompressible turbulence *only*. These quantities may not mean the same thing, or even exist in the same form, for compressible turbulence. The equation for $T(k)$ describes the rate of change of the energy with time in the inertial portion of the spectrum. Now define a new quantity, $\sigma(k) = dk/dt$, a sort of flow rate in k-space. The energy flux through a given wave number is $dE(k)/dt = \sigma(k) \, dE(k)/dk$, so that from the definition of E and S we obtain the approximation $S(k) = \sigma(k) E(k)$. This follows from the definition of $S(k)$ in Eq. (52). Now since $t = \epsilon^{-1/3} k^{-2/3}$ from dimensional analysis on ϵ, the energy dissipation rate, it follows that

$$\sigma(k) \sim \epsilon^{1/3} k^{5/3} \tag{54}$$

Notice that this means that the k-space speed depends steeply on the wave number such that large wave number (small scales) cascade most rapidly. Then we obtain

$$S(k) \sim \epsilon^{1/3} k^{5/3} E(k) \tag{55}$$

so that since $dS(k)/dk = -T(k)$ we obtain

$$\frac{d}{dk}[\epsilon^{1/3}k^{5/3}E(k)] \sim -2\nu k^2 E(k) \tag{56}$$

This is the substitution of the definition of S into the evolution equation for T. Finally, we obtain an expression for the evolution of the energy spectrum with time, one discussed by Kolmogorov in his later (1962) contribution:

$$E(k) = E_0 k^{-5/3} \exp(-\tfrac{3}{2}\alpha k^{4/3}\nu\epsilon^{-1/3}) \tag{57}$$

Here α is a universal constant and we have used E_0 as a normalization. This spectrum is essentially the same as the original one, but since we have assumed that the scaling law for E holds this isn't surprising. It is interesting that there is a characteristic wave number which emerges from all this. It is given by the exponential

$$k_\star = \left(\frac{2}{3\alpha}\right)^{3/4}\left(\frac{\epsilon}{\nu^3}\right)^{1/4} \tag{58}$$

which is the point at which the assumptions of the Kolmogorov spectrum break down. This is the critical scale for the generation of cascades.

There is yet another way of deriving useful results from these definitions. We notice that with a more schematic approach, we can define T as

$$T(k,t) \equiv 4\pi k^2 \int Q(\mathbf{k},\mathbf{k}',t)\,d\mathbf{k}' \tag{59}$$

where Q is the dissipation rate,

$$S(k,t) \equiv \int_{k'=k}^{\infty} \int_{k''=0}^{k} P(k',k'',t)\,dk''\,dk' \tag{60}$$

where this is the rate of energy transfer from $k'' < k$ to $k' > k$ in a time t. Therefore, we can write the equation for the spectral evolution of the turbulence as

$$\frac{\partial E(k,t)}{\partial t} = T(k,t) - 2\nu k^2 E(k,t) \tag{61}$$

and steady state gives $T = 2\nu k^2 E$. We thus have the condition on the dissipation rate for any system which has reached the inertial subrange—the dissipation function is fixed.

7.5.3 Astrophysical Applications

What this means for astrophysical problems is not at all clear from the derivation. Notice that we have made the assumption that the medium is incompressible. This is not at all kosher for astronomical bodies, as we have been repeatedly saying, and the assumption that vorticity is conserved, which is the primary source of the equations of motion, is clearly violated in the compressible case. However, we know that there will be a finite time for the decay of any turbulent spectrum and can therefore at least see what the effects are of the application of the theory to astrophysically interesting situations. First of all, consider the motion of a galaxy through the intergalactic medium (IGM). There will be a wake, as has been shown by a number of studies including simulations. This will have a finite lifetime for the decay, which will be given by the rate of energy dissipation. However, in the optically thin environment of the IGM, this will be due primarily to radiative processes. Thus we have a way of independently estimating the rate of energy dissipation and of fixing the scale of the lengths and times for the decay of the eddies in the wake of the galaxy. In a more spectacular application, if the energy of the eddies decays with time in the fashion we have just derived, in the expansion after the Big Bang there will also be a finite lifetime for the eddies. This will vary as $(1+z)^{-5/2}$, where z is the redshift, and this is a very steep function of the time. Thus the eddies of the primordial turbulence, if they can be approximated as incompressible, will show a rapidly decaying spectrum due to the effective cooling of the medium by the universal expansion, and the scale of the system should progressively grow smaller. Of course, you may object that we have not included the effects of gravity in this picture—and quite rightly—but the medium is initially dominated by the turbulence relic from the initial conditions and the expansion. It is only on the smaller or Jeans scale that the effects of gravity play a significant role.

It is interesting at this stage to note that, in the literature on the applications of turbulence to astrophysical environments, especially to molecular clouds, there is a simple classification based on the assumptions about the origin of the inertial range. As an example, some of the prominent ideas for the origin of turbulence are:

1. Hoyle (1953): $t_{\text{diss}} \sim t_{\text{ff}}$—that is, that the dissipation time scale is of the same order as the collapse time, which is assumed to occur on the free-fall time scale:

$$t_{\text{ff}} = (G\rho)^{-(1/2)} \qquad (62)$$

2. Low and Lynden-Bell (1976): $t_{cool} \sim t_{ff}$—that is, that the cooling time is comparable with the free-fall time, where

$$t_{cool} = \frac{3c_p \rho T}{n^2 \Lambda} \tag{63}$$

with Λ being the cooling function and c_p being the specific heat.
3. Fleck (1983): $t_{diss} \approx t_{rotate}$—that is, that the dissipation time scale, given by

$$t_{diss} \sim \left(\frac{\rho v_t^3}{l}\right)^{-1} E_{grav} \tag{64}$$

where E_{grav} is the gravitational self-energy of the cloud, and the rotational time scale, given by

$$t_{rotate} \sim \kappa^{-1} \equiv \left[\left(\frac{d\Omega}{dr} - \frac{\Omega}{r}\right)r\right]^{-(1/2)} \tag{65}$$

where Ω is the rotation frequency and κ is called the epicyclic frequency, are of the same order of magnitude.
4. Henricksen and Turner (1984): $t_{collision} \sim t_{ff}$—that is, that the collapse time for a fragment is of the same order as the collision time between fragments.

Only Fleck's mechanism makes explicit reference to any external action of the galaxy as a whole. The source for the turbulence in this case is assumed to be the large-scale shear flows, which are set up in a sizable cloud as a result of the differential rotation of the galaxy.

There is yet another source for the turbulence, but this is operative only after the cloud has begun the process of star formation. The winds and turbulence generated thereby in the vicinity of young stars are well known to be a probable source for the turbulent structuring of a cloud. That is, as soon as the stars turn on there is a complete change in the source of the turbulence and also, possibly, of the turbulent spectrum. We will discuss the role of stellar mass outflows in depositing mechanical energy into the environment, but within molecular clouds this appears to be very important as a structuring and stirring mechanism. A typical stellar wind with $\dot{M} \approx 10^{-7}$ M$_\odot$ yr^{-1} and a terminal velocity of a few hundreds of kilometers per second provides an equivalent mechanical luminosity of about 10^{33} erg s^{-1} to the cloud. This means that the effective volume that can be stirred up is about 10 pc^3 by a single star in about 10^7 years, the typical lifetime of the cloud. The typical rate of star formation in a cloud is of order a few solar masses per year, so that once these compact bodies have been formed the cloud is permanently altered. It can therefore be said somewhat differently that there appears to be a self-regulatory process involved in the maintenance of turbulence and star formation within clouds.

7.6 The Transition to Turbulence

The process by which a laminar flow develops turbulent eddies, and how they grow, is one of the current frontier areas of physics (Swinney and Gollub 1985). Ultimately, part of the problem is that there are many routes to turbulence. The development of large-scale instability in a fluid through the Kelvin–Helmholtz, or two-stream, instability is one of the most frequently encountered in cosmic bodies. The development of large-scale eddies appears to be the best way to promote both mixing and dissipation. These are observed in many cases where comparatively low-density bodies are moving rapidly with respect to one another. Explosions fragment from density gradients, often becoming Rayleigh–Taylor unstable, and this also serves as a mechanism for injecting vorticity into the surrounding medium.

7.6.1 Some Observations for Illustration

Several phenomena are readily accessible to daily observation, which well illustrate the principles we have been covering. One of the most obvious is the transition to turbulence in rapids. Notice, when you are looking at the flow, that the turbulent part of the flow is one of large shear, always (for water) accompanied by the growth of eddies. The best place to see this is in the flows around rocks, where the formation of a boundary layer and wake highlights all of the effects we have discussed. Look also at paintings of the Sung dynasty in China and the later sketches by Leonardo da Vinci of the Deluge. All of these illustrate the artist's perception of the phenomenon and serve as useful visualizations of the detailed behavior of such flows.

To see the transition for boundary layers, one of the best ways is to look at polluted water running off in a gutter or the oil-slick surface on a bank of a not too clean river. The oil will serve as an excellent marker of the flow and allow for easy visualization of the effects. The streamlines in the laminar part of the flow will begin to show both curling and interweaving as the transition is entered, until separate eddies, which begin to break up and become more diffuse in the downstream portion of the flow, become visible. After that, the fluid will become so mixed and murky that it is often very difficult to follow the subsequent development of the layer; but it will grow to a finite thickness, which will be larger than the laminar region, and then saturate. This can also be seen in using smoke along the surface of a table, or against a window, where the smoke is used as the tracer of the flow. Several excellent aeronautical photographs show the effects of both the boundary layer and the finite size of the plate—the formation of vortices at the edges of the surface which subsequently propagate as separate structures.

You can feel as well as see the effect of the turbulent boundary layer whenever you put your hand outside the window of a car, and most especially if you blow some smoke out of the window. As well as an excellent illustration of the Bernoulli equation (the pressure drops against the boundary of the car and therefore the smoke is sucked out), you can also see and feel all of the intermittency effects associated with a turbulent boundary layer. Another effect which can be observed along highways, best seen in rainstorms or fogs, is the turbulent wake behind a truck and the transition which takes place as the flow proceeds over the front of the vehicle. As a way of reducing drag, many truckers have installed a deflector over the cab of the vehicle to deflect the flow over the trailer and reduce the ram pressure resistance of the air ahead of the truck. This allows one to see both separation effects as the boundary layer is formed and the turbulent transition as the streamlines form vortices along the top of the trailer.

Small-scale turbulence is best seen in the ricelike appearance of the image of a twinkling star seen at high magnification through a large telescope. Seen through large airmass, the structure of the seeing disk of the star is an excellent illustration of scattering by a turbulent medium and also of the distribution function for the scatterers.

Probably, though, the best effects are seen in buoyant plumes, like those from smokestacks or cigarettes. Here, the laminar flow is observed near the top of the source. Especially with a cigarette (candles are intrinsically too unstable at the top of the flame to make this easy to observe independent of the source) one sees a laminar streamline rising, which becomes turbulent as it entrains more of the surrounding air. The curling of the smoke and the subsequent spreading of the plume laterally are both characteristic of the turbulent transition. Such effects have been seen in jets in radio galaxies, and it has recently been argued that, if not due to buoyancy, these structures can at least be an indication of the entrainment of the surrounding interstellar medium by the jet. Twisting modes are also sometimes observable in the plumes, as they are in the jets, and these are due to a Kelvin–Helmholtz mode which has azimuthal perturbations rather than radial ones—they represent a higher-order mode.

7.7 Compressible Turbulence: The Lighthill Process

We now look at compressible turbulence, motivated by our discussions of the incompressible case and the need for some astrophysical description of a *squishy* medium. In the case of a compressible medium, we are really examining the conditions for the generation of sound. Any compression generates mechanical waves that are transmitted through the medium and serve as the prime source for dissipation. So the problem of the energy spectrum of, and dissipation rate for, a compressible fluid is essentially one of stochastic acoustics (Lighthill 1952).

Recall that for the pressure deviation we had derived a Poisson-like equation. That is, there was a source term. This was the perturbation of the stress tensor, an important quantity in that it is nonlinear and therefore a term strongly coupled to the fluctuations in the velocity field. We did not make specific assumptions about the nature of the medium—in particular, we have not assumed that there was any particular equation of state. Now, we see that in the case of the continuity equation, allowing for the variation in the density as well, we will have

$$\frac{\partial \rho}{\partial t} = -\rho_0 \frac{\partial u_i}{\partial x_i} \tag{66}$$

where we have assumed that the medium is otherwise at rest. Already, a basic difference appears between compressible and incompressible turbulence. Variations in the velocity field change the local density. This means that the collision and redistribution of eddy vorticity will feed back into the density field. Since we have now to deal with the equation of state, the fluctuations in the density will produce local pressure changes, which will drive the emission of sound waves. Hence our earlier statement about acoustics. From the equations of motion, we have

$$\frac{\partial \rho u_i}{\partial t} + \frac{\partial \tau_{ij}}{\partial x_j} = -\frac{\partial p}{\partial x_i} = -a_s^2 \nabla \rho \tag{67}$$

which is the basic set of equations to be solved. We will look for the amplitude $\delta \rho$ such that the total energy contained within a fluctuation in the density field is translated into the kinetic energy field. This is the *Lighthill analogy*, which allows the turbulence to appear as if it is generating sound waves and then in the weak field limit derives the solution from the fluctuations in the stress thus produced. That is, the total energy will be of the order $\langle \delta\rho^2 \rangle$. First we need an equation for $\delta\rho$. This comes from taking the time derivative of both sides of Eq. (67) and substituting the continuity equation for $\partial\rho/\partial t$ on the right (remember that this term would not exist in the case of incompressible turbulence). Finally, since for the temporal fluctuations of the Reynolds stress we can write $\partial\tau_{ij}/\partial t \sim a_s(\partial\tau_{ij}/\partial x_j)$, we arrive at

$$\left(\frac{\partial^2}{\partial t^2} - a_s^2 \nabla^2\right)\rho = -\frac{\partial^2 \tau_{ij}}{\partial x_i \partial x_j} \tag{68}$$

Here the stress is provided by the velocity fluctuations (Reynolds stress) alone. The laminar portion of the flow, if there is one, does not play a role in generating turbulent fluctuations that are radiated as noise by this process. Equation (68) is a wave equation with a source term and is solved using a Green function. We assume that the energy is radiated away from the site and use the far field approximation. This assumes that the time delay is given by $\tau = t - r/a_s$ and that the measured energy flux is obtained at large distance from the source. Then $G(\mathbf{r} - \mathbf{r}')$, the propagator, is given by the delayed time. We assume also that the space derivatives in Eq. (68) can now be represented as $\partial/\partial x_i \to n_i \partial/a_s \partial t$, where n_i is a unit vector. It is important to note that this step is essential to the Lighthill analogy—that all of the spatial derivatives can be replaced by time derivatives and that therefore the wave spectrum can be directly obtained.

From Eq. (68) we see that the density fluctuations produce quadrupole radiation. The density fluctuations produce sound with a flux proportional to $\langle \delta\rho^2 \rangle a_s^3/\rho_0$. The density fluctuation is given by

$$\delta\rho \sim -\frac{1}{a_s^2} \int \frac{\partial^2}{\partial x_i x_j} T_{ij}(t - r/a_s, r) \frac{d\mathbf{r}'}{r} \tag{69}$$

Here we have used the Green function for $r^{-1} = |\mathbf{r} - \mathbf{r}'|^{-1}$. For distant regions, we see that Eq. (69) becomes

$$\delta\rho \sim \frac{1}{a_s^4} \int \frac{(r_i - r_i')(r_j - r_j')}{|\mathbf{r} - \mathbf{r}'|^3} \frac{\partial^2}{\partial t^2} T_{ij} d\mathbf{r}'$$

so that the density fluctuations give rise to a radiation pattern that depends on $\ddot{T}_{ij} n_i n_j$.

We now average over all of the fluctuations that are assumed to be stochastic. Schematically, this means that the mean square density fluctuations are given by

$$\langle \delta\rho^2 \rangle \sim \langle \ddot{T}_{ij} \ddot{T}'_{ij} \rangle / a_s^8$$

7.7 Compressible Turbulence: The Lighthill Process

This is the form we are looking for. But we still have a little more work to do. The time derivatives are proportional to v_t/L, where v_t is the turbulent velocity and L is a characteristic scale length. Each stress tensor contributes a term of order v_t^2. We therefore see that the rate of radiation is a steep function of the energy in the turbulent flow, of order v_t^8/a_s^5. Put another way, we see that the rate of radiation is proportional to $\rho v_t^3 M_t^5$, where M_t is the turbulent Mach number. For near supersonic turbulent flow, the efficiency is consequently high. The full solution is the Lighthill formula:

$$\epsilon = \mathcal{H} \rho_0 \frac{v_t^8}{a_s^5} \tag{70}$$

The numerical constant, $\mathcal{H}$, is of order 40 and was calculated by Proudman (see Lighthill 1952).

There is a steep dependence on the Mach number, primarily because of the conversion of the waves into density perturbations through the second-order effects of the stress tensor, and not because the process involves the creation of shocks. However, it is possible that the waves will steepen as they propagate and generate dissipative shocks, which is the basis of many of the coronal heating mechanisms which have been explored for the past few decades.

Various models for turbulent heating have been applied in many different places and for many different reasons. Models for coronal heating by shocks usually employ steepening acoustic waves, generated from the photospheric convection zone, that transform into shocks and dissipate in the chromosphere. Alternatively, Alfvén waves have been invoked as the energy source. The Alfvén wave mechanism has been used for heating as well as pressure support and dynamical driving for both stellar winds and the interstellar medium. These waves are incoherent, but depending on the propagation length, they may not necessarily be dissipative. This is simply the amount of energy which can be carried away from the site of the acoustic wave generation by sound waves of finite amplitude and nothing more. It says *nothing* about what happens to the medium into which the waves are moving, and therefore only a detailed examination of the dissipation process will reveal the spectrum of the heating turbulence which is finally generated by this deposition mechanism.

7.7.1 The Role of Vorticity

We have seen that mixing and vorticity are essential to the generation of and maintenance of turbulence. Shear is a source for dissipation, and it is essential that turbulence, true turbulence, be dissipative. One way of looking at this is to imagine a random collection of vortices. By random, we mean that the orientation of ω is a random variable in three dimensions. When these vortices interact, assuming that there is a viscosity, there are several outcomes. They can merge and increase the local intensity of the vorticial velocity field. This will happen if their orientations are the same. They may simply scatter off each other. Or, and this is the essential point, they may merge if counterrotating and come out with lower vorticity than initially. In fact, they may simply disappear. The energy of the vortical motions then goes into entropy and is ultimately dissipated by the fluid. For most astrophysical examples, this takes place radiatively. The role of the *enstrophy*, defined by $\int \omega^2 \, dV$, is therefore that it measures the energy latent in the chaotic vortex field.

This last point also has great consequence for cosmic fluids. Simple collapse or expansion is not a way to generate turbulence. Some vorticity must be pumped into the fluid. That is, there must be shear (now how many times have you heard that before?) in order to mix the medium and dissipate energy effectively. That is, after all, the role of viscosity and the reason the Reynolds number and the Kolmogorov scale exist. In the case of a shear flow, the Kelvin–Helmholtz instability naturally generates ω at the boundary, on all scales, and these interacting may form a cascade from the large scale to the dissipative scale. For the Rayleigh–Taylor instability, the interface develops fingers of fluid. These are buoyantly driven, either by gravity or by something else that looks like it (like a pressure gradient). Again the driving is on the large scale and there is no characteristic minimum wavelength for the development of the instability. The fingers are a natural way of pumping vorticity into the fluid because they shear relative to the background. In three dimensions, they can tangle and undergo all of the effects of vortex interactions just described.

Any instability that promotes fragmentation locally is not necessarily going to generate turbulence unless the fragments have an environment with which they are strongly coupled. The angular momentum given to the fragments by interaction is not turbulence. It will only be observed as such when there is a continuum of scales that act to transfer energy from the sources to the fluid as a whole. Here the problem of self-gravitating turbulence becomes very important. It isn't clear that the normal mechanisms for describing fully compressible homogeneous turbulence apply. The energy is fed from small length scales. The interaction with the surrounding medium is as much from fragmentation as from motion. And the essential feature is that once the fragments become self-gravitating they are not dissipative. There is an internal source of energy in this particular case, the fluid itself, and the fact that the strength of the source grows as the scale of the source contracts. The negative specific heat aspect of gravitation complicates life for fluid dynamicists yet again.

7.8 More Physical Complications

Most of the problems that you are likely to encounter in astrophysical turbulence are only loosely connected with normal experience and are even more remote from the incompressible cases most often treated in detail (here and elsewhere). In low-density media, where the mean free paths are long and the time scales for energy redistribution and loss are relatively short, local effects can rapidly cascade into global chaos. In many of the environments that appear to show turbulent structures, the motions indicated are highly supersonic. And often, the length and density scales are such that these regions may also be self-gravitating. So, just briefly, we will look at several of these complicating physical attributes of turbulent flows. The discussion will be more qualitative than some might wish, but that's the nature of the problem.

7.8.1 Supersonic Turbulence

When the turbulent region becomes supersonic, internal shocks become the primary mode for dissipation. Little is known analytically about this regime, but there is considerable aerodynamic experience with it. The medium is still highly compressible. At high Mach numbers in the flow, it is found from simulations that strongly compressed regions develop, with high vorticity, and the medium filaments rapidly into very localized regions. Some of these may still be carried in the bulk with supersonic speeds, but as they collide and the overall temperature rises within the medium the motions become progressively more subsonic. Recall that contact surfaces are formed by the intersections of shocks. These will form on collision of the turbulent eddies and locally planar fronts within a supersonic medium and as a result inject vorticity. This injection is, however, on a small scale. Unlike the Kolmogorov model, these do not necessarily proceed from the large-scale structures through a cascade to smaller scale, and it is possible that the inertial range does not exist within such media.

If a magnetic field is present, waves can be transmitted at both the magnetosonic and Alfvénic speeds, stirring the medium up and giving rise to local islands of low density and high density. The major difference between supersonic and subsonic turbulence is, however, manifested in magnetohydrodynamic (MHD) turbulence. Here initially supersonic, but sub-Alfvénic, modes propagate nonlinearly and turn into internal shocks. The energy is rapidly dissipated, but in the process localized current sheets form. These serve, through reconnection, as excellent sites for dissipation. Alfvén waves are free to move through the medium and promote cascades when they collide and scatter. Unlike the unmagnetized case, MHD turbulence has Joule heating, generated by the formation of these current sheets, for producing locally strong departures from the barotropic condition, $\nabla p \times \nabla \rho$, thereby feeding the local vorticity.

Most current work on supersonic MHD turbulence (see Montgomery 1989; Pouquet *et al.* 1991) must be performed numerically. One point that appears a very valuable one is the analogy between the helicity, $\int \mathbf{v} \cdot \omega \, d^3 x$, and the magnetic helicity, $\int \mathbf{A} \cdot \mathbf{B} \, d^3 x$. This is not just a formal one, because it indicates how the local vorticity is fed by the overall field structure (whether velocity or magnetic). The magnetic helicity already appeared in Woltjer's theorem for the generation of a force-free field, and the presence of large values of dynamical helicity is an indication of a strong interaction between the turbulence and the large-scale flow. Large values of magnetic helicity, for instance, signal the breakdown of force-free configurations and indicate that reconnection and dissipation will be important. For the dynamical helicity, it indicates that efficient interaction is taking place between vortices and again is a signal of dissipation. And one final point: in MHD turbulence, there is an additional form of pressure support so that locally, large bubbles may develop in the medium in which there is very little fluid. When these collapse, they feed their energy dissipatively into the rest of the flow. The phenomenon is still poorly understood but is very likely to have important astrophysical consequences.

7.8.2 Self-Gravitating Turbulence

Supersonic turbulence appears to be required to explain the structures and velocities observed in molecular clouds. Typically, molecular clouds have infrared emission consistent with temperatures of a few tens of degrees. In extreme cases, where star formation is obviously heating the interior of the clouds, one sees $T_{\rm rad} \approx 50$ K. The thermal motions are consequently expected to be only a few hundred meters per second, at most. But the observed velocity dispersions in molecular lines are usually a few kilometers per second, very supersonic. The source, or sources, powering these motions remains elusive. Yet some constraints can be placed on it. The motions are clearly dissipative because the clouds can radiate their turbulent energy in the infrared (to which they are optically thin). We have already discussed the ideas for the large-scale sources, but some internal sources of energy may also be important. At the densities typical of molecular clouds, Alfvén speeds are of order $0.03 \, B_\mu n_4^{-1/2}$ km s^{-1}, where B_μ is the magnetic field in μG and n_4 is the number density in 10^{-4} cm^{-3}. Therefore, for fields of order a few 100 μG, the Alfvén speeds will be of the order observed in the line profiles. These waves are certainly turbulent within the cloud and provide a pressure that contributes

7.8 More Physical Complications

to the support of the cloud through the virial theorem. Extensive two- and three-dimensional modeling of self-gravitating turbulent media is just beginning, but some consistent results are emerging. Passot *et al.* (1988) and Leorat *et al.* (1990) find that M, the Mach number within the flow, fluctuates with an rms value below unity. This is true even for models that do not include radiative losses or magnetic fields. The dispersion in M does not depend strongly on the initial conditions and can become very low, of order 0.1. Bonnazzola *et al.* (1987) have studied the onset of the Jeans instability in self-gravitating media, finding similar results.

7.8.3 Persistent Large-Scale Structures

Following the observations of large-scale structures in mixing boundary layers by Brown and Roshko (1974), many laboratory studies have pointed to the existence of *persistent* structures in turbulence. The original observations showed that the mixing in planar jets is initiated by vortex wrap-up. However, these do not disappear into microscopic dissipative scales. They persist and are advected in the flow, feeding turbulence far downstream of the source region. The so-called second Kolmogorov theory takes intermittency into account and so in a sense can be said to include some aspects of these structures. Intermittency is the portion of the spectrum at large scale (low wave number). You can think of this as the low probability that a pause in the turbulent flow occurs over a scale similar to the source scale. Coherent structures may be partially produced by such fluctuations. They are defined by Hussain (1983, 1986) as follows:

1. A coherent structure is a connected, large-scale turbulent fluid mass with a phase-correlated vorticity over its spatial extent.
2. Coherent structures are spatially mutually exclusive. Their interactions involve tearing and pairing; interactions result in new structures.
3. They may occur as solo structures (i.e., helicial vortices) or as collective structures (i.e., vortex sheets, rolls, rings). They may also connect via *braids*, regions of low vorticity.
4. They are characterized by high levels of coherent vorticity, coherent Reynolds stress, but not necessarily a high kinetic energy density, most of which is concentrated in the incoherent portion of the turbulent flow.
5. A coherent structure is *not* a wave and its advection in the flow should not be confused with wave propagation.

The primary mechanism for destruction of these structures is vortex merger and stretching. Once the structures dissipate helicity through viscous merger, often observed to occur at a large distance from the orifice, the turbulent flow becomes fully developed. Notice that many of the properties of these structures are similar to the ones we have been emphasizing for vortex filaments. The importance of these large-scale structures to astrophysical flows has yet to be explored. There are possibly some examples observed in some extragalactic radio jets, especially Hercules A and Fornax A. The problem is that these are continuum observations of dilute, possibly relativistic MHD flows observed only with continuum imaging in synchrotron emission. Little work is available for supersonic flows that display such structures.

7.8.4 Turbulent Entrainment in Jets

Jets expand by plowing up material ahead of them, some of which is "absorbed" and some of which is deflected. The jet surface exerts a force on the medium $\rho v_j^2 A$, where A is the surface area of the head. The medium in turn is acting to slow it down, so the velocity of the head into the external medium simply scales as $v_h = (\rho_j/\rho_0)^{1/2} v_j$. It is very important to note, though, that this only estimates the rate of momentum transfer and neglects some very important dynamical effects. The jet carries momentum of magnitude $\dot{M} U_1$ and a bulk kinetic energy $\frac{1}{2}\dot{M}(U_1^2 + \sigma^2)$, where σ is the random component estimated by the spectral line widths or by random motions observed directly within the jet. It has a force of $\rho_j U_1^2 A_j$ on the background gas and stagnates when the background pressure, $\rho_0 a_{s,0}^2$, is of the same order. Therefore, to ensure that the jet not stall, the Mach number for the jet must be approximately $M_j \geq (\rho_0/\rho_j)^{1/2}$. Notice that for an overdense jet any supersonic flow, and even subsonic flows, will not be stopped simply by ram pressure. But for an underdense jet (a so-called *light jet*) the flow must be supersonic in order to ensure continued propagation. But the jet is ultimately moving into a medium, and the simple dynamical interaction at the head is not the only reason for the flow to slow down. It is likely that the jet will also begin to entrain its environment.

For a subsonic jet, the only one we will deal with here, whatever the details of the mechanism, the entrainment problem can be stated succinctly. The momentum flux of the jet is conserved. Therefore, it is the mass loading from transport of environmental material that slows the jet down. To show this, consider a steady-state jet with a radius r_J and axial velocity v_J which we will also call U_1. Define the mass flux as $\rho\mu = 2\pi \int \rho v_J r \, dr$ and the momentum flux as $\rho M = 2\pi \int \rho v_J^2 r \, dr$. The rates of mass and momentum transport through the jet's periphery are, respectively,

$$\frac{d\mu}{dt} = v_J \frac{d\mu}{dz} = \alpha 2\pi R_J \rho v_J^2, \qquad \frac{d}{dz}\left(\frac{1}{2} R_J^2 v_J\right) = 2\pi \alpha R_J v_J \qquad (71)$$

7.8 More Physical Complications

Here α is the *entrainment efficiency*, a quantity that is usually measured in the laboratory. Since we have already seen that M is constant for the jet, then $v_J \sim \mathcal{M}^{1/2} z^{-1}$ and the width of the jet scales as $R_J \sim z$. For viscous jets, we therefore have a Reynolds number that scales as $M^{1/2} \rho^{-1/2} \nu^{-1}$.

First, let's expand on some of the physics to be included in the discussion. Real jets moving out into stable media rapidly become unstable. Their surfaces start to ripple and eventually disrupt. Even in the absence of strong density gradients, even in the absence of other environmental effects, the jet is Kelvin–Helmholtz unstable. For an axisymmetric jet, this takes the form of shear-generated vortex rings that surround the jet axis, mix with the environment, and ultimately contribute to the demise of the jet. For a fully three-dimensional flow, the disruption mechanism is too complicated to treat by simple arguments, but again it appears to involve both helicity and enstrophy cascade and the eventual growth of long-wavelength modes that produce twisting and kinking of the jet axis. Regardless of the details, however, the point is that simple molecular viscosity isn't likely to be the most important contributor to the viscous interactions within the jet or between the jet and its surroundings.

This point leads to another change in the equations of motion for the jet. We have already discussed the Reynolds stresses and the fact that in a turbulent medium these contribute to the pressure and energy dissipation. Now we need to include the Reynolds terms in the stress tensor. Another important point is that the head of the jet is given by the Bernoulli equation, so that we have an expression for the driving axial pressure gradient. This is, taking P_1 to be the driving pressure and U_1 to be the maximum (free-stream) velocity along the jet,

$$\frac{dP_1}{dz} + U_1 \frac{dU_1}{dz} = 0 \tag{72}$$

Now we take $\hat{z}$ along the jet axis and $\hat{r}$ radially away from the axis. The jet is assumed to be axisymmetric. Let's continue treating the fluid as incompressible, mainly because this case has been well studied in the laboratory. Astronomers don't often have the option of appealing to such experience, no matter how remote. We need to include the turbulent contributions to the stress tensor:

$$\tau_{ij} = \rho(V_i V_j + \langle u_i u_j \rangle) \tag{73}$$

We make use of the boundary layer model. This states that there is no viscous force along the direction of the motion; only shear contributes to the dissipative terms. Then we can reduce the equation of motion for the flow of a steady jet to the form

$$U\frac{\partial U}{\partial x} + V\frac{\partial U}{\partial y} + W\frac{\partial U}{\partial z} + \frac{\partial}{\partial y}\langle \delta u\, \delta v\rangle + \frac{\partial}{\partial z}\langle \delta u\, \delta w\rangle$$

$$= U_1 \frac{dU_1}{dx} + \nu\left(\frac{\partial^2 u}{\partial y^2} + \frac{\partial^2 u}{\partial z^2}\right) \tag{74}$$

Here the mean velocity components along the jet is U, (V, W) are perpendicular to the jet, and $\nabla \cdot \mathbf{U} = 0$. If the velocity fluctuations are uncorrelated, then the nondiagonal Reynolds stresses vanish. This does not happen for shear-generated turbulence; it is always the case that shear produces correlations between axial and radial components of velocity fluctuations. For axisymmetric flow, the equation for the velocity along the jet reduces to

$$U\frac{\partial U}{\partial x} + W\frac{\partial U}{\partial r} + \frac{\partial}{\partial x}(\langle \delta u^2\rangle - \langle \delta w^2\rangle) + \frac{1}{r}\frac{\partial}{\partial r}(r\langle \delta u\, \delta w\rangle) = U_1\frac{dU_1}{dx} \tag{75}$$

The momentum transfer across an axial jet is $\mathcal{M} = 2\pi \int_0^\infty \rho U(U - U_1) r\, dr$. In the presence of turbulence, we still have momentum conservation within the jet; that is, $\mathcal{M}$ is still constant. The conservation condition is

$$\frac{d}{dz}\int_0^\infty U(U - U_1) r\, dr + \frac{dU_1}{dz}\int_0^\infty (U - U_1) r\, dr$$

$$+ \frac{d}{dz}\int_0^\infty [\langle \delta u^2\rangle - \frac{1}{2}(\langle \delta v^2\rangle + \langle \delta w^2\rangle)] r\, dr = 0 \tag{76}$$

Notice that the last term is the difference between the energy densities of the random components along the orthogonal to the jet axis. Now what is happening in this momentum and mass transfer? The bulk flow is being slowed by the entrainment of material from the quiescent external fluid. We haven't included the effects of buoyancy (these probably are not very important astrophysically). The entrainment process, which we glossed over a moment ago with the constant α, is now perhaps a little clearer. The Reynolds stresses build up because a shear layer is created by viscous coupling between the jet and background. This causes the growth of vorticity at the jet boundary, already certainly unstable by the Kelvin–Helmholtz instability, and a vortex sheet begins to mix the jet and background. As the vorticity is advected along the jet axis, its amplitude grows, eventually to the point at which it wraps enough material into the flow to slow the bulk down.

7.9 Some Observational Signatures of Astrophysical Turbulence

Just looking at a medium and saying that it looks random does not mean it is turbulent. Repeatedly throughout this chapter I have tried to stress the point that turbulence is a state of the fluid, not a state of mind! It isn't just the evidence of stochasticity; there must be energetic and dynamical signatures associated with the flow. Perhaps the most important observation pointing to turbulence is the measurement of the velocity and density correlation functions. In particular, the velocity autocorrelation function, because all we ever get to observe is the (*radial*) velocity relative to our line of sight, is the measure of the energy density in the velocity field and therefore a measure of the energy in the cloud.

One of the most important signatures is the presence of larger than thermal widths for line profiles. This is seen best in emission lines from stellar chromospheres and from molecular clouds. Let's start with the second case. The thermal motions expected for clouds are estimated from the excitation temperatures observed for CO $\Delta J = 1$ transitions in the millimeter. Typically these are a few tens of kelvins. These temperatures also agree well with those derived from infrared dust emissivities and from other, more density-sensitive, molecular transitions like those of CS. Such temperatures imply motions at a few tens to hundreds of meters per second. However, the observed line widths are often tens of kilometers per second. In fact, the lines do not even always fit simple Gaussians, as would be expected from the thermal motion alone. Observations frequently show the presence of broad wings, often containing kinetic energies comparable to those of the narrower, but still superthermal, cores (Falgarone and Phillips 1990). This is taken as an indication of intermittency and large-scale turbulent flows. The profiles appear in an enormous range of molecular environments and always have essentially the same profile form.

A similar effect is observed in some stellar chromospheres. Here the temperatures are considerably higher, more in the range of 10^4 K, and the thermal velocities also higher, about 10 km s^{-1}. Yet here too line profiles are often far broader than expected from simple thermal broadening and also show the narrow cores superposed on broad wings. The mechanism for heating and energy dissipation is, of course, very different in the two cases, but the universality of the signature is a good indication that turbulence is involved. In the molecular environment, self-gravity may come into play, something certainly absent in stellar chromospheres. But both have magnetic fields that play a major role in both support and heating, and both are environments where the heating is expected to be supersonic and on the same order as the thermal time scale. In other words, both cases, as different as they appear, may have the same underlying turbulent mechanism.

Appendix A: Stochastic Functions and Their Application to Turbulence

The general methods required for the calculation of turbulence are the same as those required for the modeling of any generalized stochastic process. In the case of random numbers, we need to define several concepts which will serve as the basis for future discussion. The first of these is the *joint probability distribution* and the other is the idea of statistical independence.

Begin with the idea of joint distributions. Consider two variables, both of which may be functions of time, for instance, which have a distribution $P(u(t), v(t)) = P_{uv}(u, v; t)$, where $P_u(u; t) = \int P_{uv}(u, v) \, dv$ is the integrated probability distribution, which is the probability density distribution for the variable u integrated over the full range available for the variable v. We assume that the function $P_{uv}(u, v)$ can be normalized by taking $\int P_{uv}(u, v) \, du \, dv = 1$.

This implies that we have the full range of the function available to us and that there is not any region which will be forbidden for our calculation of the integral. In the case in which the variables are statistically independent, we can calculate a much more direct representation for the variables. In the case in which the variables are uncorrelated, the joint probability that we have just defined is simply vanishing in the limit of the complete domain of integration. There is no reason for this not to vanish, since we have variables which are being averaged over large fluctuations and therefore the integral should vanish over the range available for the variables. In the case of statistically independent variables, however, we can use the equivalent of separation of variables.

$$P_{uv}(u, v) = P_u(u) P_v(v) \tag{77}$$

where it is immediately seen that the integral also breaks into two domains. The densities, which are never in practice really measured, are defined as the probability that a variable will lie in the range of $(u, u + du)$ and $(v, v + dv)$. The cumulant, which is the probability that the variable will lie in the range less than or greater than a certain value, is more clearly defined in the physical measurement and can be used to specify the turbulence spectrum fairly precisely.

To see this, consider that the autocorrelation function is taken to be a function of the shift in the position within the cloud, or the turbid medium, or wherever. The correlation is the Fourier transform of the power spectrum, which is denoted $S(\mathbf{k})$. In the case of a homogeneous Markov process, the value of the autocorrelation function is independent of the origin of the coordinate system we have chosen, and so we can assume that the same is true in the case of fully developed turbulence. The correlation function for a velocity field is defined as the spatial or temporal average of the velocity component $v_i(\mathbf{r})$ between two positions $\mathbf{r}$ and $\mathbf{r} + \mathbf{x}$. The *Taylor hypothesis* states that for fully developed turbulence the time averages at a fixed point within the flow will be the same as the volume or surface averages so that

$$R_{ij}(\mathbf{r}) = \langle v_i(\mathbf{x}) v_j(\mathbf{r} + \mathbf{x}) \rangle \tag{78}$$

where $\mathbf{r}$ is the lag. Notice that this is a convolution, so that the Fourier transform of the function is the product of the individual velocity transforms. Empirically, in the laboratory there are several ways of obtaining this function. Observationally we rarely have the opportunity to measure more than one component of the velocity field. The radial velocity is the component of the velocity in the line of sight, and it is this one that is usually determined in the course of spatially extended observations. As we have discussed earlier in the chapter, the velocity field measured in a molecular cloud, for example, is the structure of the flow projected toward the observer. In fully developed isotropic turbulence, this will be a measure of the total energy of the flow (since the trace of the correlation function is the kinetic energy of the flow). Normally, probes are used (like hot wire anemometers) to measure the transverse and longitudinal components in the flow at different separations. The spacing between the probes is altered and the correlation function measured by taking the power spectrum of the fluctuations in the cross-correlation. However, in astronomical work we usually don't have the luxury of such measurements.

Observationally, the velocity correlation function is not simply a direct measurement of the velocity but is one weighted by the local intensity, $I_v(\mathbf{x})$. For molecular and atomic work, this is one of the basic problems. Excitation changes the determination of the correlation, and the filling factor for the dynamical tracer is extremely important. If the correlation function is large for large wave number (small spatial separation), this may be due to an intrinsic correlation or due to excitation conditions within some substructures in the flow.

One more concept needs to be mentioned, that of a long-normal distribution. Suppose we take a stochastic process that is multiplicative. By this, we mean that the process is a cascading one, the later steps depending multiplicatively on the earlier ones. Consider the realization of a stochastic process X such that $X_n = X_0 X_1 \cdots X_{n-1}$. Then to make this an additive process, we take the logarithm of X_n and thus $\ln X_n = \sum_{k=0}^{n-1} \ln X_k$. Then we can define a process which has the Gaussian properties in this new variable such that

$$P(x) = P_0 \exp\left[-\frac{(\ln x - \langle \ln x \rangle)^2}{\langle \ln x \rangle^2} \right] \tag{79}$$

The distribution that gives rise to this is called the *broken stick* process by ecologists. Imagine a stick that has a total length L. Successive breaking of the stick into lengths ϵ produces a distribution function given by Eq. (79).

Appendix B: Stochastic Differential Equations

Several results in this chapter make use of the idea that a function may not be differential but may still solve a differential equation for the motion. In a random forcing field, a particle undergoes unpredictable jumps in time, but we may be interested only in the evolution of the mean value of the variable. To show how this can be understood, let's examine for a moment the idea of a stochastic differential equation. We begin with the equation

$$\dot{v} = \alpha V(t) v \tag{80}$$

where $v(0) = a$ and $\langle V(t) \rangle = 0$. Assume that V is a random function, depending on some parameter ω that has a distribution function $f(\omega)$. By direct integration we get

$$v(t) = a + \alpha \int_0^t V(t') v(t') \, dt' \tag{81}$$

We now iterate on this solution. That is, substitute the zeroth-order term into the equation (also called Picard's method in classical differential equation theory) and average over the statistical ensemble:

$$\langle v(t) \rangle = a + \alpha^2 \int_0^t \int_0^t \langle V(t')V(t'')v(t'') \rangle \, dt' \, dt'' \tag{82}$$

Assume that the correlation in the fluctuating field V is given by $\langle V(t)V(t')v(t') \rangle \approx \langle V(t)V(t') \rangle \langle v(t') \rangle$ to separate the integrand. Now we can *differentiate* to obtain

$$\frac{d}{dt}\langle v(t) \rangle = \alpha^2 \int_0^t dt' \langle V(t')V(t'') \rangle \langle v(t'') \rangle \tag{83}$$

which is the *equation for the evolution of the mean*. If $\langle V(t)V(t') \rangle = D\delta(t-t')$, which is the definition of a Wiener process, we obtain

$$\frac{d}{dt}\langle v(t) \rangle = \alpha^2 D \langle v(t) \rangle \tag{84}$$

in the limit of $t \to \infty$. This is Bonnet's integral equation for the evolution of the mean value of v in the presence of a fluctuating driving force $V(t)$.

This can also be examined in a more general way, as was first discovered by Langevin following the pioneering work on Brownian motion by Einstein and Smolukowski in the first decade of this century. Let us start instead with the equation

$$\frac{d}{dt}y(t) = ay + b(\lambda, t) \tag{85}$$

where λ is some stochastic variable with a known distribution. Again, the integration is quite simple, yielding

$$y(t) = y(0) + \int_0^t dt' \, e^{a(\lambda)(t'-t)} b(\lambda, t') \tag{86}$$

We will assume that a is a constant, in order to simplify matters for discussion, but it can be seen immediately that if it is a stochastic function, one merely needs to expand the integral into a power series and then evaluate the moments term by term. Nonetheless, we will press on with the discussion for constant "rate" a. The second moment gives

$$\langle y(t)y(t') \rangle = \langle y(0)^2 \rangle + 2D \int_0^t \int_0^t d\xi \, d\eta \, e^{a(\eta+\xi)} \langle b(\lambda, \xi)b(\lambda, \eta) \rangle \tag{87}$$

as we had before.

Now, in the case of systems which are strongly nonlinear, we cannot use this iterative approach as it stands. The modification can be easily achieved, though, through the use of the Fourier space representation of the equation. Specifically, we know that the Fourier transform of a product is the convolution of the transforms and vice versa, so the algebra can be considerably reduced using this trick.

Appendix C: Fractals

It seems that it is nowadays impossible to write anything on turbulent fluids without mentioning *fractals* and *chaos*. It is not just a fad, though. The idea of a fractal is that it is a space-filling self-similar structure. In the sense that it is space filling, it behaves like a volume

Appendix C: Fractals

or surface. In the sense that it is self-similar and nowhere dense, it behaves like a line or a set of disconnected points. In fact, the name derives from Mandelbrot's attempt to create an appropriate *neologism* that captures the fact that while space filling, these structures behave dimensionally as if they were between points and their appropriate Euclidean dimension.

The original definition of a fractal is that it is an object whose structure persists down to some very small scale. Put another way, no matter what magnification you use it looks the same, which is to say it remains self-similar. In this sense it means that we can take the density to be a nonlinear function of the volume, or, put another way, fractals are frothy. It is the most basic property of a fractal that it looks dense only because it is certain that any line of sight will encounter the fluid. However, intrinsically, the volume is only partially filled and in fact may be very sparsely filled indeed. One simple example is to consider a cirrus-filled sky. The clouds may appear extremely dense, but when you fly through them you see they are wispy and filamentary, distributed chaotically through the upper atmosphere in such a way that no matter what direction you look in there is a filament in the way. In this way, you get the impression of the structure.

A fractal has the property that the number of elements on a scale ϵ is given by

$$N(\epsilon) = (L/\epsilon)^D \tag{88}$$

where L is the characteristic length in the medium. A way of looking at this is to imagine that you take tiles to cover a surface and then look at the number of tiles that contain completely elements of the fluid at each scale of magnification ϵ. (See Fig. 7.3.) Notice that this means that the fractal dimension of the medium is given by $D = \log N(\epsilon)/\log \epsilon$. The tiling procedure also serves to define the *Haussdorf–Besicovitch dimension* of a medium (Barnsley 1988). Consider a p-dimensional space (in the case of a cloud, for example, $p = 3$) and a coordinate system defined by a metric $d(x, y)$. The diameter of a region A is given by diam $(A) =$

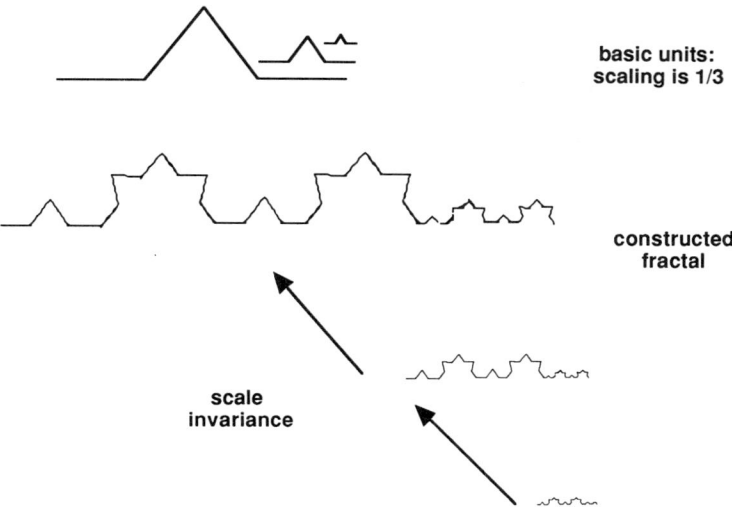

Figure 7.3 Fractal curve. The scaling between the individual segments is a factor of 3.

$\sup\{d(x, y) : x, y \in A\}$. Then the dimension of A is given by

$$\mathcal{M}(A, p, \epsilon) = \inf\left\{\sum_{k=0}^{\infty} (\text{diam}(A))^p : \{A_k\} \in \mathcal{A}, \text{ and } \text{diam}(A_k) < \epsilon \text{ for } k = 0, 1, \ldots\right\} \quad (89)$$

and the Haussdorf–Besicovitch dimension is $\mathcal{M}(A, p) = \sup\{\mathcal{M}(A, p, \epsilon) : \epsilon > 0\}$. For a fractal structure this will not be the same as the Euclidean dimension p for the medium (it will be lower than p).

This box-counting method has been developed by Sreenivasan and Mereveau (1986) (see also Sreenivasan 1991) and has been employed by Bazell and Desert (1988). (See Fig. 7.4.) However, free axisymmetric jets do not appear to show this form of scaling (Miller

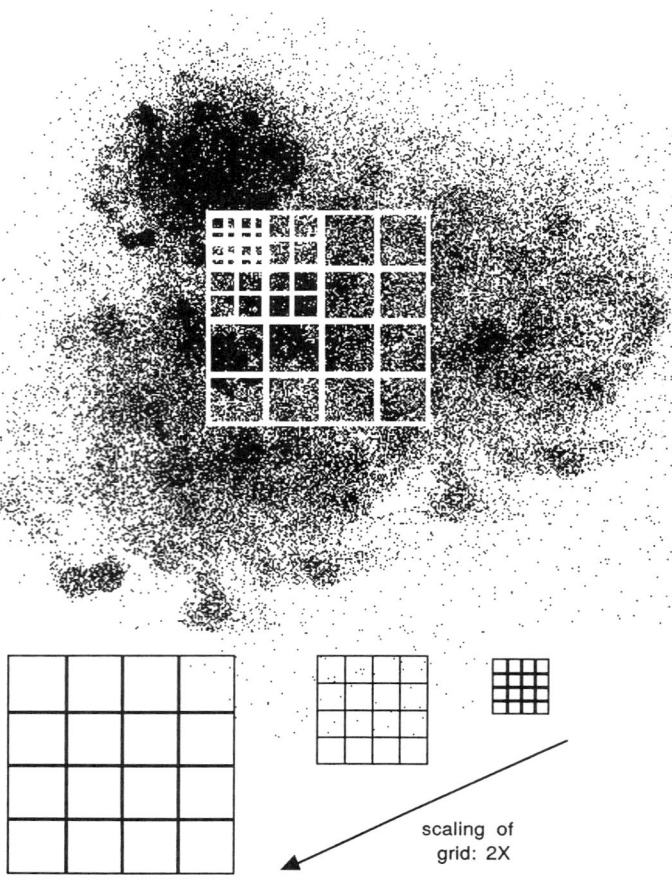

Figure 7.4 Schematic fractal grid for the analysis of a molecular cloud.

and Dimotakis 1991) and caution is advised before drawing sweeping conclusions about the fractal geometry of a surface. Also, it should be remembered that astronomical observations are generally of three-dimensional objects projected onto the surface of a sphere[5] and thus we are integrating through the fractal. This makes the filling factor appear larger than it truly is because of the probability of encountering some turbulent element. It is best to also include the velocity field, at least by making cuts through the medium at different velocities. This can be done with modern Fabry–Perot techniques and also with interferometric spectral line observations.

Suggested Readings for Fractals

Barnsley, M. (1988). *Fractals Everywhere*. San Diego: Academic Press.
Bazell, D., and Desert, F.-X. (1988). Fractal structure of the interstellar cirrus. *Astrophys. J.* **333**, 353.
Mandelbrot, B. (1974). Intermittent Turbulence in Self-Similar Cascades: Divergence of High Moments and Dimension of the Carrier. *J. Fluid Mech.* **62**, 331.
Mandelbrot, B. (1982). *The Fractal Geometry of Nature*. San Fransisco: W. H. Freeman.
Miller, P. L., and Dimotakis, P. E. (1991). Stochastic geometric properties of scalar interfaces in turbulent jets. *Phys. Fluids A* **3**, 168.
Procaccia, I. (1984). Fractal structures in turbulence. *J. Stat. Phys.* **36**, 649.
Sreenivasan, K. R. (1991). Fractals and multifractals in fluid turbulence. *Annu. Rev. Fluid Mech.* **23**, 539.
Sreenivasan, K. R., and Meneveau, C. (1986). The fractal facets of turbulence. *J. Fluid Mech.* **173**, 357.
van Buren, D. (1989). The volume filling factor of the infrared cirrus is 0.2. *Astrophys. J.* **338**, 147.

Appendix D: Nonlinear Maps and the Transition to Chaos

About a decade ago, it was realized that there is a class of extremely simple equations which display much of the behavior one has come to associate with turbulence. Specifically, the quadratic map was shown, for certain values of the control parameter of the map, to display stochastic behavior. Partly because of this, and also because of the incredible simplicity of the mapping, this has been one of the most productive (at least in the number of papers generated) areas in the recent history of physics. Because of its ubiquity in the literature, and also because of its simplicity, we will digress on it for a moment. It has virtually no astrophysical application *as such*, but it shows some of the methods which may later be used to advantage in the modeling of turbulence in cosmic environments.

---

[5] This is what the poet Omar Khayyám was talking about: "Above, around, about, below,/'tis nothing but a magic shadow-show."

The simplest nonlinear system which can be shown to have an analytic solution was first proposed by Volterra and Lotka, independently, for the modeling of a biological population with birth and death. It was the codification of Verhulst's investigations of populations in the last century. A population with a birthrate a can be written as

$$\frac{dx}{dt} = ax(1-x) \tag{90}$$

which has the analytic solution

$$x(t) = \frac{x(0)}{x(0) + (1 - x(0))e^{-at}}$$

which saturates in a time a^{-1}. The mapping is the discrete system which represents the equation. One used as a representation of the Verhulst equation is

$$x_{n+1} = ax_n(1 - x_n) \tag{91}$$

and it is this which has been the most carefully examined. The iteration is begun, for a fixed a, with some arbitrary value for x on the interval $[0, 1]$. The fixed points of the system are 0 and 1, so these should be avoided. Now as the value of a is increased, a series of bifurcations takes place. First, for $a < 3.5$ there is only one root. As this value is crossed, two roots appear which are separated from the first—a bifurcation has been seen. Then, as one continues to increase a, a doubling process takes place. This continues until one is near $a = 4$, at which point odd numbers of solutions appear and the system begins to go chaotic. That is, there do not appear to be any fixed points or stable iterates; the values continue to change on each iteration of the map. The value $a = 4$ is the upper limit to the range of the control parameter, the birthrate. This is one which you can easily try out even on a pocket calculator, and it will prove on such a device to be even more interesting than if done on a computer. The roundoff error acts like a noise to the calculation and simulates a stochastic perturbation in the deterministic system. This causes the transition to chaotic behavior to occur earlier than it would for a higher-accuracy calculation. Thus, one can simulate the effects in the dynamical map of noise.

Why is this system thought of as an effective model for the turbulent transition even though it is so amazingly simple? The most direct answer is historical. In the first edition of Landau and Lifshitz's book *Fluid Mechanics*, Landau included a discussion of a possible picture for the onset of turbulence. It is that the fluid passes through a series of bifurcations in either frequency of oscillation or wave number of the critical solutions. The fact that the transition is observed to take place through a cascade of multiple, interacting states as the Rayleigh number is increased is one which forces the intuitive picture that Landau was arguing for. Recent experimental work on turbulent convection shows that this idea of successive bifurcation is an adequate, although by no means complete, description of the process of turbulence. In addition, the onset of chaos in optically bistable systems, such as lasers, has been shown to be modeled in similar fashion.

The reason for the applicability of this model to fluids is that the Navier–Stokes equation has a nonlinear term—the inertial or convective term—which is responsible for the strong coupling between the temperature gradient and the velocity field. It was Lorenz who showed that there is also a set of nonlinear ordinary differential equations which behaves in the same way for compressible convection.

References

Bartlett, M. S. (1965). *An Introduction to Stochastic Processes.* Cambridge University Press.

Bonazzola, S., Falgarone, E. Heyvaerts, J., Pérault, M., and Puget, J. L. (1987). Jeans collapse in a turbulent medium. *Astron. Astrophys.* **172**, 293.

Brown, G. L., and Roshko, A. (1974). On density effects and large structure in turbulent mixing layers. *J. Fluid Mech.* **64**, 775.

Canuto, V. M., Goldman, I., and Chasnov, J. (1988). Turbulent viscosity. *Astron. Astrophys.* **200**, 291.

Chandrasekhar, S. (1951). The invariant theory of isotropic turbulence in magnetohydrodynamics. *Proc. R. Soc. London* **A204**, 435 [(1950) *ibid.* **A207**, 301.]

Dickman, R. L. (1985). Turbulence in molecular clouds. In *Protostars and Planets, II.* (p. 150). D. Black and M. S. Matthews, eds. Tucson, Ariz.: University of Arizona Press.

Falgarone, E., and Phillips, T. G. (1990). A signature of intermittency of interstellar turbulence: The wings of molecular line profiles. *Astrophys. J.* **359**, 344.

Falgarone, E., Boulanger, F., and Duvert, G., eds. (1991). In *Fragmentation of Molecular Clouds and Star Formation: IAU Symp. 147.* Dordrecht, The Netherlands: Kluwer.

Fleck, R. C. (1983). A note on compressibility and energy cascade in turbulent molecular clouds. *Astrophys. J. (Lett.)* **272**, L45.

Forster, D., Nelson, D. R., and Stephen, M. J. (1977). Large-distance and long-time properties of a randomly stirred fluid. *Phys. Rev. A* **16**, 732.

Gardiner, J. (1983). *Handbook of Stochastic Processes.* Berlin: Springer-Verlag.

Henricksen, R. N., and Turner, B. E. (1984). Star cloud turbulence. *Astrophys. J.* **287**, 200.

Hoyle, F. (1953). On the fragmentation of gas clouds into galaxies and stars. *Astrophys. J.* **118**, 513.

Hussein, A. K. M. F. (1983). Coherent structures—Reality and myth. *Phys. Fluids* **26**, 2816. [See also Hussein, A. K. M. F. (1986). Coherent structures and turbulence. *J. Fluid Mech.* **173**, 303.

Kida, S., and Orszag, S. A. (1990). Enstrophy budget in decaying compressible turbulence. *J. Sci. Comput.* **5**, 1.

Kolmogorov, A. N. (1941). *Dokl. Akad. Nauk SSSR* **26**, 115. Local structure of turbulence in incompressible flow. [See also Kolmogorov, A. N. (1962). A refinement of previous hypotheses concerning the local structure of turbulence in a viscous incompressible fluid at high reynolds number. *J. Fluid Mech.* **13**, 82.

Kraichnan, R. H. (1967). Inertial ranges in two-dimensional turbulence. *Phys. Fluids.* **10**, 1417.

Leorat, J., Passot, T., and Pouquet, A. (1990). Influence of supersonic turbulence on self-gravitating flows. *Mon. Not. R. Astron. Soc.* **243**, 293.

Lighthill, M. J. (1952). On sound generated aerodynamically. I, general theory. *Proc. Roy. Soc.*, **211**, 564.

Low, C., and Lynden-Bell, D. (1976). The minimum Jeans mass or when fragmentation must stop. *Mon. Not. R. Astron. Soc.* **176**, 367.

Montgomery, D. (1989). Magnetohydrodynamic turbulence. In *Lecture Notes on Turbulence: NCAR-GTP Summer School* p. 75. J. R. Herring and J. C. McWilliams, eds. Singapore: World Scientific Publishers.

Oboukhov, A. M. (1962). Some specific features of atmospheric turbulence. *J. Fluid Mech.* **13**, 77.

Passot, T., Pouquet, A., and Woodward, P. L. (1988). On the plausibility of Kolmogorov-type spectra in molecular clouds. *Astron Astrophys.* **197**, 228.

Pedretti, C. (1980). *Leonardo da Vinci: Nature Studies from the Royal Library at Windsor Castle.* New York: Johnson Reprints. [See also Pedretti, C. (1980). *The Codex Hammer (Formerly the Codex Leicester) by Leonardo da Vinci.* London: Christie, Manson, and Woods.]

Perault, M., Falgarone, E., and Puget, J. -L. (1985). *Astron. Astrophys.* **152**, 371; 1986, *Astron Astrophys.* **157**, 1399.

Pouquet, A., Passot, T., and Leorat, J. (1991). Numerical simulations of turbulent compressible flows. In *Fragmentation of Molecular Clouds and Star Formation: IAU Symp. 147.* E. Falgarone, F. Boulanger and G. Duvert, eds. Dordrecht, the Netherlands: Kluwer.

Rott, N. (1990). Note on the history of the Reynolds number. *Annu. Rev. Fluid Mech.* **22**, 1.

Scalo, J. (1987). Theoretical approaches to interstellar turbulence. In *Interstellar Processes* (p. 349). D. J. Hollenbach and H. A. Thronson, eds. Dordrecht, The Netherlands: Kluwer. [See also Scalo, J. (1985). Fragmentation and hierarchical structure in the interstellar medium. In *Protostars and Planets*, Vol. II (p. 201). D. Black and M. S. Matthews, eds. Tucson, Ariz.: University of Arizona Press.]

Swinney, H. L., and Gollub, J. P., eds. (1985). *Hydrodynamic Instabilities and the Transition to Turbulence.* New York: Springer-Verlag.

CHAPTER 8

Outflows and Accretion

> *But don't panic; base 8 is like base 10, really, if you're missing two fingers.*
> Tom Lehrer

8.1 Introduction

The aim of this chapter is twofold. The first is to show something of the theory of outflows, whether from stars or from any other source such as galactic nuclei, and to introduce some of the methods for the diagnosis of the condition and its dissection. More on the observational techniques will be found in the chapter on diagnosis of mass flows. The second aim is more pedagogical: to show that accretion isn't really all that different from outflow.

The first part of this chapter develops some of the basic dynamical tools for the spherically symmetric case and will serve as an extension for the inclusion of rotation, magnetic fields, and other departures from sphericity. Magnetic field effects and angular momentum transfer in and by winds are dealt with next. Then we change the point of view, looking at flows coming from the outside in, and examine some aspects of accretion. We will also deal with problems related to accretion disks. Finally, we look at the complication of cooling flows, extended mass distributions, and accretion in clusters of galaxies.

First, though, a quick plunge into the cold waters of history.

8.2 Historical Overview of Winds, Especially from Stars

The problem of recognizing that stars can lose mass is more complicated than one might initially think. The first emission line stars were recognized quite early in the period of visual spectroscopy; in fact, Secchi included them in his class V (bright line stars) in the 1850s. These included γ Cas and several other well-known Be stars. Maury, following her discovery of the binary nature of β Lyr and of its anomalous emission line spectrum, also pointed to several stars in the southern hemisphere, like γ Vel, which display strong emission (Clerke 1902). In the 1850s and 1860s the application of spectroscopy to the Sun showed that the outer layers, the chromosphere being observed only in eclipse, were seen in emission, in contrast to the Fraunhofer spectrum of the photosphere, which appeared to obey Kirchhoff's law quite well. So there were already seeds for the eventual recognition that the temperature gradients of the outer layers of stars can be nonmonotonic. But it is a long way from the recognition that the peripheral regions of stars can be hot to the determination that this material is moving.

Several events contributed to this. First, the observations prior to the turn of the century were made with very low resolution spectrographs—often objective prisms [as in the Henry Draper (HD) catalogue effort] or visual spectrographs. Emulsions were slow and many sites were located at sea level or in murky climates, so there were not too many hours during which observations could be made. But with all of these problems, many types of stars now known to possess strong outflows were identified. In particular, Wolf and Rayet described the stars now named after them. The peculiar spectrum of P Cyg was observed. The distinction between nebular and stellar absorption lines was recognized and, by the turn of the century, the Harvard scheme had been applied to a large enough number of stars so that the anomalous few could be recognized easily and isolated for more detailed study.

It is in this climate that we find the paper by Schuster (1905) discussing the transfer of radiation (in a foggy atmosphere) by scattering. Ignoring the effects of mass motions, he discussed the fact that in an extended atmosphere scattering would produce emission lines and that the Wolf–Rayet (WR) and related objects might be due to such an environment. Schuster's essential point was that the integrated intensity weighted over the scattering function, $\frac{1}{2}\int_{-1}^{1} I(\mu')p(\mu, \mu')\,d\mu'$, will produce emission if it exceeds S, the source function for the photosphere, as will certainly happen in the case of an extended atmosphere. K. Schwarzschild, at the same time, introduced the idea of a reversing layer and, in the assumption of thermodynamic equilibrium, produced emission lines by an inversion in the temperature gradient (by extension of the same mechanism by which the lines are formed in the photosphere). Neither questioned the idea of mechanical equilibrium and stationarity, although Schwarzschild did consider the conditions under which the medium would be convectively unstable.

The possibility that stars change mass was ignored through most of the

8.2 Historical Overview of Winds, Especially from Stars

early period of stellar interiors work, largely because it was a difficult enough problem to solve the equations of equilibrium. Polytropic models, including eventually radiation pressure, were produced by Jeans and Eddington (see especially Eddington 1927, a work still definitely worth reading). The latter eventually, by the late 1920s, recognized that pulsational equilibrium is possible even if the star is globally in mechanical equilibrium, a result thoroughly developed for adiabatic stars by Rosseland in the 1930s. In the meantime, Shapley had computed the profile expected for a pulsating star. It was N 1901 Per (= GK Per) which first made the idea of mass loss palatable and paved the way for Beals' work (1931) on the spectrum of P Cyg. There are a few odd digressions during this period, prior to the 1950s, especially Chandrasekhar's solution for the line profile in an expanding atmosphere (described best in Chandrasekhar 1945), but otherwise the subject did not come up again until Kuiper's work on β Lyr and O. Struve's synthetic model for the line profiles in the Be stars. D. McLaughlin and S. S. Huang generalized the discussion of interpreting line profiles for rapidly rotating stars, demonstrating that the Be star phenomenology could be unified by assuming a circumstellar disk. All of the work on mass loss was, however, confined largely to explosive or pathological cases.

It was not until Greenstein and Deutsch demonstrated the existence of the mass loss in α Her that it was made forcibly clear to astronomers that mass loss is a general, not freak, phenomenon among stars. The comparison between the line profiles observed for giants and supergiants and the strange stars like P Cyg gradually forced a unification into the discussion of mass loss. Beal's (1951) generalized discussion of line formation in moving atmospheres gave the general name "P Cygni profile" to the characteristic signature of mass loss. Sobolev (1960) developed a general theory for line formation using the large velocity gradient approximation which has proved to be one of the most robust techniques ever developed in radiative transfer, especially as extended by Castor (1970) (see also Kuan and Kuhi 1975 for a useful bibliography and discussion of the general problem of line formation).

The work on mass loss took a different theoretical turn for the Sun during the Second World War when Edlen recognized the lines of Fe XIV in the optical solar spectrum and a temperature of greater than 10^6 K was assigned to the inner corona (the outer corona, which shows a reflected solar spectrum, is the F corona and due to zodiacal dust). Two models vied for supremacy during the period prior to the first interplanetary monitoring platform (IMP) satellite launch: Chamberlin (1961) and Parker (1958, 1963). The former is a breeze model, under the assumption that the corona can cool conductively, and it predicted that the density of material at the

Earth's orbit is about 30 cm^{-3}. The latter is a fully hydrodynamical treatment, assuming initially that the expansion is isothermal, and came far closer (of order a few cm^{-3}) to the proper density and predicted a velocity of about 400 km s^{-1}, which Parker also generalized to the case of a stellar wind in addition to the Sun. So by the end of the decade, the idea that stars can have low-level supersonic mass loss was well accepted. The Be stars remained a problem, but their connection to the shell stars, which are eruptive episodic mass losers, suggested that the mass loss need not be steady and could be due to rotational ejection.

The discovery of strong P Cyg profiles in the ultraviolet spectra of ζ Pup and γ Vel with terminal velocities of order 3000 km s^{-1} from sounding rocket observations seems to have been the critical trigger in starting work on radiatively driven winds. Within one year, Paczynski, Underhill, Lucy and Solomon (1970), and Michaud (1970) had invoked radiation pressure to drive mass motions, although only Lucy and Solomon and Michaud developed the theory of line transfer, both influenced by Milne's (1927) paper on radiative force transferred through spectral lines.

The final stage in this line of development was the generalization of the Parker solution for radiative driving by Castor, Abbott, and Klein (1975) allowing for the line formation as a function of depth in the atmosphere and including the effects of velocity gradients on the radiative driving in its most detailed form. There are many other strands to this cloth which we will examine as the subjects come up—this short rendition of the historical development is by no means exhaustive. The theory is still evolving and deepening, and it is useful to keep in mind how the paradigms have altered in the past few decades in order to anticipate where some of the assumptions, now deeply imbedded in the theory of mass loss, can be extended or altered. So to understand the basic structure of the problem, let us examine the most general mathematical model for a steady outflow from a star—the Parker solution for an isothermal wind.

8.3 The Isothermal Wind Problem

8.3.1 Escape of an Atmosphere: Subsonic Evaporation

Consider a spherical body with a mass M and radius R surrounded by a tenuous atmosphere. This atmosphere cannot be completely stationary in the sense that there will always be the few particles which leak out at speeds above the escape velocity. For since the layer can be, for the moment, assumed to be in thermodynamic equilibrium, there will always be some particles in the distribution function, $f(v)$, whose speeds lie above

8.3 The Isothermal Wind Problem

v_∞ so that there will be a net current:

$$J = \frac{1}{2}n_0(t) \int_{v_\infty}^{\infty} f(v)v^3 \, dv \tag{1}$$

where $n_0(t)$ is the density as a function of time. Assume that the mass of the body is not altered substantially by the loss of the particles so that v_∞ is time independent, and also assume that collisions can always repopulate the tail of $f(v)$. Also assume that the distribution function is stationary and that flow is diffusive. This latter point is simply stating that the dynamical equations are not needed to calculate the location of the atmosphere and that the particle loss can be accounted for by the continuity equation alone. Then

$$\frac{\partial n_0}{\partial t} + \frac{1}{r^2}\frac{\partial r^2 J}{\partial r} = 0 \tag{2}$$

or, integrating over the volume,

$$\dot{M} = -4\pi R^2 J \tag{3}$$

This equation implies that there is always some loss from the atmosphere, which increases as v_∞ is approached by the mean velocity of the distribution function, the thermal speed v_{th}. As a rule of thumb, if the thermal speed is of the same order as the escape velocity the atmosphere will be in dynamical flow and no static state will be possible. So

$$T_\star \approx \frac{\mu}{\mathcal{R}}\frac{GM}{R} \approx 4 \times 10^7 \frac{M/M_\odot}{R/R_\odot} K \tag{4}$$

is an approximate critical value (to within an order of magnitude) for the heating of an atmosphere to the point where it will become largely unbound on the dynamical time scale. Of course, this is an overestimate of the required temperature because this value would imply that the material would be explosively unstable. But if we are even an order of magnitude away from this, there will be some net loss of the atmosphere simply because it is hot. This is the so-called Jeans escape mechanism, first discussed by J. H. Jeans in his book on *The Dynamical Theory of Gases*.

8.3.2 Dynamical Mass Loss

The problem we are then faced with is to find under what conditions an atmosphere will be unstable to a flow but stable to explosive loss. Let us start with the equations that will come to dominate the future discussions. We first know that regardless of the driving mechanism, for compressible flow the continuity equation expresses mass conservation:

$$\frac{\partial}{\partial t}\rho + \nabla \cdot \rho \mathbf{v} = 0 \tag{5}$$

where ρ is the mass density and $\mathbf{v}$ is the velocity of the bulk flow (here taken by averaging over the thermal distribution function). The equation for momentum conservation, in the presence of a gravitational field, is

$$\rho\left(\frac{\partial}{\partial t} + \mathbf{v} \cdot \nabla\right)\mathbf{v} = -\nabla p - \rho \nabla \Phi \tag{6}$$

where Φ is the gravitational potential, p is the pressure (supplied by some equation of state), and we assume that the potential is not a function of time. The energy equation is

$$\frac{\partial}{\partial t}\rho E + \nabla \cdot \rho \mathbf{v}\left(E + \frac{\gamma}{\gamma-1}\frac{p}{\rho} + \Phi\right) = 0 \tag{7}$$

neglecting loss terms other than adiabatic losses due to expansion. Notice that we have explicitly excluded forces other than gravitation in the momentum equation and have assumed that the material moves without heating or cooling above the source region. Now we choose an equation of state, $p = a_s^2 \rho$ and a_s, the sound speed, is assumed to be constant; this is the isothermal wind problem. This is a very special choice of equation of state. It assumes that the energy balance is fixed and that the velocity dispersion of the particles is constant with distance as the wind expands. Yes, we will find that here we have a dynamical outflow, called a *wind*, but this is still to come.

Choose the flow to be spherically symmetric. Also, take the mass of the atmosphere to be negligible so that the gravitational field is that of a point source. This last assumption is very important; without it we cannot write a full hyperbolic system of equations. We can take the flow to be $dE/dt = 0$ so that the energy in the flow will be shown eventually to be constant. Finally, and here we are on shaky ground without further examination, we *assume that the flow is steady, that is, that the outflow is independent of time.* Then

$$v\frac{dv}{dr} = -\frac{a_s^2}{\rho}\frac{d\rho}{dr} - \frac{GM}{r^2} \tag{8}$$

which has a simple analytic solution:

$$\frac{1}{2}v^2 + a_s^2 \ln \rho - \frac{GM}{r} = \mathscr{C} \tag{9}$$

8.3 The Isothermal Wind Problem

where the constant is a function of the streamline we choose. Notice that this is essentially the Bernoulli equation, which states that the pressure drop in a flow is driven by the acceleration of the flow. The continuity equation is $\dot{M} = 4\pi r^2 \rho v$, so that

$$\frac{1}{2}v^2 - a_s^2 \ln v - 2a_s^2 \ln r - \frac{GM}{r} = \mathscr{C} \tag{10}$$

If the flow is polytropic, the solution is

$$\frac{1}{2}v^2 + \frac{\gamma}{\gamma - 1}\frac{p}{\rho} + \Phi = \mathscr{C} \tag{11}$$

In either case, there is an infinite family of solutions, depending on the Bernoulli constant. Most of the solutions are double-valued. For solutions that have $v = 0$ at $r = 0$, only one solution passes to infinity with finite velocity, as we will see from analyzing the differential equation. There is also a family of solutions which don't vanish at the origin, and the structure of the solution in the vicinity of the critical point is a separatrix or X-type critical point.

There is far more physics imbedded in the isothermal wind model than would seem to be the case. Suppose we rewrite the wind equation in the following form:

$$(v^2 - a_s^2)\frac{dv}{dr} = \frac{2a_s^2}{r} - \frac{GM}{r^2} \tag{12}$$

and make an additional, very strong, assumption—we want to look only at those solutions which are able to pass to infinity at finite speed. If we are to have a wind, we must have material actively driven to escape from the star at the stellar surface. The gravitational acceleration decreases with increasing distance in a spherically symmetric outflow, so if the initial driving is large enough the acceleration of gravity may be incapable of slowing the material down at any radius sufficiently to bring it to rest at a large distance from the star. This implies that the flow must be monotonic and that the velocity gradient must be positive definite at all radii; we can also consider that the flow will consequently be guaranteed to be time independent. Thus there is a critical point in the flow at which the pressure gradient balances the gravity:

$$r_\star = \frac{GM}{2a_s^2} \tag{13}$$

at which, if $dv/dr \neq 0$, we have $v_\star = a_s$. That is, at the critical distance from the star, the flow becomes transonic. If the velocity gradient is

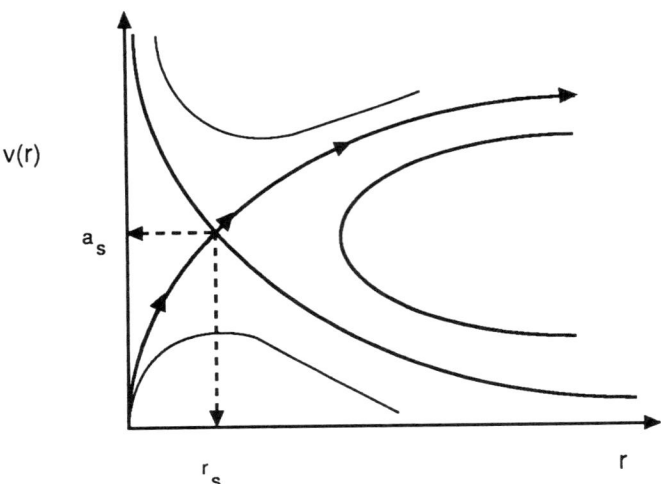

Figure 8.1 Velocity field for the Parker solution. The sonic radius ($v = a_s$) is indicated as r_s.

positive, then the consequent flow, farther from the sonic point, is hypersonic for all distances. The flow accelerates outward, incapable of reaching a static state at any distance from the star, and a rarefaction wave moves back into the star, becoming stable when the sonic point is established. (See Fig. 8.1.) Notice that the critical solution is determined by the ability of the pressure to build up in the flow as a result of the retardation of the flow from gravity (see the digression on the de Laval nozzle analogy).

In fact, the real condition that governs the flow is the fact that $\dot{M}$ remains constant with distance, so it is not the nonlinearity of the momentum equation which is important but the fact that there is a steady-state relation between the density and velocity at every point in the flow. The source of material literally manages to push its way to infinity, and this in either a polytropic or isothermal wind guarantees that the matter escapes from the confining gravitational field.

As an alternative solution, we could choose to have the velocity gradient vanish at $r_\star$. Then the flow must reach a maximum and coast to a stop with large distance. The flow, since it never reaches the supersonic state or escapes from the star, establishes an extended atmosphere which is ultimately a slowly expanding medium—a *breeze solution* of the sort discussed by Chapman. The wind solution was found by Parker as early as 1956, although most of the discussion is contained in Parker (1963). All other solutions to the equations of motion other than the critical solution are (1) not monotonic (so that there will be an extremum in the velocity

8.3 The Isothermal Wind Problem

field) and (2) not necessarily single valued (so that the solution cannot be steady state).

8.3.3 Wind Tunnels and Stellar Winds—The de Laval Nozzle Analogy

We can add that any driving force which is sufficiently strong in its initial acceleration will produce the same general behavior of the outflow. To see this, let us write the most general kind of force ρa instead of gravitation, and allow for the area to vary for a flow tube in a fashion different from a spherically symmetric flow:

$$\rho v \Sigma = \text{constant} \tag{14}$$

and

$$v \frac{dv}{dr} = -\frac{1}{\rho} \left(\frac{\partial p}{\partial \rho} \right)_s \frac{d\rho}{dr} + a \tag{15}$$

so that we obtain

$$(v^2 - a_s^2) \frac{dv}{dr} = a_s^2 \frac{d}{dr} \ln \Sigma + a \tag{16}$$

and here we have taken Σ and a to be implicit functions of the radius. Notice that the form of the equation is unchanged from the isothermal solution but that, depending on the precise form of the cross section and the acceleration, we may have multiple solutions for the critical point $r_\star$. Notice, however, that the sound speed is not taken to be a constant in this treatment but is allowed to be a function of distance as well. For an isothermal flow, if there is no retarding force (as in a wind tunnel, where the walls provide the only reaction force) the supersonic transition takes place at the extremum of the cross section. This is the principle of a wind tunnel or of a de Laval nozzle (a nozzle with a taper that narrows to a very thin throat and then flares out rapidly producing a very high velocity jet, first developed by de Laval for use in steam engines about a century ago).

This analogy is often used for the description of the flow in a gravitational field, but it is really not relevant. The only important point that it demonstrates is that the retardation of the flow is due to the gravitational "walls" and that there is a reservoir pushing material through the "throat" of the "nozzle." Really, the continuity equation is the most critical determining factor in this flow—the statement that the flow must be steady state is the fact that forces the behavior of the velocity field at the critical point! Drop this assumption and the entire solution is changed. Waves become possible which can become unstable, and these we shall later face head on. But for now, let's look in more detail at what this steady-state solution means.

The boundary conditions are relatively simple. We know that the solution must pass through the critical point. The velocity must be zero at the stellar surface and finite and large at infinity. Since the isothermal wind is governed by a first-order equation, the solution is fixed by $r_\star$. It is continued both inward and outward from that point. Thus the solution is completely specified. But this is only one possible solution, because there may be a more complex topology at the sonic point if we go to a more general equation of state.

Notice what we have assumed in order to achieve this solution. First, the motion is steady state (how many times can this be emphasized?). Then the star is assumed to be static—no rotation or pulsation. The heat sources are not specified. The sound speed is taken to be constant, or at least simple. Finally, there is no magnetic field. But what we have gained is dramatic. For, once we know the solution at the critical point, we can continue it down to the stellar surface, and also if we know the density at the sonic point, we can specify the mass loss rate uniquely. The entire solution can be nondimensionalized so that we have solved the problem once and for all.

8.3.4 Effect of a Temperature Gradient

The energy equation must be included if we have temperature variations in the flow. This is because both convective (that is, due to fluid motions) and conductive losses and transport (due to temperature gradients) can change the pressure and therefore the driving of the wind near the base and near the critical point. Notice that the assumption of the isothermal equation of state means that the driving of the outflow depends uniquely on the velocity field, there is a one-to-one feedback between changes in the velocity and the density, and therefore we obtain a simple topology for the flow as a function of mass loss rate near the critical point.

Let us drop the assumption of constant a_s. Then

$$\frac{\partial T}{\partial t} + \mathbf{v} \cdot \nabla T = \nabla \cdot \kappa \nabla T \tag{17}$$

Now assuming that the only mode of energy transport is conduction, for a static atmosphere (one which is only boiling off in Jeans escape), $\nabla \cdot \kappa_C \nabla T = 0$ or

$$\frac{1}{r^2}\frac{d}{dr}\left(\kappa_0 r^2 T^{5/2}\frac{dT}{dr}\right) = 0 \tag{18}$$

8.3 The Isothermal Wind Problem

which gives the structure of a static corona and completely determines the density distribution. If flow is possible, the conductive losses must be included in the energy conservation equation. The equation of state is explicitly a function of temperature:

$$p = \frac{\mathcal{R}}{\mu} \rho T \tag{19}$$

so that

$$v\frac{dv}{dr} = -c_v \frac{\gamma}{\gamma-1}\left(\frac{dT}{dr} + \frac{d\ln\rho}{dr}\right) - \rho\frac{GM}{r^2} \tag{20}$$

If we have chosen a polytrope, then $T \sim \rho^{\gamma-1}$, so again we have a direct relation between the density and velocity through the continuity equation:

$$T = T_0\left(\frac{r_\star^2 a_s}{r^2 v}\right)^{\gamma-1} \tag{21}$$

What is usually done to solve this equation? We have direct measurements for the Sun and, from ultraviolet satellite observations of other stars, of the temperature gradient through the chromosphere and corona. From phenomenological models, one can place limits on the gradient in the temperature in the vicinity of the photosphere. These are incorporated into the solar wind solution but cannot be used (yet—Voyager will take a while to get there) for extrasolar systems. So we need another prescription for solving the general wind case.

Let us look at a scaled solution, due to Roberts (1971; see also Roberts and Soward 1972; Freeman and Johnson 1972), which takes $\tau = T/T_0$ for the temperature, $\psi = mv^2/kT_0$ for the kinetic energy, and $\lambda = GMm/kT_0 r$ for the inverse scaled radius; T_0 is the temperature at the base of the corona, and $m = m_p/2$. Then

$$\frac{1}{2}\frac{d\psi}{d\lambda} = \frac{1 - 2\tau/\lambda - d\tau/d\lambda}{1 - \tau/\lambda} \tag{22}$$

for the momentum equation and

$$\frac{1}{2}A\tau^{5/2}\frac{d\tau}{d\lambda} = \epsilon_\infty + \lambda - \frac{1}{2}\psi - \frac{5}{2}\tau \tag{23}$$

where ϵ_∞ is the energy per particle at infinity and we have defined, using κ_0 as the coefficient of thermal conductivity, the constant

$$A \equiv \frac{8\pi GMm\kappa_0 T_0^{3/2}}{k^2 \dot{M}} \tag{24}$$

We use the polytropic equation of state to close this system. The solution to these equations is specified by the scaled values for the temperature. Only one solution passes through λ_c, which is found from the momentum equation:

$$\psi_c = \tau_c, \qquad 1 - \frac{2\tau_c}{\lambda_c} = \left(\frac{d\tau}{d\lambda}\right)_c \tag{25}$$

so that the temperature gradient must be specified by the solution to the energy equation (which can be found in terms of the temperature at the critical point). This is really as far as we can go. The value of A must be determined by direct observation, since an infinite family of solutions exists for this pair of equations in terms of this parameter.

8.4 Driving Stellar Mass Loss

The main problem for late-type stars is, why, given the fact that the luminosities of main sequence stars are insufficient to drive mass loss, do these objects show outflows? More directly, why do these stars possess outer atmospheres which have been heated to temperatures in excess of 10^6 K? The fact that all stars with effective temperatures less than about 7000 K show chromospheres and consequently show coronae, and that these stars are also convective, seems to point to at least a portion of the heating being the conversion of mechanical to radiative energy through the generation of waves in and above the convective zone. In addition, there is substantial evidence for magnetic activity in late-type stars, certainly from direct measurements on the Sun but also from flare star observations, spot activity, rotational modulation of emission lines, and time variations in the chromospheric activity. The expectation is that, in the near future, the energy budget of coronae and chromospheres will be accounted for at least by some form of dynamo mechanism.

In early-type stars, there are several mechanisms which may play a role in mass loss. These include the role of pulsation and magnetic fields. Perhaps time-dependent magnetic fields also come into play; this is not presently known.

In both early and late-type stars, the presence of a binary companion exacerbates the inability of the star to remain at constant mass because the perturbation to the gravitational potential always lowers the surface gravity. However, the complication is that this is not a spherical perturbation, and the additional role of angular momentum makes a more complete treatment of the geometry an essential feature of the discussion.

8.4 Driving Stellar Mass Loss

8.4.1 Radiation Pressure as a Driving Force

The solar problem is very hard. Not only do we have to come up with some way of heating the lower solar atmosphere in a way that ensures that we reach the critical condition somewhere near the stellar surface, but the real solar wind is a great deal more complicated than our simplistic model allows for. We will return to this later, but for now let's look at what happens if the radiation is strong enough to transfer substantial momentum to the atmosphere.

To begin with, we know from solutions of the interior equation that there is no stable solution for a self-gravitating sphere (a polytrope, to be sure, but still a good guide) when the sole means of pressure support is due to radiation. This is because the equation of state is critically squishy and does not rise sufficiently rapidly on compression to prevent the star from reaching zero radius. A critical luminosity must therefore be present in this system. Radiation pressure is due to scattering, as well as absorption, when the photon directions are changed, and therefore there is a transfer of momentum from the radiation field to the atoms in the atmosphere. When the optical depth is large and the radiation diffuses outward, the radiative acceleration is small. But when the optical depth decreases, the radiation field becomes progressively more anisotropic (stated otherwise, the Eddington factor departs from 3, its asymptotic value).

Now we can examine the effect of the scattering. The momentum carried by the radiation field is $h\nu/c$ per photon. If the flux is F_ν then the rate of momentum transfer is

$$\nabla p_{\text{rad}} = \frac{\pi}{c} \rho \int F_\nu \kappa_\nu \, d\nu \qquad (26)$$

where κ_ν is the cross section, or absorption coefficient, for the photons. If scattering is gray, then $\kappa_\nu = \sigma_T$, where $\sigma_T^\nu = 0.4$ cm^2 g^{-1} is the Thomson scattering cross section. (For low-energy photons; we will need to return later to the problem of very high energy photons and the generalization of this process to the Compton effect.) Consequently, the critical condition for the radiation pressure is that the effective gravity vanishes:

$$\nabla p_{\text{rad}} = \rho \mathbf{g} \qquad (27)$$

Assuming that the medium is optically thick, the radiation pressure is given by a simple function only of temperature, $p = \epsilon/3 = \frac{1}{3}aT^4$, and the radiative transfer equation is simply diffusive with $F = \kappa_{\text{rad}} \nabla T$, κ_{rad} being the radiative conductivity. When $p = p_{\text{rad}}$ in the optically thick limit we obtain the radiative polytrope. However, for the single scattering case, in the optically thin limit, we get the well-known Eddington limit:

$$L_{\text{Edd}} = \frac{4\pi GMc}{\sigma_T} \approx 3 \times 10^4 \frac{M}{M_\odot} L_\odot \qquad (28)$$

when the radiative flux is strong enough to support the matter completely. This is different from the polytropic solution because, as is well known from stability theory, such a star would not be strictly stable. Instead, this critical luminosity will cause the atmospheric layers to be mechanically unable to achieve hydrostatic equilibrium. If the star is completely radiation supported, the star is unstable. Surprisingly, it wasn't recognized that this really implies that the star loses mass. Instead, it was assumed that this represents the upper limit for stars on the main sequence.

We can look again at the isothermal wind equation, only this time we generalize the acceleration to include both gravity and radiation:

$$(v^2 - a_s^2)\frac{dv}{dr} = -(g - g_{\text{rad}}) + \frac{2a_s^2}{r} = -\frac{GM}{r^2}(1 - \Gamma) + \frac{2a_s^2}{r} \qquad (29)$$

where we now write, for the simple case of electron scattering, $\Gamma = L/L_{\text{Edd}}$. If the radiative opacity of the medium is simply scattering, and this could also include dust, the atmosphere will still have a simple critical point when the effective gravity vanishes. This point moves into the star as the luminosity approaches the Eddington luminosity, so inevitably the mass loss rate goes up, being controlled by the density at the sonic point. The sonic point eventually reaches the center of the star in the case when $L = L_{\text{Edd}}$. Most of the analysis is, therefore, unchanged by this alteration of the pressure. In effect, we have decreased the stability of the atmosphere by the addition of radiation—hardly a surprise since we have added a highly anisotropic source for momentum which can couple very efficiently to the material in the atmosphere.

But we have been proceeding with a very simplified analysis. What if we make the opacity dependent on frequency? Then the most effective driving will occur when F_ν peaks where κ_ν does. Since the opacity from the continuum is not an effective avenue for communicating a kick to the constituent atoms, this leaves lines as the most likely causative agents whereby the matter is destabilized. Line opacity depends on the relative population of the scattering or absorbing level, n_i, the population of the ion, the oscillator strength f_{ij} for the transition, and, in principle, the phase function (see Chapter 10). In general, the larger the oscillator strength the larger the kick, so the lines with the strongest absorption are likely to be resonant lines of dominant ion states. This automatically forces us into the

8.4 Driving Stellar Mass Loss

ultraviolet, where most of the resonance lines reside and also where the greatest momentum is transported by the photons.

But one line is not sufficient as a model for the radiative acceleration, in part because it cannot transfer enough momentum and in part because we cannot ignore the myriad of weak lines that dominate the absorption spectrum of any ionic state in the ultraviolet. The real absorption coefficient is a frequency average over all of the available ion states. We must also solve for the ion fraction as a function of position in the atmosphere, but this is another complication that we shall examine later. For now, assume that

$$g_{\text{rad}} \to \sum_i x_j \int_0^\infty \kappa_{\nu,j} F_\nu \, d\nu \quad (30)$$

summed over all of the states j with relative populations x_j. This is trivial in the weak line limit and if the ion fraction is constant, but neither is likely to be true. For one thing, if the line is strong enough, the flux is depressed in the line and only the deepest layers feel the radiation pressure, not the ones where the effect will cause a flow to begin. But if the atmosphere starts to move ever so slightly, the higher layers in the atmosphere will see progressively more of the continuum from the deeper layers as the line core shifts into the wing of the underlying line. Thus a large velocity gradient is an essential consequence of radiative driving—the additional avenue open to the gas for increasing the radiation pressure is to accelerate more promptly near the photosphere. To put it the other way around, the driving will saturate at some value of the velocity gradient because there will be some height at which the atmosphere simply runs out of driving ions yet has desaturated the lines to the greatest extent possible.

Here we can cheat and say that, even if we don't have a clear idea of how it works, the radiative driving depends on the local conditions and also on the velocity gradient so that

$$g_{\text{rad}} \sim \left(\frac{dv}{dr}\right)^\alpha \quad (31)$$

and, amazingly enough, we have a truly nonlinear first-order equation of motion even in the isothermal state. To solve for the motion, we must first iterate to the sonic point assuming that we have constant α, also called the force multiplier. Since $dv/dr > 0$ for all r,

$$r_\star \sim \frac{GM}{2a_s^2}\left(1 - \Gamma_0 \left(\frac{dv}{dr}\right)^\alpha_{v=a_s}\right) \quad (32)$$

evaluated for $v = a_s$. If we assume a velocity field in advance we obtain a unique relation beween the mass loss rate and the force multiplier, such that the larger α becomes, the greater the mass loss and the steeper the velocity gradient will be. The problem is complicated by the fact that we need to know the velocity gradient in the sonic point as well as the velocity. At least we can make use of the condition that v increases outward and is monotonic, but the fact that α is not an integer introduces wholly new structure at the critical point. An analytic solution is possible for the velocity field but this may not be realistic if there is a large ionization gradient in the atmosphere (even if the flow is isothermal the continua become optically thin at different stages, affecting the variation in the driving force and the ions responsible for the acceleration).

If the medium is optically thin, the photon scatters once (or even less often), which means that the radiation can kick the atoms only once. The force obtained from this process is easily computed, since in the optically thin limit the flux is also as anisotropic as it gets. The momentum carried in the stellar radiation is easily calculated, so the number of photons, integrated over the efficiency of the coupling, gives the acceleration. That is, there is a direct link between the flux and the mechanical energy of the wind:

$$\dot{M} v_\infty^2 \sim L \qquad (33)$$

If the radiation is transferred in lines, then the efficiency (measured relative to electron scattering) is $\eta = \langle \kappa \rangle / \sigma_T$, where the average is taken over the stellar flux distribution. In general, $\eta > 1$, so that the radiation transfers considerable momentum to the gas. If the optical depth is increased so that the lines are optically thick, the effect of the radiation is to additionally destabilize the atmosphere. Scatterings in the line core serve to multiply the momentum transfer by some factor which is essentially the number of scatterings (which is approximately $\tau_L^{1/2}$, where τ_L is the line core optical depth). Thus, assuming that the radiation can escape because the continuum is optically thin, the trapping and the resultant velocity gradient serve to increase the efficacy of the photons as drivers for the outflow.

The resulting scaling law is not, therefore, precisely linear with the luminosity of the star. It depends on the precise spectrum of the radiation and on the state populations. In general, the parameters expected to govern the outflow are T_{eff} and g, the surface gravity, as well as L (see, e.g., Howarth and Prinja 1989).

The reason for this digression is that the effects of radiative driving on a wind are virtually identical to those of a heated atmosphere at the level of the basic equations of motion. The thermal structure of the medium can be

8.4 Driving Stellar Mass Loss

handled separately, and often is, from the line formation and the driving. One of the aims of the so-called *unified atmospheres approach* is to treat the structure of the outer stellar layers as a continuous medium, going from the inner envelope through the wind smoothly.

We can add a simple remark here about the effect of adding the radiation pressure. Recall that the primary effect is the reduction of g, the surface gravity, by an amount $(1 - \Gamma)$. Once the flow is optically thin, this is a constant. So the sonic point is moved closer to the star, as is the Alfvén point. The topology of the critical points is also altered by the presence of radiation pressure. In one sense, it is easy to see this because the Alfvén and sonic points are affected differently. For example, we might be able to have a flow which exceeds the Eddington limit but is still unable to escape the star because of confinement by the magnetic field. The field also alters the flow in preventing spherically symmetric outflow from occurring and even possibly funneling the flow entirely out the magnetic poles. Finally, whether the flow is rotating or not there must be an Alfvén point, so even for spherically symmetric initial conditions there must be some point at which the radial flow meets the Alfvén critical condition.

Radiation pressure is spherically symmetric, so the driving for a rotating wind is initially spherical. But, in a rotating frame, this immediately translates into differential motion between the polar and equatorial flows. The polar region is able to expand freely while the equatorial flow feels a progressively stronger centrifugal barrier. This can be seen from the one-dimensional isothermal form:

$$(u^2 - a_s^2)\frac{1}{u}\frac{du}{dr} = \frac{2a_s^2}{r} - \frac{GM}{r^2}(1 - \Gamma(\rho,r)) + \frac{J^2}{r^3} \tag{34}$$

Here the critical velocity is still $u = a_s$, but there are two such points because the equation for r_s depends on J. The same feature could have been seen in the equation for the magnetic rotating wind if we were to set the field to zero and look only at the added terms for the angular momentum, ignoring the magnetic flux transport in the wind. Thus a disk is expected to form for a rotating wind. In the case in which this disk is supported by radiation pressure, we have the prospect of a radiation-dominated torus, now a familiar feature of the theory of active galactic nuclei. Radiation pressure from the disk may support the structure vertically and the wind, now slowed to a breeze solution, forms a shroud around the central star while blowing out the poles and perhaps off of the surface of the disk. But for more discussion of this, you will have to wait a bit longer.

8.5 Magnetic Winds

8.5.1 Rotation and Magnetic Fields

The problem of including magnetic fields, and rotation, in the equations of motion is almost as old as wind theory. Again, it is an example of the simultaneous discovery of a solution by two groups, Weber and Davis (1967) and Mestel (1968), but from two totally different points of view. As an aside, a remark about the temper of the times. The magnetic spindown problem had been examined by Deutsch (1957) for the case of a rotating dipolar Ap star. When the pulsars were first observed, Pacini (1967) obtained the basic result and proposed the electromagnetic braking solution for pulsars. Goldreich and Julian (1969) proposed the solution for the structure of the circumstellar environment around a pulsar; Michel (1991) was one of the first to stress the effect of inertia and the light cylinder and also the Alfvén point in the outflowing particle wind. These solutions are all relevant to the overall problem of the acceleration of mass by the presence of a strong magnetic field in a rotating object and the effect that this mass loss has on the rotation of the star. The electromagnetic problem is easier, in the sense that we aren't transferring mass anywhere but only radiating Poynting flux $c\mathbf{E} \times \mathbf{B}/4\pi$ and through this losing angular momentum. Recall that for the pulsar magnetosphere, the potential drop between the pole and equator is generated largely by the rotation of the dipole and so depends on the angular momentum of the star. Hence, loss of electromagnetic flux is equivalent to loss of angular momentum, and the star, on a long time scale, spins down. The problem of the rate of spindown of the Sun and of solar-type stars became interesting following the initial period of surveys of field stars for distribution of rotational velocity along the main sequence and once it became possible to measure the *in situ* angular momentum loss from the solar wind. Magnetic fields were added to Be star models in an attempt to explain the stability of the disk structures, and the general problem of the origin of the solar system seemed to demand the inclusion of magnetic effects in the early, mass-losing, protostellar environment.

8.5.2 Basic Physics

We will start with the basic assumptions of *any* model. The first is that the plasma in the environment of the stellar atmosphere is perfectly conducting. This may be thought of as a way of saying that there are no internal fields other than those which are maintained by the flow itself, coupled to the externally imposed magnetic field. The conductivity may,

8.5 Magnetic Winds

however, be anisotropic (because of the magnetic field effects on a very low density medium) and this assumption may eventually break down in the flow, but not near the star. The second is that the flow is electrically neutral (which follows) and for our purposes that the protons and electrons are thermalized. Again, the assumption of perfect conductivity is essentially the same as saying that the collisional coupling is so rapid between the species in the wind that the velocity distributions of the species are identical. The most important assumption we will have to make is that the flow is steady state. As we found in the isothermal wind solution, this will drive all of the dynamics. It cannot be stated too often that *if we change the rate of mass loading of the flow with time, its effects will propagate through all of the various physical processes involved with the flow*. The assumption that the mass loss is steady is *the* key assumption of the whole process. Any solution we obtain will necessarily enforce self-consistency, but this is because of the underlying structure of the mathematical machine created by the physical assumptions. Only in the case of extremely low-amplitude rapid variations in the mass loss can we include time dependence in the steady-state scenario. Otherwise, we have to solve the fully time-dependent equations of mass loss, and that we won't do here.

We begin with the equations for the electrodynamics of the flow, making the assumption that there is a bulk current $\mathbf{J}$:

$$\mathbf{E} = -\frac{1}{c}\mathbf{v} \times \mathbf{B} \tag{35}$$

$$\mathbf{\nabla} \times \mathbf{E} = 0 \tag{36}$$

$$\mathbf{\nabla} \times \mathbf{B} = \frac{4\pi}{c}\mathbf{J} \tag{37}$$

supplemented by the divergence conditions $\mathbf{\nabla} \cdot \mathbf{E} = 0$ and $\mathbf{\nabla} \cdot \mathbf{B} = 0$. Then for the equation of motion:

$$\rho\mathbf{v} \cdot \mathbf{\nabla}\mathbf{v} = -\mathbf{\nabla}p - \rho\frac{GM}{r^2} + \frac{1}{c}\mathbf{J} \times \mathbf{B} \tag{38}$$

which, of course, we could generalize by adding a factor $(1 - \Gamma)$ to the gravitational term, or directly by adding $\mathbf{\nabla}p_{\text{rad}}$ to the pressure gradient (whichever seems more congenial to you).

Now we have to make some hard decisions. First, do we work with a static or dynamic magnetic field? That is, the flow will be assumed slow enough that it is dominated by the magnetic field near the surface of the star and out to very large distance. If this is the case, then we can impose a field structure and do not have to think about self-consistency. Such is the case with pulsars, for instance, where a strong dipole field is provided by the neutron star and the surrounding plasma simply falls into line. The second decision concerns the motion of the flow near the surface. Is the underlying star rotating or not? In order to discuss the most general solution, we will take the case of a field which is not terribly strong, so that the full MHD effects will manifest themselves, and we will also assume that the star is a relatively slow rotator. Again, all of this will be in the time-independent picture.

The azimuthal equation of motion is

$$\rho \frac{u}{r} \frac{\partial}{\partial r} rv = \frac{B_r}{r} \frac{\partial}{\partial r} rB_\phi \tag{39}$$

because we assume that the flow is at least axisymmetric. This assumption can be dropped but here it makes life *much* easier. In this and subsequent discussions, u and v represent the radial and azimuthal velocities, respectively, and the other symbols are obvious. The mass loss rate is

$$\dot{M} = 4\pi r^2 \rho u \tag{40}$$

which is the conserved quantity that results from the continuity condition, as always. That we are taking this as a constant, in both space and time, is the central assumption.

8.5.3 The Structure of Magnetic Outflows

From the induction equation,

$$\nabla \times \mathbf{E} = 0 \rightarrow \frac{\partial}{\partial r} r(uB_\phi - vB_r) = 0 \rightarrow r(uB_\phi - vB_r) = \text{constant} \tag{41}$$

Here we have a very important result. The magnetic field in the azimuthal direction is the result of the dragging of field lines in the outflow due to conservation of angular momentum by the wind fluid. Put another way, the induction equation serves as the condition that the outflow, as it spirals outward, wraps the field and causes a back-reacting torque on the central star. As a result, there will be a net angular momentum transport in the wind. This can be seen also from the fact that another way of writing the induction equation solution is

$$\mathbf{v} = \kappa \mathbf{B} \tag{42}$$

where κ is a scalar function. In other words, the flow aligns along the magnetic field lines. If the field is predominantly dipolar, then meridional as well as radial flows will be possible. If the velocity field has an azimuthal component, so will the magnetic field—so that the induction of a torus is a

8.5 Magnetic Winds

necessary consequence of the equations of MHD. How this relates to the equations of motion can be seen from the angular momentum equation,

$$\rho u(rv)' - B_r(rB_\phi)' = 0 = (rv)' - \frac{B_r r^2}{\rho u r^2}(rB_\phi)' \qquad (43)$$

with the prime denoting d/dr. Then using $r^2 B_r = $ constant (from the flux conservation equation) and $r^2 \rho u = $ constant we obtain

$$J = r(v - \eta B_\phi) \qquad (44)$$

where J is the angular momentum, which is constant, and

$$\eta = 4\pi r_0^2 B_{r,0}/\dot{M} \qquad (45)$$

is also a constant with r_0 and $B_{r,0}$ being the conditions at the stellar surface (or anywhere else in the flow for that matter). This is *the* central equation for the magnetic wind problem, the one which expresses the dynamical coupling between the axisymmetric flow and the magnetic field.

We now define a new variable, one which is directly related to the speed with which a wave moves in a magnetic field and the one which determines the time scale for establishing hydrostatic equilibrium in a magnetic field, the Alfvén speed:

$$v_A = \frac{B_r}{(4\pi\rho)^{1/2}} \qquad (46)$$

so that the Alfvénic Mach number is

$$M_A^2 = \left(\frac{u}{v_A}\right)^2 = \frac{4\pi\rho u^2}{B_r^2} \qquad (47)$$

and choose some radius r_A to be the position at which the flow speed exceeds the Alfvén velocity. At this radius, it will not even be possible to reach stasis in the field, and at larger distance the energy density in the magnetic field is exceeded by the kinetic energy density in the flow. This last point is important for understanding the nature of the transition point where $M_A = 1$. Prior to reaching this radius, although accelerating outward, the material of the wind is still essentially dominated by the local magnetic field and structured by it. This is the regime where the flow increases angular momentum because of this constraint, thus torquing the emitting star down. At distances greater than r_A, the field is dominated by, and consequently structured by, the outflow. Since from this point angular momentum in the matter is strictly conserved, the material spirals outward carrying the magnetic field with it.

We now have to look back at the origin of the flow, the surface of the mass-losing body. For simplicity, we start the flow off with two specific conditions. First, we assume that there is no initial acceleration. The flow starts from rest. This is not a very important assumption, though; it simply changes some of the algebra but you can see that it will only quantitatively change the mass loss rate and the density at a given radius, not the whole qualitative structure of the solution. Next, we assume that the material is initially corotating with the surface. This assumption is more important because it means that we have no excess angular momentum other than $r_0^2 \Omega$ for the matter as it enters the flow. Under these two assumptions, we write the induction equation as

$$r(uB_\phi - vB_r) = -v_0 r B_r = -\Omega r^2 B_r \tag{48}$$

Now we can solve for the angular velocity, v, by first writing

$$B_\phi = (v - \Omega r) B_r / u \tag{49}$$

which states that the magnetic field will be wrapped up by the outflow, and then we have for the azimuthal velocity

$$v = \frac{L - \Omega r^2 / M_A^2}{r(1 - 1/M_A^2)} \tag{50}$$

Finally, scaling to the Alfvén radius, which we can do by assuming this as our reference level for the flow, we obtain

$$B_\phi = -B_r \frac{\Omega r}{u_A} \frac{r_A^2 - r^2}{r_A^2 (1 - M_A^2)} \tag{51}$$

for the azimuthal magnetic field and

$$v = \frac{\Omega r}{u_A} \frac{u_A - u}{1 - M_A^2} \tag{52}$$

for the angular velocity of the wind. We need these two quantities to calculate the rate of angular momentum transfer by the mass loss. The magnetic field provides a measure of the moment arm, and the azimuthal velocity provides the momentum loss rate. Together, they come into the torque. (See Fig. 8.2.) Notice that here the velocity and the radius can be *scaled* by the Alfvén distance, which we do not know quantitatively in advance. But we do know that $r = r_A$ when $M_A = 1$, so the angular velocity and the azimuthal field are both finite at this point. In fact, we would normally have to solve the full equations of motion, but you already know from the Parker solution that there is a way around this stringent requirement. We have assumed that the flow is stationary, so we also know that

8.5 Magnetic Winds

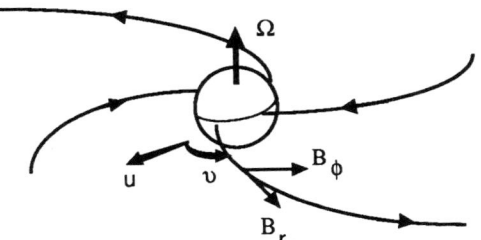

Figure 8.2 Spiraling magnetic field lines of the Weber–Davis solution.

the net energy flux is

$$F = \rho u r^2 \left[\frac{1}{2} V^2 + \frac{\gamma}{\gamma - 1} \frac{p}{\rho} - \frac{GM}{r} + \frac{B_r B_\phi}{4\pi \rho} \frac{\Omega r}{u} \right] \quad (53)$$

The total velocity, V, is $(u^2 + v^2)^{1/2}$, and the last term represents the Poynting flux carried by the field with a tilt angle $\Omega r/u$. But recall that $\rho u r^2$ is a constant, so we have an equation which is an *algebraic* expression of the solution to the equation of motion analogous to the Parker solution. If we choose a particular value of $F/\dot{M}$, we can solve for the radial velocity of flow with distance once and for all. You see that the reason is simply that we have equations for v and B_ϕ in terms of r/r_A and u/u_A. Thus we have a scaled solution in a dimensionless radial velocity M_A as a function of the scaled radius x. The final solution is

$$\frac{1}{2} v_A^2 M_A^2 + \frac{\gamma}{\gamma - 1} \frac{p_A}{\rho_A} M_A^{-2(\gamma-1)} - \frac{GM}{r_A x} + \frac{\Omega r_A^2}{2x^2} \left[1 + \frac{(2M_A^2 - 1)(x^2 - 1)}{r_A^2 (M_A^2 - 1)} \right]$$
$$= \text{constant} \quad (54)$$

We have the free parameters v_A and r_A for the quantitative solution. For an isothermal wind, we have

$$\nabla \cdot \rho \mathbf{v} \left(\frac{1}{2} v_p^2 + \frac{1}{2} \Omega r^2 \sin^2 \theta + a_s^2 \ln \rho - \frac{GM}{r} \right) = -\nabla \cdot \frac{c}{4\pi} (\mathbf{E} \times \mathbf{B}) \quad (55)$$

where θ is the latitude on the rotating star. The left-hand side represents the advected energy flux and the right-hand side is the Poynting flux in the magnetic field. This last equation was derived by Mestel; it transforms into Eq. (54) when we include the variation of a_s with distance and take the electric field in the perfectly conducting flow from $-(1/c)\mathbf{v} \times \mathbf{B}$.

In general, then, we have two conditions for the flow. The first is that the flow must become supersonic at some point, and the second is that it must pass through the Alfvén speed at some distance. The azimuthal velocity is very small compared with the radial flow speed and the B_ϕ field becomes progressively stronger at the expense of the radial field (since the divergence condition ensures that the total magnetic flux is conserved). For the Sun, for which this solution was first obtained, we have the advantage of being imbedded in the flow and consequently *we* serve as an additional boundary condition—the flow must match the velocity at 1 AU. This gets us around not knowing what the value is for the absolute value of r_A; it can be determined by the algebraic condition that the mass flux is conserved and measured at the Earth's orbital radius and that the initial radial velocity is zero at the solar radius $R_\odot$.

8.5.4 Angular Momentum Loss—The Spindown Problem

To say that algebra determines the structure of the flow, however, seems more Platonism than physics. Instead, you might look at it another way. We assume that the star is rotating and that the flow is steady. Angular momentum is conserved and immediately couples the motion in ϕ with that in r. Since we also have a magnetic field threading through the medium, this coupling is enhanced by the fact that the field must be divergenceless and is completely coupled to the flow. This coupling increases the efficient torquing applied to the star, enormously extending the moment arm above the stellar surface and thus providing "more kick per proton" in the angular momentum lost per unit mass. A simple estimate of this effect can be derived from essentially dimensional arguments. Let's take a dipolar magnetic field at the surface of the star and assume that there is no distortion of the field out to the Alfvén point. Then when the corotation speed is the same as the Alfvén speed, we have

$$v_A = \frac{B_0 r_0^3}{r_A^3 (4\pi \rho_A)^{1/2}} = r_A \Omega \tag{56}$$

Now use the condition that $\dot{M} = 4\pi r_A^2 \rho_A u(r_A)$, where u is the radial velocity, and assume that the radial velocity at the Alfvén point is approximately the same as the terminal velocity, $u(r_A) \approx u_\infty$. The small error we are likely to make by this assumption does not seriously affect the estimate, as we will find out in a moment. Then a good estimate for the radius of the Alfvén surface is

$$r_A \approx \left(\frac{B_0^2 u_\infty}{\dot{M} \Omega^2} \right)^{1/6} r_0 \approx 8.2 \times 10^{-3} B_{0,G}^{1/3} U_\infty^{1/6} \dot{m}^{-1/6} P_d^{1/3} r_0 \tag{57}$$

8.5 Magnetic Winds

where the magnetic field is now a gauss, the mass loss rate is in $M_\odot$ per year, the terminal velocity is in kilometers per second, and the rotation period P_d is in days. For a massive star, say about 10 $M_\odot$, with a rotation period of about 2 days (arbitrary, but not out of line with observed rates), a mass loss rate of 10^{-8} $M_\odot$ yr^{-1}, and a strong surface magnetic field of order 5 kG (which is observed in a number of upper main sequence helium strong stars), the escape velocity is about 1000 km s^{-1}, so the Alfvén radius is at about 10 stellar radii. For the Sun, with a period of about 1 month, a mass loss rate of about 10^{-14} $M_\odot$ yr^{-1}, and a surface field of order 1 G, the Alfvén radius lies at about 16 $R_\odot$, a good guess. Notice that as the mass loss rate decreases, the Alfvén radius moves outward. This is because the lower density ensures that the magnetic field can control the flow farther out because of the relatively low kinetic energy density in the wind; eventually, however, the wind dominates because the magnetic energy density is falling off like r^{-6} for a dipole.

The effect on the angular momentum of the star is profound. This enormous increase in the moment arm means that the rate of angular momentum loss is a factor of at least $(r_A/r_0)^2$ greater than would have been expected from the same mass loss rate. For the Sun, for instance, this means that the spindown time is about a factor of 50 shorter than would be expected without the field being present. Also, the spindown rate can be seen to scale as

$$\tau^{-1} = \frac{1}{J}\frac{dJ}{dt} = \frac{2}{3K}\frac{\dot{M}}{M}\left(\frac{r_A}{r_0}\right)^2 \sim \frac{\dot{M}^{2/3}}{M}\left(\frac{B_0}{\Omega}\right)^{2/3} \tag{58}$$

Here K is the numerical factor for the central concentration needed for the moment of inertia of the star. As an aside, one of the commonly quoted laws for the spindown of solar-type stars, the so-called Skumanich law, $\Omega \sim t^{-1/2}$, is based on the coupling of the rotation rate to the dynamo-generated magnetic field and is a very special example of the most general case. Notice that the rate of loss of angular momentum goes up rather slowly with the mass loss rate.

An additional effect of the magnetic field is, for a rotating star, to set up a centrifugal wind even if the matter is otherwise not driven by heating or radiation (see below for a digression on radiation pressure, but not just yet). Assuming that we have a perfectly conducting fluid, there is an internal electric field set up by the magnetic field which draws matter off the surface and causes the material to drift outward. The matter at the equator feels the greatest centrifugal acceleration (recall that the outflow is set up merely because of the presence of a high enough pressure at the surface anyway) and is trapped in an extended magnetosphere. The Alfvén point represents a dead zone for the flow in that the field lines here are transverse to the flow for a strongly dipolar field. This can be seen by considering the equation for a poloidal–toroidal decomposition for the magnetic field, $\mathbf{B} = \mathbf{B}_p + \mathbf{B}_\phi$. The flow still satisfies the condition that $v_p = \kappa \mathbf{B}_p$, so there is a kind of Parker condition along the field lines. This means that the matter collects in a disk at the magnetic equator, which is constrained to corotate with the stellar surface and drift slowly outward because of centrifugal acceleration.

8.5.5 What Might a Magnetic Wind Look Like?

It is sufficient to state that the primary effect of the presence of a magnetic field is that the wind cannot be spherically symmetric. Whether the star is rotating or not, the material streaming out of the magnetic poles will be able to escape the star more easily than that confined to the magnetic equator. Thus the flow will, at large distance, look jetlike. The field lines which close within the Alfvén point, however we estimate such a distance, will trap the outflow and permit the formation of a magnetosphere, while those which are open at that distance will connect to infinity and permit free streaming. A critical magnetic latitude exists for the star, θ_c, estimated by

$$\theta_c = \sin^{-1}(r_A/r_0)^{-1/2} \tag{59}$$

for a dipole field which opens as the Alfvén point moves closer to the stellar surface. One effect of a magnetic field is therefore to confine the escaping material in space and to alter the amount of material at the terminal velocity. The trapped matter corotates (approximately) with the surface out to r_A, so one would expect a broad emission line with a width Ωr_A and an undisplaced narrow absorption component superimposed. From the jet, seen against the photosphere, the profile should have a substantially reduced emission component (irrespective of the magnetospheric component) and a broad blueward extended absorption profile. In other words, the profile does not look anything like a normal P Cygni profile (see Chapter 10).

8.6 Winds within Winds

8.6.1 Spherically Symmetric Wind Collisions in Astrophysics

The idea that stars lose mass during their lifetimes is, as we have seen, not at all surprising, but the fact that they may do this at different rates, depending on their evolutionary state, brings up some interesting questions. During the final stages of protoplanetary nebula evolution—that is,

8.6 Winds within Winds

after the asymptotic giant branch formation of a hot semidegenerate core—the slow wind which had characterized the red giant is replaced by a fast wind from the hot core, which is ejected at nearly the Eddington luminosity and rams into the base of the old giant envelope. The details of this process are not well understood, but we will use a simple analysis to show how to understand the evolution of the dynamical structures obtained in this scenario. Some of the first work on this problem is that of Pickel'ner (1968, 1973) on the collision of a stellar wind with the surrounding nebula (H II region) during the last stages of star formation. The more specific application of the problem to planetary nebulae is by Kwok, Purton, and Fitzgerald (1978). In the standard model, the old wind is from a very high mass loss stage, but with a very low velocity. The total mechanical energy is, however, fixed by some fraction of the Eddington limit for the star at the time of the ejection, so that there is a limit on the mass loss rate for this *superwind* stage:

$$\tfrac{1}{2}\dot{M}_{SW}v_{SW}^2 < L_{Edd} \sim M_\star \tag{60}$$

so that assuming there is a limit of $\dot{M}_{SW} \sim L$ where L is the stellar luminosity, then the velocity of the superwind is limited by the M/L ratio of the progenitor (with lots of proportionality constants). Since this is likely to be a low number, one would expect that the wind is slowly expanding.

For some reason, not now well understood, it is assumed that the fast wind turns on at a stage well into the development of the superwind. Empirically based, this assumption has yet to be a natural consequence of any models. But for the moment, let us look at the consequences of having the speed and mass loss rate change for the outflow. If the slow wind turns off and is then hit by a fast wind, a shock develops on the inner surface of the superwind shell, at which the fast wind undergoes a strong deceleration. The outer part of the superwind remains unaffected for some time as the fast wind plows up a shock within the slowly expanding envelope. A second shock develops where the fast wind, having been shocked, finally comes to equilibrium with the background superwind material and forms a cooling layer. This is the matter running ahead of the inner shock, as it would in front of a piston driving into the old wind. This outer shock is the outer boundary of the observed H II region and is the one likely responsible for the shock emission from molecular species. Beyond it, there should be considerable cool material from the old wind.

8.6.2 Similarity Solution: Stalled Shock Propagation

The shocked wind expands in a stalled state, as we have already discussed. That is, the pressure is in equilibrium. To see this, one need only compute the pressure from the similarity solution:

$$P \sim ML^{-1}t^{-2} \approx \frac{\dot{M}_{SW}}{v_{SW}t^2} \tag{61}$$

which states that the pressure at the working surface of the shock goes down as t^{-2}. Now notice that this also means that

$$\rho v^2 \sim \dot{M}vr^{-2} \sim P \sim \dot{M}_{SW}v_{SW}^{-1}t^{-2} \tag{62}$$

where $\dot{M}$ and v are the mass loss rate and speed for the fast wind. This leads directly to the equation of motion for the expansion of the shock:

$$r(t) \approx \left(\frac{\dot{M}vv_{SW}}{\dot{M}_{SW}}\right)^{1/2} t \tag{63}$$

and therefore the shock speed is given directly by dr/dt. For instance, if the mass loss ratios are about a factor of 100 and the wind speeds are 100 and 2000 km s^{-1} for the superwind and fast wind, respectively, the shock speed is about 45 km s^{-1}.

Consider what happens if the mass loss rate is a function of time. We have assumed that γ is constant, which is quite reasonable, and that the envelope is the result of mass loss from an earlier epoch, which also had constant properties. If $\dot{M}$ increases, the stagnation pressure increases. Then

$$v_{\text{shock}}(t) \approx \frac{1}{2}\left(\frac{\dot{M}vv_{SW}}{\dot{M}_{SW}}\right)^{1/2} t\left(\frac{\ddot{M}}{\dot{M}} + \frac{\dot{v}}{v} + \frac{2}{t}\right) \tag{64}$$

so that if the mass loss rate increases with time, due to the increase in the temperature of the core, the shock accelerates. The critical time scale is the rate of increase of the terminal velocity of the wind and the rate of change of $\dot{M}$. Taking these as power laws, we see that, if $\dot{M} \sim t^a$ and $v \sim t^b$, then

$$v_{\text{shock}}(t) \sim (a+b+2)v_{\text{shock},0}t^{(a+b)/2} \tag{65}$$

where $v_{\text{shock},0}$ is the initial velocity of the shock front into the surrounding material.

8.6.3 Similarity Solution: Constant Mechanical Luminosity

Suppose instead that the mechanical luminosity of the wind remains constant. Assume also that the pressure is given by the dynamical pressure of the fast wind on the slow shell, but now take $L_{\text{wind}} = \frac{1}{2}\dot{M}v^2$ to be constant. This means that both $\dot{M}$ and v can vary in time. Then the expan-

8.6 Winds within Winds

sion of a shock in the stalled state will be $r(t) \sim (L_{\text{wind}}/P)^{1/3} t^{1/3}$. We have in hand the approximation for the pressure in the slow wind, so that

$$r(t) \sim \left(\frac{L_{\text{wind}} v_{\text{SW}}}{\dot{M}_{\text{SW}}}\right)^{1/3} t \qquad (66)$$

thereby proving that a stalled shock by any other conserved name is still a stalled shock—the velocity is constant with time. The difference between the two solutions is that the proportionality constant is different. But the *scaled* dynamical solution is the same.

A more thorough examination of the problem begins with the equation of motion for the shell created by the interaction of the fast wind with that of the red giant:

$$M_s(t)\frac{dV_s}{dt} = 4\pi r^2(\rho_{\text{FW}}(v - V_s(t))^2 - \rho_{\text{SW}}(v_{\text{SW}} - V_s(t))^2) \qquad (67)$$

The mass of the shell, as a function of time, is given by

$$M_s(t) = \int_{R_\star + v_{\text{SW}} t}^{r} \frac{\dot{M}_{\text{SW}}}{v_{\text{SW}}} dr + \int_{r}^{r_{\text{CPN}} + vt} \frac{\dot{M}}{v} dr$$

$$= \left(\frac{\dot{M}_{\text{SW}}}{v_{\text{SW}}} - \frac{\dot{M}}{v}\right) r(t) - (\dot{M}_{\text{SW}} - \dot{M})t - \left(\frac{\dot{M}_{\text{SW}}}{v_{\text{SW}}} R_\star - \frac{\dot{M}}{v} r_{\text{CPN}}\right) \qquad (68)$$

where $R_\star$ is the radius of the red giant at the time the wind was initiated and r_{CPN} is the current radius of the central star. This is the solution that Kwok, Purton, and Fitzgerald first obtained. It is worth adding that the factors of the form $\dot{M}/v$ are just $4\pi r^2 \rho$, so the integrals are simply the fraction of the mass of the shell coming from the wind of the red giant plus the fraction coming from the CPN wind.

Notice that the similarity solution gives the same result for the velocity of the shock that we have previously obtained. The problem is that the similarity method must also be applied in detail to the structure of the shock subsequent to the wind collision, and this is a more difficult problem. But, as in the chapter on similarity solutions and the Sedov method, once λ and μ are known it is possible to calculate directly the exact solution to the problem. The boundary condition at the shock surface is given by the Rankine–Hugoniot relations. Consider a wind coming out of a star impinging on an envelope with a constant density. A shock forms at the intersection of the two regions, so that the postshock flow in the stationary frame is $u_2 = (2/(\gamma + 1))v$ where v is the wind velocity. It is assumed that the shock is strong and that the wind is moving slowly. Then the velocity at the boundary of the shock, u_2, serves as the input for the similarity solution of the postshocked gas.

The mechanical luminosity of the wind is given by

$$L_{\text{wind}} = \frac{1}{2}\dot{M}v^2 = \epsilon L_{\text{CPN}} = \frac{dE}{dt} + p\frac{dV}{dt} = \frac{d}{dt}(2\pi r^3 p) + 4\pi r^2 \dot{r} p \qquad (69)$$

where V is the volume and E is the internal energy $c_V T$, with c_V the specific heat. The pressure is given by

$$p = \rho_{\text{SW}}(\dot{r} - v_{\text{SW}})^2 + \frac{\dot{M}_{\text{SW}}(r - v_{\text{SW}}t)\ddot{r}}{4\pi v_{\text{SW}}r^2} \qquad (70)$$

so that the mass of the shell is

$$M_s(t) = \frac{\dot{M}_{\text{SW}}}{v_{\text{SW}}}(r - v_{\text{SW}}t) \qquad (71)$$

The detailed development of the planetary nebula from this interaction has been discussed by several authors (see Frank et al., 1990, for details), and it is essentially the expansion of a Strömgren sphere driven by the dynamical expansion of the envelope. As the wind reaches progressively lower-density regions of the shock, its thickness increases and the ionized matter in the interior fills a progressively larger fraction of the volume. (See Fig. 8.3.)

Following its passage through the first shock, the CPN wind (the fast wind; the terminology gets muddled here) is heated so that $T \approx (v/3)^2$ using the Rankine–Hugoniot relations. The temperatures are typically very high, greater than about 10^7 K, and this may place the gas above the temperature at which cooling is efficient. Thus, as the material flows through the shock, and as it cools, it remains hot until it falls well below 10^6 K. The LyC emission from this gas runs ahead of the shock as a precursor, so the emission from Lyα should surround the shock and extend farther into the cool gas than the fully ionized region and should be important in heating dust in the vicinity of the hot shocked wind. Infrared emission may show some trace of the structure of the shock due to this heating.

8.6.4 Rayleigh–Taylor Instabilities at Wind Interfaces

The shock front is also Rayleigh–Taylor unstable. We will discuss this at greater length in the chapter on instabilities, but here it is worth

8.6 Winds within Winds

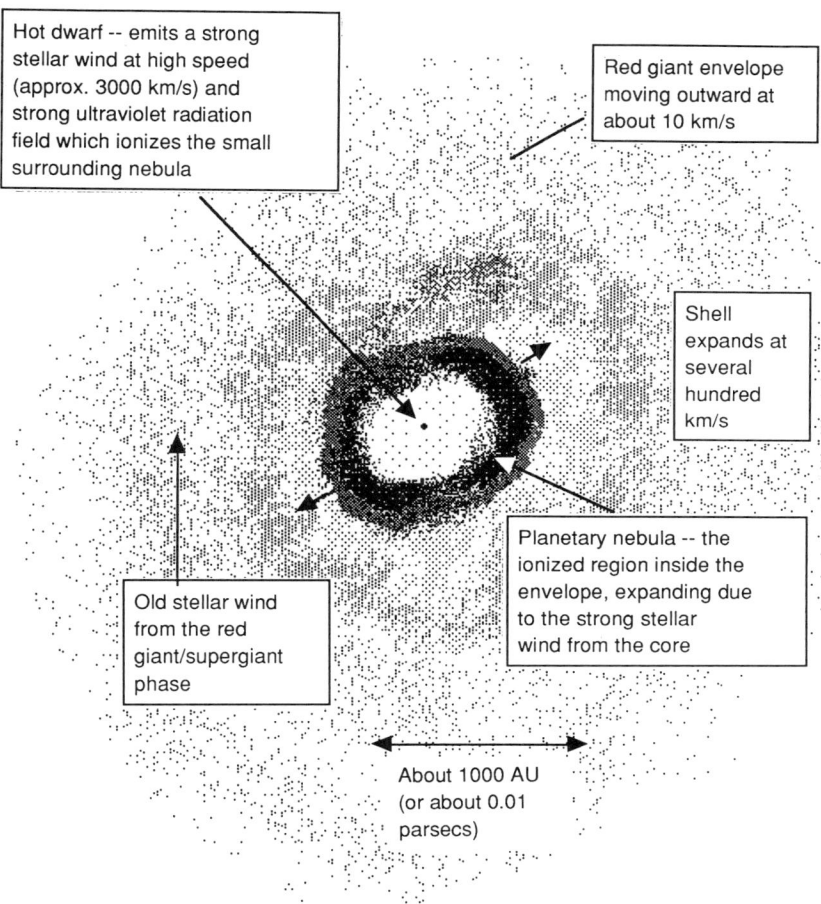

Figure 8.3 Formation of a planetary nebula.

reflecting on why the front might become unstable and form structures. Suppose a dense fluid rests above a lighter one in a gravitational field. Then if a perturbation occurs at the interface, and some of the material from the two layers is interchanged, the denser fluid gives up more potential energy than can be made up by the upward displacement of the lighter fluid. The result is that more of the lighter fluid must be displaced and the layers interchange completely and on a dynamical time scale. At the wind interface, we can neglect gravity. But there is an acceleration present at the shock front. Let's briefly examine what this might do. If the acceleration is a, the change in the potential energy is $\delta E_p = (\rho_2 - \rho_1) a \delta z$ and the total kinetic energy gained is $\delta E_K = (\rho_2 + \rho_1)(\delta z)^2 \tau^{-2}$. Here the rate of growth is τ^{-1} and δz is the displacement. Then the growth time scale is given by

$$\tau^{-1} = \left(\frac{\rho_2 - \rho_1}{\rho_2 + \rho_1}\right)^{1/2} (ak)^{1/2} \tag{72}$$

where $k \sim z^{-1}$ is the wave number for the disturbance (assuming that the wavelength is of order z). The instability is guaranteed if the density gradient is in the opposite direction to the local acceleration. Now by noting that any acceleration is locally equivalent to gravitation (this is after all the basis of general relativity), we can replace a by ΔP, the pressure jump across the shock front at the wind–wind interface. We could also use g_{eff}, the effective gravity including radiation pressure. Let's concentrate on the first choice. Across a shock, which moves with a speed v_Σ into the surrounding gas, the pressure is given by

$$p_2 = \frac{2}{\gamma + 1} \rho_1 v_\Sigma^2 \tag{73}$$

For a perfect gas this is larger on the postshocked side and the pressure jump is $p_2 - p_1 \approx p_2$. The interface is unstable to the formation of knots and blobs.[1] As we have already discussed, as these rise, they tend to show induced vorticity and develop the characteristically mushroom-shaped structures so familar from both violent explosions (like nuclear blasts) and simulations of supernova envelopes. This same condition will certainly prevail during the formation of planetary nebulae, and the wind–wind collision should generate stringy and bloblike structures because of this instability. If optically thick, they will shadow the material in the slow wind, and the result is that low ionization regions should be mixed into the H II region. One expectation is that in any of the blastlike expansions, be they Wolf–Rayet winds, planetary nebulae, or supernova remnants, it should be possible to see the effects of the wind collision preserved in these structures.

---

[1] An example of stability analysis for these fronts is given in Ryu and Vishniac (1988).

8.7 Accretion Disks in Astrophysics

8.7.1 Some Observational Motivations

We have invoked binary systems frequently in this chapter, and it is therefore appropriate to end it with a discussion of a most interesting confluence of rotation and viscosity, namely accretion disks around massive objects. But first, before launching into an extended exposé of the properties of the disks themselves, let's examine the conditions under which rotating accretion flows may arise.

For many binaries, especially ones of long period, this is precisely what is observed. The more evolved star really is the more massive. In the case of Algol (β Persei), an extremely well-studied star and the first discovered eclipsing binary, just the opposite is observed. Algol consists of a G giant and a B main sequence star. The mass ratio is ≈ 3, *but in favor of the main sequence B star*. The orbital period is short, less than 3 days. The light curve data and modeling the equilibrium shapes of the stars show that the red subgiant completely fills its Roche surface. This is the limiting surface for tidal interaction, given approximately by

$$R_{\rm RL} \approx 0.49 a q^{2/3} [0.6 q^{2/3} + \ln(1 + q^{1/3})]^{-1} \tag{74}$$

where a is the semimajor axis given by Kepler's law $GM = \omega^2 a^3$ for circular orbits, q is the mass ratio, and M is the total mass of the system.[2] The reason for the peculiar mass of the red giant is that it has been significantly altered by mass loss from the binary system and by mass transfer onto the main sequence star. The fact that this is still going on, in both this and related *semidetached* systems (the term comes from the fact that only one of the stars is in contact with the critical surface), means that accretion flows onto the companion have played a role in the orbital dynamics and that this has fed back into the stellar evolution through the alteration of the mass and boundary conditions on the stars. The observation, for a number of these stars, of emission lines which are formed in a Keplerian disk surrounding one of the components adds fuel to the argument, although in Algol it does not appear that an extensive accretion disk is observed.

---

[2] See Paczynski (1971), Eggleton (1983), and Mocnoski (1984) for further discussion of techniques for fitting equivalent volumes for tidally distorted stars. These approximation formulae are also helpful for estimating potential surface distortions for calculating tidally driven currents.

8.7.2 Flow through the Inner Lagrangian Point

First, a binary system, that is, a close system, is one that is not spherically symmetric. The presence of the companion star, as well as the rotation of the mass-losing star due to spin–orbit coupling, produces immediate departures from sphericity. To see what happens to the mass transfer at the inner Lagrangian point, also called L_1, we need to consider the flow of material in a potential that switches sign at some point in the flow (see Fig. 8.4).

Let's look back at the spherical case for a moment. The mass loss is driven by the combined effect of pressure gradient and retardation due to gravity. At some point, where the outward driving becomes strong enough relative to gravity, the material coasts at the sound speed and then accelerates as the gravitational acceleration continues to fall off. In other

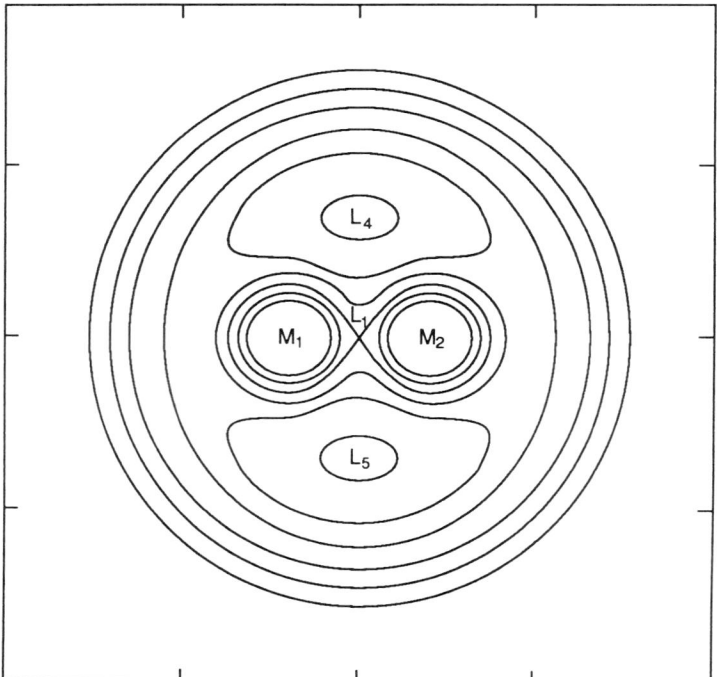

Figure 8.4 Calculated Roche surfaces for equal mass stars. The Roche surface shows an x-type critical point at L_1. The L_4 and L_5 points are located along the y axis.

8.7 Accretion Disks in Astrophysics

words, the reason the effective gravity vanishes is that at some point, g is balanced by ∇p. But what about other possible cases? We've already seen that if g is balanced by ∇p_{rad}, or that at some point $g(1 - \Gamma)$ vanishes, then the material becomes supersonic and a strong wind. Both of these depend on the presence of a pressure gradient to do the job of producing a sign change in the acceleration. The alternative is to say that the gravity itself reverses sign, something impossible for a single star but normal for a close binary. Put differently, imagine that the potential is taken to be

$$\Phi(\zeta) \approx \Phi_0 + \left(\frac{\partial \Phi}{\partial \zeta}\right)_{L_1} \zeta + \frac{1}{2}\left(\frac{\partial^2 \Phi}{\partial \zeta^2}\right)_{L_1} \zeta^2 + \cdots \quad (75)$$

in the vicinity of the L_1 point. The problem can be rendered one dimensional if we assume that we look at the flow only in the vicinity of the Lagrangian point and that the system is not so rapidly rotating that the Coriolis deviation of the flow is large compared with rectilinear flow. This means that $\Omega < u'$, where the prime denotes the spatial derivative of the velocity.

The fact that the equation of motion can be written as

$$u\frac{du}{d\zeta} = -a_s^2 \frac{d \ln \rho}{d\zeta} - \frac{\partial \Phi}{d\zeta} \quad (76)$$

means that we can apply the same condition that we had in the spherical Parker wind solution. Take a look at the flow through a small region around the L_1 point. The equipotentials on either side of L_1, along the line of centers, give a local critical point to the flow. This is because there is a local maximum in the gravitational field. Perpendicular to the line of centers the gradient has a local minimum. (See Fig. 8.5.) The L_1 point is a saddle point in the gravitational potential; the gravitational acceleration changes sign on crossing this point. For the case of one-dimensional flow, this has the same effect as the changing gravitational acceleration relative to the pressure gradient. The critical condition for stream formation is similar to the Parker solution; that is, $u = a_s$ at L_1. Thereafter, as the gravitational field increases toward the secondary, the flow accelerates. As material is forced through this point, it has the same effect as the passage through the $g_{\text{eff}} = 0$ point in a spherical wind. The pressure gradient does not vanish, so the material is accelerated and the sound speed is reached, after which the flow is ballistic toward the secondary. Stream formation is important because it transfers material with high specific angular momentum toward the accreting star. Once in the vicinity of the companion, the matter forms an accretion disk, the details of which we shall now discuss.

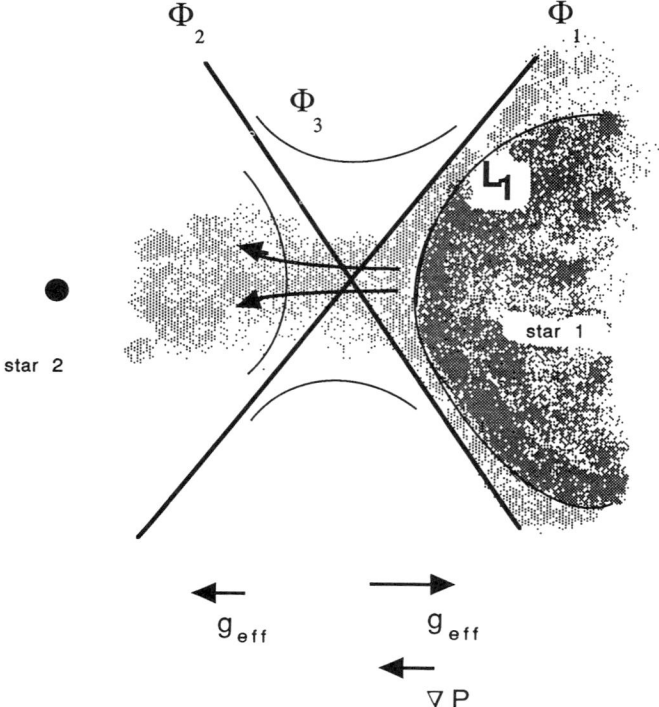

Figure 8.5 Flow through the L_1 point in a semi-detached close binary system undergoing mass transfer. Equipotentials are indicated by Φ_{j1} with the critical (Roche) equipotential indicated by Φ_1 and Φ_2. The directions of the effective gravity, and of the presure gradient in the flow, are indicated.

8.7.3 Some Consequences of Mass Transfer

The observation of tidal distortion leads immediately to important hydrodynamic consequences. The existence of the Roche surface in general, and the limiting radius in particular, is the result of the three-body problem. The inner Lagrangian point, L_1, is the point along the line of centers at which the effective gravitational acceleration vanishes. However, since the pressure does not vanish even though the gravitational acceleration does, matter will be forced to exit through this region and begin to flow to the other star. In effect, we have set up the de Laval nozzle problem from Chapter 1. The variation in the effective gravity acts like the nozzle (although without material walls) to accelerate the flow through the sonic point and eventually to hypersonic speeds. The matter carries some net angular momentum because the L_1 point is generally not at the center

8.7 Accretion Disks in Astrophysics

of mass, and therefore the deviation of the flow and its acceleration toward the companion produce an accretion disk.

The stream must eventually rid itself of this excess angular momentum before it can accrete onto the companion even for direct impact. Several mechanisms are available, probably all of which operate somewhere in the universe. One is turbulent viscosity. That is the one we shall mainly deal with here. Another is magnetic breaking. If the material forms a disk that becomes Kelvin–Helmholtz unstable at the boundary of a stellar magnetosphere, blobs may be formed that accrete onto the companion. The process is certainly not well understood but can be simulated for neutron star accretion and is well established as a scenario. For direct impact, the stream may submerge and pump angular momentum into a deeply generated boundary layer.

The final mechanism is spiral shocks. Since the matter falls into the disk with excess angular momentum and drives spiral waves in the disk, these may form stable circulating structures that serve to deviate the flow onto the companion and dissipate energy and momentum. (See Fig. 8.6.) Presently, however, the details of the accretion process are the most schematic parts of accretion disk theory. This is a pity, because these details contain virtually all of the essential physics.

If the mass-accreting star is a compact object, like a white dwarf,

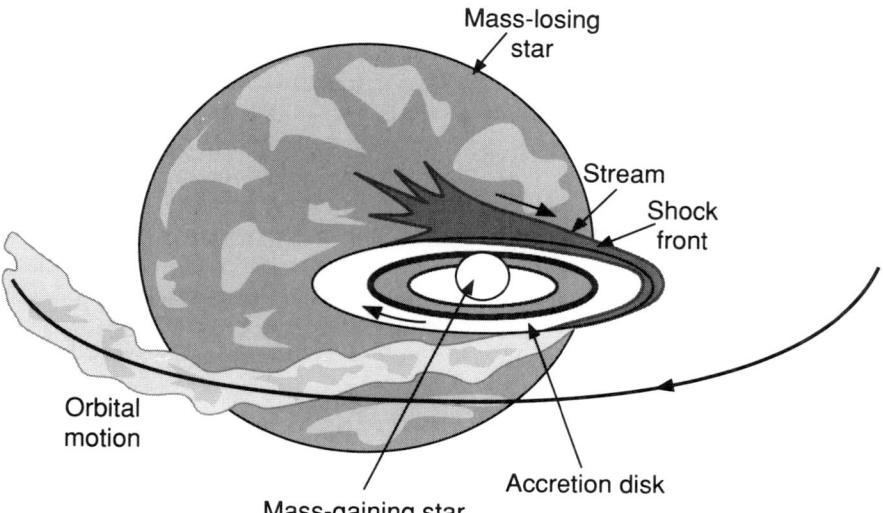

Figure 8.6 Edge-on presentation of mass transfer in a close binary system. The standing shock at the stream–disk interface and the mass loss from the disk are shown.

neutron star, or black hole, the disks reach temperatures considerably higher than for main sequence stars. The simple reason is that the gravitational well around the accreting object is very deep and the energy source for the dynamics is consequently greater. Any dissipative processes fed by the global circulation will therefore find a very large reservoir of energy which can be tapped, and the resulting emission of radiation can take place in the ultraviolet or even the x-ray. It was the latter wavelength region, observed with satellites like Uhuru and Einstein, that first signaled the presence of accretion disks around neutron stars like Hercules X-1 = HZ Her and black holes like Cygnus X-1 = HD 226868. The added discovery that the optical counterparts of these and other galactic x-ray sources are spectroscopic binaries and the observation of optical emission lines which tracked the compact star clinch the argument for accretion powering the radiation.

The observation of strong emission lines arising in a very compact region in the cores of active galaxies like quasars and Seyferts also indicates that accretion can occur on a scale of extremely massive but otherwise "single" collapsed objects. These galaxies have emission line widths indicative of velocities of order 1000 to 10^4 km s^{-1} coming from a region less than 1 pc across. The temperature of the regions, indicated by the presence of extremely high radiation rates for x-rays, argues that accretion flow around a black hole is the likely source of the observed luminosity.

In light of these observations, and because the range of physics required for an understanding of such flows touches on virtually all aspects of astrophysical hydrodynamics, we will discuss accretion disk theory at some length.

8.7.4 Heating the Disks: Dissipation and Viscous Torques

We first take up the question of the effect of the generation of energy by the shearing within the disk due to its differential rotation. Recall that the viscous energy dissipation rate is given by

$$\frac{d\epsilon}{dt} = -\frac{1}{2} T_{ij} \sigma_{ij} \tag{77}$$

where T_{ij} is the stress tensor and σ_{ij} is the shear. The shear for an axisymmetric system is given by $\sigma_{r\phi} = \partial v_\phi/\partial r - v_\phi/r = r(d\omega/dr)$. The negative sign in the second term comes from $\partial \hat{\phi}/\partial \phi$. In light of the previous discussion, the disk is Keplerian, and since the angular frequency is given by $\omega_K(r) = (GM/r^3)^{1/2}$, the shear is $-\frac{3}{2}\omega_K$. Since the shear varies with radius for such a disk, so does the rate of energy generation. If the stress

8.7 Accretion Disks in Astrophysics

tensor is proportional to the shear, then $\dot{E} = \eta r^2 (\partial \omega / \partial r)^2 \sim \eta \omega_K(r)^2$. This produces a relatively steep dependence of the local luminosity of the disk on radial distance, falling at larger distance from the central body.

The accretion disk will be assumed to be thin. It means that a vertical average through the disk, assuming that it is uniform vertically and approximately one pressure scale height in thickness, provides a reasonable first guess for the disk structure. The disk thickness is determined from the equation of state, since we know the coupling between the pressure and the stress tensor. It would perhaps be more correct to say that we *assert* this coupling, but it is at least intuitively correct. The pressure gradient in the disk counteracts the acceleration of gravity in the vertical direction. Assume that the disk does not contribute to the local gravitational acceleration g. Then if g is due only to the central object, there is a vertical component g_z near the symmetry plane:

$$g_z = -\frac{GM}{r^2}\frac{z}{r} = -\frac{v_\phi^2 z}{r^2} \tag{78}$$

Here r is the radial distance from the central object. We have introduced an important assumption here, that the disks are *geometrically* thin. The vertical support is therefore given by

$$\frac{dp}{dz} = -\rho g_z \tag{79}$$

If the disk is vertically isothermal (and this is something that we will have to examine later) the equation of state can be written as $p = \rho a_s^2$ and we get a simple scaling law for the thickness,

$$h = \left(\frac{a_s}{v_\phi}\right)^2 r \tag{80}$$

The pressure (and, in this case, density) scale height is the thickness of the disk.

Now, in the case of accretion around a compact object, the inner boundary is a sink for the accretion flow. That is, we do not really need to treat the details of the inward-directed flow but instead use only the assumption of continuity of the flow with time to provide the rate of inward transport of mass. The assumption is that, in response to the presence of viscosity in the turbulent flow in the disk (due to the α parameter) there is a slow but steady inward drift of mass. This is embodied in

$$\dot{M} = 2\pi r h \rho v_r \tag{81}$$

where v_r is the radial velocity of the material, and the flow is directed inward. The area is assumed to be the integration over a vertical column.

The equation for conservation of angular momentum is

$$\rho v_r \frac{1}{r}\frac{\partial}{\partial r}(rv_\phi) = \tau \tag{82}$$

where τ is the torque per unit volume.[3] Here we see the use of the continuity equation: it provides the connection between the mass flux and the disk scale height. Therefore, it governs the temperature gradient, $\Delta T/h$, and ultimately regulates the way that the disk gets rid of the energy it generates internally by friction.

We are therefore presented with a very pretty picture of equilibrium, one where rate of energy generation by shear stress is balanced by a change in the disk thickness and a change in the rate of inward drift of mass. But we have actually made a critical *implicit* physical assumption here. In order to solve these equations, the Shakura–Sunyaev approach is to assume that the radial flows are very slow in comparison with either the rotational velocity or the sound speed. In fact, in order to be able to neglect shocks in the flow and to maintain homogeneous subsonic turbulence throughout the disk (the thin disk approximation) we must maintain slow inward drift. On the other hand, this is not self-consistently treated in that it may be possible within physically imposed models to achieve very high flow velocities. This must be checked after the fact.

If (and that is a big *if*) we ignore this question, we can nonetheless solve for the local energy generation. This follows from the slow drift assumption *and* the condition that the disk is always Keplerian so that the gradient in the angular velocity is functionally specified. If the disk is anywhere self-gravitating, for example, or if it is imbedded in some peculiar gravitational potential within which it moves, the form of the angular velocity will not be so simple; but here we have $\partial rv_\phi/\partial r = \frac{1}{2}v_\phi$. This means that $a_s/h = v_\phi$ relates the torque and the scale height of the disk.

Look now at the rate of energy generation. Since we can take the stress as proportional to the strain, we have $L = \nu\sigma_{r\phi}^2$ as an approximation for the local rate of energy generation. But this is only an approximation; ν isn't defined except as a phenomenological constant. But ignoring this for a moment, we see that the flux, that is, the energy per unit area, is given by $F \sim GM\dot{M}r^{-3}$, so that locally we can estimate the temperature to vary as $r^{-3/4}$. The outer part of the disk should be cooler than the inner part, so different portions of the spectrum are sensitive to contributions from different portions of the disk. In a way, this behavior of the temperature

---

[3] *Torque* and *couple* are used on opposite sides of the Atlantic to denote the same force. This should save some confusion when approaching the literature.

8.7 Accretion Disks in Astrophysics

permits imaging of the disk (see Horne 1985). The problem is complicated by the effects of the central region, both the accretion boundary layer and the central body; reprocessing of radiation from the central parts of the accretion disk by its periphery is an especially troublesome phenomenon.

This picture provides basic scaling arguments, but let's examine the energy generation process a bit more closely. The work done by the stress must also be included in any reasonable disk model—it is the reason for the accretion in the first place. This work term is equal to $2\pi r^2 \omega T_{r\phi}$. Then the energy release per unit area, that is, the internally generated flux, is

$$F_{\text{internal}} \equiv \frac{1}{4\pi r^2} \frac{dE}{dt} = \frac{1}{4\pi r} \frac{d}{dr}\left[\dot{M}\left(\frac{1}{2}v_\phi^2 - \frac{GM}{r}\right) - 2\pi r^2 \omega T_{r\phi}\right] \quad (83)$$

This expression simplifies considerably because of the Keplerian motion assumption. For example, the first two terms combine to give the rate of gravitational energy release from the accreting matter, $-\frac{1}{2}GM\dot{M}/r$, and the third term is also just a multiple of GM/r. The exact answer is that the internally generated energy flux is given by

$$F_{\text{internal}} = \frac{3}{8\pi} \frac{GM}{r^3} \dot{M} \quad (84)$$

The total disk luminosity is ultimately derived from accretion, unless some nuclear processing occurs in the disk, or magnetic reconnection (both of which have been suggested for different accretion environments). In general, this means that the rate of energy generation is some multiple of what would be derived from infall of the matter from infinity,

$$L_{\text{acc}} = \frac{1}{2}\chi \frac{GM\dot{M}}{R} \quad (85)$$

where χ is an efficiency of energy release. In fact, here we have $\chi = 3$; the viscosity transports energy within the disk as well as the local release of gravitational energy, making the disk considerably brighter than we would have originally guessed. The rate of energy generation is balanced by the vertical conduction of flux by an optically thick medium. Here we again use the thin disk approximation and assume that the medium has a vertical temperature gradient of T/h. Then the flux, $F = K_{\text{rad}}\hat{z} \cdot \nabla T$, is given by

$$F = -K_{\text{rad}} \frac{dT}{dz} \sim -\frac{4acT^4}{3\kappa\rho h} \quad (86)$$

where κ is the radiative opacity.

Now have the basis for the separation of the disk into different phenomenological regimes. The disk is divided into a set of zones depending on the question of state and the source of the opacity. In the regime in which the temperature is highest, radiation pressure dominates. Here $p = p_{\rm rad} = \frac{1}{3}aT^4$. Then either electron scattering or thermal bremsstrahlung is the opacity source. In the innermost region of the disk, where the shear-generated energy is greatest, the temperature is so extreme that scattering and radiation pressure dominate the medium. The slowly moving outer parts of the disk are more inflated vertically and cooler. Here the temperature is low enough so that gas pressure dominates over radiation pressure and the opacity is from thermal absorption (and perhaps ionization). It is easy to see how this must be the case. Since the vertical component of the gravity scales as the inverse cube of the distance from the central source, the disk becomes progressively thicker in the outer parts, which reduces (at fixed input of energy) the temperature at the surface. The slower rate of energy generation guarantees that the temperature falls off faster than inversely with increasing distance. Some additional examples of disk structures, including self-gravity, are given in Shore and White (1982).

8.7.5 Stirring the Disks: The Turbulence Parameter

Most of the luminosity observed from systems with circulatory accretion flow comes from the disk itself. This means that the x-ray emission which is observed from the central regions of compact binaries and active galaxies must come from the very inner part of the accretion disk. It is also known that this region will be thermally unstable, depending on the cooling function for the disk. There is also a serious problem here, which arises as a result of the detailed calculation of a model for the α parameter. In the case of ordinary turbulence,

$$\eta = \tfrac{1}{3}\rho v_{\rm T} \qquad (87)$$

where $v_{\rm T}$ is an estimate of the turbulent velocity. The value of α can therefore be compared with the result for the determination of the viscous coefficient in both the gas-dominated and radiation-dominated regimes of the disk and, unfortunately for the model, the α required for the radiation is much higher in the assumed value than results from the viscosity.

In general, the argument goes, it isn't possible to specify the detailed turbulent flow field throughout the disk. Instead, it is probable that the kinetic energy density at any point in the material will be some fraction of the pressure. The turbulent pressure will ultimately come to some multiple of the thermal energy density, ρa_s^2. Tangled magnetic fields may also contribute to the turbulent viscosity, contributing terms of order $\rho v_{\rm A}^2$, where $v_{\rm A}$ is the Alfvén speed. Ultimately, these fields are probably limited

8.7 Accretion Disks in Astrophysics

by buoyancy and dissipation to having approximately the thermal energy density. So a good first guess would seem to be that if the turbulence is fully developed and isotropic, the shear stress tensor is given by

$$T_{r\phi} = \alpha p \tag{88}$$

This is the parameterization first examined in detail by Shakura and Sunyaev (1973) and Novikov and Thorne (1973). You have here the basis of almost all of the work done on accretion disks in the past two decades! An assumption so simple yet so sweeping in its implications that we have to spend more time on what it means.

The viscosity is therefore a function of temperature. By Eq. (87) we have already assumed a form that depends on the scale height, $\nu = \alpha a_s h$, where both h and a_s are local variables. The coefficient α is usually assumed to be a constant, but it need not be. Other forms are available for the viscous coupling, most taking into account the vertical temperature gradients and solving for some analog of convection within the disk (Meyer and Meyer-Hofmeister 1983; see also Pringle 1981). Again, notice that once the rate of viscous energy generation is chosen, the entire structure of an equilibrium thin disk is fixed. Unless we alter a physical assumption, like allowing for time dependence or studying the departures from locally Keplerian circulation, the entire model is a done deed.

8.7.6 Time-Dependent Accretion Disks: I. Basic Equations

For an axisymmetric disk flow, where $\partial/\partial\phi \to 0$, the structure depends on only two variables, the radial distance from the accreting object r and the vertical distance from the plane of symmetry z. Let's examine as skeletal a problem as possible. For a disk in centrifugal support, radial motions are merely a perturbation. However, turbulent disks are dominated by viscosity. It is responsible for the heating and for the vertical pressure balance. But such a disk, left to its own devices, will slowly change effective shape and internal mass distribution because of this viscous transfer of angular momentum. Without any outside perturbations, the disk would slowly spread out as material diffused inward toward the central accreter and the angular momentum lost from this material was transferred outward to the periphery of the disk. Quantifying this process is not terribly difficult in light of the tools we have at hand, so let's see what happens.

Assume again that $\omega = \omega_K \sim r^{-3/2}$ so that the specific angular momentum varies as $j \sim r^{1/2}$. In the presence of viscosity, gases in different annuli exchange angular momentum through the shear. The shear is given by $\sigma = r \, d\omega/dr$ and assuming that the stress is proportional to the strain gives

$T_{r\phi} \sim \eta \sigma_{r\phi}$. The torque is the volume integral of the shear times the moment arm:

$$\tau = r \cdot 2\pi r h \cdot \eta r \frac{\partial \omega}{\partial r} = -3\pi h(r) r^2 \omega_K \tag{89}$$

Even though you know the form of the torque, and so have an idea of what the ultimate answer looks like for the angular momentum evolution, let's proceed in a general way without supplying the viscosity prescription just yet. The assumption of Keplerian circulation disposes of the radial momentum equation; all radial motions are approximately in equilibrium. We need the continuity equation for the disk:

$$\frac{\partial \rho}{\partial t} + \frac{1}{r}\frac{\partial}{\partial r}(\rho r v_r) = 0 \tag{90}$$

where we have suppressed the z direction because of midplane symmetry. The disk is assumed to be in radial centrifugal equilibrium. This means that any perturbations in the circulation velocity are ignored for this first pass and $v_r \ll v_\phi$. It also means that $\omega = \omega_K$. The full conservation equation for angular momentum is

$$\frac{\partial}{\partial t}\rho v_\phi + \frac{1}{r}\frac{\partial}{\partial r}(r\rho v_r v_\phi) = \frac{1}{r}\frac{\partial}{\partial r}rT_{r\phi} \tag{91}$$

where the stress tensor $T_{r\phi}$ is not yet specified. Dropping the first term and using the continuity equation,

$$\rho v_r = \frac{\partial r T_{r\phi}/\partial r}{\partial r v_\phi/\partial r} \tag{92}$$

Since Keplerian motion has been assumed, the only unknown is the stress tensor.

Let's examine this last equation in a bit more detail. It states that the advection of angular momentum drives changes in the density of the disk. It is the density that we are interested in, not the change in time of the angular momentum. In fact, that is assumed to remain constant. This last point may be a bit confusing. How is it possible to transfer mass consistently and not deal with the alteration in the angular velocity? The answer is that we have assumed a disk configuration for which the time scale for exchange of angular momentum is short, where the central source dominates the disk structure. If the disk were to become self-gravitating, the problem would have a very different character. We would have to solve for the change in the circulation and the change in the disk mass simultaneously, a very difficult problem. But that isn't the disk we're studying. As

8.7 Accretion Disks in Astrophysics

long as the central accreter dominates, it is perfectly alright to ignore the alteration in ω with time. However, we cannot ignore the fact that ω is a function of radial distance in the disk, because it is this dependence that determines the shear rate and consequently drives all of the disk evolution through the viscous coupling between adjacent zones.

We have in hand an equation for the temporal evolution of the disk surface density because we substitute Eq. (92) back into the continuity equation (90) and integrate over the vertical direction. In fact, once you have performed the algebra, you'll notice that this looks like a diffusion equation. To make this more apparent, look at the surface density $\Sigma = \int \rho \, dz$ and assume that the viscosity is not constant vertically or radially. In other words, take $\langle \mu \rangle = \int \eta \, dz$. Recall that η actually contains the density and that the stress tensor depends on α and also on the local pressure, so both of these are functions of radial distance from the central accreter. Including Keplerian motion explicitly in the angular frequency, you will arrive at

$$\frac{\partial \Sigma}{\partial t} = \frac{3}{r} \frac{\partial}{\partial r} \left[r^{1/2} \frac{\partial}{\partial r} (r^{1/2} \langle \mu(r,t) \rangle) \right] \qquad (93)$$

This is an equation for the surface density that depends only on the physical properties of the turbulence. The factor 3 is due to the Keplerian form of the angular frequency, and the same is true for the $r^{1/2}$ terms. For other angular momentum laws the details will change, but as long as the frequency is a power law there will be no qualitative change in the nature of the solution.

Now we have some freedom here. The viscosity is something that connects the surface density and the state variables P and T. It therefore will also govern the rate of mass accretion, turning around the argument that gave us Eq. (93) in the first place. Once again, as long as the rotation law for the disk remains fixed, the entire problem hinges on the choice of the viscous coupling. If, on the other hand, we attempt to solve the disk structure self-consistently, this will not be a choice with such simple and immediate consequences. But taking $T_{r\phi} = \alpha p$ gives

$$\mu = \alpha \rho a_s h \qquad (94)$$

so that the vertically integrated viscosity is $\langle \mu \rangle = \alpha a_s h \Sigma$. The coupling between the temperature and surface density, mass accretion rate, and opacity are responsible for the behaviors observed in model thin α-disks. We will not detail all of the pathologies of such disks. Many of the references are contained in Pringle (1981), Frank, King, and Raine (1985), and Belvedere (1990). Here we will stay confined to more general questions.

Now what happens as the viscosity is turned on? Assume that the disk starts out as a thin ring. As the viscous coupling is increased, the region of the disk over which angular momentum transport is possible increases and the ring expands to compensate. That is, there will be an attempt to minimize the gradients in the surface density. Therefore, in time, the material in any narrow zone should be diffusively spread out over the whole accessible space and the central object should grow in mass. The basic idea behind disk instabilities is that the mass stored in the disk is a function of time and that if there were a large increase in the mass of the disk, it would take a finite time for that matter to make its way inward to the accreter. Consider for a moment what this might mean to the thermal properties of the disk. If the temperature goes up in some portion of the disk, one would expect that the disk would expand in the vertical direction and therefore the density, and the surface density, should drop. If the opacity then drops, the matter will cool and fall back down and there will be no large-scale transport of the disk gas inward. On the other hand, suppose we have a disk which starts to heat up and which then sees its opacity drop. Then the disk never expands.

8.7.7 Time-Dependent Accretion Disks: II. Disk Stability

One of the most important aspects of accretion disks is their radiative equilibrium. Unlike other forms of accretion, accretion disks are normally expected to be optically thick. This has been built into the equations for the equilibrium, of course, but it is essential that such disks can actually be constructed. In effect, accretion disks are like very flat stars and therefore one might expect that they would be unstable to pulsations in the same way that stars are. But in stars we know that pulsation is governed by the ratio of specific heats and opacity. These are both important in the convection zone and in regions of partial ionization. It is therefore not surprising that the same physical process can take place in accretion disks. In fact, to continue the "flat star" analogy one step further, disks are actually much like one-zone models used for pulsation studies. The one difference is that the gravitational acceleration is vertical but changes with radial distance from the central accreter and the geometrical thickness of the layer also changes.

We now rewrite the equations for disk radiative equilibrium:

$$\mathbf{F} \cdot \hat{\mathbf{z}} = F_z = -\frac{4ac}{3\kappa\rho} T^3 \frac{dT}{dz} + \frac{1}{4} c_p T \rho (gh)^{1/2} (\nabla - \nabla_{\text{Ad}})^{3/2} \left(\frac{l}{h}\right) \quad (95)$$

The first term is the radiative diffusion equation. It's the second term that is the most interesting (see also Chapter 9). First, the difference in the temperature gradients influences the total flux (this is well known). But the scale height enters through the mixing length. Normally we wouldn't think twice about this, but here it means that the mixing length varies with distance from the central star. All this means that the thermal time scale is a function of motion in the disk and that global relaxation is not possible if the disk has been disturbed. Depending on the local opacity and temperature, the disk may either damp or pulsate at different annuli on very different time scales. The changes in the vertical structure feed back into the driving (see, e.g., Saio *et al.* 1987).

8.7.8 Boundary Layers in Accretion Disks

When matter circulates around an accreter (in a binary system or elsewhere), it must eventually lose angular momentum and fall onto the surface of the star. Since it is generally true that the underlying star is rotating more slowly than the accretion disk, there must be a larger shear between the stellar surface and the disk. Therefore, one expects that a boundary layer will appear about the accreter and that the appearance of this layer will be important for the observational properties of the system. In this section, we will discuss some ways of calculating the structure of this region. It must be added, however, that there is at present no complete theory for the description of this region.

The rotational velocity at the surface of the star is given by $v_{\phi\star}^{(\text{rotate})} = 5.1 \times 10^6 R_\star P^{-1}$ cm s^{-1}, where $R_\star$ is the stellar radius in $R_\odot$ and P is the period in days, while the circulation velocity for the accretion disk is $v_{\phi\star}^K = 4.4 \times 10^7 (M_\star/R_\star)^{1/2}$ cm s^{-1}, where $M_\star$ is the stellar mass in $M_\odot$. These are useful orders of magnitude to bear in mind, at least for stars. It is generally true that the rotational velocity of the accreter is less than the Keplerian velocity. (See Fig. 8.7.) This is especially true for synchronous rotation. Shear *at the stellar surface* is of the same order as the Keplerian velocity. This is an enormous burden to place on the fluid, slowing down over a short distance to become synchronous with the surface. If the surface is rotating rapidly enough, the matter may never actually accrete. Instead, it may either blow out of the system or accumulate in a ringlike structure. Such situations are not entirely unrealistic for many interesting applications (a possible exception is the accretion of matter by newly formed neutron stars in close systems). It is possible that the accreter may be rotating at nearly the Keplerian velocity, so that virtually no boundary layer is formed.

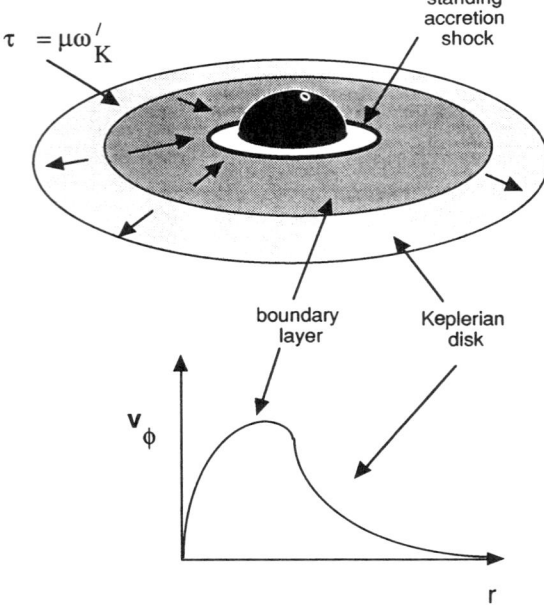

Figure 8.7 A boundary layer in the inner portion of an accretion disk. The velocity curve shows that the outer disk is Keplerian, while at some critical distance from the star the flow attempts to become corotating with the surface.

The luminosity of material which is falling into the gravitational potential of the accreter is

$$L_{BL} = \frac{1}{2}\frac{GM\dot{M}}{R_\star} = 4\pi R_\star \delta_{BL} \sigma T_{BL}^4 \qquad (96)$$

Here σ is the Stefan–Boltzmann constant and δ_{BL}, the thickness of the boundary layer for which the temperature is T_{BL}, is to be calculated. The temperature of an optically thick layer is

$$T_{BL} = \left(\frac{g\dot{M}}{8\pi\sigma}\right)^{1/4} \delta_{BL}^{-1/4} \qquad (97)$$

For an optically *thin* disk, the virial temperature is the maximum achievable value:

$$T_V = \frac{3}{8}\frac{GM m_P}{8kR_\star} \geq T_{BL} \qquad (98)$$

Here k is the Boltzmann constant and m_P is the mass of the proton. The thickness of the disk is fixed by the thermal and kinetic energies of the region around the accreter, and the boundary layer thickness is

$$\delta_{BL} \sim \left(\frac{a_s}{v_{\phi\star}^K}\right)^2 R_\star = \frac{a_s^2}{g} \tag{99}$$

This prescription has been used for compact objects and represents an equilibrium between the rate of generation of energy by the circulation and the source energy. Any estimate for the Reynolds number shows that the boundary layer is turbulent. Also, the layer is unstable by the Rayleigh criterion (see Chapter 9) because the velocity field now has an inflection point. The estimate of the shear-generated viscosity is thus

$$\eta_t \sim \frac{\langle\rho_{BL}\rangle\langle v_t^2\rangle}{(\partial v_\phi^K/\partial r)_\star} \tag{100}$$

where v_t is the turbulent velocity, and this will be assumed to be subsonic. Then the turbulent velocity is of order $\alpha^{1/2}a_s$ and it follows that

$$\eta_t \sim \frac{2}{3}\langle\rho_{BL}\rangle\alpha\frac{a_s^2}{v_\phi^K}R_\star \tag{101}$$

The thickness of the boundary layer becomes $\delta_{BL} = \frac{2}{3}\alpha^{1/2}h < h$ and the local heating is given by $\Gamma = \eta_t(v_\phi^K/\delta_{BL})^2$. For a recent discussion of numerical models, see Kley (1991).

8.8 Spherical Accretion

The essential similarity of the infall and outflow solutions derives from the simplicity of the assumption—only pressure drives the flow outward, only gravity constrains it, and the flow is in steady state and mass loss rate (or accretion rate) is constant. This statement will be our coda for the remainder of this chapter.

8.8.1 Mass Infall and the Parker Solution

We have gone through much effort to understand the Parker solution and what an isothermal or polytropic outflow looks like. But we make use of this solution in a totally different context, one which is equally important as a stellar wind but of perhaps wider application—the accretion of material in a gravitational field. The first detailed discussion of this problem was by Bondi (1952) and elaborated in a series of papers by Mestel (1954). We want to be sure that the accreting matter does not hit the surface of the star going at infinite speed; we place the interior condition that the stuff settles hydrostatically onto the surface of the accreting body. This is a convenience—if the central body happens to be a black hole there is no possibility for hydrostatic equilibrium if the matter falls freely through the event horizon. But ignoring this case, we would be surprised indeed if, for instance, the gas lost by a companion star didn't settle quasi-statistically onto the surface of an accreting star in a binary system. Assuming that the material starts from rest at large distance and falls inward, the fact that the velocity gradient switches sign means that the accreting material goes through a shock which stands off at some distance from the stellar surface and which has a very high temperature. The shock, if optically thin, reaches the *virial* temperature, which is the energy released by falling from large distance onto the surface of a body and coming to rest:

$$T_V = \frac{GM}{2kr_s} \tag{102}$$

which is independent of the mass accretion rate. If it is optically thick, the shock will look instead like a blackbody at the local temperature given by the luminosity in the shock:

$$T_s = \left(\frac{L}{4\pi\sigma r_s^2}\right)^{1/4} \tag{103}$$

where the luminosity is given by

$$L = \frac{GM}{r_s}\dot{M} \tag{104}$$

The equation of motion for the accreting gas will, by now, look quite familiar:

$$u\frac{du}{dr} = -\frac{1}{\rho}\frac{dp}{dr} - \frac{d\Phi(r)}{dr} \rightarrow (u^2 - a_s^2)\frac{1}{u}\frac{du}{dr} = \frac{2c_s^2}{r} - \frac{d\Phi}{dr} \tag{105}$$

where Φ is the gravitational potential and the second form of the equations assumes the accretion to be isothermal. As for the outflow solution, $u = a_s$ when $r = r_s$. There is a twist here, however, because if the accretion flow is not isothermal, a_s is determined by the conditions at the shock front and therefore we have a condition for the mass accretion rate which is $\dot{M} = 4\pi r_s^2 \rho_s a_s(r_s)$. The condition that the flow is at rest at infinity constrains the solution that passes through the sonic point. This doesn't mean that there will necessarily be a shock at this position, only that the speed is greater than the local sound speed. Only if the velocity field turns to

8.8 Spherical Accretion

decelerating flow—in other words, if the flow starts out at too high a speed to settle down quietly onto a hydrostatic core—does the matter certainly pass through a shock. For purposes of discussion, but also because it is physically reasonable in stellar and even galactic accretion problems and the simple inverse of the stellar wind case, we assume that the flow settles down onto the surface at rest. So the flow must start out with some conserved mass flux which chooses one of the branches of the wind solution, moves inward, goes through the shock, and dissipatively jumps to another branch.

We supplement the momentum and mass conservation equations by the condition for energy conservation:

$$\frac{1}{r^2}\frac{d}{dr}\left[\rho r^2 u\left(\frac{\gamma}{\gamma-1}\frac{p}{\rho}+\frac{1}{2}u^2+\Phi\right)\right]=\frac{\partial \epsilon}{\partial t}=-\alpha\rho^2 T^{1/2} \quad (106)$$

where ϵ is the energy density, α is a constant characteristic of free–free radiation, and we have explicitly chosen bremsstrahlung as the mode of energy loss. Heating is affected by the shock compression. Once we know the position of the sonic point and have chosen a boundary condition at infinity, that is, $a_s(\infty)$, we can solve for the run of temperature and velocity throughout the flow. And once we have specified these, we also know the mass accretion rate. This is the basis of a cooling flow—that the matter cools and loses dynamical support, after which it accretes onto the central galaxy. If the flow isn't isothermal or adiabatic, then the equation for the temperature feeds back into the dynamics through the sound speed (or the pressure) and the rate of dissipation becomes critically important.

A very interesting technique was derived by Hunt (1975), who looked at the mapping of the equations of accretion into oblate spheroidal coordinates with an extended gravitational potential (this solution makes use of the fact that the Galaxy, the object of the study, can be modeled as a spheroid potential). The technique could be readily generalized to include the effects of shocks, which are not included in the model.

8.8.2 Spherical Accretion: Solution for Polytropic Flows

Let's take the simplest case—spherical accretion without radiative terms. As we have seen, the momentum equation is actually the same as for the solar wind problem, but as we will see there is a twist. The continuity equation gives $C = r^2\rho v = $ constant. The radial equation of motion is

$$v\frac{dv}{dr}=-\frac{GM}{r^2}-\frac{1}{\rho}\frac{dp}{dr} \quad (107)$$

For this gas, we have to select an equation of state, so assume a polytropic flow, $p = K/\rho^\gamma$. Now the sound speed is $a_s = K\gamma\rho^{\gamma-1}$. Rewriting Eq. (107) using C gives

$$(v^2 - a_s^2)\frac{dv}{dr} = v\left(\frac{2a_s^2}{r} - \frac{GM}{r^2}\right) \tag{108}$$

which has a critical point, $v = a_s$, at a radial distance $r_\star = GM/2a_s^2$. This is the most important part of the calculation, as with the solar wind. We now know where the flow will turn supersonic, if it does at all. The energy condition for this polytropic flow is

$$E = \frac{1}{2}v^2 + \frac{\gamma}{\gamma - 1}K\rho^{\gamma-1} - \frac{GM}{r} = E_\infty \tag{109}$$

Assume that the matter starts falling in from rest with a total energy E_∞. Then from energy constant and knowing the matching conditions at $r_\star$

$$r_\star = \left(\frac{5 - 3\gamma}{\gamma - 1}\right)\frac{GM}{4E_\infty} \tag{110}$$

Now you see why $r_\star$ is so vital. The mass accretion rate is a constant; therefore it can be fixed by its value at $r_\star$, namely $\dot{M} = 4\pi r_\star^2 \rho_\star a_s$. But the sound speed is a function only of the density, and we can determine this from the energy equation since we know $r_\star$. So the mass accretion rate depends only on the mass of the accreter and the energy that the material started with from large distance:

$$\dot{M} = \pi GM^2\left(\frac{2(\gamma - 1)E_\infty}{5 - 3\gamma}\right)^{(5-3\gamma)/2(\gamma-1)} \tag{111}$$

Note that this accretion rate is independent of E for $\gamma = 5/3$. We have thus derived the solution for a spherically accreting flow (Bondi 1952). (See Fig. 8.8.) It assumes that the flow goes through a sonic point, but it also makes the critical assumption that there are no added sinks for energy. There is no thermal conduction, no radiative loss (other than what is required to thermostat the material). There is no shock at the critical point. In fact, that is how it was derived. If the flow starts out with some value different from E_∞, a given mass, or has rate accretion different from the one required by Eq. (109), the flow will have to pass through an accretion shock or it will never become supersonic (see also Scharlemann 1978). The former condition is frequently encountered. The latter is the analog of the stellar breeze solution in the Parker case. It is hydrostatic and we need not bother with it anymore.

8.8 Spherical Accretion

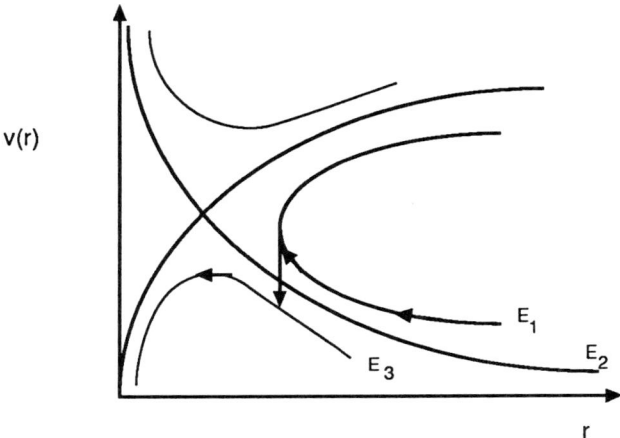

Figure 8.8 Inverse Parker-type solution for accretion. Streamlines with different initial energies E_∞ are indicated. The standing shock is indicated by the transition to the subsonic branch.

The Eddington limit also plays a role for accretion. Notice that the gravitational acceleration has been assumed to be from the mass M. But you know that by allowing for radiation pressure in the optically thin, spherically symmetric limit, we reduce this effective mass by a factor $(1 - L/L_{\text{Edd}})$. So the Eddington luminosity is also an effective accretion limit—if we dump matter onto the star at a rate so high that we would generate a luminosity above L_{Edd}, the flow becomes unstable and highly time dependent. Steady-state flow is not possible under such circumstances.

8.8.3 Accretion within a Wind

Binary systems provide excellent environments in which to study both disk and nonspherical accretion. As we have seen, mass transfer can take place between components in a close binary if one of the stars fills its Roche surface. There is, however, an alternative mode of mass transfer that occurs if the mass-losing star does not completely fill its critical equipotential. This is the *wind accretion* mode. Consider matter coming from infinity with some initial velocity v_∞, in this case the terminal velocity of the wind from the loser. Once in the vicinity of the companion, this material is gravitationally deflected and may accrete. The simple statement of the critical condition is that anything with a terminal velocity which exceeds the escape velocity at some distance from the companion will be able to freely escape the system, although deflection may occur forming a trailing wake. This point requires a moment's reflection. Recall that the wind is supersonic. Therefore, any deflection amounts to the introduction of oblique shocks in the wind. This both dissipates kinetic energy in the wind and further refracts the flow outward. The net result is the formation of a hot trailing region behind the accreter, a phenomenon also well reproduced in numerical simulations (see, e.g., Livio *et al.* 1986). The condition for capture can be written as

$$\frac{1}{2} v_\infty^2 = \frac{GM}{r_\star} \tag{112}$$

where now $r_\star$ is the critical radius. Therefore, the gravitational cross section is approximately

$$\sigma_g \approx \frac{4\pi G^2 M^2}{v_\infty^4} \tag{113}$$

and the mass accretion rate is

$$\dot{M}_\star \approx \rho_\infty \frac{\pi G^2 M^2}{m_p v_\infty^3} \tag{114}$$

where m_p is the mass of the proton and ρ_∞, the density in the outflow from the companion at the distance of the secondary, is given by $\dot{M}/4\pi a^2 v_\infty$. Here a is the separation of the components. Therefore

$$\dot{M}_\star \approx \left(\frac{G^2 M^2}{m_p a^2 v_\infty^4} \right) \dot{M} \tag{115}$$

where the expression in parentheses can be thought of as an accretion efficiency, ϵ. Since the accretion luminosity is given by $GM\dot{M}_\star/R$, where R is the radius of the accreter, we can compute the total output from a wind accreter in terms of the mass loss parameters for the loser alone, with the addition of the M/R ratio for the companion. (See Fig. 8.9.)

There is another very powerful constraint that can be used to delimit the properties of the mass-gaining star. The accretion cannot exceed the Eddington rate, which is that required to provide the Eddington luminosity. Therefore it follows that

$$\frac{G^2 M^2 \kappa}{4 m_p c v_\infty^4 a^2 R} < 1 \tag{116}$$

or, put in scaled variables, $M > 1.7 \times 10^{-4} (v_\infty/1 \text{ km s}^{-1})^2 (a/1 \text{ AU})(R/R_\odot)^{1/2} M_\odot$. The temperature of the infalling matter can be determined

8.8 Spherical Accretion

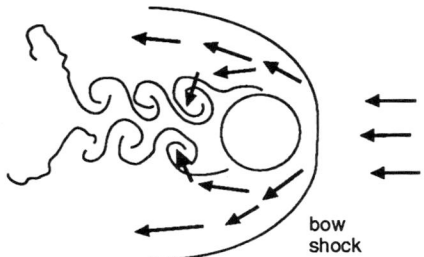

Figure 8.9 Wind accretion in a close binary system. The formation of an accretion wake is a consequence of the gravitational attraction of the mass gainer. The bow shock is shown as Σ. The flow may also show a vortex sheet trailing the accreter; recent models seem to find this in the form of a wake-flipping instability (Livio 1991).

directly from observation and, if optically thin, provides a direct constraint on M/R. Thus we can determine many of the basic properties of the accreter in the case of wind accretion which are not directly obtained from disk accretion. If the star is able to form an accretion disk, additional energy release takes place prior to the last plop of the material onto the stellar surface.

If the star has a strong magnetic field two additional processes may be observed. First, the accretion is given by the ability of the magnetosphere to capture the wind, which may be larger than the gravitational radius. This follows from the fact that for slow highly ionized winds, the local Alfvén speed may be greater than the escape velocity for the star. Therefore the accretion condition is given by the condition that $v_A = v_\infty$ at the capture point, which in turn depends on the density through

$$v_\infty = \frac{BR^3}{2\pi^{1/2} r^3 \rho_\infty^{1/2}} \tag{117}$$

for a dipole field. Thus, the accretion cross section is

$$\sigma_B = \pi \left(\frac{BR^3 a}{v_\infty^{1/2} \dot{M}} \right)^{2/3} \tag{118}$$

You might try to calculate the following: What is the condition that the Alfvén accretion cross section exceeds the gravitational cross section? What happens if the star accreting matter in a disk has a magnetic field? The field may be small enough to produce a negligible increase in the accretion cross section, but it may dominate the flow. In the vicinity of the star, should an accretion disk be formed, the local Alfvén speed is usually smaller than the Keplerian circulation speed at the magnetospheric boundary.

8.8.4 Time-Dependent Flows

Let's now add time dependence for this spherically symmetric infall. First the continuity equation becomes

$$\frac{d\rho}{dt} = -\rho \frac{1}{r^2} \frac{\partial}{\partial r} r^2 v \tag{119}$$

Then the momentum equation for the radial flow:

$$\frac{dv}{dt} = -\frac{1}{\rho} \frac{\partial p}{\partial r} - \frac{\partial \Phi}{\partial r} \tag{120}$$

Finally, adding the complication of heating and cooling terms to the usual energy equation we get

$$\frac{de}{dt} = -(e + p) \frac{1}{r^2} \frac{\partial}{\partial r} (r^2 v) - \rho v \frac{\partial \Phi}{\partial r} - v \frac{\partial p}{\partial r} \tag{121}$$

The latter can also be written explicitly in terms of the temperature

$$\frac{1}{r^2} \frac{d}{dr} \left[\rho v r^2 \left(\frac{\tfrac{5}{2} kT}{\mu m_p} + \frac{1}{2} v^2 - \frac{GM}{r} \right) \right] = -\Lambda(\rho, T) \tag{122}$$

where Λ is the energy loss rate per unit volume. Parametrically, this loss rate is assumed to be of the form

$$\Lambda(\rho, T) = -\alpha_L \rho^2 T^{1/2} \tag{123}$$

where α_L is a constant and the rate is assumed due to thermal bremsstrahlung. Incorporating the perfect gas equation of state gives

$$\rho v \frac{dv}{dr} + \frac{d}{dr} \left(\frac{\rho k T}{\mu m_p} \right) + \rho \frac{GM}{r^2} = 0 \tag{124}$$

for the momentum equation. Here e is the energy density (the energy per unit mass is E and the mass density is ρ as usual), p is the pressure, v is the radial velocity, and Φ is the gravitational potential. We need to supplement these equations with two additional equations, one for the pressure and one for the gravitational field.

8.8 Spherical Accretion

8.8.5 Cooling Flows in Clusters of Galaxies

Observational Motivations

After the first x-ray observations of clusters of galaxies were completed during the 1970s, using HEAO-A and Einstein (both x-ray satellites), it was realized that the cooling time for the gas observed in the cluster potential is short in comparison with the Hubble time. That is, since one knows that the temperature of an optically thin plasma can be directly determined from the bremsstrahlung (or free–free emission) spectrum, the energy content, E, of a volume of gas can be directly observed. The luminosity is therefore a direct measure of the cooling time through $\tau_{cool} = E/L$. The luminosities observed for many rich clusters are high, yielding $\tau_{cool} < \tau_{Hubble}$, the latter being the canonical cosmological time scale of about 2×10^{10} years. The logical conclusion seems to be that this gas cannot remain hydrostatic forever and simply cool, because it is bound within the gravitational potential of the cluster. The matter has no choice, unless some heating source is present, than to fall inward toward the center of the cluster and accrete onto whatever galaxy might be located there. This is the *cooling flow* model. It leads to a very interesting picture, one which we will examine here as an example of an accretion flow but one which contains a serious problem as yet unresolved.

To begin with the problem, it is that the rate of mass accretion implied by the cooling is exceedingly high, of order 10 to 1000 $M_\odot$ yr^{-1}, and one which would have an obvious effect on the mass-accreting galaxy. Stars should form in this newly supplied galaxy, and there should also be an increase in the mass of the galaxy with time. If the rate were merely a few tens of solar masses per year, the galaxy might more than double its mass over the age of the universe. Further, the increase in the gas content should mean that newly formed stars should be present in these galaxies, and these should show up in both the ultraviolet and the infrared. Neither of these has yet been observed. The pain this has evoked among theorists in this field is considerable. Many highly ingenious models have been proposed for altering the nature of the stars formed in such cooling flows in order to hide them from detection. So one should keep in mind, while we discuss the standard picture which has emerged for this phenomenon, that as in most of astrophysics there are still foundational questions lurking in the wings which need to be addressed at some stage.

More Complications: Extended Mass Distributions

Normally, in the case of a stellar wind, we can assume that only the central potential plays any role in structuring the flow. There are some cases, especially in galactic and galaxy cluster–scale environments, where we must assume that the self-gravitation of the flow is important or that the mass of the accreter is sufficiently spatially distended that we will have to deal with changes in the mass, as well as the radius, in computing Φ. In other words, we have to have some equation which spells out how $M(r)$ is to be calculated. In the case of clusters of galaxies, the amount of accreting gas is so great that we may have to consider that the mass accretion rate is not a constant with distance. Take

$$\frac{dM_c(r)}{dr} = 4\pi r^2 \rho_g \tag{125}$$

where the galaxy density distribution, $\rho_g(r)$, can be modeled using a polytropic or King-type law. The reason we need this complication is that the mass in the cluster is extended, so we require a mass model in order to compute the gravitational acceleration $-GM(r)/r^2$. Because the potential is extended, we do not necessarily have a solution for the accretion that must undergo a shock. The gravitational acceleration decreases as the flow falls deeper into the galaxy or cluster potential and therefore may never pass through the sound speed. It may settle quiescently into the central body, or may undergo shock compression within the central galaxy. We have also not included in this discussion the possibility of wind or particle heating of the flow. If the galaxy generates a wind due to star formation activity, the flow can be slowed as it approaches the accreter. All of these are discussed in Sarazin (1988) and references therein.

References

Beals, C. S. (1931). The contours of emission bands in novae and Wolf–Rayet stars. *Mon. Not. R. Astron. Soc.* **91**, 966. [See also: Beals, C. S. (1951). *Publ. Dom. Astrophys. Obs.* **9**, 1.]

Bondi, H. (1952). **112**, 195: On spherically symmetric accretion. *Mon. Not. R. Astron. Soc.* **112**, 195. [See also Hoyle, F., and Lyttleton, R. A. (1941). On the accretion theory of stellar evolution (and references therein). *Mon. Not. R. Astron. Soc.* **101**, 27.]

Castor, J. (1970). Spectral line formation in Wolf–Rayet envelopes. *Mon. Not. R. Astron. Soc.* **149**, 111.

Castor, J., Abbott, D., and Klein, R. (1975). Radiation-driven winds in Of stars. *Astrophys. J.* **195**, 157.

Chamberlin, J. W. (1961). Interplanetary gas. III. A hydrodynamic model of the corona. *Astrophys. J.* **133**, 675.

Chandrasekhar, S. (1945). Line formation in an expanding atmosphere. *Rev. Mod. Phys.* **17**, 138.

Clerke, A. M. (1902). *A Popular History of Astronomy During the Nineteenth Century*. London: Adam and Chas. Black.

Cowie, L., Ostriker, J. P., and Stark, A. A. (1978). Time-dependent spherically symmetric accretion onto compact x-ray sources. *Astrophys. J.* **226**, 1041.

Davidson, K., and Ostriker, J. P. (1973). Neutron star accretion in a stellar wind: Model for a pulsed x-ray source. *Astrophys. J.* **179**, 585.

Deutsch, A. J. (1955). The electromagnetic field of an idealized star in rigid rotation *in vacuo*. *Ann. Astrophys.* **18**, 1.

Eddingtion, A. S. (1927). *The Internal Constitution of the Stars*. Cambridge: Cambridge University Press.

Eggelton, P. (1983). Approximations to the radii of Roche lobes. *Astrophys. J.* **268**, 368.

Frank, A, Balick, B., and Riley, J. (1990). Stellar wind paleontology: Shells and halos of planetary nebulae. *Astron. J.* **100**, 1903.

Freeman, N. C., and Johnson, R. S. (1972). A note on stellar winds and breezes. *Proc. R. Soc. Lond.* **A329**, 241.

Goldreich, P., and Julian, W. (1969). Pulsar electrodynamics. *Astrophys.J.* **157**, 869.

Horne, K.(1985). Images of accretion discs—I. The eclipse mapping method. *Mon. Not. R. Astron. Sco.* **129**, 213.

Howarth, I. D., and Prinja, R. K. (1989). The stellar winds of 203 galactic O stars: A quantitative ultraviolet study. *Astrophys. J. Suppl.* **69**, 527.

Hunt, R. (1975). Accretion of intergalactic gas by a realistic model for the galaxy and its consequences. *Mon. Not. R. Astron. Soc.* **173**, 465.

Jeans, J. H. (1916). *The Dynamical Theory of Gases*. Cambridge: Cambridge University Press.

Kley, W. (1991). On the influence of viscosity on the structure of the boundary layer of accretion disks. *Astron. Astrophys.* **247**, 95.

Kwan, J., and Kuhi, L. (1975). P cygni stars and mass loss. *Astrophys. J.* **199**, 148.

Kwok, S., Purton, C., and Fitzgerald, P. (1978). On the origin of planetary nebulae. *Astrophys. J. Lett.* **219**, L125.

Livio, M., Soker, N., de Kool, M., and Savonije, G. J. (1986). On accretion of angular momentum from an inhomogeneous medium. *Mon. Not. R. Astron. Soc.* **218**, 593.

Lucy, L. B., and Solomon, P. (1970). Mass loss by hot Stars. *Astrophys. J.* **159**, 879.

Lubow, S., and Shu, F. H. (1975). Gas dynamics of semidetached binaries. *Astrophys. J.* **198**, 383.

Lynden-Bell, D., and Pringle, J. (1974). The evolution of viscous discs and the origin of nebular variables. *Mon. Not. R. Astron. Soc.* **168**, 603.

Mestel, L. (1954). The influence of stellar radiation on the rate of accretion. *Mon. Not. R. Astron. Soc.* **114**, 437.

Mestel, L. (1968). Magnetic braking by a stellar wind—I. *Mon. Not. R. Astron. Soc.* **138**, 359.

Michaud, G. (1970). Diffusion processes in peculiar A stars. *Astrophys. J.* **160**, 641.

Michel, F. C. (1991). *Theory of Neutron Star Magnetospheres*. Chicago: University of Chicago Press.

Milne, E. A. (1927). Selective radiation pressure and the structure of a stellar atmosphere. *Mon. Not. R. Astron. Soc.* **87**, 697.

Mocnoski, S. (1984). Accurate integrations of the Roche model. *Astrophys. J. Suppl.* **55**, 551.

Novikov, I., and Thorne, K. (1973). In *Black Holes*. C. de Witt and B. de Witt eds. New York: Gordon and Breach.

Pacini, F. (1967). Radio emission from a neutron star. *Nature* **216**, 567.

Paczynski, B. (1971). Close binary star systems. *Ann. Rev. Astron. Astrophys.* **9**, 183.

Parker, E. N. (1958). Dynamics of the interplanetary gas and magnetic fields. *Astrophys. J.* **128**, 664. [See also Parker, E. N. (1963). Dynamical properties of stellar coronas and stellar winds. I. Integrations of the momentum equation. *Astrophys. J.* **139**, 72.

Pickel'ner, S. B. (1968). Interaction of a stellar wind with a diffuse nebula. *Astrophys. Lett.* **2**, 97.

Roberts, P. H. (1971). Transformation of the stellar wind equations. *Astrophys. Lett.* **9**, 79.

Roberts, P. H., and Soward, A. M. (1972). Stellar winds and breezes. *Proc. R. Phil. Soc.* **A328**, 185.
Ryu, D., and Vishniac, E. (1988). A linear stability analysis for wind-driven bubbles. *Astrophys. J.* **331**, 350.
Saio, H., Cannizzo, J., and Wheeler, J. C. (1987). A linear thermal stability analysis for the vertical structure of alpha model accretion disks. *Astrophys. J.* **316**, 716.
Scharlemann, E. T. (1978). The fate of matter and angular momentum in disk accretion onto a magnetized neutron star. *Astrophys. J.* **219**, 617.
Schuster, A. (1905). Radiation through a foggy atmosphere. *Astrophys. J.* **21**, 1.
Shakura, N. I., and Sunyaev, R. A. (1973). Black holes in close binary systems: Observational consequences. *Astron. Astrophys.* **23**, 1.
Shore, S. N., and White, R. L. (1982). Self-gravitating accretion disk models for active galactic nuclei: Self-consistent α-models for the broad line region. *Astrophys. J.* **256**, 390.
Sobolev, V. (1960). *Moving Envelopes of Stars*. Cambridge: Cambridge University Press.
Weber, E. J., and Davis, L., Jr. (1967). The angular momentum of the solar wind. *Astrophys. J.* **148**, 271.

Suggested Readings

Stellar Winds

Brandt, J. C. (1970). *Introduction to the Solar Wind*. San Fransisco: W. H. Freeman.
Conti, P., and Underhill, A., eds. (1988). *O Stars and Wolf-Rayet Stars*. NASA SP-497.
Hjellming, R. M., and Gibson, D. M., eds. (1985). *Radio Stars*. Dordrecht, The Netherlands: Reidel.
Garmany, C. D. (1990). *Properties of Hot Luminous Stars*. San Fransisco: ASP Press.
Johnson, H. R., and Querci, F. R., eds. (1986). *The M-type Stars*. NASA SP-492.
Jordan, S., ed. (1981). *The Sun as a Star*. NASA SP-450.
Mihalas, D., and Mihalas, B. (1984). *Foundations of Radiation Hydrodynamics*. Oxford: Oxford University. Press.
Parker, E. N. (1963). *Interplanetary Dynamical Processes*. New York: Wiley Interscience.
Underhill, A. B., and Michalitsianos, A. G., eds. (1985). *The Origin of Nonradiative Heating/Momentum in Hot Stars*. NASA CP-2358.

Accretion Disks

Belvedere, G., ed. (1990). *Magnetic Fields and Accretion Disks in Astrophysics*. Dordrecht, the Netherlands: Reidel.
Bertout, C., Collin, S., La Sota, J-P., and Tran Thanh Van, J., eds. (1991). *Structure and Emission Properties of Accretion Disks*. Paris: Editions Frontières.
Cordova, F., ed. (1988). *Multiwavelength Astrophysics*. Cambridge: Cambridge University Press.
Frank, J., King, A., and Raine, D. (1985). *Accretion Power in Astrophysics*. Cambridge: Cambridge University Press.
Mauche, C., ed. (1990). *Accretion-Powered Compact Binaries*. Cambridge: Cambridge University Press.
Meyer, F., Duschl, W. J., Frank, J., and Meyer-Hofmeister, E., eds. (1989). *Theory of Accretion Disks*. Dordrecht, The Netherlands: Kluwer.
Pringle, J. (1981). Accretion disks in astrophysics. *Ann. Rev. Astron. Astrophys.* **19**, 137.

Sellwood, J. A., ed. (1989). *Dynamics of Astrophysical Accretion Disks*. Cambridge: Cambridge University Press.
Shu, F. H., and Lubow, S. H. (1981). Mass, angular momentum, and energy transfer in close binary stars. *Ann. Rev. Astron. Astrophys.* **19**, 277.

Accretion and Cooling Flows

Forman, W., and Jones, C. (1982). X-ray imaging observations of clusters of galaxies. *Ann. Rev. Astron. Astrophys.* **20**, 547.
O'Dea, C. P., and Uson, J. M., eds. (1986). *Radio Continuum Processes in Clusters of Galaxies*. Green Bank, West Virginia: NRAO.
Sarazin, C. (1990). Cooling flows and x-ray emission in early-type galaxies. In *The Interstellar Medium in Galaxies* (p. 201). H. A. Thronson and J. M. Shull, eds. Dordrecht, The Netherlands: Kluwer.

CHAPTER 9

Instabilities

> *Et tu, Brute? Then fall, Caesar.*
> Shakespeare, *Julius Caesar, Act I, Scene 2*

9.1 Introduction

In the case of the fluid equations, the idea that the flow can remain steady is a nice fairy tale. In fact, in general, experience forces us to admit that the solutions we have found thus far are quite far from being realistic. The flows of everyday life may not be stable. We are all familiar with wave phenomena, turbulence, and all of the possible departures from steady Bernoulli flow in between. It is therefore necessary now to look at the cause, and cure, of some of the more important astrophysical instabilities which can take place in the presence of gravitational fields and external media.

One historical point is in order before we begin. Many of the instabilities we will be discussing were already discovered during the nineteenth century. It was the bias of the Victorian scientists that they were generally looking for *stability* and thus many of these solutions appeared pathological to them. It may be an indication of just how much the view has changed that the onset of an instability is now met with delight. It reflects the evolution of our understanding of the dynamic nature of astrophysical environments and of the transient nature of many of the states and structures we observe.

9.2 Waves

9.2.1 Sound Waves as a Prototypical Instability

Imagine that a fluid of uniform initial density ρ_0 is sitting in equilibrium and is perturbed by a slight amount $\delta\rho$. The equations describing the development of this perturbation, under the condition that it is extremely small in amplitude, are

$$\frac{\partial \delta \rho}{\partial t} + \rho_0 \frac{\partial v}{\partial x} = 0 \qquad (1)$$

$$\frac{\partial v}{\partial t} = -\frac{1}{\rho_0} \frac{\partial \delta p}{\partial x} \qquad (2)$$

Here we have taken δp to be the perturbation on the pressure. Now take the approximation that $p = p(\rho)$ so that the second term of Eq. (2) becomes

$$\frac{\partial \delta p}{\partial x} = \left(\frac{\partial p}{\partial \rho}\right)_0 \frac{\partial \delta \rho}{\partial x} \qquad (3)$$

We see that we have two first-order equations, which we can combine into a single second-order equation:

$$\frac{\partial^2 \delta \rho}{\partial t^2} = a_s^2 \frac{\partial^2 \delta \rho}{\partial x^2} \qquad (4)$$

We have just written down the familiar wave equation for a density disturbance. It is a hyperbolic equation that has two characteristics, and which will have a speed a_s. This speed is simply given by the derivative of the pressure with respect to the density. So far, all of this should be familiar. The solution is well known to be

$$-\omega^2 + a_s^2 k^2 = 0 \qquad (5)$$

This is a wave which propagates without dispersing and which will not grow in time. For all values of k, there is a solution to the flow. There is no *onset* of the disturbance. *Any* wave, no matter how small in amplitude, will succeed in moving through the medium. Adding a velocity field to the original medium complicates the problem a bit, but the basic physics does not change. An alternative way of stating what we have just found is that for all real values of k, there are only *real* values for the frequency—thus the wave is a *simple* wave and the characteristics have a unique slope of a_s.

9.2.2 The Jeans Instability and Self-Gravitation: Density Waves on the Cheap

Now let's consider what happens if we add gravity. The case of an *externally imposed* gravitational field is well understood intuitively. In the presence of this field, the pressure acts like a restoring force and we get a gravity wave. Such waves are as familiar as the sight of water running over rough pavement after a rainstorm. However, in a cosmic context, there is an additional possibility—the medium may be self-gravitating. In this case, we must add an additional equation, namely the Poisson equation, to the ones already used. To show what happens, we note that the Poisson equation is prelinearized, so to speak, in that a perturbation in the density feeds immediately into a perturbation in the potential. There is no need to look at the orders of the terms; they balance exactly. That is, taking Φ to be the perturbation in the gravitational potential:

$$\frac{\partial^2 \Phi}{\partial x^2} = -4\pi G \delta \rho \tag{6}$$

We write the equation of motion as

$$\frac{\partial v}{\partial t} = -\frac{a_s^2}{\rho_0} \frac{\partial \delta \rho}{\partial x} - \frac{\partial \Phi}{\partial x} \tag{7}$$

and the continuity equation as

$$\frac{\partial \delta \rho}{\partial t} + \rho_0 \frac{\partial v}{\partial x} = 0 \tag{8}$$

Now, we *assert* that all of the perturbations vary as $\delta q \sim \exp i(kx - \omega t)$ for all quantities $q(x, t)$. Therefore our system of equations is

$$-k^2 \Phi + 4\pi G \delta \rho = 0, \qquad -i\omega \delta \rho + ik\rho_0 v = 0 \tag{9}$$

$$-i\omega v + ik\frac{a_s^2}{\rho_0} \delta \rho + ik\Phi = 0 \tag{10}$$

Eliminating Φ, we obtain an *algebraic* equation for the growth of a perturbation in *either* of the other quantities:

$$-\omega^2 + k^2 a_s^2 - 4\pi G \rho_0 = -\omega^2 + a_s^2(k^2 - k_J^2) = 0 \tag{11}$$

This is the basic equation that J. H. Jeans first derived at the turn of this century. This dispersion relation describes a compressible fluid and in this sense is like a longitudinal sound wave.

Notice, however, that this is *not* a simple sound wave. If we set $\omega \to 0$ we get a critical condition for the wave number:

$$k_J = \left(\frac{4\pi G \rho_0}{a_s^2}\right)^{1/2} \tag{12}$$

which is called the Jeans wave number, or using its inverse, the Jeans length. This is the characteristic size of a gravitationally bound perturbation. The mass associated with this disturbance is simply (neglecting the

9.2 Waves

geometry-specific volume factors for a moment)

$$M_J \sim \rho_0 k_J^{-3} \sim \rho_0^{-1/2} T^{3/2} = \left[\frac{3}{4\pi}\left(\frac{5\mathcal{R}}{2G}\right)^3\right]^{1/2}\left(\frac{T}{\mu}\right)^{3/2} \rho^{-1/2} \quad (13)$$

Again, we have used the isothermal representation for the sound speed. If a disturbance grows to a size greater than the Jeans mass, as this factor is called, the perturbation will become an isolated, self-gravitating blob relative to the background. This is, then, the *minimum* mass that a perturbation must achieve if it is to begin to grow further. Notice that we have taken a plane wave and moved it through the medium. What we are setting up, then, is the case of a slab with a longitudinally propagating compressional wave. Such a situation might be envisioned to arise in the case of a disk in which there is turbulence, and sound wave generation, and in which the waves can reach some critical wavelength λ_J. What a great way to make planets, you are probably thinking. In fact, this is one of the most likely scenarios for both planet and star formation.

A Dimensional Analysis of the Jeans Instability

There is an even easier way, using dimensional analysis, to see what is happening. Take the dimensional form of the hydrostatic equation:

$$\frac{P}{R} = \frac{GM\rho}{R^2}$$

and of the force equation

$$\frac{R}{t^2} = \frac{GM}{R^2}$$

If we define the time scale from the second equation as the free-fall time, t_{ff}, we get

$$t_{ff} = (G\rho)^{-1/2} \quad (14)$$

which is correct to within a constant of order unity. We can define a sound travel time for the medium to be

$$t_{sound} = \frac{R\rho}{P} = \frac{R}{a_s}$$

and therefore we can form a dimensionless ratio of the two time scales. This represents the competition between pressure and gravity. Such a parameter is

$$\mathcal{J} = t_{sound}/t_{ff} = (G\rho)^{1/2} R a_s^{-1} \quad (15)$$

The parameter is essentially the Jeans criterion. If the free-fall time for the medium is shorter than the sound travel time, pressure will be swamped and the medium will be inherently unstable to collapse. This is the essential meaning of the Jeans length—if the medium is sufficiently compressed, it will be unable to support itself. If the sound speed is sufficiently low, or if the density is sufficiently high, the same thing is true.

9.2.3 The Rotating Self-Gravitating Slab Problem

Since the effect of centrifugal acceleration in the potential is to stabilize the medium by reducing its effective gravity for collapse toward the axis, we can adumbrate a dispersion relation for the rotating case pretty quickly. Take the surface density to be Σ, integrated over, say, one scale height. Now in the rotating frame there is a natural frequency given by 2Ω, coming from the Coriolis term. Since this acts like pressure in the plane, but is scale free (every parcel, no matter how far from the axis, knows it is in a frame rotating with this frequency), we have

$$\omega^2 = 4\Omega^2 - 2\pi G \Sigma_0 |k| + a_s^2 k^2 \tag{16}$$

We know that the absolute value of the wave number must enter, because the term we have adumbrated for the surface density doesn't care whether we integrate from the top of the plane through to the bottom or the other way around (put differently, the plane is the symmetry of the system and gravity vanishes at $z = 0$). This assumes that centrifugal acceleration is not important (relatively slowly rotating system, in fact nearly geostrophic). The net effect of rotation is to delay the onset of the instability when $\Omega t_{\rm ff} \gg 1$. For differentially rotating self-gravitating slabs, the Jeans instability is governed by a parameter first described by Toomre (1964):

$$Q \equiv \frac{\kappa a_s}{\pi G \Sigma_0} < 1 \tag{17}$$

where the frequency for differential rotation is replaced by the epicyclic frequency, κ.

9.2.4 The Jeans Instability and the Virial Theorem

Since the body is self-gravitating, we can look at the virial theorem and how it relates to a Jeans mass. We will also include, schematically, the effects of turbulence. The relevance of the Jeans instability for star formation is obvious. Since we have already noted that turbulence appears to play an important role in the support of molecular clouds, the sites of star formation, it is important to see whether the stability criteria can be

9.2 Waves

changed for a mass just at the Jeans limit. The gravitational self-energy of a body with radius λ_J is

$$E_{\text{grav}} = -\frac{3}{5}\frac{GM_J^2}{\lambda_J} \sim -\frac{3}{5}\left(\frac{4\pi}{3}\right)^{1/3}\lambda_J^5 G\rho^2 \tag{18}$$

Its thermal energy is

$$E_{\text{thermal}} = \frac{3k}{2\mu}TM_J = \frac{3k}{2\mu}\left[\frac{3}{4\pi}\left(\frac{5\mathcal{R}}{2G}\right)^3\right]^{1/2}\left(\frac{T}{\mu}\right)^{5/2}\rho^{-1/2} \sim \frac{3}{2\pi}\lambda_J^5 G\rho^2 \tag{19}$$

By the virial theorem, the system is required to have $2E_{\text{thermal}} + E_{\text{grav}} = 0$ to maintain equilibrium. The comparison of these two quantities shows that the system is marginally unstable and that, as we have expected, even the global constraints agree that λ_J is the scale on which collapse is assured for an isothermal, uniform-density layer.

Now suppose instead that the body is supported by turbulence. We don't know its mass or its radius, but we know that the virial theorem can be used to place conditions on them. If the unknown radius is called R_c and the mass is $M_c = (4\pi/3)\rho R_c^3$, the virial theorem gives the following condition for collapse (that is that $d^2I/dt^2 < 0$):

$$2E_{\text{turb}} + 3M_c a_s^2 - \frac{3}{5}\frac{GM_c^2}{R_c} < 0 \tag{20}$$

[see Chapter 1, Eq. (25)]. Also, we have assumed that gas pressure ($3\int p\,dV \approx 3\rho a_s^2 V$) is added along with the turbulence. The turbulent energy depends on the spectrum of the turbulence. The steeper the energy spectrum, the more unstable the medium becomes. Magnetic fields can also be included in the same way as the pressure, adding a term $3\int P_{\text{mag}}\,dV \sim 3B^2V/8\pi$ to the virial and with the same sign as the thermal term.

9.2.5 Bifurcation for the First Time

The Jeans instability presents us with a first example of the most common property of the other instabilities of astrophysical interest and a new buzzword for your vocabulary: *bifurcation*. To illustrate what this is, we should compare the sound wave solution with the one we have just obtained. For an instability to grow, the frequency should be either purely imaginary or complex, with the imaginary part being negative. (*Why?*) In the sound wave solution, *there are only real roots for* $\omega(k)$. In the Jeans case, on the other hand, once we have crossed the critical wave number there are *two* complex solutions, one of which has a negative imaginary part. The other solution dies away—we can safely ignore it—but the growing mode is a signal of impending disaster for the linear solution. In light of the trivial dimensional analysis we have just been through, you should now be able to see why. The medium simply cannot adjust its pressure structure rapidly enough to prevent collapse of the blob, with the inevitable nonlinearities that attend such a situation.

This kind of bifurcation behavior is characteristic of systems which have intrinsic scale lengths. In our case, this is the Jeans length. But, in fact, it can also be thought of as resulting from the fact that the system has two characteristic *time scales*. The sound travel time and free-fall time form a dimensionless critical parameter which when of order unity indicates the onset of an instability. The simplicity of the example should only serve to emphasize that such behavior can sometimes be anticipated before the completion of a detailed analysis. In fact, it is even more interesting perhaps because it occurs in a system which *contains a parabolic character due to the introduction of the Poisson equation.*

9.3 Convection

9.3.1 The Rayleigh–Bénard Problem

The Basic Physical Picture

We will begin with a kitchen-level experiment. Imagine a pot of water, our classic incompressible fluid, sitting on a flame. The processes which are transferring heat within the pot are well understood, at least intuitively, by any amateur chef. Heat is conducted from the outer surface to the fluid, and then within the fluid. The characteristic time scale for this process is determined by the heat conduction coefficient and the size of the pot. As the fluid at the bottom becomes hotter, it may be gaining heat at a rate faster than that at which it will transfer that additional energy away. In addition, since there is a thermal response of the water to the added heat—it begins to expand—there is also induced buoyancy. As the water gets hotter, there comes a point at which it begins to become sufficiently buoyant that it will start to churn at the bottom of the pot. Noise will be generated, and ripples will appear on the upper surface of the fluid. Finally, when it is sufficiently hot, the water will start to boil. Generally the onset is sudden, global, and violent (or at any rate chaotic). The transfer process of convection takes over from conduction, and the temperature gradient is stabilized. The fluid motions continue as long as the heat source is applied. Eventually, we have soup, turn off the heat, and the medium (after a short lag time) settles down.

All of the essential features of convection are demonstrated by this little experiment. The characteristic time scales, the length scales, and the fact that buoyancy is critically involved are all vital features of the process. In fact, as a next step in the delineation of processes of importance in astrophysics, convection can be seen as a member of that class of instabilities which occur because of the presence of an external gravitational field. For this reason, we shall use it as a model of all such instabilities and beat it to death before moving on to some of the more recently studied effects.

The Basic Equations

As usual, we have relatively familiar equations to start with. To make life easy from the start, we will take the medium as incompressible and examine the *Rayleigh–Bénard* case of convection. This means that we can ignore all of the time derivatives in the density and write the continuity equation as $\nabla \cdot \mathbf{v} = 0$. This guarantees that all of the fluid motions we excite will be circulatory. This also means that we have a known form for the velocity potential. We can represent the velocity as $\mathbf{v} = \nabla \times \boldsymbol{\psi}$ where $\boldsymbol{\psi}$ is a vectorial velocity potential. Now we take the equation of motion as

$$\left(\frac{\partial}{\partial t} + \mathbf{v} \cdot \nabla\right)\mathbf{v} = -\frac{1}{\rho}\nabla p - \mathbf{g} + \nu \nabla^2 \mathbf{v} \tag{21}$$

where $\mathbf{g}$ is the gravitational acceleration and ν the viscosity. We have not yet linearized the equation, but we already notice that the problem is parabolic in nature because of the inclusion of the diffusion term.

If we assume that the fluid is initially at rest, we can drop the nonlinear term from the start. Thus we see that a new operator appears for the first time:

$$L_q = \frac{\partial}{\partial t} - q \nabla^2 \tag{22}$$

where q is an arbitrary diffusion coefficient. We will have recourse to this representation frequently in the next few sections. We will have a problem—the density *and* pressure are still represented in our equation and we have left out the driving term for the potential instability. There is no equation of heat transfer. For the Rayleigh–Bénard problem, this is essentially the conduction equation, but taking the convective derivation into account. We write it as

$$\left(\frac{\partial}{\partial t} + \mathbf{v} \cdot \nabla\right)T = \kappa \nabla^2 T \tag{23}$$

where κ is the heat conduction coefficient.

Pausing here for a moment, you will notice that several time scales are already appearing in this problem. The thermal conduction time,

$\tau_{\text{cond}} = l^2/\kappa$, where l is the scale length of the system, is characteristic of the heating. There is a viscous time scale, $\tau_{\text{visc}} = l^2/\nu$, on which the momentum is dissipated. There is also a *third* time, the buoyancy time scale, which is given by the velocity of a rising blob and the scale length. It is this one which is a little tricky. Remember that Archimedes made a fool of himself running naked through the streets of Syracuse after realizing the existence of this effect—you won't have to do this to show that you have gotten the physics down right—but it is easy to see that it is related to the gravitational acceleration. We will see momentarily that it is *this* time scale which is the crucial one for the convection problem.

We will assume, as we well know from daily experience with water, that the density of the fluid changes with temperature. While not knowing precisely how, we will take a simple law of the form (called Boussinesq)

$$\frac{\delta\rho}{\rho_0} = -\alpha\frac{\delta T}{T_0} \tag{24}$$

Here α is the coefficient of expansion of the fluid. We now have a closed system because, as we'll see in a minute, we can get rid of the pressure by a mathematical artifice. Now assume that we have an initial temperature gradient placed across the layer of fluid of magnitude $\Delta T/l$. We will further assume that this remains constant, that there is no feedback from any motions of the medium into the temperature gradient. This is, clearly, a very weak assumption, but it at least makes sense as a starting point. With this approximation in mind, we now proceed with our analysis of the velocity field.

To recapitulate, the fluid is assumed to be initially at rest, of uniform density, and with a sufficiently small scale length that the variations in g are negligible. Now we take

$$\rho = \rho_0 + \delta\rho, \quad \mathbf{v} = 0 + \mathbf{v}, \quad T = T_0 + \frac{\Delta T}{l}z + \delta T \tag{25}$$

The linearized equations can be shown to be

$$L_\nu \mathbf{v} = \frac{1}{\rho_0}\nabla p - \frac{\delta\rho}{\rho_0}g\hat{\mathbf{z}} \tag{26}$$

$$L_\kappa \delta T = -v_z \frac{\Delta T}{T_0} \tag{27}$$

and

$$\nabla \cdot \mathbf{v} = 0 \tag{28}$$

9.3 Convection

Before removing the troublesome p, stare at the equations and analyze them dimensionally. It is easy to see that we can change our variables to dimensionless ones by taking a characteristic time scale to be the viscous time τ_{visc} and the length to be l. We also substitute in the representation for the velocity

$$v_x = \frac{\partial \psi}{\partial z}, \quad v_z = -\frac{\partial \psi}{\partial x} \tag{29}$$

Take the curl of both sides of the system of Eqs. (21)–(23) to remove ∇p. We thus obtain, using subscripts as a convenience (from here on, *subscripts will indicate partial derivatives with respect to the subscript*),

$$L_\nu \nabla^2 \psi = \alpha g \left(\frac{\delta T}{T_0} \right)_x \tag{30}$$

$$L_\kappa \delta T = \psi_x \frac{\Delta T}{l} \tag{31}$$

Now we can make use of that moment's reflection on dimensions. Scaling the time and length variables, we can write Eqs. (30) and (31) as

$$L \nabla^2 \psi = \frac{\alpha l^3 g}{\nu} \left(\frac{\delta T}{T_0} \right)_x \tag{32}$$

and

$$\left(\frac{\partial}{\partial t} - \frac{\kappa}{\nu} \nabla^2 \right) \delta T = \psi_x \frac{\Delta T}{\nu} \tag{33}$$

Now substituting in the derivative of the temperature from Eq. (27), we finally obtain the equation describing Rayleigh–Bénard convection:

$$\left(\frac{1}{\sigma} \frac{\partial}{\partial t} - \nabla^2 \right) L \nabla^2 \psi - R \psi_{xx} = 0 \tag{34}$$

We have also introduced a new dimensionless number of great physical importance, the *Rayleigh number*:

$$R = \frac{\alpha g l^3 \Delta T}{\kappa \nu} \tag{35}$$

We have also absorbed the steady-state density and temperature into our value of α in order to make it more compatible with standard notation. We could have done this earlier, of course, but here it is a little more convenient. The auxiliary number, alternately called the *Prandtl number* or

Schmidt number,

$$\sigma = \kappa/\nu \tag{36}$$

has also appeared through this process of dimensional analysis. It has the role of contrasting the rate at which energy is redistributed by thermal conduction with the rate of viscous transport. Since both of these processes are essentially diffusive, the Prandtl number measures the comparative importance of the various sources of diffusive transport in a hydrodynamic system. When the Rayleigh number is large, it implies that the buoyancy time scale is short. When the Prandtl number is large, it implies that heat conduction is more important than viscosity. In general, the Prandtl number is held to be of order unity (it is for most fluids of interest), while, as we shall see, the Rayleigh number has certain specific critical values—in fact, we are faced with the Rayleigh number acting like an eigenvalue of the convection problem. We therefore suspect, even at this stage, that its value will somehow depend on the choice of both geometry and boundary conditions, which to the point have remained unknowns. More on this after a brief commercial message.

A Quick Analysis of the Rayleigh Number

As advertised, two time scales have been needed to analyze the problem. We were forewarned, though, of the presence of a third—the buoyancy time. Now we can pull the Rayleigh number apart and see how it enters. If a blob is buoyant, it is because it has a density contrast relative to its surroundings. If this contrast is $\delta\rho$, it will feel an Archimedian force $\delta\rho g$. The rise time will therefore be $\tau_{\text{buoy}} = (l/\delta\rho g)^{1/2}$. Given our three time scales, we can form a dimensionless parameter by taking

$$\mathcal{P} = \frac{\tau_{\text{visc}} \tau_{\text{cond}}}{\tau_{\text{buoy}}^2} \sim \frac{\alpha g l^3 \Delta T}{\nu \kappa}$$

the Rayleigh number. In other words, the ratio of the buoyancy to the geometric mean "heat–viscous" time scale is proportional to $1/\sqrt{R}$.

Boundary Conditions and What They Mean for Convection

A simple analysis will suffice to set up the picture. Consider Eq. (34) for the Rayleigh–Bénard equation. If we take the approximation that the motion is steady state (in other words, we ignore the explicit time dependences and look for the critical solution) and make the assumption that the solution is periodic in space, we can substitute

$$\psi(x, z) \sim e^{i(kx + mz)} \tag{37}$$

9.3 Convection

for the potential; then we get

$$R = \frac{(k^2 + m^2)^3}{k^2} \tag{38}$$

The Rayleigh number will thus be a function of the horizontal and vertical wave numbers. We can ask a simple question: *what is the minimum value that R can attain*? If we were to take the Rayleigh number to be a function of k and m, we would find that it does indeed have a minimum value. The easiest solution is to look at it in terms of k. We get, by differentiation with m fixed,

$$k_{\min} = \frac{m}{\sqrt{2}} \tag{39}$$

The critical value of the Rayleigh number, R_c, is then given by

$$R_c = \tfrac{27}{4} m^4 \tag{40}$$

even without knowing what the value of m, the vertical wave number, is. In fact, another way of putting this is that the *length scale* on which the convection will occur is of order $R_c^{-1/4}$. One thing should occur to you from this little exercise. The value of R is quite sensitive to small changes in the characteristic length introduced by the boundary conditions, and so one must be careful to choose them correctly. Since the critical temperature gradient, for example, is linearly proportional to the value of R_c, it will also be quite sensitive to changes in the value of the vertical wave number. This could affect estimates of when the soup on the stove will be done, or whether a star can be convective. Some thought will be necessary.

We will have to look at the physical conditions, not just the numerical ones, in order to set up these equations. Since we have taken the layer to be thin, there is for now no need to consider viscous effects or boundary layers. The fluid motion at the surface of the medium can be either free or constrained. If *free*, there is a chance for oscillations of the fluid, waves, etc. and only surface tension helps to keep the medium stable. If *constrained*, the fluid velocity at the boundary must vanish. We have taken the horizontal wave number to be a quantity which we shall determine using periodic conditions. This is something that follows from the *implicit* assumption of a horizontally infinite thin layer. Had we taken it to be something closer to the kitchen pot, walls would have played a more important role. We can now state a useful "theorem": *stars do not have walls, nor do they have corners or edges*. Let us look at the two cases a little more closely for just a moment. First, we will consider the simplest case, of free boundaries at the top and bottom of our layer, and then go on to both the constrained and mixed conditions. It should be remembered, all

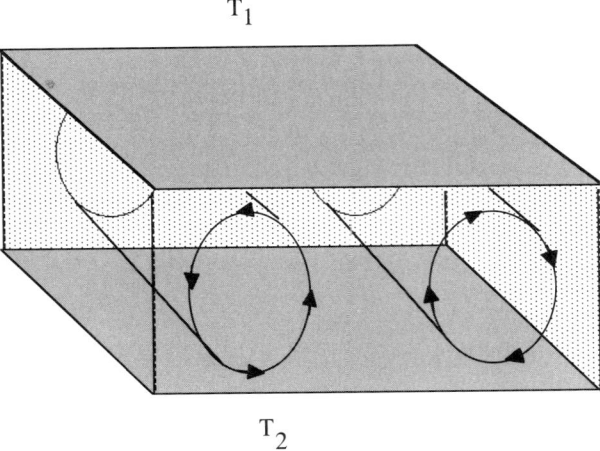

Figure 9.1 Rayleigh–Bénard convection.

through this analysis, that the *details* are meant merely to show you how to fit the exact numbers into the machinery you already have at your command.

Take first the case of a fixed or, using the more normal terminology, rigid boundary. In this case, all perturbations vanish at the surface so that δT, ψ, and $v \rightarrow 0$ at the surface. If we look at what this means for the velocity potential derivatives, we see that it also means that no net flux is transported across the surface. Thus the divergence also vanishes there, but we already have this assured by the conditions of the Rayleigh–Bénard problem. How many boundary conditions do we need? There are three variables, and the equation is sixth order—we require three conditions at each surface. (See Fig. 9.1.) The flow is circulatory and we can make the assumption that the roots of the conditions will be of order $n\pi$. Therefore, for the Rayleigh number we get

$$R_c = \frac{27\pi^4}{4} n^4 \tag{41}$$

You see, we are refining things further. It is now easy to guess that the critical value of R will be of order 10^2 to 10^3. The solution depends on whether n is even or odd. For the odd mode, $n = 1$, $R_c \approx 6.5 \times 10^2$. This is the lowest available mode. As the Rayleigh number is increased, higher-order modes start to appear and the convective pattern becomes extremely complicated.

Stellar Convection Using R_c

Suppose we ignore the compressibility, viscosity, and conductivity—just for a moment (set them equal to scaling constants for now)—and instead concentrate on the effects of the temperature gradient, radius, and gravity. Then we can write the critical Rayleigh number schematically as

$$R_c \sim M_\star R_\star^2 \nabla T \tag{42}$$

where $M_\star$ is the stellar mass, $R_\star$ is the stellar radius (not to be confused with the Rayleigh number—the Latin alphabet is occasionally too limiting), and ∇T is the *mean* temperature gradient for the star. The equation of radiative transfer gives us another estimate of the temperature gradient:

$$\nabla T \sim \frac{L \kappa_{\text{rad}}}{R_\star^2 T^3} \tag{43}$$

Here κ_{rad} is the radiative opacity and T is the temperature in the region (the envelope) of the star which we are studying. Note that *this temperature may also be the T_{eff} of the star and so not scale simply with $M_\star$ and $R_\star$*. Now notice that there must be a critical value of the temperature gradient for the star to be convective. At the same time, there is also a condition on the gradient given by the fact that the star has been assumed to be *radiative*. The result of equating these two terms is that there must therefore be, for a star of a given mass and radius, a *critical luminosity* at which convection will succeed radiation as the primary mode of energy transport. To see this, merely equate the two representations of ∇T and get

$$L_c \sim \frac{R_\star^3 T^3}{M_\star^2 \kappa_{\text{rad}}} \tag{44}$$

That is, at each mass, if the opacity of the medium is increased the star becomes more *unstable* to the onset of global convection. As another example, we can choose main sequence stars. For dwarf stars the radius varies as $M_\star^{1/2}$ and the mass–luminosity relation gives $L \sim M_\star^n$ ($n \approx 3$). Thus the critical condition for the variation of the temperature with mass is $M_{\star,c} \sim T^{6/7}$.

If we are looking not at the entire star, but only the envelope, we can gain even more insight into the problem. The scale length can simply be set at the *pressure scale height* H_p, which is given by

$$H_p = \frac{\mathcal{R} T}{\mu g}$$

where $\mathcal{R}$ is the gas constant and μ is the mean molecular weight. So stated, the Rayleigh number becomes

$$R_c \sim \frac{T^4 \nabla T}{g^3}$$

Now, for a moment, make the assumption that we hold the mass and temperature of the star fixed. The density then scales approximately as $g^{-3/2}$ so that we can substitute in the luminosity for $T^3 \nabla T$:

$$L_c \sim g^{1/2}/\kappa_{\text{rad}}$$

The higher the opacity or the lower the surface gravity, the lower the critical luminosity, all other things being the same. This implies, again, that giants and supergiants will be more unstable to convection than their main sequence counterparts. It can also be said that equations of this sort can be generalized to give a critical luminosity as a function of placement in the Hertzsprung–Russell diagram or as a function of spectral type.

As an aside on this last point, it should be pointed out that the Hayashi track can be understood using this "global Rayleigh number" approach. Hayashi tracks are the paths in the HR diagram followed homologously by fully convective stars. They are so named because of their discovery by C. Hayashi and his collaborators in the early 1960s. These are stars for which the luminosity is sufficiently high at a given temperature and gravity that they must develop deep convective envelopes. In addition, it is found that these stars are typically red giants, so they are quite cool. Since low effective temperature implies high opacity, the effect is enhanced. The useful point to notice is that we can determine that such stars must exist, regardless of the details of the evolutionary tracks, simply on the basis of the Rayleigh criterion for convection. Many other amusing relations, many of which are observationally interesting, can be obtained from this simplistic kind of playing around. One activity which you should never be loath to try is simply looking at the interrelationships among the variables characterizing these problems. As Sheridan once wrote, "*Prudence, like experience, must be paid for.*"

A Remark on Chaos

It is important now to think about what happens as we begin to exceed the Rayleigh number's critical value. If you think again back to the pot of boiling water, you will recall that the flow pattern is nothing like the smooth, streamlined view that I've been trying to pass off on you here. There are bubbles, turbulent eddies, convergences, and divergences which shift constantly over the face of the fluid. In short, the picture is a mess. In fact, the picture is *chaotic*. Recently, this word has taken on an aura of mystery and wonder among physicists. In large measure, the growth industry in chaotic dynamical systems is the result of the discovery of a scale invariance which occurs in simple discrete representations of dynamical systems (Feigenbaum, 1980). It also stems from the work of Prigogine on nonlinear representations of open thermodynamic systems. Not surprisingly, then, this interest should relate back to the Rayleigh–Bénard problem. Why? Simply because we see that chaos has a generic avenue of growth—successive bifurcation eventually breaking down into odd harmonics, which by their interaction merge into a power spectrum of the form usually associated with noise. The transition is rapid, if not discontinuous, and is always seen in the kinds of systems that describe convection. As a matter of fact, the first such system to be considered in detail was that of Lorenz (1963), who was describing a nonlinear model for compressible atmospheric convection and found that a deterministic set of three ordinary nonlinear first-order differential equations describing the density, velocity, and temperature fluctuations behaved stochastically. The solutions became astonishingly sensitive to initial conditions, and trivial (noise-like) perturbations in those conditions were found to induce unpredictable trajectory deviations in the dynamical system.

9.3.2 Mixing Length Theory

Introduction to the Idea of Mixing Length

Let us begin with a picture you have likely seen before. Imagine driving on a highway during a fog. If you take your eyes cautiously from the road and look at the flow of the foggy air over the cars near you, you will notice that wakes are being formed behind them—a turbulent region of finite extent is seemingly attached to each auto. It is the finite extent which is important to note. The turbulence blends into the surroundings at some point, and the wake "softly and suddenly vanishes away." It was Prandtl and von Kármán who first noted that the phenomenon of wake extent implies that some characteristic scale length, typical of dissipation, is involved in turbulence and who seem to have been the first to dub this the *mixing length*. However, the idea of applying it to convection, originally due to Unsöld and Biermann and later elaborated on by Vitense, has nothing *directly* to do with the original hydrodynamic idea of mixing length. It was motivated by the basic observation that any turbulence has a finite distance over which it freely travels before dissipating, but the formalism intended for the problem was totally different. The reason for this difference is that the turbulence of the convection problem is driven by the temperature gradient, while that observed in wakes and boundary layers is strictly driven by shear. The idea, though, that the turbulence will have a characteristic mixing length is basic to any turbulence dissipation model.[1]

Actually, the idea that turbulence is important in the transport of energy is an ancient one in astrophysics, tracing back to the stability analysis by Schwarzschild at the turn of the century. We will first set up the idea of an instability by "blob" analysis, in analogy with the Rayleigh–Bénard problem, and then go on to see whether we can construct an entire theory of energy transport on the basis of such a picture.

The Schwarzschild Criterion

Picture a duck sitting calmly on a pond (and please don't ask why a duck!). If we say that the bird is buoyant, we mean that if we depress him a bit by pushing from above, he will bob back to the surface and, ignoring his agitation, bounce up and down for a while neutrally. If we have one, on the other hand, who is not well preened and therefore not waterproof, and push down on him, he may sink. Now think of a blob which is hotter than its surroundings. It will begin to rise, since we already know that its density will be lower than that of the medium and it will thus be buoyant. It will continue to rise until it reaches a level in the medium at which it is neutrally buoyant again. On the other side, if the blob is pushed down, it will sink until it reaches a point at which the density again allows for stable balance. As in the duck case, we are assuming that the motion is without heat transfer—that the motion is adiabatic.

Now we have the criterion at hand. In a gas, there will be a density gradient resulting from the fact that the gas is one with an adiabatic structure (completely optically thick). Call this density $\nabla \rho|_{ad}$. This then implies that the medium is characterized by $\nabla T|_{ad}$. Therefore, if $\nabla T > \nabla T|_{ad}$, the medium will be unstable. This is the Schwarzschild criterion. It is assumed that once the blob has buoyancy, it will continue to move until it suddenly vanishes into the surrounding medium, dumping all of its energy locally. (See Fig. 9.2.) The criterion is thus a *local* one and one which assumes that small perturbations simply wash out—that if the adiabatic gradient is exceeded by even a small amount, the action of convection thus set up is to push the value back to $\nabla T|_{ad}$. This is the same statement that a geophysical fluid dynamicist would clothe with the statement that the Brunt–Väisälä frequency is positive in the case of stable atmospheres.

---

[1] I thank E. Bohm-Vitense for information on the origin of the mixing length theory.

9.3 Convection

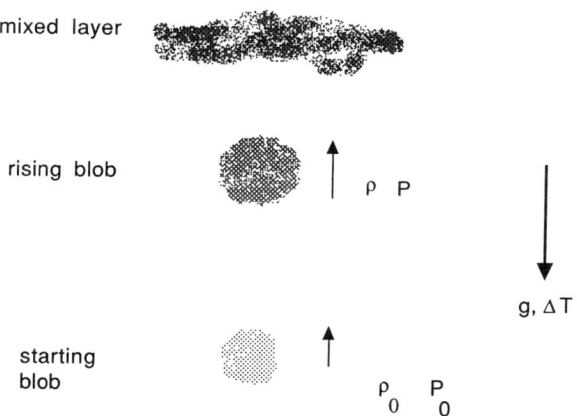

Figure 9.2 The basic picture for mixing length theory.

Some Additional Physics

We can take this picture further. Look at the force balance on the blob. The buoyancy force will, in the neutral case ($\omega = 0$), give

$$\rho v^2 = \delta \rho g l \tag{45}$$

The blob will be slowed by the ram pressure against the background to a velocity v. If we take the pressure inside and outside the blob to balance (recall that *pressure equilibrium does not imply the absence of motion*), then

$$\frac{\delta \rho}{\rho} = -\frac{\delta T}{T}$$

This is essentially the same as the Rayleigh–Bénard problem with $\alpha = 1$. But notice that now we are making the problem explicitly compressible. Even if some of the formalism is the same, the underlying motivations are very different. Mixing length theory is an essentially compressible treatment of the turbulent transport of heat, one that is a local theory and as such much closer in spirit to Kolmogorov theory than anything encountered in the Rayleigh–Bénard problem or Boussinesq theory. We take δT to be the excess of the internal temperature over the environmental value, so that

$$\delta T = (\nabla T - \nabla T|_{\text{blob}})l \tag{46}$$

where l is now the mixing length. The velocity is therefore given by
$$v \sim g^{1/2}(\Delta \nabla T)^{1/2}l \tag{47}$$
The heat flux is given by ρv^3 so that we get
$$F_c \sim g^{3/2}(\Delta \nabla T)^{3/2}l^2 \tag{48}$$
This is, however, only the first step. We know that the heat transported by convection will be
$$F_c = c_p \rho v \delta T \tag{49}$$
where c_p is the heat capacity at constant pressure. In the case of fully developed turbulence, which is what we have been using to this point, we obtained a particular scaling with g and the excess temperature gradient, even without the substitution of the scale height for the mixing length. In the case of the more traditional approach to the theory for astrophysical applications, where it is assumed that the turbulence is weak, the heat capacity of the medium comes into play.

The essentials of the treatment remain unchanged. The velocity is given as before in terms of the gravity and excess gradient, but the flux clearly scales as
$$F_c \sim c_p g^{1/2}(\Delta \nabla T)^{3/2}l^2 \rho = \frac{1}{\sqrt{32}} c_p T (gH_p)^{1/2}(\Delta \nabla)^{3/2}\left(\frac{l}{H_p}\right)^2 \tag{50}$$
where $\nabla \equiv d \ln T / d \ln p$. Thus, if we ask at what point the two theories yield the same result, we see that this depends only on the gravity and that it will occur at some critical length scale $l_\star \sim c_p g^{-1}$. If we take the mixing length to be a multiple of the scale height, this should yield, at a fixed temperature, a unique value of the gravity $g_\star$ at which the fully developed turbulence takes over. Now a problem is how to decide on the superadiabaticity of the fluid. If the medium is radiative, we can ask whether the gradient required to transport heat by radiation is steeper than the adiabatic gradient. This will happen in any region where the opacity is great enough. In the deep layers of a fluid, the radiative transport is essentially diffusive because the optical depth is sufficiently large that the mean free path for a photon is approximately the same as the collisional mean free path. Then
$$\mathbf{F}_{rad} = -\frac{4ac}{3\kappa\rho} T^3 \nabla T = \frac{4\pi ac}{3\kappa\rho} \frac{T^4}{H_p} \nabla_{rad} \tag{51}$$
where a is the Stefan constant. In our present notation, this means that $\nabla_{rad} \sim F$. Should the opacity become large enough that $\nabla_{rad} > \nabla_{Ad}$, then calculations proceed by assuming that the convection is efficient and the temperature gradient immediately becomes adiabatic. In an optically thin

9.3 Convection

region of the fluid, however, it is possible for some of the energy to escape as radiation and therefore our discussion of the convective instability must be modified. Because of this escape, the temperature gradient, $\nabla \equiv d\ln T/d\ln p$, for the blob will not be the same as *either* the radiative or the adiabatic gradient. We know that the total flux is carried by convection and radiation for the outer layers of the body so that

$$F_c + F_{rad} = F_{total} \tag{52}$$

where F_{total} is set by the luminosity of the body in the deep interior. Now we know that the convective flux can be written as $F_c = F_{total}(1 - \nabla/\nabla_{rad}) = \mathcal{A}(\nabla - \nabla_{blob})^{3/2}$, where $\mathcal{A}$ is a function of the local temperature and pressure through the scale height and ∇_{blob} is the convective cell's temperature gradient. Therefore:

$$\mathcal{A}\nabla_{rad}(\nabla - \nabla_{blob})^{3/2} = \nabla_{rad} - \nabla \tag{53}$$

The efficiency of energy transport compares δT in the blob with the energy lost by radiation during the motion:

$$\gamma = \frac{\nabla - \nabla_{blob}}{\nabla - \nabla_{Ad}} \tag{54}$$

and this depends on the optical depth of the medium and the rate of rise of the blob. If the radiative time scale is short compared with the rise time, the medium will lose energy mainly from optically thin escape of photons, while if it is optically thick, the convection will be very efficient (and asymptotically approach the adiabatic condition). Therefore γ is given approximately by the ratio $v_c c_p \delta T/F_{rad}$, which we used at the start to set up the equations for convective transport. Since F_{rad} depends on $\delta T/l$, where l is the mixing length, we have an equation for the efficiency that is proportional to $(\nabla - \nabla_{blob})^{1/2}$ and thus a way to calculate the radiative flux from the blob. By using this and the total flux condition, the values of ∇ and ∇_{blob} can be separately determined. This is the essential feature of the mixing length theory. Keep in mind that this is a strictly local theory for energy transport. In that sense, there are no boundary conditions and no large-scale, global motions. On the other hand, it isn't really a turbulence theory either because there is no characteristic scale for the dissipation that is connected in any way with the velocity.

Molecular Weight Gradients

We know that there may be a variation in the mean molecular weight as the blob moves. There is a possibility that recombination will set in or that the blob has a different chemical composition than the surroundings. In this case, there is an additional factor to include in the density variation at constant pressure:

$$\frac{\delta\rho}{\rho} = -\frac{\delta T}{T} + \frac{\delta\mu}{\mu}$$

where μ is the mean molecular weight. Now taking the approximation that $\mu = \mu(T)$, we can write this equation as

$$\frac{\delta\rho}{\rho} = -Q\frac{\delta T}{T} \tag{55}$$

where we now define Q to be

$$Q \equiv 1 - \left(\frac{\partial \ln \mu}{\partial \ln T}\right) \tag{56}$$

It may happen that the hotter material has a different composition than the neutral layer into which it is moving. This frequently happens in stellar interiors. A region undergoing nuclear burning may, as we have seen, have so great a luminosity from what is essentially a pointlike region in the core that the medium around it is driven to become convective. The burning region will have a higher μ than its surroundings, as a result of nucleosynthesis, and thus $Q \neq 1$. Also, even if the medium is convectively stable, large gradients in the mean molecular weight can produce a slow mixing, called *semiconvection*. This instability is driven by an entropy gradient and is not a mode of heat transport. Rather, it redistributes mass so as to smooth out the molecular weight discontinuities.

The Mixing Length and How to Choose It

The most critical problem with the mixing length approach is the freedom to choose a representation of l. Normally, one argues that the blob of hot fluid will rise until it mixes with the background without exchanging heat. Therefore, the buoyancy is limited by the pressure gradient. The length scale which characterizes the mechanical structure of the atmosphere, if one wants to think of it this way, is the pressure scale height. The usual choice is consequently $l = \alpha H_p$. Here, α is conventionally taken to be a constant (usually between 1 and 2). However, studies of two-dimensional convection by Dupree (1977) show that it is possible to parametrize α as a function of temperature. The reason for going to this formalism is the following.

It is usually the case that the opacity becoming high in the envelope is the trigger for convective motion. As we showed in the previous section, according to the Rayleigh criterion, an opacity increase decreases the stability of the atmosphere. Therefore, a portion of the stellar envelope (or any other environment) in which the opacity dramatically increases causes a steepening of the temperature gradient required to push flux through, which leads the matter to the brink of instability (at some level). Such regions as the ionization zones for hydrogen and helium are especially good for this, since the opacity is very high in partially ionized regions. The presence of an additional absorber near the peak of the radiation field, like H^- in solar-type stars, is another example. Molecules are generally not too efficient in this regard, because they are normally not abundant enough to severely affect the radiative transport except in the outer portions of the atmosphere. In general, these regions will be unstable from other, more deeply seated, envelope opacity sources.

The prescription for the temperature-dependent mixing length is provided by Dupree (1977) and is

$$\log \alpha(T) = 0.03 + 4 \times 10^{-8}(T - 11{,}000)^2 \tag{57}$$

From examining the temperature dependence it seems that the mixing length goes through a minimum in the middle of the hydrogen partial ionization zone. In such regions, the specific heats are also changing, as is the mean molecular weight per electron, so there are several factors which contribute to making the convective motions a bit more complex to treat in such regions than in the fully developed case. This is a field in which computational capabilities are rapidly expanding. It should be possible to provide scaling laws for the mixing length from completely compressible three-dimensional calculations in years ahead. Considerable progress has been made in numerical simulations, but such general scaling laws have yet to be realized.

The Nusselt Number

If we have a value for the convective flux, which we see is given by the solution to the equations of motion that we have outlined above, we can form yet another dimensionless number—the Nusselt number—which is a measure of the efficiency of convection. Put another way, it is the rate of convective flux transport compared with that which we would have obtained by conduction. If we have c_p as the heat capacity, again, and l as the characteristic scale length, then for a temperature gradient ΔT we have

$$N = \frac{F_c l}{c_p \Delta T} \tag{58}$$

called the *Nusselt* number. It is not a quantity which we can specify in advance of the solution of the eigenvalue problem, unlike the Prandtl number, for instance. In fact, like the Rayleigh number, it is the product of the calculation. In effect, in any convective theory, it is the efficiency factor which tells us how effective the fluid motion is at transferring heat. In classical mixing length theory, we might say that this is the factor which takes into account the drag and geometric properties of the rising blob.

9.3.3 Some Effects of a Magnetic Field

Here, for the sake of illustrating the effects of adding in external perturbations, we shall include the effects on Boussinesq convection of a magnetic field. There is another reason for studying this problem. *Thermohaline* instabilities have become increasing important in astrophysics (Kippenhahn *et al.* 1980). These are also known as *double-diffusive* instabilities and are the result of a fluid having more than one characteristic diffusive constant. For instance, in the case of a magnetic field the diffusivity of the field itself, due to ohmic dissipation, competes with the viscous force in the presence of buoyancy and two different time scales appear in the problem. For this reason, the magnetic convection problem serves as an excellent introduction to a wider class of instabilities.

We start with the basic equation for the magnetic Lorentz force $\mathbf{J} \times \mathbf{B}$:

$$(\nabla \times \mathbf{B}) \times \mathbf{B} = -\tfrac{1}{2}\nabla B^2 + \mathbf{B} \cdot \nabla \mathbf{B} \tag{59}$$

and the equation for the evolution of the magnetic field is written as

$$\left(\frac{\partial}{\partial t} - \eta_M \nabla^2\right)\mathbf{B} = \nabla \times (\mathbf{u} \times \mathbf{B}) \tag{60}$$

where η_M is the magnetic diffusivity. This latter equation is also the standard form of the dynamo equation with the effects of current dissipation and the Lorentz force included. The perturbed equations of motion are now

$$\left(\frac{\partial}{\partial t} - \nu\nabla^2\right)\mathbf{u} = -\frac{1}{\rho}\nabla\delta p - \frac{\delta\rho}{\rho}\mathbf{g} - \frac{1}{\rho}\nabla\mathbf{B}\cdot\mathbf{b} + \mathbf{B}\cdot\nabla\mathbf{b} + \mathbf{b}\cdot\nabla\mathbf{B} \tag{61}$$

$$\left(\frac{\partial}{\partial t} - \kappa\nabla^2\right)\theta = -\mathbf{u}\cdot\nabla T \tag{62}$$

$$\left(\frac{\partial}{\partial t} - \eta_M\nabla^2\right)\mathbf{b} = \mathbf{B}\cdot\nabla\mathbf{u} \tag{63}$$

9.3 Convection

This system of equations is considerably more complex in appearance than the one for standard convection, partly because of the addition of the equation for the magnetic field but primarily due to the nature of the vector coupling between the magnetic and velocity fields. Notice, for example, that the source term for the magnetic field is a direct nonlinear coupling of the unperturbed field and the velocity, of the same sort as we have for the temperature. It is important to note that the addition of a magnetic diffusivity also has the effect of introducing several new time scales to the problem. For instance, we can compare the rate of magnetic dissipation with the growth rate for the velocity perturbation, or with the buoyancy time scale. In words, we can describe what we expect to happen and then examine the system formally in more detail.

If the velocity perturbation is due to buoyancy, then the temperature gradient is still the driving term. However, the growth of the velocity field counterbalances the tendency of the field to decay by providing a generating term for the magnetic field which consequently increases the energy of a nonthermal component of the system. As the velocity grows, so does this term, and it serves to act as a damping agent for the field. The presence of a dynamo term also has the effect of increasing the stability of the medium against convection and of introducing a possible double-diffusive instability.

Double-diffusive instabilities have been much discussed in the literature in recent years. It is known that a signature for the onset of chaos in dynamical systems is the appearance of several, rather than two, time scales in the system. In this sense, we can use the magnetic convection problem also as an illustration of what geophysicists call the "salt finger" or *thermohaline* instability. The origin of this term is interesting, in that it is an example of an empirical test you can perform yourself of the nonlinear system. In the ocean, when you are swimming, there are times you encounter isolated patches of extremely cold, fresh water. The contrast is usually made more marked by the fact that you will normally have just exited from extremely saline warm water. The reason for the instability is connected with the double effects of salinity on the molecular weight of the water and of buoyancy. Warm water wants to rise, of course, but it can retain a higher salinity, which makes it heavier. The opposite is true for the cold fluid. The salt also wants to diffuse between the two fluids, and the competition of this with the rate of thermally generated rise is what gives rise to the instability. (See Fig. 9.3.) Much the same is true for the magnetic convection problem—the magnetic fluid is cooler at the same pressure and therefore a competition is set up between the effects of the temperature gradient and the magnetic field.

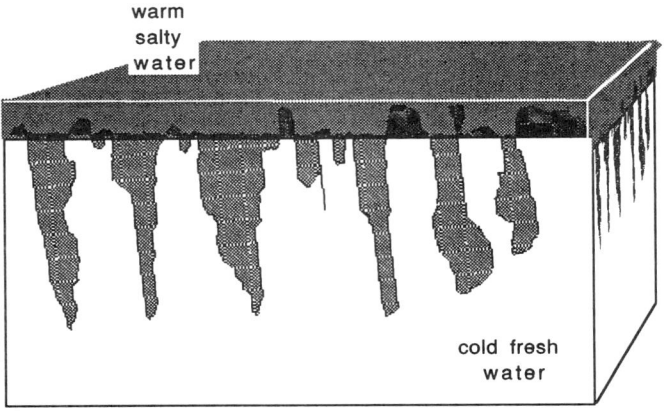

Figure 9.3 Magnetic, or *thermohaline*, convection.

The Hartmann Number

The system of equations describing the magnetic convection problem is

$$L_{\eta_M} b_x = B_0 \psi_{,zz} \tag{64}$$

$$L_{\eta_M} b_z = -B_0 \psi_{,xz} \tag{65}$$

$$L_\kappa \theta_{,x} = \psi_{,xx} \frac{\Delta T}{d} \tag{66}$$

$$L_\nu \nabla^2 \psi = -\alpha g \theta_{,x} - \frac{B_0}{\rho}[b_{x,xx} - b_{z,xz}] \tag{67}$$

Combining these equations using the methods which by now should be familiar gives the following sixth-order linear equation for the velocity potential:

$$L_\kappa L_{\eta_M} L_\nu \nabla^2 \psi = -\alpha g \frac{\Delta T}{d} L_{\eta_M} \psi_{,xx} - \frac{B_0^2}{\rho} L_\kappa \nabla^2 \psi_{,zz} \tag{68}$$

We can now scale the equation as usual to obtain in dimensionless final form

$$\nabla^8 \psi = \frac{\alpha g d^3 \Delta T}{\kappa \nu} \psi_{,xx} + \frac{B_0^2 d^2}{\eta_M \nu} \nabla^4 \psi_{,xx} \tag{69}$$

This provides the dispersion relation:

$$(k^2 + s^2)^4 = Rs^2 + Qk^2(k^2 + s^2)^2 \tag{70}$$

which yields the *cubic* for the stability condition of minimizing the Rayleigh number:

$$4(k^2 + x)^2 x - (k^2 + x)^3 - 2Qk^2 x + Qk^2(k^2 + x) = 0 \tag{71}$$

where $x = s^2$. This has the possibility of multiple roots, which is the signature of the double-diffusive phenomenon and the source of a multiple time scale instability. The value of Q is the controlling parameter, being the measure of the effect of the magnetic field on the stability of the medium, and is called the *Hartmann* number. It has also, following the detailed investigations by Chandrasekhar, been named after him. It is the measure of the retarding effect of the magnetic field on thermal convection, and this can be generalized to the case in which there is also a molecular weight gradient in the medium. Therefore, in the case of a magnetic dynamo, which involves the formation of strong toroidal fields by wrapping and buoyancy, it is possible that a filamentation or fingerlike instability can develop.

9.4 Pulsation as an Instability

The problem of pulsation is closely related to that of convection, in that it is a mechanical instability which is produced in response to the instability of the medium to remain in equilibrium and transfer energy by radiation or conduction. It will be the aim of this section to expand the discussion of the previous part of this chapter to the point of providing a simple solution to the pulsation equations, and to show you how it is that convection and pulsation are intimately related within real stars.

To begin with, let us consider a simple picture. We have a region of the star which is especially opaque, for whatever reason, to the transfer of radiation. The medium heats up as a consequence, and in keeping with even the simple interpretation of the Boussinesq conditions given earlier, the buoyancy will increase and the region will begin to expand. Assume that there is a source for the gravitational field which remains unaltered by this change in the zone structure—that there is a core mass which provides the mechanical arena in which this play will be performed. Now, if the zone expands and cools in the process and the opacity goes down, the region will cool and recontract. (See Fig. 9.4.) This is what is meant by the basic model of pulsation. In order to see what happens more clearly, let us take a *one-zone model* approach—assume that there is only one thin zone in which there is no gradient in the physical properties of the star which has a mass ΔM and therefore which has properties dependent only on t, the time. This basic picture, first elucidated by Eddington in the 1920s, is still

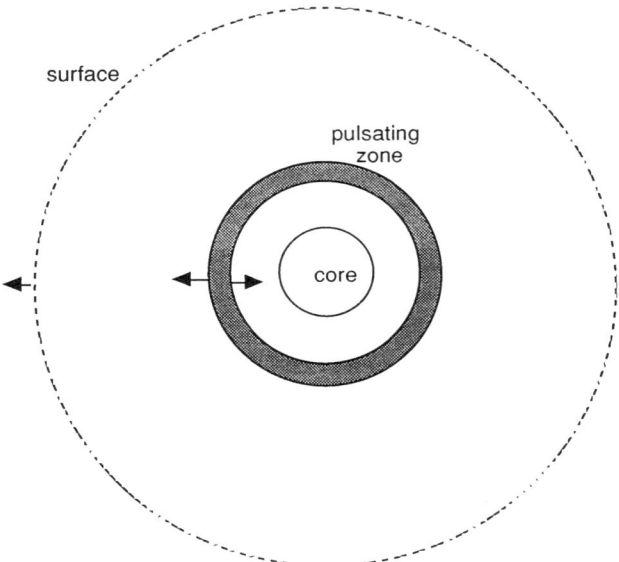

Figure 9.4 Physical picture of the one-zone pulsation analysis.

the basic picture to be kept in mind when approaching the problem of pulsational instability.

Before we start on the problem in earnest, let's digress to a simple treatment. We know from the beginning of this chapter that a self-gravitating fluid mass has a characteristic dynamical time scale which is a multiple of the free-fall time:

$$t_{\rm ff} = \frac{1}{(G\rho)^{1/2}} \sim R^{3/2}$$

This is also the fastest mechanical time scale available in the medium. Assume that all stars which are unstable on this time scale have a temperature T so that the luminosity is $L = 4\pi R^2 \sigma T^4$ but have the same mass M. Then the luminosity is related to the time scale by

$$L \sim t^{4/3} \to M_{\rm V} \sim \frac{10}{3} \log P$$

where P is the period and $M_{\rm V}$ is the absolute magnitude. We assume as well that there is a color term (dependent on the effective temperature), but for the moment we ignore this. Quite simply stated, we have a period–luminosity relation for a dynamically unstable star. This was one of the

9.4 Pulsation as an Instability

discoveries of greatest import at the beginning of the modern era of astrophysics, since it provides a way of determining the absolute magnitude of a star knowing *only* what type of variable it is and what the period is. The zero point of the relation is *still* controversial, but it is possible to determine observationally. We will discuss this further below. For the moment, however, it should serve as a justification for even being interested in the problem.

The basic equations of structure now come into play. The mass equation is inverted in order to solve the equations; we take the mass M to be the independent variable. The equation is written as

$$\frac{\partial r}{\partial M} = \frac{1}{4\pi r^2 \rho} \tag{72}$$

We will follow the motion of a zone of specified mass, not the mass as a function of radius and time. The equation of hydrostatic equilibrium is

$$\rho \ddot{r} = -\frac{\partial P}{\partial r} - \frac{GM}{r^2} \rho \tag{73}$$

where we have neglected the effect of viscosity (more on this later in the discussion of turbulence in pulsation). The equation of radiative transfer is also crucial to our discussion, since it defines when there is no avenue for the photons other than to be absorbed by the medium resulting in an increase in the local temperature. To study this equation, we must back up for a moment.

The equation for the opacity which we shall use in this section is $\kappa = \kappa(\rho, T)$. Then

$$\frac{\delta \kappa_0}{\kappa} = \kappa_T \frac{\delta T}{T_0} + \kappa_P \frac{\delta P}{P_0} \tag{74}$$

and also the equation of state $P = P(\rho, T)$ of the form

$$\frac{\delta P}{P_0} = \frac{\delta \rho}{\rho_0} + \frac{\delta T}{T_0} \tag{75}$$

where we have *assumed* the equation of a perfect gas (of course!). Let us begin the analysis with the statement of the equilibrium condition:

$$-\frac{dP_0}{dr} = \rho_0 \frac{GM}{r_0^2} \tag{76}$$

We have neglected the acceleration in this state so that we can define a set of variables which we shall use in the rest of this analysis. These are l, the luminosity perturbation; ξ, the radius change; Π, the pressure change; λ,

the density perturbation; and τ, the temperature perturbation. Thus, for example, we have
$$r^2 \rho\, dr = r_0^2 \rho_0\, dr_0$$
$$\Pi = \lambda + \tau$$

We have the equations, then, for the physical variables:
$$P = P_0(1 + \Pi), \qquad \rho = \rho_0(1 + \lambda), \qquad T = T_0(1 + \tau),$$
$$r = r_0(1 + \xi), \qquad L = L_0(1 + l) \tag{77}$$

Now, we will solve for the successive orders of the perturbed equations. We begin with the momentum conservation equation written in terms of the mass. We have already considered what happens if we write this in terms of the radius. We get

$$\frac{\partial P}{\partial M} = -\frac{1}{4\pi r^2}\left(\frac{GM}{r^2} + \frac{\partial^2 r}{\partial t^2}\right) \tag{78}$$

and take the perturbation to get, to first order in the perturbing variables (as usual we seek the linearized result),

$$\frac{\partial P_0}{\partial M} = -\frac{GM}{4\pi r_0^4} \tag{79}$$

$$\frac{\partial \Pi}{\partial M} = \frac{1}{4\pi r_0 P_0}(\sigma_0^2(4\xi + \Pi) - \ddot{\xi}) \tag{80}$$

Here we have defined the basic frequency of the system to be the inverse of the free-fall time:

$$\sigma_0^2 \equiv \frac{GM}{r_0^2} \tag{81}$$

It should be clear that the natural vibration of a gaseous mass must be related, if it is self-gravitating, to the free-fall time (recall the example of the Jeans instability with which we opened this chapter).

When we linearize the radius equation, we get (remember, this is the inverse of the mass equation)

$$\frac{\partial \xi}{\partial M} = -\frac{1}{4\pi r_0^2 \rho_0}(3\xi + \alpha\Pi - \delta\tau) \tag{82}$$

where we define the two moduli α and δ as the compressional modulus with pressure and the expansion modulus with temperature:

$$\alpha \equiv \left(\frac{\partial \ln \rho}{\partial \ln P}\right)_T, \qquad \delta \equiv -\left(\frac{\partial \ln \rho}{\partial \ln T}\right)_P \tag{83}$$

9.4 Pulsation as an Instability

The definition of α is that of the ratio of specific heats, γ, for a polytropic gas. Here we allow for a more general equation of state and don't force the equivalence just yet.

The linearization of the transport equation provides the connection between the temperature gradient and the luminosity:

$$\frac{\partial \tau}{\partial M} = (l - 4\xi + \kappa_P \Pi + (\kappa_T - 4)\tau) \frac{\partial \ln T_0}{\partial M} \tag{84}$$

The luminosity equation is a little more time consuming. We consider that the basic equation is

$$\frac{\partial L_0}{\partial M} = 0$$

We here assert that there is no local energy generation. This is the statement that we treat only the variation in the outer envelope of the star and that the core luminosity does not change during the pulsation. If the star has significant core pulsation, this will not be true in general. For our purposes, however, we will ignore the possible nonlinear effects that are likely to change the rate of core burning.

The second law of thermodynamics states, for our purposes, that the luminosity generation is a measure of the entropy generated by the nuclear processes. Recall that in the discussion in the introductory chapter, we derived several results from the second law for the connection between the internal energy, E, the pressure, volume, and specific heat. We know that

$$T\, dS = DQ = c_p\, dT + \left(P + \left(\frac{\partial E}{\partial V}\right)_T\right)\left(\frac{\partial V}{\partial P}\right)_T dP \tag{85}$$

so that

$$\frac{\partial L}{\partial M} = \epsilon - \frac{DQ}{dt} \tag{86}$$

Here we take $\epsilon = \epsilon(T, P)$ to be the nuclear energy generation rate. We therefore obtain

$$\frac{\partial l}{\partial M} = -c_p T_0 \dot{\tau} + \left(P_0 + \left(\frac{\partial E}{\partial V}\right)_T\right)\frac{\alpha}{\rho_0 L_0} \dot{\Pi}$$

To further simplify the calculation we define two new variables K_1 and K_2 which are the coefficients of the above equation:

$$\frac{\partial l}{\partial M} = -K_1 \dot{\tau} + K_2 \dot{\Pi} \tag{87}$$

We now make an essential approximation in order to solve this system of equations. We notice that much ground can be covered with the simple approximation of a one-zone model. That is, we assume that we have a thin zone, somewhere in the star, which has vanishingly small perturbations. That is, we assume that with the exception of the luminosity generation within the zone, all of the mass gradients vanish within the zone. Taking the luminosity in a zone to have a perturbation equation:

$$l = \tfrac{1}{2}(l_0 + l_1)$$

where l_1 and l_0 are the upper and lower boundary values of the luminosity fluctuations. We also assume that the zone has a mass $\delta M \equiv m$ so that

$$\frac{\partial l}{\partial M} = \frac{l_1 - l_0}{m} \rightarrow \frac{2l}{m} \tag{88}$$

on making the assertion that the luminosity at the lower boundary is constant (there is assumed to be an enormous heat engine sitting at the center of the star).

The spirit of the derivation we are now going to present is that of a local analysis. You have encountered this throughout this chapter, but here it is most important. As in the convection treatment, and that of the Jeans instability, we are interested in the conditions under which the instability will develop, not necessarily what the detailed spatial dependence of the amplitude will be. This will not, therefore, be useful in giving us any feeling for *how* the pulsation occurs within the star but will tell us what the likely frequencies will be. This analysis can be generalized, and we will do that at the end of this section, but for the moment it is important to keep in mind that we do not intend to detail the radial dependence of the amplitude.

The final equation for the radial perturbation is

$$A\frac{d^3\xi}{dt^3} + B\frac{d^2\xi}{dt^2} + D\frac{d\xi}{dt} + E = 0 \tag{89}$$

Here the coefficients are defined by

$$A = \frac{K\sigma_0\delta + \alpha C}{K\delta\sigma_0^3} \tag{90}$$

$$B = -\frac{1}{\sigma_0^2\delta}(\kappa_P\delta + \alpha(\kappa_T - 4)) \tag{91}$$

$$D = \frac{(4\alpha - 3)C - 4K\sigma_0\delta}{K\sigma_0\delta} \tag{92}$$

9.4 Pulsation as an Instability

and

$$E = \frac{1}{\delta}(4\delta(1 + \kappa_P) + (4\alpha - 3)(\kappa_T - 4)) \tag{93}$$

The adiabatic approximation takes $K \to 0$ so that both D and B vanish. The equation then becomes

$$\frac{d^2}{dt^2}\dot\xi = -\frac{4\alpha - 3}{\alpha}\dot\xi \tag{94}$$

The radial perturbation has the form $\xi \sim e^{i\omega t}$. We therefore see some immediate consequences. First, if the star has $\alpha < 4/3$, the perturbation is intrinsically unstable. This is a basic result in pulsation theory. Since we can identify α with the ratio of specific heats, γ, this result means that if the ratio falls below that for a relativistic gas the medium is catastrophically unstable. It also means that the characteristic frequency becomes progressively smaller as one lowers the ratio (makes the medium intrinsically more compressible) so that *anything* that lowers the ratio of specific heats contributes toward destabilizing the medium. In fact, from the analysis of the individual terms we see that there are two cases of instability that can be identified. If $\sigma_0^3 B > 0$, the medium is *dynamically* unstable, while if $K\sigma_0^3(AB - E) > 0$, the medium is said to be *secularly* unstable.

The final one-zone pulsation equation that we arrive at is simple in form. It is strictly *time dependent*, since we have carefully removed all dependence of the perturbed quantities on mass. The equation is third order in the time derivative of the radial perturbation, since the equations from which it derives form a third-order system. The second thing we note is that there will consequently be a very interesting behavior: the system will display multiple roots and critical values of the period. In fact, this third-order system yields both real and complex roots and these are a bifurcation set—precisely as in the case of convection.

9.4.1 The Hurwitz–Routh Criterion

It states that if the equation for the frequency s is

$$s^3 + a_1 s^2 + a_2 s + a_3 = 0 \tag{95}$$

then

$$a_2 > 0 \to \text{dynamical stability}$$

$$a_3 > 0 \to \text{secular stability}$$

$$a_1 a_2 - a_3 > 0 \to \text{pulsational stability}$$

Assume that we have a cubic for the frequency and that the third-order dynamical system has been solved using $x \sim \exp \lambda t$. Now for $\lambda = \lambda_r + i\lambda_i$ we want to ensure that all $\lambda_r \le 0$. Therefore in a cubic of the form $\lambda^3 + a_1\lambda^2 + a_2\lambda + a_3 = 0$, we know that $\text{Re}(a_1) > 0$ and $\text{Re}(a_3) > 0$ simultaneously ensures stability (the real part is taken only for completeness; in general we don't have to worry about that). Now a_2 is the sum of the pairs of roots $\lambda_1\lambda_2 + \lambda_2\lambda_3 + \lambda_1\lambda_3$. If two of the roots are conjugates (that is, if all of the coefficients are positive) then stability is ensured if $a_3 < a_1 a_2$ (von Kármán and Biot 1940).[2] The point of this set of equations is that it allows the determination of the natural frequencies of a one-zone pulsator and delineates the regions of stability for a medium which is thermally and mechanically unstable.

9.4.2 Another Approach: Lagrangian Variational Method

Another approach, distinct from the one used in the previous section, is to look at the variations in the moving frame. In a real sense, we have used this many times before, but it is useful to see what happens in the context of pulsation theory. To begin with, by taking the mass to be a constant (and not making the one-zone approximation which we have been discussing), we have a new way of writing the continuity equation:

$$\frac{\partial \rho}{\partial t} + \frac{\partial \rho u_i}{\partial x_i} = 0 \rightarrow \frac{d\rho}{dt} + \rho \frac{\partial u_i}{\partial x_i} = 0 \tag{96}$$

which on integration gives

$$\Delta \rho + \rho \frac{\partial \xi_i}{\partial x_i} = 0 \tag{97}$$

This is the Lagrangian variation of the density. In other words, it is the variation seen in the comoving frame. We can generalize the operator Δ then by taking

$$\Delta = \delta + \xi_i \frac{\partial}{\partial x_i} \tag{98}$$

---

[2] There is a simple theorem which can be demonstrated elegantly using the ideas of catastrophe theory (Poston and Stewart 1978; Stewart 1981) which applies to the analysis of the roots of the frequency cubic. One point is to note that an equation of the form $f(x) = ax^3 + bx + c$ has the critical point $f'(x_\star) = f''(x_\star) = 0$. This is the point at which there are an extremum and an inflection point simultaneously, and from this we get the following conditions for the critical unstable solution: $a_1^2 - 3a_2 = 0$, and $a_1^3 - 27a_3 = 0$. Now consider the equation $f(x) = x^3 = a_2 x + a_3 = 0$. Here we expect the existence of unstable solutions. So $4a_2^3 + 27a_3^2 = 0$ is the critical solution.

9.4 Pulsation as an Instability

where δ is the Eulerian variation (what we have been calling the ordinary variation). It is important to note that although the Eulerian variation commutes with the various partial derivatives, the Lagrangian does not. In the comoving frame, we will assume that the pressure is given by a polytropic equation of state (that is, we will assume that the pulsation is adiabatic):

$$\frac{\Delta p}{p} = \Gamma_1 \frac{\Delta \rho}{\rho} \tag{99}$$

Combining this with the continuity equation, we find that the pressure Lagrangian variation can be written as

$$\Delta p = -\Gamma_1 p \frac{\partial \xi_i}{\partial x_i} \tag{100}$$

The equation of motion is written as

$$\ddot{\xi}_i = -\Delta\left(\frac{1}{\rho} \frac{\partial p}{\partial x_i}\right) + \Delta \frac{\partial \Phi}{\partial x_i} \tag{101}$$

The potential, Φ, is derived from the Poisson equation:

$$\frac{\partial^2 \Phi}{\partial x_i \, \partial x_i} = -4\pi G \rho$$

We could have, of course, begun with the Eulerian perturbation and then generalized it, but it is interesting to see what happens if we turn this around. Now take the Lagrangian perturbation of the pressure gradient, as an example:

$$\Delta\left(\frac{1}{\rho} \frac{\partial p}{\partial x_i}\right) = -\frac{\Delta \rho}{\rho^2} \frac{\partial p}{\partial x_i} + \frac{1}{\rho} \Delta \frac{\partial p}{\partial x_i} \tag{102}$$

which becomes

$$\Delta(\ ) = -\frac{1}{\rho}\left(\frac{\partial \Delta p}{\partial x_i} + \frac{\partial \xi_j}{\partial x_i} \frac{\partial p}{\partial x_j}\right) + \frac{\Delta \rho}{\rho^2} \frac{\partial p}{\partial x_i} \tag{103}$$

We now substitute the equation for the polytropic equation of state to remove Δp and the continuity equation for $\Delta \rho$ to obtain an equation which depends only on the displacement ξ:

$$\Delta(\ ) = -\frac{1}{\rho} \frac{\partial \xi_j}{\partial x_j} \frac{\partial p}{\partial x_i} + \frac{1}{\rho} \frac{\partial}{\partial x_i}\left(\Gamma_1 p \frac{\partial \xi_j}{\partial x_j}\right) + \frac{\partial \xi_j}{\partial x_i} \frac{\partial p}{\partial x_j} \tag{104}$$

The analysis can now proceed using the Poisson equation as the only other equation we will require. The perturbation in the gravitational field can be written as

$$\frac{\partial \delta \Phi}{\partial x_i} = -4\pi G \rho \xi_i \qquad (105)$$

so that the Lagrangian variation of the potential is

$$\Delta \frac{\partial \Phi}{\partial x_i} = -4\pi G \rho + \xi_j \frac{\partial^2 \Phi}{\partial x_i \partial x_j} \qquad (106)$$

The equation we arrive at is therefore

$$-\omega^2 \xi_i = \frac{\partial}{\partial x_i}\left(\Gamma_1 p \frac{\partial \xi_j}{\partial x_j}\right) - \frac{\partial \xi_j}{\partial x_j}\frac{\partial p}{\partial x_i} + \frac{\partial \xi_j}{\partial x_i}\frac{\partial p}{\partial x_j}$$
$$- \rho \xi_j \frac{\partial^2 \Phi}{\partial x_i \partial x_j} + 4\pi G \rho^2 \xi_i = L_{ij}\xi_j \qquad (107)$$

To remove the potential entirely from the problem, it is merely necessary to recall the equation of hydrostatic equilibrium which defined the pressure gradient in the first place. This is the equation for the linear, adiabatic radial oscillation equation.

It is important to note that this equation is one which allows us to find the radial dependence of the oscillation as a function of position in the star. It is also an eigenvalue equation, in that not all modes will satisfy the boundary conditions. These are simply that the oscillation should vanish at the center and be finite at the surface of the star. Since the density is assumed to vanish at R, the outer radius of the star, we see that Δp vanishes there as well.

The final, radial, form of the equation is

$$\frac{d}{dr}\left(\frac{\Gamma_1 p}{r^2}\frac{d}{dr}(r^2 \xi_r)\right) - \left(\frac{4}{r}\left(\frac{dp}{dr}\right) - \omega^2 \rho\right)\xi_r = 0 \qquad (108)$$

Since from the polytropic equation of state we know that we can write Δp as a linear function of p, it follows that we merely require that the amplitude be finite at the surface. Given a functional form, then, for $p(r)$, we can find the amplitude as a function of position in the star, and the eigenvalue as well. You will get a rest now, since it is merely important that you have seen how the treatment proceeds for this equation. It is only a set of approximations which then allow the solution of the equation in detail. The essential feature of the linear adiabatic equation is that we don't use *any* information about the energy generation or transfer—it is assumed that the amplitude is small and that there is no change in the thermal properties of the medium during the pulsation. Historically, the equation we have arrived at was derived by Rosseland and Eddington in the early part of this century. The fact that the assumptions do not require us to consider anything about the energy transfer accounts for the early derivation of the equation—it is a purely classical result which could just as easily have been written down by Darwin, Emden, or Rayleigh had they chosen to consider the problem of radial pulsation of polytropes. Interestingly enough, none of these investigators seem to have been interested in the problem. It only became central to astrophysics after the discovery of the period–luminosity relation by Levitt and Shapley for Cepheid variables in the Large Magellanic Cloud.

9.5 Thermal Instability

Although not strictly hydrodynamical in origin, there are a number of thermal instabilities of importance in astrophysics. In a sense, the problem of pulsation highlights the coupling between mechanical and thermal time scales. If a medium is optically thick, and the cooling time is very long, any motions may take place adiabatically. That is, it is possible that they do but not necessary. If, on the other hand, the cooling time is very short we are in trouble. The medium may change its pressure very rapidly because of sudden onset of cooling and in the absence of any additional heat source may start to collapse. This collapse is usually due to two effects in astrophysics. One is self-gravity. If the mass of the body is great enough, and the local density is large enough as well, contraction can cause the body to pass over the limit of Jeans' stability. This is one of the standard scenarios for the collapse of molecular clouds at the onset of star formation. The other cause is that the pressure of the background medium is finite. This means that a drop in pressure locally due to a rapid drop in temperature leads to a large pressure deficit very rapidly and the region implodes. Such an event is analogous to the collapse of a bubble, a very complicated hydrodynamic event.

The basic outline of a thermal instability calculation is to start with the equations of thermal balance:

$$\rho T \frac{dS}{dt} = c_v \frac{dT}{dt} - \frac{p}{\rho} \frac{d\rho}{dt} \qquad (109)$$

so that the equations for the instability can be written in a very general form. First, notice that the instability will not involve gravity. It may, but need not, involve effects like rotation, stratification, magnetic fields, multiple compositions, interfaces, etc. It is actually the simplest instability we can write down, except perhaps for sound waves. In a way, you can look at it as lying somewhere between sound waves and pulsation in terms of complexity. Radiation comes into play in this instability, and in that sense it is like the mechanism by which sound waves are expected to damp.

The chief culprit in this instability is the *loss term*, $\mathscr{L}$. A function of ρ and T, this term expresses the imbalance between the radiative losses, Λ, and heating rates, Γ, of the medium. For instance, photoionization is a heating term. Hard photons impinging on a gas ionize it and inject energetic electrons into the gas. These free electrons subsequently collide with other gas particles and redistribute their kinetic energy. Ultimately, this heats the gas. This input of energy may, however, have an unfortunate consequence. For some processes, the rate of radiative loss is a decreasing function of temperature. This produces a cooling catastrophe. The cooler the gas becomes, the greater its cooling function becomes. Thus after the heating is turned off, it is possible that nothing stops the gas from cooling to a temperature below that of its environment. Such a cooling curve is known for low-density plasmas. Below about 10^4 K, it behaves like $\Lambda \sim T^2$. But above this, the temperature increase leads to a lower cooling rate; in fact, $\Lambda \sim T^{-1/2}$ until about 10^7 K (see review by Rosner, Tucker, and Viana 1978).

As in the film *Casablanca*, when the police chief gives the order to "round up the usual suspects," we write down the usual equations, supplemented by the heat transport equation. This time, however, we write them down as their modal "aliases" for the linearized system. It's simplest to assume that the medium is initially homogeneous and at rest. Then take the problem as a local one, away from all boundaries. This means replacing all of the growth terms by $q(\mathbf{x}, t) = q_0 + q_1 \exp(\lambda t + i\mathbf{k} \cdot \mathbf{x})$ and assuming that all perturbations higher than first order vanish. First write the continuity equation

$$\frac{d\rho}{dt} + \rho \nabla \cdot \mathbf{v} = \lambda \delta\rho + i\rho_0 \mathbf{k} \cdot \delta\mathbf{v} = 0 \tag{110}$$

and then include the momentum equation:

$$\rho \frac{d\mathbf{v}}{dt} + \nabla p = \lambda \rho_0 \delta\mathbf{v} + i\mathbf{k}\delta p = 0 \tag{111}$$

The thermal balance equation completes the set. However, it is important to note that the equilibrium condition for this equation is that $\mathscr{L}(\rho_0, T_0) = 0$ and that we then expand the loss function using the density and tempera-

9.5 Thermal Instability

ture as the primitive thermodynamic variables:

$$\frac{p}{\gamma - 1} \frac{d}{dt} \ln(p\rho^{-\gamma}) + \rho \mathscr{L} - \nabla \cdot K \nabla T = -\lambda \frac{p_0}{\rho_0} \delta\rho$$

$$+ \frac{\lambda}{\gamma - 1} \frac{p_0}{T_0} \delta T + \rho_0 \left(\frac{\partial \mathscr{L}}{\partial \rho}\right)_T \delta\rho + \rho_0 \left(\frac{\partial \mathscr{L}}{\partial T}\right)_e \delta T + k^2 K \delta T = 0 \quad (112)$$

Here K is the *thermal* conductivity. The most important new terms to appear in this problem are the two derivatives of the loss function. They can be either positive or negative and therefore are extremely important in governing the onset of the instability. (See Fig. 9.5.) A critical point is reached if $\lambda = 0$, because this means that the medium may become overstable (we have already used this for the computation of the Rayleigh number). So for this case we get

$$\left(\frac{\partial \mathscr{L}}{\partial T}\right)_\rho - \frac{\rho_0}{T_0}\left(\frac{\partial \mathscr{L}}{\partial \rho}\right)_T = 0 \quad (113)$$

in the absence of conduction. The effect of conduction is that it introduces a characteristic length scale into the system. Because of the sound speed $(\gamma p_0/\rho_0)^{1/2}$, once K appears there is also a typical time scale for the onset

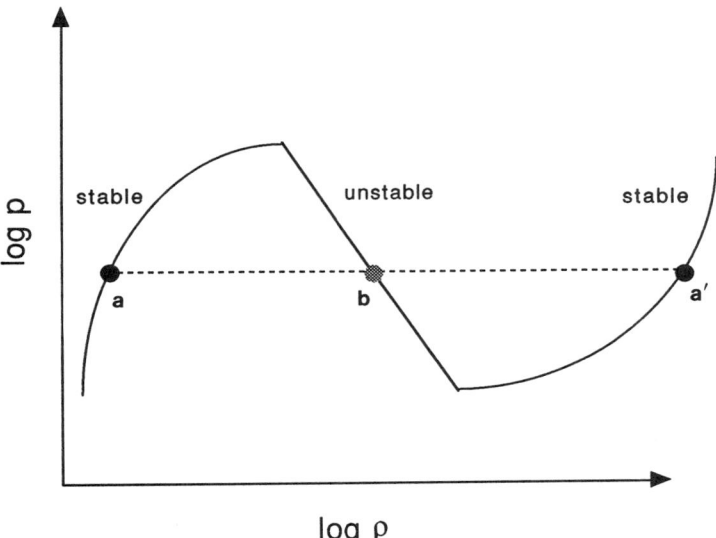

Figure 9.5 Condition for thermal instability of a two-phase medium. The phases are in mechanical equilibrium at constant pressure (a, a'). Phase b is unstable.

of the instability. It is no longer local, and the dispersion relation reflects this.

We have already discussed the Hurwitz–Routh criterion in the context of pulsation; thermal instability will seem almost familiar. You should continue the derivation of the general dispersion relation, just to confirm that you get a cubic and that one of the conditions for instability is the one we have just written down for the critical point. In fact, there is another instability even in the absence of conduction, $(\partial \mathscr{L}/\partial T)_\rho < 0$, but this is less important for most of the interesting astrophysical problems involving thermal instabilities.

There are many examples of this instability, and oddly enough not all of them involve thermal processes for the losses. For instance, a medium undergoing strong synchrotron radiation is also subject to this instability, even though the emission mechanism isn't thermal at all. If the magnetic field strength is proportional to some power of the density, the strength of the synchrotron source will increase if the density is increased. Even though this does not feed back into the temperature directly, it does drive the collapse of the medium if the emissivity increases with increasing density. (See Fig. 9.6.) Of course, there is a catch to all this. The opacity of the medium must enter at some stage. If the material becomes optically thick at some stage in the collapse, the loss term ceases to increase with density and may even drop. This shuts off the instability and provides a characteristic scale to the problem once again. It is the same as in the Jeans instability. The reason is actually hidden in the definition of the opacity—like the conductivity it introduces a characteristic length and time when

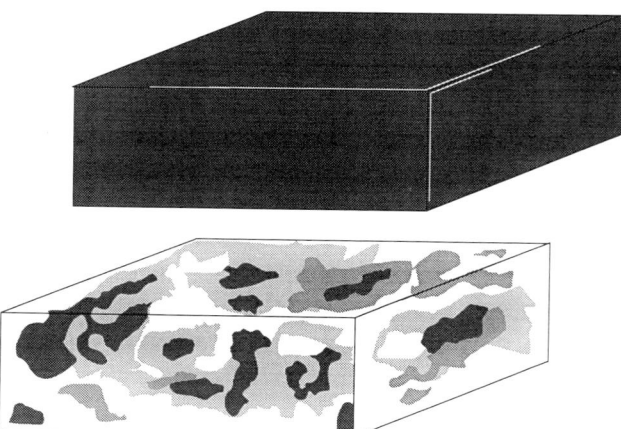

Figure 9.6 Thermal instability in an optically thin medium, illustrated by an initially uniform slab condensing into knots. The density is indicated by the grey scale.

combined with the sound speed. The historical role played by this instability was in the first explanations of why cold clouds and hot diffuse gas cohabit the interstellar medium. The basic idea was that if these regions are in pressure equilibrium, there must exist some density and temperature range at which the cooling function is locally minimized. The broad range of conditions seemed to lead naturally to phase transitions if the diffuse gas is compressed. An analogy of this process is invoked now for accretion disks, where the heating is internally generated by viscous torques and the cooling is radiative and convective and provided through the surface of the disk.

9.6 A Selection of Important Instabilities

9.6.1 Rayleigh–Taylor Instability: Buoyancy and Density Discontinuities

The interaction of buoyancy and diffusion, as represented in the case of the Rayleigh–Bénard problem, is even more dramatically shown by the problem of fluids which are stratified by density and composition. The classic expression of this problem is the so-called *Rayleigh–Taylor instability*, but the more recent work on double-diffusive convection makes the problem even more acute and interesting. In this section, we will review the problem for superimposed fluids of different density and then discuss the problem in an astrophysically interesting context. We will then discuss the double-diffusive problem in a simple way. Keep in mind that the methods of this section are among the most general for the analysis of instability generation at the interfaces between fluids of any that can be used for astrophysical problems.

The Basic Physics of the Problem

Start with a hot cup of coffee, or tea if you'd rather, and pour some very cold cream into it. If you have ever used a glass cup, you have seen a lovely sight. If not, keep the picture going in your mind. The fat content of the cream is normally high. Therefore, the cream has a lower density than the surrounding coffee (which is, after all, just caffeinated water). In the case of the very cold cream, however, the density is higher than normal (*remember the Boussinesq approximation?*) and therefore it quickly sinks. In time, the cream will begin to heat up, just due to conduction (forget mixing and entrainment in the falling column for the moment). The density drops, and suddenly the cream layer finds itself *beneath a fluid of higher mean density*! The cream now becomes buoyant, and starts to rise. In displacing the incompressible fluid above it, it releases more potential energy than it carries and therefore even more fluid will be displaced.

Eventually, the fluid layers invert, and the cream will mix (turbulence is generated in this case, but you can also think of the fluids as simply changing places). (See Fig. 9.7.) The more traditional example of water and paraffin is not as familiar from the kitchen, although it does show the inversion effect better. The basic physical point is that the fluids interchange due to this buoyancy-driven instability, because the density gradient is negative—that is, the heavier fluid is on top. This is one of the easiest ways of setting up the physical situation leading to the Rayleigh–Taylor instability I know of and also has the advantage that one can consume the experiment, which is equally stimulating on being drunk, once you're done.

Dimensional Analysis of the Interface Conditions

Let us now look a bit more carefully at the boundary conditions. We have set up an interface, call it Ξ, across which we have assumed pressure equilibrium. Therefore, $\delta p|_\Xi = 0$. Then the fluid being displaced downward across Ξ gives a potential energy term $-\rho_1 g \Delta z$, while for the fluid on the underside of the layer the potential is $+\rho_2 g \Delta z$. It is assumed that a small cough sets up the ripple, so that the energy input is minimal. The conservation condition for the energy gives

$$(\rho_1 + \rho_2) \Delta z^2 \omega^2 \sim (\rho_1 - \rho_2) g \Delta z \tag{114}$$

so that we get, for the frequency,

$$\omega^2 \sim \left(\frac{\rho_1 - \rho_2}{\rho_1 + \rho_2}\right) g k \tag{115}$$

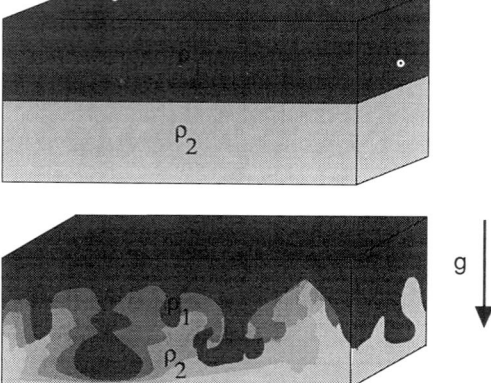

Figure 9.7 Example of the Rayleigh–Taylor instability. The overlying fluid has density $\rho_1 > \rho_2$. The gravitational acceleration, **g**, is shown. Mixing produces fingers that eventually invert the density distribution and, in the absence of diffusion, reestablish the stable interface.

9.6 A Selection of Important Instabilities

We find that *any* fluid which is set up with a density inversion will inevitably wind up undergoing the Rayleigh–Taylor instability. Of course, this is a very simplistic derivation of the condition, but the essentials are preserved. As in the Rayleigh–Bénard case, the growth time shows two roots, but the negative one dies away in time and never concerns us.

More Detailed Derivation by Perturbation Analysis

The medium is assumed to be at rest, incompressible and irrotational. We take the conservation equations to be

$$\mathbf{v} = \nabla \phi \tag{116}$$

where the velocity potential is denoted as ϕ. The conservation of momentum is

$$\rho \frac{D\mathbf{v}}{Dt} = -\nabla p - \rho g \tag{117}$$

If we take the pressure to be constant across the interface, $p_1 = p_2$, then

$$-\rho_1 \left(\frac{\partial \phi_1}{\partial t} + \frac{1}{2}[\nabla \phi_1]^2 + g\xi \right) = -\rho_2 \left(\frac{\partial \phi_2}{\partial t} + \frac{1}{2}[\nabla \phi_2]^2 + g\xi \right) \tag{118}$$

at the interface. We also have the condition that

$$\frac{\partial \xi}{\partial t} + \frac{\partial \phi_1}{\partial x} \frac{\partial \xi}{\partial x} - \frac{\partial \phi_1}{\partial z} = 0 \tag{119}$$

Now if the boundary has no velocity, that is, if there is no net velocity across Ξ, we get

$$\frac{\partial \xi}{\partial t} = \frac{\partial \phi_1}{\partial z} = \frac{\partial \phi_2}{\partial z} \tag{120}$$

If we take the boundary to have a form like a simple wave,

$$\phi_1 \sim \exp[i(kz - \omega t)],$$

then we obtain

$$\omega^2 = \left(\frac{\rho_1 - \rho_2}{\rho_1 + \rho_2} \right) gk \tag{121}$$

precisely as we found before heuristically.[3]

---

[3] A more detailed analysis of the compressible case including relativistic effects is given by Allen and Hughes (1984).

Some Astrophysical Applications

Several examples of the Rayleigh-Taylor instability have been shown to occur in cosmic environments. Perhaps the most celebrated is the Parker instability. This occurs in the interstellar medium when the mass of cold gas is supported by the pressure exerted by magnetic fields and cosmic rays. The fluid, if that is the appropriate word, of the magnetic field is intrinsically lighter (and also more compressible) than the gas, and therefore the situation is intrinsically unstable by the discussion above. The form of the instability is the formation of dense knots of gas in the medium, with the magnetic field presumed to squirt out between them. The result is the formation of large-scale loop structures between what should be local clouds. Material trapped along the field lines will funnel toward the plane, so the knots will grow in time—eventually becoming full-scale clouds. While the detailed observational consequences of this picture have not been studied in great detail (*how would you go about observing this?*), the existence of large-scale structure in neutral hydrogen above the plane was initially taken as evidence for the presence of the instability in the medium. Many of these structures are now attributed to the formation of "superbubbles" by the combined effects of supernovae and expanding H II regions in the disk. But the role of the light gas support of the more ponderous interstellar gas is still of considerable interest to molecular cloud formation and star formation.

In the explosion of a supernova, the envelope is low in density and quite hot. The blast wave should become Rayleigh-Taylor unstable, mixing material from the envelope into the blast and also causing knots to appear early in the evolution of the remnant. Detailed hydrodynamic calculations have confirmed the existence of this effect, and it is now taken as one of the mechanisms operating early in the evolution of the expanding shell. A similar mechanism is acting in the formation of the interface between H II regions and molecular clouds, and between blast waves and clouds at the boundary of the cloud. We have also met with this manifestation of the Rayleigh-Taylor instability in the chapter on stellar winds and accretion flows.

A final application is a bit more subtle and also relates to the double-diffusive effect. A class of main sequence stars, the Ap and Bp stars, shows abundance enhancements of some of the heavy elements and strong magnetic fields. The model which has been most successful to date for explaining this phenomenon is that of radiative diffusion, first suggested by Milne in 1927 and later expanded upon by Michaud in 1970 and Vauclair in 1975 (see Vauclair 1983). The idea is that in the presence of a strong magnetic field, turbulence and large-scale circulation currents can be suppressed, and radiation pressure can cause selective separation of ions in

9.6 A Selection of Important Instabilities

the atmosphere. The resulting concentration of the atoms in a thin layer is due to the saturation of the radiative driving. It is a stellar wind that never quite makes it. The layer thus formed is denser than the underlying photospheric gas and therefore the layer should be Rayleigh–Taylor unstable on time scales shorter than the main sequence lifetime of the star. However, the details of the process have yet to be looked into. Some preliminary discussion[4] shows that the layer will initially compact, then become unstable and mix to some finite thickness, at which point the growth of the instability and the diffusive motion from the radiative driving eventually balance. Instead of mixing completely, blobs or "fingers" form in the gas, which are enhanced relative to the surroundings.

9.6.2 Kelvin–Helmholtz Instability: Shear and Velocity Discontinuities

Let us assume that the layers are in motion relative to each other. We can take, as the simplest possible case, the scenario in which there is no gravity acting but instead there is only shear. Also, we continue to ignore diffusive effects. The shear generates vorticity on all scales because $k \Delta U$ is finite on all scales. The enstrophy density in the flow, $\rho \omega^2$, is derived from the shear $\rho U (\Delta U)^2 k^2 \approx \rho U^2$ and the kinetic energy in the vorticity grows at the expense of the shear flow. For each side of the interface we have a $\hat{z}$ velocity due to the displacement ζ:

$$v = \frac{d\zeta}{dt} = \left(\frac{\partial}{\partial t} + U_j \frac{\partial}{\partial x}\right)\zeta, \tag{122}$$

assuming that the bulk uniform flow is only in the $\hat{x}$ direction on either side of the discontinuity. Here j is 1 or 2, depending on which side we are on.[5]

---

[4] Lin, J. 1980, Ph. D. thesis, Columbia University. This is the only available analysis. See also Childress et al. (1975).

[5] Let's look physically at what is happening here. When the interface is displaced upwards, it forces the overlying flow to deviate upwards over the perturbation. This means that the pressure in the flow drops, just as in the airfoil case (recall the discussion in Chapter 3). The boundary thus experiences lift. Of course, the situation is mirror symmetric across the interface. The downward displaced boundary feels a downward-directed lift. The ultimate cause of this force is that the fluid is incompressible and follows Bernoulli flow. The penetrating boundary is then sheared by the respective flow and rolls up into a vortex sheet. Recall that for incompressible flow any velocity perturbation results in a pressure perturbation of opposite sign. The interface is also unstable for compressible flows because of this same lift. If there is a magnetic field in the problem, the fluid acts as if there is a surface tension (much like Kelvin's original problem, the generation of waves on the ocean surface by wind) and there is also a characteristic speed, the Alfvén speed v_A.

From the perturbation of the equation of motion we obtain

$$\frac{\partial v}{\partial t} + U\frac{\partial v}{\partial x} = -\frac{1}{\rho}\frac{\partial p'}{\partial z} \qquad (123)$$

where ρ is a constant on either side of the interface. The technique for solving for the motion of the shear interface is the same one encountered for the Rayleigh–Taylor problem. Pressure equilibrium is assumed across the interface so that $p'_1 = p'_2$. The displacement is assumed to have the same wave number in x and z but of the form $e^{\pm kz}$. Then substituting for $\zeta(x, z, t)$ we obtain

$$\omega^2(\rho_1 + \rho_2) + 2(\rho_1 U_1 + \rho_2 U_2)\omega k + (\rho_1 U_1^2 + \rho_2 U_2^2) = 0 \qquad (124)$$

so that we obtain

$$\frac{\omega}{k} = \frac{(\rho_1 U_1 - \rho_2 U_2) \pm (U_1 - U_2)(\rho_1 \rho_2)^{1/2}}{\rho_1 + \rho_2} \qquad (125)$$

Here U_2 has been taken as negative, but the choice is purely a matter of convenience. Notice that if the densities are equal across the interface, we can still have an instability because the velocity shear does not vanish. This configuration is unconditionally unstable (see Fig. 9.8). Unlike the

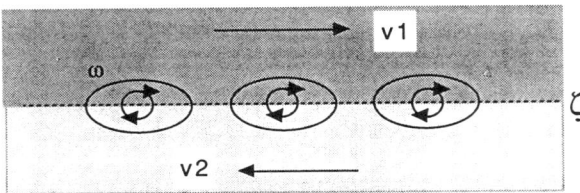

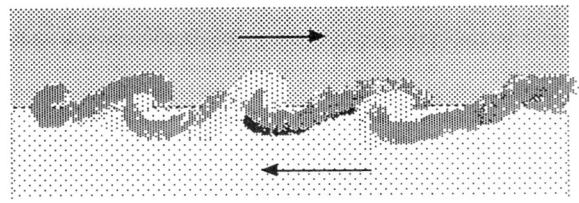

Figure 9.8 The origin of the Kelvin–Helmholtz instability. Shear is generated at an interface, ζ, between two uniformly moving incompressible fluid layers. The resultant mixing takes place in a vortex sheet.

9.6 A Selection of Important Instabilities

Rayleigh–Taylor instability, the Kelvin–Helmholtz instability does not go away in the absence of gravity. *This instability is driven by shear, not buoyancy.* The effects of surface tension have been discussed since the first study by Kelvin, who was dealing with the problem of the generation of waves by a wind shear at the surface of the ocean, but in astrophysical environments it plays no role. Helmholtz and Kelvin both pointed out that the continuous nature of fluids forces them to erase the discontinuity through the generation of a vortex sheet and that the size of the largest vortex is limited only by the magnitude of the velocity discontinuity.[6] In an astrophysical context, the most prominent use of this instability has been in studying the structure of radio jets in active galaxies. The boundary shear is sufficiently great for galactic jets, moving outward in the interstellar medium of the system, that waves and twists should develop on the surfaces of the jets. These appear as kinks in the radio maps of the sources (see Hardee 1979, Ferrari *et al.* 1982, Birkenshaw 1984 in the General Bibliography). It is also possible that the formation of Herbig–Haro objects, which appear to be associated with *bipolar mass outflows*, often observed from protostellar environments, may be due to this instability generated within the jets (see Cohn 1983 in the General Bibliography).

9.6.3 Shear Instabilities: Rayleigh Criterion for Velocity Gradients

Begin with an incompressible inviscid fluid, so we can neglect the Navier–Stokes diffusion term, and assume that the velocity field can be represented as $\mathbf{v}(\mathbf{x}, t) = U(z)\hat{\mathbf{x}} + \mathbf{v}'(\mathbf{x}, t)$. We write the perturbation as

$$v'_x = \psi_{,z}, \qquad v'_z = -\psi_{,x} \tag{126}$$

assuming an incompressible fluid. Now separate the perturbation using $\psi(\mathbf{x}, t) = \phi(z) \exp i(kx - \omega t)$ so that

$$\psi_{,zt} + U\psi_{,xz} - \psi_{,x} U' = -p_{,x} \tag{127}$$

$$\psi_{,xt} - U\psi_{,xx} = p_{,z} \tag{128}$$

---

[6] A very interesting and readable discussion of induced vorticity in the presence of a discontinuity is provided by Kelvin in his article "On the Doctrine of Discontinuity in Fluid Motion" (*Mathematical and Physical Papers*, **4**, 215). This paper also has a very nice discussion of planar and cylindrical jets issuing from orifices into stationary fluid and a brief discussion of wakes (the "dead water" phenomenon) and starting vortices. The problem was of immense practical importance in hydraulics and therefore received considerable attention at the end of the nineteenth century.

This pair of equations becomes, on combination and substitution of the formal representation of the perturbation,

$$(-\omega + kU)\phi'' - \frac{U''}{U - v_p}\phi = 0 \tag{129}$$

This equation is sometimes referred to as the *Orr–Sommerfeld* equation. Here, we use the phase velocity, $v_p = \omega/k$ to make dimensional consistency in the terms. Let the frequency be complex, so that $\omega = \omega_r + i\omega_i$, and $v_p = v_{p_r} + iv_{p_i}$ and substitute this into Eq. (129). The final term in Eq. (129) is

$$\text{(last term)} = \frac{(U - v_{p_r}) + iv_{p_i}}{(U - v_{p_r})^2 + v_{p_i}^2} U''\phi \tag{130}$$

Let's concentrate on the imaginary part of this term. This is the one that ultimately drives the instability. If we take the complex conjugate of ϕ and then take the integral over z, which is across the direction of the flow, we obtain

$$\int (|D\phi|^2 + k^2|\phi|^2)\, dz + \int \frac{(U - v_{p_r})U''}{(U - v_{p_r})^2 + v_{p_i}^2} |\phi|^2\, dz$$
$$+ i\int \frac{U''v_{p_r}}{(U - v_{p_r})^2 + v_{p_i}^2}\, dz = 0 \tag{131}$$

It is obvious that although the real terms *can* cancel, the imaginary term does not go away *unless* the integral itself vanishes. Notice that the integrand is symmetric *except* for the second derivative of the velocity. Therefore, *the value of U'' must change sign if the flow is to satisfy the equation.* There must be an inflection point in the flow if the stability is to be assured. This is the Rayleigh criterion for a shear flow. A monotonic velocity field will be stable against the formation of a turbulent shear flow.

This criterion is also extremely important for the evolution of boundary layers, for which it was originally intended. If there is an inflection point, it is possible for the flow to be stable. Therefore, one suspects that the layer will grow until it is possible for the inflection to appear, and then the medium will stabilize with a turbulent boundary layer of finite thickness having been formed. Note, as in the discussion of turbulence, that if the layer is not laminar the kinematic viscosity no longer applies and we have to resort to a full statistical treatment of the turbulence. One point is that, depending on precisely how the circulation behaves in the inner part of an accretion disk, the Rayleigh criterion may or may not be violated. Notice

9.6 A Selection of Important Instabilities   365

that it does not depend on viscosity, just shear. For an accretion disk boundary layer, it is essentially an extension of the angular momentum criterion we have already established.

Shear Stability of a Free Jet

We have been discussing incompressible jets because these provide the best comparison with laboratory work. But astrophysical analogs are probably quite different. Compressibility dominates the structure of cosmic jets. The continuity equation must be solved explicitly, and we do not have the option of employing streamlines in the solutions of the fluid equations (since $\nabla \cdot \mathbf{v}$ doesn't necessarily vanish). Jets observed in the laboratory, even compressible ones, have a finite opening angle, albeit a small one. This means that we are faced with a choice, whether to express the motion in cylindrical coordinates (using z as the width of the jet) or spherical coordinates (using θ to give the opening angle, which can be kept constant). In light of the discussion in the chapter on turbulence, we choose the latter representation.

Take the jet to be expanding with some constant opening angle θ_0. The inviscid fluid equations now have to be written in spherical coordinates. For the moment, to maintain the condition that the jet is three-dimensional, we will keep the azimuthal terms. These deformations are responsible for helical waves that, if they grow nonlinear, will distort the jet sufficiently to produce twisting and kinking about the axis and eventual disruption of the jet. We will stay, however, in the linear regime. The continuity equation is

$$\frac{d\rho}{dt} + \rho \nabla \cdot \mathbf{v} = \frac{\partial \rho}{\partial t} + \frac{1}{r^2}\frac{\partial}{\partial r}(r^2 \rho v_r) + \frac{1}{r \sin \theta}\frac{\partial}{\partial \theta}(\sin \theta \, \rho v_\theta)$$

$$+ \frac{1}{r \sin \theta}\frac{\partial}{\partial \phi}(\rho v_\phi) = 0 \qquad (132)$$

There are now three equations of motion:

$$\frac{d}{dt}v_r - \frac{v_\phi^2 + v_\theta^2}{r} = -\frac{1}{\rho}\frac{\partial p}{\partial r}$$

$$\frac{d}{dt}v_\theta + \frac{v_\phi v_\theta}{r} - \frac{v_\phi^2 \cot \theta}{r} = -\frac{1}{\rho r}\frac{\partial p}{\partial \theta}$$

$$\frac{d}{dt}v_\phi + \frac{v_\phi v_r}{r} + \frac{v_\theta v_\phi \cot \theta}{r} = -\frac{1}{\rho r \sin \theta}\frac{\partial p}{\partial \phi} \qquad (133)$$

Here to save space we have abbreviated the inertial terms in spherical coordinates:

$$\frac{d}{dt} = \frac{\partial}{\partial t} + \mathbf{v} \cdot \nabla = \frac{\partial}{\partial t} + v_r \frac{\partial}{\partial r} + \frac{v_\theta}{r} \frac{\partial}{\partial \theta} + \frac{v_\phi}{r \sin \theta} \frac{\partial}{\partial \phi}$$

Notice how the cross-coupling between the coordinates introduces the motions that serve to destabilize the jet. A variation in the azimuthal speed alters the centrifugal terms in the radial equation, for example, and this change in the radial motion feeds back into the bending and twisting terms. To make the analysis simpler, assume that θ_0 is small so that $\sin \theta \approx \theta$ and $\cot \theta \approx 1/\theta$. Now assume that the jet is perturbed by some small disturbance. All of the quantities can be assumed to be of the form $q(r, \theta, \phi, t) = q_0(r) + q_1 F(\theta) \exp i(-\omega t + kr + m\phi)$, taking F to be a function only of θ and q_1 as a constant amplitude. The analysis proceeds without the need for the energy equation only because we are going to assume an isothermal jet. This simplifies the problem tremendously since it provides an equation of state. We can use either an isothermal or polytropic law for p to close the system of equations. We do not make any assumptions about the initial velocity of the jet with respect to its background, only that it is moving at constant radial velocity.

The analysis is considerably simplified by prelinearizing the equations, noticing that the time dependence provides a frequency, ω, which we will be able to use later to determine the eigenfunctions and eigenfrequencies. We assume that the motion is initially along the axis,

$$\mathbf{v}_0 = (u, 0, 0)^T \qquad (134)$$

and that all opening angles are initially small. We will also now write all of the unperturbed quantities (functions only of r) as q_0. Then we have the perturbation equations:

$$-i\omega \rho + \rho_0 \left[\frac{1}{r^2} \frac{\partial}{\partial r}(r^2 v_r) + \frac{1 - \frac{1}{2}\chi^2}{r\chi} \frac{\partial}{\partial \chi}(\chi v_\theta) + \frac{imv_\phi}{r\chi} \right] = 0 \qquad (135)$$

$$-i\omega v_r = -\frac{1}{\rho_0} \frac{\partial p}{\partial r} + \frac{\rho}{\rho_0^2} \frac{dp_0}{dr} - \frac{\rho}{\rho_0} u \frac{du}{dr} \qquad (136)$$

$$-i\omega v_\theta = -\frac{1 - \frac{1}{2}\chi^2}{\rho_0 r} \frac{\partial p}{\partial \chi} \qquad (137)$$

$$\left(-i\omega + \frac{u}{r}\right) v_\phi = -\frac{imp}{\rho_0 r \chi} \qquad (138)$$

9.6 A Selection of Important Instabilities

Here we have replaced $\sin\theta$ with χ to further simplify the notation. The reason for concentrating on the linearized equation is that we can reduce it immediately to a single, second-order, ordinary differential equation for the bending-type modes (the $\hat{\theta}$ modes). One very important assumption we will now make is that $u = \text{constant}$. This means that we drop the unperturbed advective term in the radial momentum equation. To keep the problem most general, especially for the Kelvin–Helmholtz instability, would mean including the possibility of entrainment. The difficulty we face, however, is the absence of a proper theoretical treatment for this process, so it would only have added a formal term to the equations without adding any physical prescription for how to model the entrainment. So to proceed, we substitute the individual velocity components into the continuity equation to obtain

$$-\frac{i}{\omega}\left[ikp_1 + \frac{p_0'}{\rho_0}\right]\left(ik + \frac{2}{r}\right)F + \frac{ip_1}{\omega}\frac{1 - \frac{1}{2}\chi^2}{r^2\chi}\frac{d}{d\chi}\left[\left(1 - \frac{1}{2}\chi^2\right)\frac{dF}{d\chi}\right]$$

$$-\left[\frac{m^2 p_1}{r^2\chi^2(u - iw)} - iw\frac{\rho_1}{\rho_0}\right]F = 0 \quad (139)$$

To remove the pressure perturbation requires closure through an equation of state. We assume the jet is barotropic, $p = p(\rho)$, and isothermal so that

$$\frac{\partial p}{\partial x_i} = \frac{dp}{d\rho}\frac{\partial \rho}{\partial x_i} \quad (140)$$

Had we assumed a polytropic jet, then

$$\frac{1}{\rho}\frac{\partial p}{\partial x_i} = K\gamma\rho^{\gamma-1}\frac{\partial \rho}{\partial x_i} \quad (141)$$

The perturbation equations would not be substantially altered by the polytropic choice since $a_s^2 = \gamma p_0/\rho_0$. The major problem would be including cooling and entrainment, which we explicitly ignore. Now assume that the opening angle is sufficiently small that $\chi^2 \to 0$, and we arrive at a form for the final equation:

$$\frac{\omega^2}{a_s^2}F + \frac{2}{r}\left(ik - \frac{1}{\rho_0}\frac{dp_0}{dr}\right)F + ik\left(ik - \frac{1}{\rho_0}\frac{dp_0}{dr}\right)F$$

$$+ \frac{1}{r^2}\left(1 - \frac{1}{2}\chi^2\right)\frac{1}{\chi}\frac{d}{d\chi}\left\{\chi\left(1 - \frac{1}{2}\chi^2\right)\frac{dF}{d\chi}\right\} - \frac{m^2}{r^2\chi^2 i\omega(1 + iu/\omega r)}F = 0 \quad (142)$$

This is the full version for the perturbation $F(\chi)$. The equation can be further simplified to read

$$\frac{1}{\chi} \frac{d}{d\chi} \chi \frac{dF}{d\chi} + \left(1 - \frac{\beta(r)}{\chi^2}\right) F = 0 \qquad (143)$$

where now we have the coefficient

$$\beta(r) = r^2 \left[\omega^2 - a_s^2 k^2 + \frac{2ika_s^2}{r} + \frac{i\omega a_s^2 m^2}{ru}\right] \qquad (144)$$

Notice that the perturbation equation is formally a Bessel equation of complex argument. The imposed boundary conditions determine the precise eigenvalues, and these are functions of distance, r, along the axis of the jet. Asymptotically, at large distances from the central source, the axial perturbations have damped wavelike solutions.

Now it is no longer too hard to see what is physically happening here. To obtain an eigenvalue solution for the jet spreading in $\hat{\theta}$, we need the requirement that both of the auxiliary functions are constant. This means that the eigenvalues are functions of r but not of θ or ϕ explicitly. We now see that the constant spreading angle will cause the azimuthal wave number, $2\pi m^{-1}$, to be a function of distance along the jet (*why?*). As the jet expands, this helicity grows, driven by the shear at the surface. It is usually assumed that the spreading is slow and that the jet is not overpressured relative to its surroundings because the hydrodynamic time scales are too short for thin jets for their propagation to be stable if overpressured. But there is also a possibility that the jet will be surrounded by a cocoon, generated by the deflection of the environment at the working surface, and this region confines the flow. In this case, the boundary conditions of simple slip (inviscid) will not work. The classes of instability of a jet were first stated by Rayleigh. The axisymmetric pinching modes were called *varicose*, while those that produce nonsymmetric, wavelike disturbances that propagate at the boundary were called *sinusoidal*.[7]

---

[7] Rayleigh found that for cylindrical incompressible subsonic jets the two modes have the growth rate (advected with the flow as we have written above)

$$\Gamma = \frac{(\tanh kz)^{1/2}}{1 + \tanh kz}$$

Here Γ denotes a growth rate with time, ku in our notation. There is an excellent discussion of the problem in Rayleigh (1896), *Theory of Sound* **2**, 380. In the discussion of instability of jets, he states that "In the ideal case of abrupt transitions of velocity, constituting vortex sheets, in frictionless fluid, the motion is always unstable, and the degree of instability increases as the wave-length of the disturbance diminishes." Rayleigh's Chapter 21, Vortex Motion and Sensitive Jets, still makes excellent reading and contains many valuable insights.

9.6 A Selection of Important Instabilities

Magnetospheric Boundary Shear

Accreting neutron stars often possess strong magnetic fields. As a quick application of the Kelvin–Helmholtz instability, imagine that we have a disk circulating about a neutron star with a local velocity v_K. The outer magnetospheric boundary is determined from the condition that $v_A \sim v_K$, where v_A is the local Alfvén speed determined by the surface magnetic field strength and the local density. In general, the shear will be large across this boundary because $v_A = \Omega_\star r_A$ is the condition for maintaining matter rigidly within the magnetosphere, where $\Omega_\star$ is the stellar rotation frequency. The ratio $\Omega_K/\Omega_\star$ goes as $r_A^{-3/2}$. In general, this ratio is very much greater than unity, and the boundary is unstable.

9.6.4 The Richardson Criterion and the Brunt–Väisälä Frequency: Density Stratification

In the introduction to this chapter, we discussed sound waves using an equation of state $P(\rho)$. Let us take the simple case of a medium which has a density gradient (in one dimension):

$$\frac{\partial^2 \xi}{\partial x^2} = -\frac{g}{\rho}\left(\frac{\partial \rho}{\partial x}\right)\frac{\partial^2 \xi}{\partial x^2} \tag{145}$$

where we have simply taken the medium to be a harmonic oscillator. The natural frequency of this medium is

$$N^2 = -\frac{g}{\rho}\frac{\partial \rho}{\partial x} \tag{146}$$

which is called the Brunt-Väisälä frequency. It is also sometimes referred to simply as the *buoyancy frequency* (Turner 1973). This is the time scale for the oscillation of the atmosphere and in the case of an isothermal medium is the same as the ratio of the acceleration to the pressure scale height (this is something you should derive for yourself). Notice that this is the frequency for the instability of a medium with a *continuous* density gradient, which of course we could have found by dimensional analysis (the ratio of the free-fall time scale to the buoyancy time scale is dimensionless). The mechanical response of an atmosphere, independent of the details of its thermodynamic state, will be given by this frequency. It is most important in pulsation problems. We have already looked at this problem many times. It is just that sometimes, you seem to be seeing the same physics in different settings or clothing and don't recognize it.

What Richardson introduced was shear flow which has a continuous velocity gradient as well, and showed that the following is a dimensionless quantity:

$$\text{Ri} = N^2 \frac{1}{(\partial v/\partial x)^2} \tag{147}$$

and that Ri describes the condition for the stability of the flow in the presence of the vertical density stratification. This is the *Richardson number* as conventionally defined. In simplest dimensional form, as defined by Batchelor, this can be written as

$$\text{Ri}_0 = \frac{gl}{U^2} \tag{148}$$

where l is the scale length of the medium and U is the velocity scale. This latter number is sometimes called the global Richardson number. It applies to inviscid fluids and must be compared with the Reynolds number when taking the viscosity into account. If the number is high, the medium will be unstable to the formation of shear-generated waves which can transport energy vertically. In the case of meteorologically and oceanographic models, the *Froude number*, which is essentially the inverse square root of the Richardson number, is often met with.

9.6.5 Taylor Instability: Rotation and Vorticity

What if the fluid rotates? This is a most central question, since the rotation can itself create turbulence. Viscous fluids can, if the shear is great enough, become unstable and give rise to a host of interesting effects. We will concentrate only on the one which was first recognized by G. I. Taylor: that differential rotation can create an instability. If the rotation is in rectilinear coordinates, then the planar motion is given by

$$\frac{D\mathbf{v}}{Dt} + 2\mathbf{\Omega} \times \mathbf{v} = \boldsymbol{\nabla}\Phi + \nu \nabla^2 \mathbf{v} - \frac{1}{\rho}\boldsymbol{\nabla}p \tag{149}$$

You will notice that there is now a new time scale which is characteristic of the fluid—the rotation period. You know full well, of course, what happens to the medium when it becomes unstable to convection. The fluid elements will move in complex spirals, which are dominated by the Coriolis terms whch have now appeared in the equations of motion. The potential, in principle, contains terms of the order of Ω^2, but these have been neglected in the slow rotation approximation. We can write the equations in component form as

$$\dot{u} + 2\Omega v = F_x + \nu \nabla^2 u, \qquad \dot{v} - 2\Omega u = F_y + \nu \nabla^2 v \tag{150}$$

so that

$$(i\omega + \nu k^2)^2 + 4\Omega^2 = 0 \tag{151}$$

is the critical condition for the eigenvalue problem in the absence of external forces. We have the possibility for the introduction of a new dimensionless number. Noting that the factor of 4 comes in from the epicyclic frequency (see Section 9.2.3), we see that there is a number of the order of $\Omega t_{\text{viscous}}$ which has the form $\mathcal{N} \sim \Omega l^2/\nu$, where we have in effect replaced the acceleration by gravity with the centrifugal acceleration. The analogy with the Rayleigh number comes from the consideration of the acceleration as $\Omega^2 l$ so that

$$\mathcal{T} = \frac{4\Omega^2 l^4 \Delta T}{\kappa \nu} \tag{152}$$

is called the *Taylor* number. The dependence on the scale length is different, but the basic bifurcation behavior of the solution is quite similar. In fact, the point of this argument is that one can relate the convection problem without rotation to one with rotation by the inclusion of an *effective gravity* term in the equations of motion for the fluid.

9.7 Exeunt

We here end the discussion of instabilities. This is not a complete survey by any means. New ones are frequently found now as a result of numerical simulations of complex flows. Many of the basic assumptions that lead to analytic results, like dispersion relations, are no longer valid for fully compressible flows in complicated geometries. So this chapter should be ended on a note of caution. Just because a medium can become unstable doesn't mean that it necessarily will unless all of the conditions are met. And, conversely, just because a stability analysis indicates that a fluid is stable doesn't mean that it will be.

References

Allen, A. J. and Hughes, P. A. (1984). The Rayleigh–Taylor instability in astrophysical fluids. *Mon. Not. R. Astron. Soc.* **208**, 609.
Baker, N. (1966). Simplified models for cepheid instability. In *Stellar Evolution*. R. F. Stein and A. G. W. Cameron, eds. New York: Plenum.
Birkoff, G. (1962). Helmholtz and Taylor instabilities. *Proc. Symp. Appl. Math.* **13**, 55. [See also Lin, C. C., and Benney, D. J. (1962). On the instability of shear flows. *Ibid.*, 1.]
Busse, F. H. (1978). Non-linear properties of thermal convection. *Rep. Prog. Phys.* **41**, 1929.

Childress, S., Levandowsky, M., and Spiegel, E. A. (1975). Pattern formation in a suspension of swimming microorganisms. *J. Fluid Mech.* **69**, 591.

Dupree, R. G. (1977). The theoretical red edge of the RR Lyrae gap. IV. Convective steady state models. *Astrophys. J.* **215**, 620.

Feigenbaum, M. J. (1980). Universal behavior in nonlinear systems. *Los Alamos Sci.* **1**, 4.

Field, G. B. (1965). Thermal instability. *Astrophys. J.* **142**, 531.

Jeans J. H. (1928). *Astronomy and Cosmogony*. Cambridge: Cambridge University Press; reprinted by Dover.

Kippenhahn, R., Ruschenplatt, G., and Thomas, H. -C. (1980). The time scale of thermohaline mixing in stars. *Astron. Astrophys.* **91**, 175. [See also Kippenhahn, R. (1974). Mixing in stars. In *Late Stages of Stellar Evolution: IAU Symp. 66* (p. 20). R. J. Tayler, ed. Dordrecht, The Netherlands: Reidel.]

Lorenz, E. N. (1963). Deterministic non-periodic flows. *J. Atmos. Sci.* **20**, 130.

Normand, C., Pomeau, Y., and Verland, M. G. (1977). Convective instability: A physicist's approach. *Rev. Mod. Phys.* **49**, 581.

Parker, E. N. (1965). Cosmic rays and their formation of a galactic halo. *Astrophys. J.* **142**, 1086.

Poston, T., and Stewart, I. (1978). *Catastrophe Theory and Its Applications*. London: Pitman. [See also Gilmore, R. (1981). *Catastrophe Theory for Scientists and Engineers*. New York: Wiley-Interscience.]

Proctor, M. R. E., and Weiss, N. O. (1982). Magnetoconvection. *Rep. Prog. Phys.* **45**, 1317.

Roberts, P. H. (1966). On non-linear Benard convection. In *Non-equilibrium Thermodynamics, Variational Techniques, and Stability* (p. 125). R. J. Donnelly, R. Hermann, and I. Prigogine, eds. Chicago: The University of Chicago Press.

Rosner, R., Tucker, W. H., and Viana, G. S. (1978). Dynamics of the Quiescent Solar Corona. [See especially the appendix on the cooling function.] *Astrophys. J.* **220**, 643.

Shandarin, S. F., and Zeldovich, Ya. B. (1989). The large-scale structure of the universe: Turbulence, intermittency, structures in a self-gravitating medium. *Rev. Mod. Phys.* **61**, 185.

Spiegel, E. A. (1971). Convection in stars. I. Basic Boussinesq convection. *Annu. Rev. Astron. Astrophys.* **9**, 323. (1972). Convection in stars. II. Special effects. *Ibid.* **10**, 261.

Stewart, I. (1981). Applications of catastrophe theory to the physical sciences. *Physica D* **2D**, 245.

Toomre, A. (1964). On the gravitational instability of a disk of stars. *Astrophys. J.* **139**, 1217.

Vauclair, S. (1983). Atomic diffusion and abundance gradients in stellar atmospheres. In *Astrophysical Processes in Upper Main Sequence Stars*. B. Hauck and A. Maeder, eds. Geneva: Geneva Observatory. [See also Vauclair, S., and Vauclair, G. (1982). Element segregation in stellar outer layers. *Ann. Rev. Astron. Astrophys.* **20**, 37.]

von Kármán. Th., and Biot, M. (1940). *Mathematical Methods for Engineering Applications*. New York: McGraw-Hill.

CHAPTER 10

Diagnosis of Astrophysical Flows

> *But now we see through a glass darkly.*
> Corinthians I

10.1 Introduction

Thus far in this book we have discussed the problem of flows from a rather abstract point of view. The working astronomer, however, is faced with very complex observational problems, frequently without recourse to the kinds of information one would have in the laboratory. Astronomy is fundamentally an observational science and as such is not one in which controlled experiments are easily performed. We have neither time nor budget to go around building stars and planets in order to let them evolve.[1] We will examine a few test cases to see what are the signatures of different hydrodynamic phenomena. The list will not be comprehensive, just representative, in the interest of economy.

First of all, speaking in the first person for a moment, I must make a confession. I am an unreconstructed spectroscopist. It is my bias that relatively little direct information about astronomical flows can be gained solely from imaging. Imaged celestial sources, when they show resolved motion at all, usually show it only after considerable times. More often, these sources show changes in brightness or morphology that *can be interpreted* as motion. However, line photons can delineate the flow because of the Doppler effect; this unambiguously shows if the material is in motion and usually even in what direction it is moving relative to the

---

[1] See, however, Adams, D. (1979), *A Hitch-hiker's Guide to the Galaxy* (New York: Harmony) for an alternative view.

observer. This is not generally true for continuum processes, which, by their very nature, simply redistribute the energy of one region of the spectrum into another part of the continuum. Emission or absorption lines have definite rest wavelengths and intrinsic widths. As a rule, the mechanisms involved in line formation in many cosmic environments are moderately well understood. So we can use line profiles, and their variations, to tell us much of what we need to know about a gas. The relative intensity of various lines yields information about the ionization processes, and the energetics of the flow, while the line profiles themselves reveal the density and velocity structure of the flow region.

The purpose of this chapter is not to serve as a course on radiative transfer or on stellar atmospheres. There are wonderful books available in those areas and you should look to them for more details. Rather, look on this chapter as representing the analog of the one usually found in hydrodynamics books on "flow visualization" and its purpose will be clearer.

10.2 Radiative Transfer on the Cheap

Light is the basic tool for the analysis of celestial objects. Any wavelength may be important, but the basic physical processes by which the light is created are the same from one environment to another. There are two fundamental problems to be dealt with. One is how the photons are created, and here the properties of the body are the ones we are after. The other is what effect any intervening medium has on the propagation of the signal. This latter may include the problem of travel time effects for extended bodies, or the fluctuations introduced in the signal by refractive changes or motions in the intervening medium.

Although it will appear as if stars are the concern of this chapter, this is not a correct impression. The theory of radiative transfer was first developed for stars because they are points. You see, we observe spectra from these unresolved objects (at least mostly unresolved; only a few of the very nearest and largest have been imaged directly using interferometry). To infer something from these data about the surface or interior conditions requires the construction of models of the emitting body and its environment. If the source isn't resolved, we don't have the option of pointing the spectrograph at different places and obtaining diagnostic measurements over a range of environments. In a single spectrum, we see the product of all physical effects produced by the visible source, all contributing simultaneously, but with different weights, to the integrated spectrum. So we need to develop some sort of theoretical machine whereby we can infer something more detailed about the conditions within the emitting source.

10.2 Radiative Transfer on the Cheap

In the simplest case, we assume that the body is comparatively large with respect to the layer that is transparent enough for us to observe. This is the *plane parallel* approximation. It says that there is a *geometrically thin* atmosphere on top of a more opaque surface. The radiation we see emerges from this lower *photosphere* and is altered, as a function of wavelength, by the absorption, emission, and scattering from the higher regions. We also generally have to assume that the light is initially emitted from below. But if the source isn't isolated, this won't be true. Sometimes, as with planets, the radiation comes mainly from outside, reflecting off different layers in the atmosphere. Even in the case of an accretion disk, the radiation may come only in part from internal processes. Sometimes the disk is also illuminated from the central source, which is distant from the region of the disk we are observing and only radiatively affected by it.

So to make this discussion more concrete, let's derive the equation of radiative transfer. Consider radiation of intensity I_ν at frequency ν incident on a layer of gas. The radiation is assumed to be produced by a source whose luminosity is determined independently of any properties of this thin layer, an assumption that may have to be modified in some cases but is sufficient for our purposes. This atmospheric layer has a density ρ and an absorption coefficient κ_ν. The intensity of radiation emergent from this layer will be diminished by extinction and augmented by local re-emission and redistribution by scattering of external photons into the line of sight. Also, a variation in the frequency by redistribution due to motion in the medium augments the extinction, or increases the scattered intensity. The goal of radiative transfer theory is to express this succinctly.

If the photon passes through a medium that absorbs, the intensity will be diminished by thermalization. True absorption is the process by which the energy of a photon is redistributed throughout the gas and ultimately re-emitted at some arbitrary wavelength consistent with the flux distribution. As complicated a statement as this is, it is the core of what is meant by local thermodynamic equilibrium (LTE). Any photon can be either scattered or absorbed (or if the density is low enough or the cross section small enough, it can be ignored too!). Scattering alters the direction of the radiation, so that if we stand within the scattering region, the solid angle subtended by the source appears to have increased.

Another way of looking at this is the halo that appears around streetlights in a fog. The light appears to be coming not just from the lamp itself but also from a region around it. This is radiation that would have missed the observer were it not for the intervening scattering medium, the fog. The intensity of the light surrounding the lamp is increased because some of the photons that would have missed the observer (having been emitted with the "wrong" incident angle) are redirected toward the observer. However, the radiation is also decreased along any direction pointing directly to the source, and there may be a small secondary scattering effect even for the redirected photons. In short, there are two combined effects. One is that the intensity of the direct beam decreases; the other is that the local density of photons is increased over what it would have been along a given direction. But scattering is a conservative process, so what one line of sight loses is gained in some other direction.

Assume for the moment that the medium is at rest. The mean intensity J_ν characterizes the scattering process we have just discussed. It represents the average radiative intensity over all solid angles and is a scalar quantity. That is its formal definition, equivalent to the integral

$$J_\nu \equiv \frac{\int I_\nu \, d\Omega}{\int d\Omega} \quad (1)$$

The integrated intensity represents the overall energy density of radiation at any point in space and is a scalar quantity. The flux is defined as the normal, cosine-weighted component of the incident radiation,

$$F_\nu = \frac{\int I_\nu \mathbf{n} \cdot d\Omega}{\int d\Omega} \quad (2)$$

For a gas in thermal equilibrium, $\nabla \cdot \mathbf{F} = 0$. Now the change in the intensity of the radiation at any frequency along any arbitrary line of sight is given by

$$\mathbf{n} \cdot \nabla I_\nu = -\kappa_\nu \rho I_\nu + \rho j_\nu \quad (3)$$

where j_ν is the volumetric emission coefficient. The first term is the rate along a path of removal of the radiation from the line of sight, and the second is the rate at which photons appear along a path. Either of these terms can have contributions due to scattering or from actual thermal or nonthermal emission or absorption. That is, the photons can appear from "nowhere" or disappear into the "nowhere," or they can be contributed along any line of sight by redirection. Either one causes the intensity *along some specific direction* to change.

The equation of transfer is essentially a statement of *probability*, like the Vlasov and Boltzmann equations, and is the analog of those evolution equations for photons. The *source* function is the rate of contribution of photons to any line of sight. It is defined as the ratio of emissivity to absorption of the medium.

$$S_\nu \equiv j_\nu / \kappa_\nu \quad (4)$$

and is equal to the Planck function, $B_\nu(T)$ if the medium is locally in thermal equilibrium (LTE). Strong departures from LTE result when

10.2 Radiative Transfer on the Cheap

radiative excitation and de-excitation dominate over collisions in defining level populations. In low-density, low-opacity media, especially in the interstellar medium, this is often true. But for many cases we can approximate the transfer in such a way as to allow for a solution of the line profile and level populations.

In general, then, the source function can be approximated as composed of two terms, one due to thermal emission and the other due to scattering (and other non-LTE phenomena) for coherent processes:

$$S_\nu = (1 - \epsilon)J_\nu + \epsilon B_\nu(T) \tag{5}$$

where ϵ is the ratio of the collisional de-excitation rate to the radiative transition probability, C_{ji}/A_{ji}. Collisions cause the source function of the line to equilibrate with the thermal continuum, ultimately suppressing emission or absorption to yield a continuum consistent with the local Planck function. The first term therefore represents the scattering contribution, and the second is the local thermal emission from the gas.

Scattering is not a thermalizing process. It does not increase the temperature of the medium, and it changes the radiative energy density without necessarily producing a change in the temperature. To repeat something from a moment ago, it is also a strictly conservative process. In an optically thick medium, the radiation would ultimately appear isotropic and we would not see lines formed purely by scattering. Velocity gradients change all this. Once the medium isn't static, even a region that would have been completely opaque at line center begins to suffer losses and photons stream out of the layer. A single spectral line has a unique value for ϵ at any density and temperature. But because of atomic properties, different lines may have different ϵ values under the same thermodynamic conditions.

The equation of transfer is actually three-dimensional. (See Fig. 10.1.) It reduces to two-dimensional form for spherical or axisymmetric cases:

$$\mu\frac{\partial I_\nu}{\partial r} + \frac{1-\mu^2}{r}\frac{\partial I_\nu}{\partial \mu} = -\kappa_\nu \rho (I_\nu - S_\nu) \tag{6}$$

For a plane parallel medium, it is even simpler to write down the equation of transfer because the line of sight maintains constant inclination with respect to the atmosphere for each ray. Therefore, the problem is *essentially* one-dimensional with vertical distance z:

$$\mu\frac{dI_\nu}{dz} = -\kappa_\nu \rho (I_\nu - S_\nu) \tag{7}$$

The direction cosine is treated as an auxiliary variable, constant for each line of sight, and ultimately averaged over when calculating the emergent

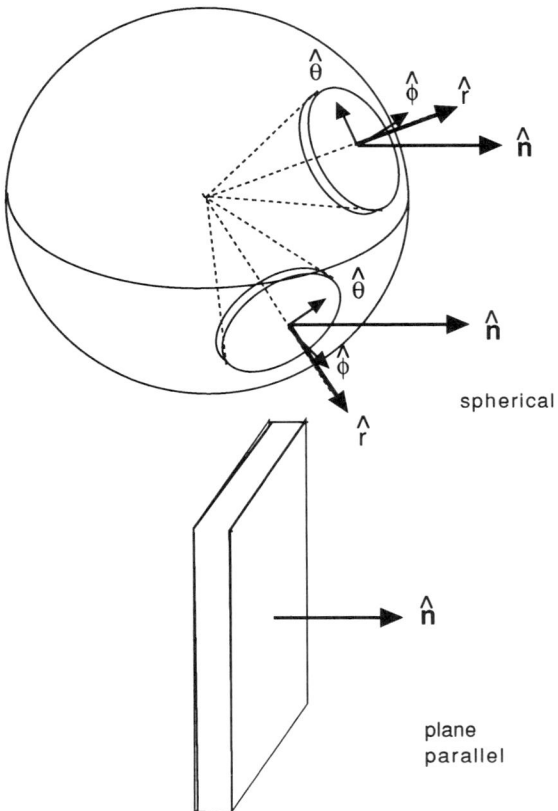

Figure 10.1 Coordinates for plane parallel and spherical atmospheres. An oblique ray encounters infinite optical depth for horizontal transfer in the plane parallel case but always sees a finite atmosphere for the spherical case.

spectrum. However, you notice that there is an extra term in the spherical equation that is missing in the plane parallel approximation. This term, acounting for curvature, behaves like an additional loss term for the radiation. Since a spherical atmosphere with finite radius has a boundary, and there are no infinitely long paths through such an atmosphere, the outer layers cool more rapidly than the plane parallel model would predict. In other words, something as simple as altering the radius of curvature of the atmosphere can alter the emergent radiation pattern. And this for a body we can't even resolve!

10.2 Radiative Transfer on the Cheap

We will begin with a plane parallel medium, which works assuming that we have a small layer which, even if it is part of a spherical object like a star, is thin compared with the radius of the star (that is, the radius of curvature is extremely large, approaching infinity). For a one-dimensional ray, we can create a new coordinate, one that gives a measure of depth to the medium *as perceived by the photons*. This is the *optical depth*, defined by

$$\tau_\nu = \int_z^0 \kappa_\nu \rho \, dz' \tag{8}$$

Then the equation of transfer becomes

$$\mu \frac{dI_\nu}{d\tau_\nu} = I_\nu - S_\nu \tag{9}$$

The source function for thermal emission is given by $B_\nu(T)$ and won't concern us just now. However, the source function for scattering is given by the mean intensity, which is now provided by the solution to the transfer equation:

$$S_\nu^{(\text{scat})} = J_\nu(\tau_\nu) \tag{10}$$

In fact, in an optically thick medium, the mean intensity is also equal to the Planck function, provided collisions dominate.

10.2.1 P Cygni–Type Profiles: Observations

At this point, let's stop for a moment to reflect on what we have written down. We sit on the outside of the star, or any flow. In order to probe the interior of any medium, the only information we *ever* have is provided by the photons that emerge from within the medium. Usually, we don't have the luxury of being able to spatially resolve the medium. We see only the integrated properties, summing up the contributions over the whole surface area of the body. Each point within the volume may provide some light, but not every one does. And in order to answer many of the basic questions about the medium, we need to be able to take a velocity field provided by the solution to the equations of fluid motion and use it with the radiative transfer to obtain the requisite information about the internal density and temperature and abundance.

A Hot Steady-State Stellar Wind: Mel 42 in the LMC

Examine Figure 10.2. Here we see the profiles for several ultraviolet lines in an O3f star in the Large Magellanic Cloud, Melnick 42. The profiles are strikingly different. The terminal velocity of the C IV reso-

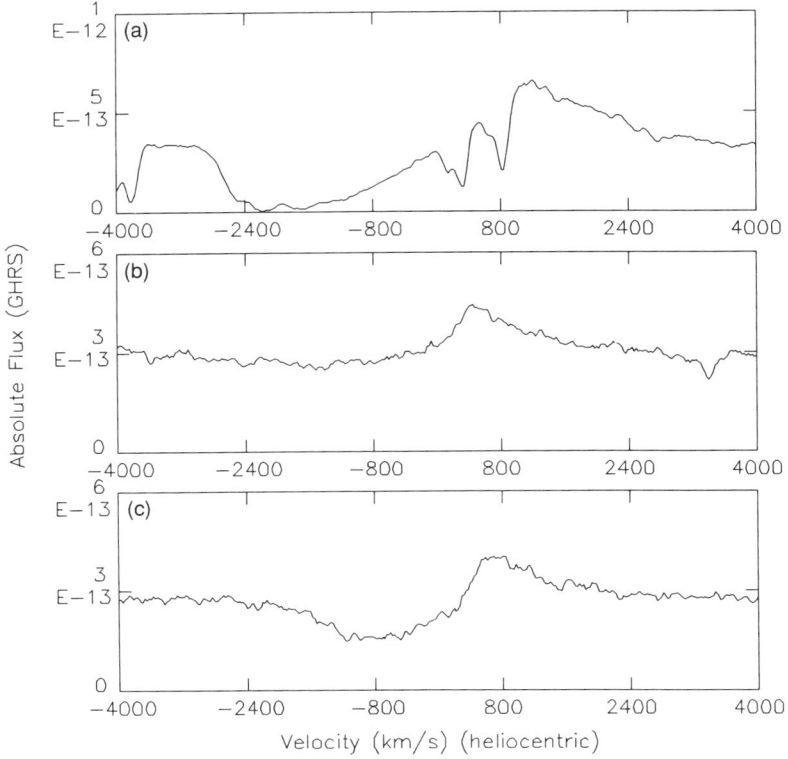

Figure 10.2 Goddard High Resolution Spectrograph (GHRS) small-aperture spectrum of the O3f star Melnick 42 in the Large Magellanic Cloud. (a) C IV λ1550Å; (b) He II λ1640Å; (c) N IV λ1718Å.

nance line at wavelength 1550 Å is about 3000 km s^{-1}. It shows a deep absorption trough and abrupt, well-defined blue edge. On the other hand, the He II 1640 Å line shows only the slightest hint of an absorption component, although the emission profile of this nonresonance line is decidedly asymmetric. The N IV 1718 Å line, also formed from an excited state, displays a profile lying between these two extremes.

Clearly, the lines are formed very differently with depth. However, using only one of them and accepting the profile at face value could lead to completely erroneous conclusions about the stellar wind dynamics and thus compromise any interpretations of the stellar properties. An equally dramatic example is provided by a comparison between the optical and ultraviolet line profiles of stars with strong stellar winds. The optical hydrogen Balmer lines, all formed from excited states, typically show much

10.2 Radiative Transfer on the Cheap

lower terminal velocities than those observed on the resonance lines in the ultraviolet and also stronger emission components.

Explosive Mass Loss: Nova LMC 1990 No. 1

Even more dramatic examples are provided by nonstationary flows. One good case in point is provided by novae, in which shell ejection occurs ballistically with a linear velocity law imposed on the outflow. An example of the evolution of the C IV profile during the outburst of nova LMC 1990 No. 1 is shown in Figure 10.3. The strong P Cyg absorption initially extends to a radial velocity of nearly 10^4 km s^{-1}. The absorption is broad and saturated, even at this low resolution (about 500 km s^{-1}). As the shell expands, the absorption narrows and the emission line strength increases.

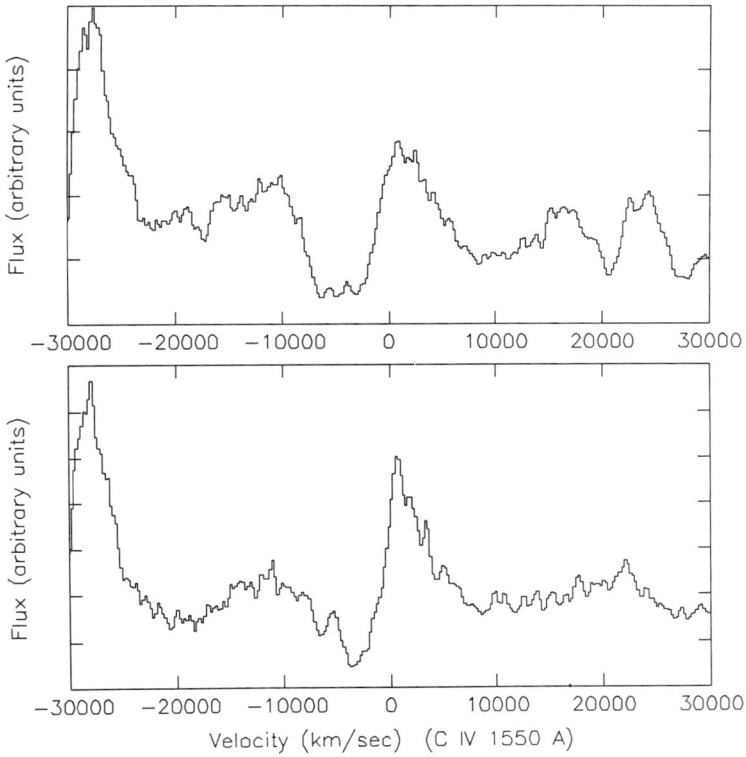

Figure 10.3 Spectral evolution of Nova LMC 1990 No. 1 during the first month of outburst. The low resolution spectra were obtained with the International Ultraviolet Explorer satellite (IUE) during Jan. 1990. Top spectrum was obtained about one week before the bottom spectrum. First spectrum was during UV maximum of this fast nova.

The reason for this is that the outer layers of the shell, moving fastest, decrease in density and reveal the lower layers, which are moving more slowly. Again, in a ballistic (explosive) ejection occurring at an instant in time, the velocity gradient is constant, so the line profile evolves almost self-similarly as it becomes progressively more optically thin. Finally, the optical depth becomes so small and the solid angle subtended by the expanding *pseudophotosphere* becomes so small that the line profile flips completely into emission.

An enormous amount of information is provided by this time sequence of profiles, like peeling away an onion skin. Once the lines are optically thin, it is possible to determine the abundances in the ejecta and compare them with the predictions of nucleosynthesis models. The dynamics preserve information about the ejection process and energetics that are not accessible by any other observation. Finally, the velocity structure of the line profiles may signal the breakup of the shell, the occurrence of secondary ejections, or the collision between the expanding shell and material in its near environs. And all of this from an unresolved point source!

10.2.2 Escape Probabilities

Now, refreshed from these encounters with nature, let's return to our discussion of the line profiles.

The emergent line profile is quite sensitive to the presence of flows. This is also true for the continuum, although not to the same degree. The line optical depth is reduced at any wavelength by a velocity gradient. This can be understood as follows. Consider an atmosphere with an underlying continuum. If the optical depth at line center is τ, then the probability for a photon to escape from the layer in question along a direction $\hat{s}$ with a directional cosine μ is

$$P_{\text{esc}}(\tau, \mu) = e^{-\tau/\mu} \tag{11}$$

and from the two layers is the product $P_{\text{esc}}(\tau, \mu) P_{\text{esc}}(\tau', \mu')$. Now the average over all possible lines of sight for a given optical depth provides the net escape probability, which we will denote as β. This probability is given for monochromatic light:

$$\beta(\tau) = \int_{-1}^{1} e^{-\tau_\nu/\mu} d\mu = \frac{1}{\tau_\nu}(1 - e^{-\tau_\nu}) \tag{12}$$

This is the estimate for the rate of escape of photons from the medium. You can see that if the optical depth vanishes, the escape probability approaches unity. On the other hand, it falls to zero only as $1/\tau$, not too rapidly.

In the presence of a velocity gradient, however, the upper layer is shifted relative to line center. This decreases the opacity at line center, but increases it in the blueward line wing, from which the photon would previously have had a higher chance of escaping. The price paid for this is that the upper layer now absorbs against the blueward (or redward) continuum relative to line center and the emergent line, which may have the same equivalent width, has a reduced central depth. If the shift is actually larger than the intrinsic width of the line profile, then the photons that are absorbed by the upper layer will never have passed through the lower layer line core; thus the emergent line is determined by the conditions in each layer independent of what the underlying (or overlying) layers are doing.

The feature of the atmosphere or flow that produces both absorption and emission in the line is that the source function for the line in a given layer may be lower or higher than that in the deeper layer. If the source function is lower, photons are absorbed at the local frequency ν and either scattered or lost through collisional de-excitation of the upper level. Scattering diminishes the number of photons along the line of sight, although the rate of scattering depends mainly on the column density of the target atoms and not on their excitation conditions. The absorption process ultimately thermalizes the photons, redistributing their energy through collisions throughout the continuum, and depends on the local thermal state of the gas. Thus the absorption line depends on the column density to the continuum source, but the emission portion of the line depends on both the optical depth through the medium and its geometry. It is also important to remark that the equivalent width of a scattering line remains constant since the photons are never thermalized.

10.2.3 Kinematic Approach

Let us begin with a radially expanding atmosphere whose outflow velocity depends only on the radius. To simplify matters still further, assume also that the velocity law is known in advance. Actually we have cheated here, because specifying the velocity law assumes that we have already solved the dynamical equations for the medium. This assumption is quite a leap of faith because as we have seen the driving force for the wind matter depends on the radiation pressure, which in turn depends on the transfer of momentum through spectral lines, which in turn depends on the velocity gradient. You see, the problem is a convoluted one and to solve it self-consistently is extremely difficult. But we don't necessarily need all of the physical information to attack the problem of diagnosing the structure of winds. In fact, one thing that is often done is to seek some phenomenological representation for the wind properties and to then study the systematics of the parametric fits.

Let me be more precise about what this means, because it affects our whole discussion. The physical problem of starting and maintaining flows in cosmic environments is terribly hard. In fact, it appears in many cases to be impossible to solve from first principles. So in the absence of an exact solution, we do what physical scientists always do—we specify the problem with as small a set of parameters as possible. This is a very general statement, of course, but it is also profoundly important. In astronomical observation, we must be selective about what we consider data. There is often so much, won at so hard a cost (observing time, data reduction, waiting for the sky to clear or the radiation to decrease) that we must grab what we can and what we can be sure of getting every time we make a similar observation. For instance, it is useful to have an entire line profile to work with, but usually this isn't available—either from the literature or in a form that can be published—so we instead boil the profile down to a few numbers. For example, we may want to specify the centroid velocity and a few moments of the profile [the dispersion or full width at half-maximum (FWHM), or the full width at zero intensity, or both], the skew or some measure of the line asymmetry, and perhaps even the absolute strength of the profile (if available). Often even these parameters are not available, so we sometimes resort to the equivalent width. Now in fact we have just made a parametric statement, specifying the results of some form of fit of the profile, or at least making a consistent set of measurements.

An alternative approach is to specify those parameters for a characteristic type of line that immediately bear some relation to the properties of the star that produced the profile. For example, for a P Cygni profile, we need the terminal velocity, v_∞, and perhaps some measure of the "blackness" of the line at the blue edge of the adsorption trough. We may also want the ratio of the equivalent widths of the blue and red portions of the profile, the centroid velocity (weighted over the absorption and emission components), and perhaps some measure of the width of the emission (Castor and Lamers 1979; Castor, Lutz, and Seaton 1981). Again, we have assumed that there is some connection between these readily observable properties of the profile and the physical conditions that produced them. It is also assumed that we will be able to go from the models to the observations in a consistent way.

Often, we are lucky. The models produce results that can be quantified directly in the same way that the observations were and that also mean the

10.2 Radiative Transfer on the Cheap

same thing. On the other hand, often the models don't agree with each other. One treatment of the problem gives an obviously different answer when compared with another treatment of the same physical conditions. We will see, in the case of stellar winds, that this discrepancy between different methods often results from very different assumptions about the internal physics and that these differences are actually important.

For instance, suppose we have a slowly expanding medium which is spherically symmetric and we embed a small star within this medium. The solid angle of the scattering region is very large compared with the size of the solid angle subtended by the continuum source. Therefore the absorption line will have a small total depth and small equivalent width compared with the broader emission line. However, the absorption extends all the way to the terminal velocity. The smaller the core, the weaker the absorption relative to the surrounding emission.

This form of the profile is especially well known for resonance lines, for which the excitation is not important and only the ionization fraction and the velocity gradient determine the strength of the line core. The terminal velocity is the point at which the profile optical depth abruptly changes from larger values to zero, and often forms a very sharp blue edge to the absorption trough for resonance lines (see Fig. 10.4). The determination of the terminal velocity is especially important because it is set by the mass and radius of the star (since it is related to the escape velocity for a radiatively driven wind) and thus serves as an independent check on the stellar properties. For main sequence stars, the mass and radius are directly related and therefore the terminal velocity is a function of only the mass (and to some extent the chemical composition) of the star.

A P Cygni profile is a unique signature of an outflow, although the dynamics of the flow may not uniquely reflect in that profile. The ability of a photon to escape the core of the profile is determined by both the local density and temperature conditions (which determine the opacity at line center in the rest frame) and the gradient in the velocity. While for a steady-state medium the density is related directly to the radius by the velocity field (and that is specified independently in most problems) through the mass loss rate, the absorption coefficient depends on the ionization state of the gas as well as the local mass density. It is therefore the problem of computing a radiative flow to determine both the solution to the equations of ionization *and* statistical equilibrium at each point in the comoving frame, and then to compute the transfer of a photon through the medium.

Usually this cannot be done exactly. Treatments have been attempted in recent years, called *unified stellar atmospheres*, that couple the comoving

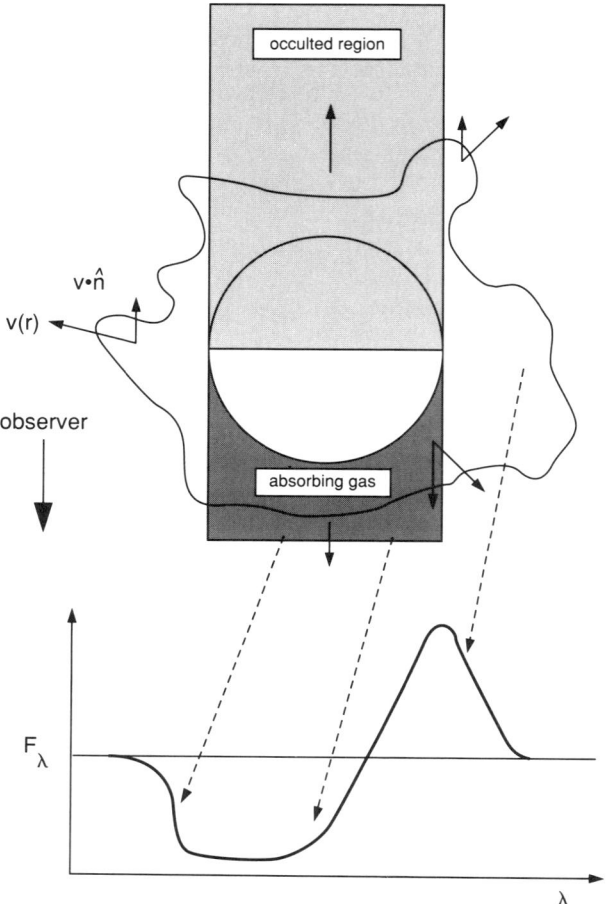

Figure 10.4 Formation of a P Cygni profile in an expanding stellar wind. Regions of formation of individual parts of the line profile are marked.

frame treatment of the radiative transfer with the calculation of line blanketed model atmospheres (see, e.g., Kudritzki *et al.* 1988). But the treatment has limited successes (albeit impressive ones) and is tremendously computer intensive (at least in this part of the decade). Instead, let's look at a simplified version of the problem.

We assume that the opacity at line center is fixed by specifying some ionization parameter as a function of radius for the ith species, $X_i(r)$, and the dilution factor for the radiation field $W(r)$. The optical depth at line

10.2 Radiative Transfer on the Cheap

center is

$$\Delta \tau_{0,i} = \kappa_{0,i} X_i(r) \rho(r) \Delta l \qquad (13)$$

where Δl is the path length. The path length is determined by the diffusion of photons out of the line core. For a static atmosphere, the photon random walks its way out toward the wings of the line, whence it escapes from the medium. In a moving medium with an internal velocity gradient, the scale for this diffusion is set by velocity shear. If the medium has a large enough gradient, ∇v determines l. The natural width of the line, the thermal speed v_{th}, sets the lower bound for the velocity gradient along the photon's path. Therefore:

$$\Delta \tau_{0,i} = \kappa_{0,i} X_i(r) \rho(r) \frac{\Delta v}{dv/dl} \qquad (14)$$

and $\Delta v = \Delta \nu_D c / \nu_0$, where $\Delta \nu_D$ is the Doppler width of a line with rest frequency ν_0. For a large velocity gradient along a line of sight, the lower layers of the atmosphere contribute strongly to the emergent line profile and may even form emission at the local blueshift (or redshift) corresponding to that layer. In a decelerating flow with a monotonic velocity gradient the reduction in the velocity gradient causes a rapid increase in the line opacity *even though the density may have dropped so far as to render an equivalent static line profile completely optically thin.* The velocity gradient is given along each ray. The velocity law, however, is given in terms of the distance from the center of the star. So we need to transform between these in order to make any sense out of the problem. Call $z = r\mu$. Then the radial gradient transforms into one along the line of sight by

$$\frac{dv_z}{dz} = (1 - \mu^2) + \mu^2 \frac{d \ln v(r)}{d \ln r} \qquad (15)$$

In what follows, we will use both z/r and μ, whenever one is preferred to the other for clarity.

This method for calculating line transfer has many names. In molecular line work, for either the interstellar medium or stellar envelopes, this is called the *large velocity gradient* or *LVG* approximation (Kwan 1978). In works on stellar winds it is usually called the *Sobolev* approximation (Sobolev 1960). It is also know as the *on-the-spot* approximation and even as the *escape probability* method (Castor 1970; Castor and Lamers 1979). Always the basic assumption is the same, that the bulk flow has so large a gradient, and so large a shear, that the mean free path of a photon is very large compared with a static atmosphere. Essentially, it states that, except for the immediate vicinity of the line-emitting region, two distant regions of the atmosphere are completely decoupled from each other.

The process of line formation is detailed along each line of sight through the atmosphere. Since the velocity of a layer relative to the observer, the radial velocity, is a function of both angle and distance from the central star, there exists a map that takes each point in the line profile and associates it with one direction through the stellar atmosphere. To see this, consider a spherically symmetric, outwardly accelerating atmosphere. The distance along a ray at a distance R to the central star is $z = (r^2 - R^2)^{1/2}$ for a fixed impact parameter. The radial velocity is given by $v_{rad} = \mathbf{v} \cdot \mathbf{n}$, where $\mathbf{n}$ is the unit vector that lies along z. For a spherical outflow, v_{rad} is bilaterally symmetric, identical across the meridional plane passing through the star and changing sign in the plane of the sky. Therefore, the entire back half of the expanding atmosphere is radiatively decoupled from what happens in the front half, at least within the line. Certainly, in a complicated atmosphere this is not true because differential shearing of a dense line spectrum may cause lines rather distant from each other in frequency to come into contact suddenly even if the lines are formed in very distant portions of the atmosphere that are otherwise mechanically distinct. For this reason, we normally resort to treating one line at a time! But obviously, the star knows something we don't, since it has to solve the problem for all of its photons for a real spectrum.

Along a specific line of sight through the atmosphere, only one point along a radial velocity surface is contributing to the line profile, but all points along that surface ultimately make some contribution. That is to say, all parts of the wind are sampled by every part of the line profile, but they are radiatively decoupled from each other and each contributes along only one line of sight. That is why for every point on the line profile we must ultimately sum up all of the contributions from every line of sight.

10.2.4 The Ray Approximation for Line Formation

Now assume that we stand on the outside of the radiative region. The simplest physically interesting case is of a monotonic outflowing velocity field, the surfaces of which are symmetric about the line toward the source of the outflow and asymmetric about the plane of the sky (see Fig. 10.5). We cannot resolve this region, so we must integrate along cuts through the flow. These are at impact parameters R relative to the line toward the central object. Lines of constant R are in the $\hat{z}$ direction. The problem is to see how the density and temperature vary along surfaces of constant radial velocity and then to sum up the contribution of each line of sight through the atmosphere to a fixed point on the line profile.

Let's shift around the definition of the viewing angle so that we can stand on the outside more easily. The equation of transfer along a ray

10.2 Radiative Transfer on the Cheap

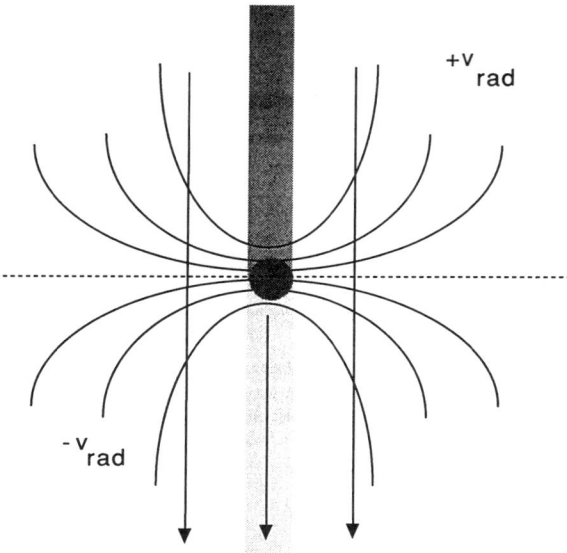

Figure 10.5 Iso-velocity contours for an accelerating mass outflow.

$z = -\mu r$ is

$$\frac{\partial I_\nu(R, z)}{\partial z} = k_\nu(R, z)(I_\nu(R, z) - S(r)) \qquad (16)$$

where we have replaced the normal equation of transfer, which depends on μ (the direction cosine), and the depth l. The optical depth along a ray is defined by

$$\tau_\nu(R, z) = \int_{-\infty}^{z} \kappa_\nu(R, z') \, dz' = \int_{\infty}^{z} \kappa_0(r) \phi\left(\nu + \frac{\nu_0}{c} \frac{v(r)}{r} z' - \nu_0\right) dz' \qquad (17)$$

Here we have defined a new function, ϕ, to describe the line profile, whose *functional form* is assumed to remain invariant with depth. The second term in the profile function is the correction for the radial velocity of the matter at radius r from the central star. The optical depth at line center is defined by

$$\tau_0(r) = \frac{\pi e^2}{mc} r \frac{(n_l/g_l) - (n_u/g_u)}{(\nu_0 v(r)/rc)} (gf)_{lu} \qquad (18)$$

where g_i is the statistical weight of the ith level (lower or upper), N_i is the population of the ith level, and f is the oscillator strength. The level populations are given by the solution of the statistical equilibrium equations and we will not discuss this further (see especially Mihalas 1978). Usually one tries to work with resonance lines, in part because they are strong and in part to get around the problem of excitation, so the ground state population, N_l, is all you need to know and one of the parameters of the fit to the line profile. Let's concentrate on how the line profile reflects the details of the originating flow. For convenience, we define an integrated profile function as

$$y(x) = \int_{-\infty}^{x} \phi(x') \, dx' \tag{19}$$

(see Castor 1970). This is a way keeping track of where you are in the line profile. The problem is often keeping a clear eye on whether you are talking about a point in the flow or a point in the line. The formal solution to the intensity along a ray is given by

$$I_\nu(R, z) = \int_{y(z)}^{1} S(r(z')) \exp(\tau_\nu(R, \infty)[y(z) - y(z')]) \tau_\nu(R, \infty) \, dy(z') \tag{20}$$

for rays with $R > r_{core}$ or for rays with $z > 0$ (from the front side of the flow) and

$$I_\nu(R, z) = \int_{y(z)}^{y(-z_{core})} S(r(z')) \exp\{\tau_\nu(R, \infty)[y(z) - y(z')]\} \tau_\nu(R, \infty) \, dy(z')$$
$$+ I_c \exp[\tau_\nu(R, \infty)(y(z) - y(-z_{core}))] \tag{21}$$

for impact parameters intersecting the core $R < r_{core}$ and with $z < 0$. Here, to keep the notation for the integrals simpler, we have used $r(z') = (R^2 + z'^2)^{1/2}$ and z_{core} for the line of sight that is intercepted by the central body. You see that what we have done is to change from the optical depth as the independent variable to the profile function. This also emphasizes the essential feature of the line formation—that the profile contains a map of the velocity field and that each frequency corresponds to some depth of formation within the velocity field.

10.2.5 Calculating the Source Function

To solve for the source function of a line, we must know the integrated intensity J. This is given by

$$J_\nu = \frac{1}{2r} \int_{-r}^{r} dz \int_{0}^{\infty} d\nu \, \phi\left(\nu + \frac{\nu_0}{c} \frac{v(r)}{r} z - \nu_0\right) I_\nu(R(z), z) \tag{22}$$

10.2 Radiative Transfer on the Cheap

We will also assume that the profile function, which measures the emission probability for a photon at any frequency relative to line center, is heavily weighted toward the peak of the line; more on this point momentarily. The integrated intensity results from continuum photons that penetrate the line core with a probability β_c and those that are trapped as a result of scattering with a probability $1 - \beta$. Therefore:

$$J_\nu = (1 - \beta)S_\nu(r) + \beta_c I_c \tag{23}$$

We must specify the temperature of the flow before starting the calculation in order to know B_ν. This can be achieved by taking an equilibrium value, consistent with the photospheric continuum, and reducing it by the dilution factor $W(r)$ given by

$$W(r) = \frac{1}{2}\left[1 - \left(1 - \left(\frac{r_{\text{core}}}{r}\right)^2\right)^{1/2}\right] \tag{24}$$

Here the unified atmosphere approach is so powerful because with it, it is possible to compute the thermal state of the flow self-consistently. Then, if necessary, the Sobolev approximation can be used to obtain the emergent line profile. The escape probabilities provide only an approximate method for computing the line profile, and there is always some assumed physical input because it is not exact.

Now we can make use of the escape probability that we have previously defined. We go back to the profile function. As we said a moment ago, assume that most of the line opacity originates in the line core. This means that rather than an exact treatment of the line profile, we can assume some approximate form for $\phi(\nu)$. This is where the analytical approximations come into play for escape probability calculations. Depending on whether Gaussian form (either weak line, thermal broadening, or macroturbulent dominant) or Voigt form (including intrinsic line width and full Stark and van der Waals broadening mechanisms) is assumed for the intrinsic line profile, you will get different analytic approximations for the escape probability. However, the point is that the formal representation of the line profile must be specified in advance with this treatment. The velocity gradient alters the definition of the optical depth and through it alters the escape probability. What we have done is to change the optical depth at line center to $\tau_0/(1 + \mu^2[d\ln v/d\ln r - 1])$ so that, including the profile function, we get

$$\tau_\nu(z) = \int_{-\infty}^{z} \frac{\tau_0(r[z'])}{1 + \mu(z')^2[d\ln v(r)/d\ln r - 1]} \phi_\nu(z')\, dz' \tag{25}$$

and then the line escape probability can be written as

$$\beta = \frac{1}{2} \int_{-1}^{1} d\mu \int_{0}^{1} \exp\left\{\frac{-\tau_0(r)(1-y)}{1+\mu^2[d\ln v(r)/d\ln r - 1]}\right\} dy \tag{26}$$

and the escape probability for the continuum is given by

$$\beta_c = \frac{1}{2} \int_{-1}^{(1-r_{core}^2/r^2)^{1/2}} d\mu \int_{0}^{1} \exp\left\{\frac{-\tau_0(r)}{1+\mu^2[d\ln v(r)/d\ln r - 1]}\right\} dy \tag{27}$$

We have now arrived at an expression for the line source function, the goal of all of this for the calculation of the intensity, assuming that the stellar continuum is diluted with distance through $W(r)$:

$$S(r) \approx \frac{(1-\epsilon)\beta W I_c + \epsilon B_\nu}{(1-\epsilon)\beta + \epsilon} \tag{28}$$

Here again ϵ is the rate of collisional de-excitation and $\beta_c \approx W\beta$. This is a function of temperature and again one of the reasons why the kinematic approach requires some formal temperature distribution in advance of the profile calculation. When collisions are important, and the escape probability is reduced because of the optical depth and low velocity gradient, the line disappears and is replaced by a "photosphere" (the pseudophotosphere to which we have already alluded). The local emission from the line increases as the collision rate is reduced and as the escape probability for the line is also increased by a decrease in the optical depth of the line profile. Once we have the intensity, we integrate over impact parameters R to obtain the emitted flux. Recall that the total flux is $F_\nu = 4\pi \int_0^\infty I_\nu(R,-\infty) R\, dR$ and then we have to substitute the intensity obtained from the foregoing analysis. The line profile is specified in terms of the residual flux, $(F_\nu - F_c)/F_c$, where $F_c = 4\pi r_{core}^2 I_c$ is the flux of the continuum.

10.2.6 Decelerating Flows

If the velocity is increasing with distance from the radiation source, even if the gradient is decreasing with distance, line formation can be thought of as integrating along a surface of constant radial velocity. There is no spatial coupling between regions along that surface. It is simply that each line of sight makes a unique contribution to any point on the profile. (See Fig. 10.5.) If the flow is decelerating with distance, two points in different parts of the atmosphere actually couple at the same position in the line profile (see Fig. 10.6; a similar problem is encountered with rotating winds, Fig. 10.7). This is due to a topological property of the surfaces of constant radial velocity for the two cases. If the flow is accelerating, these

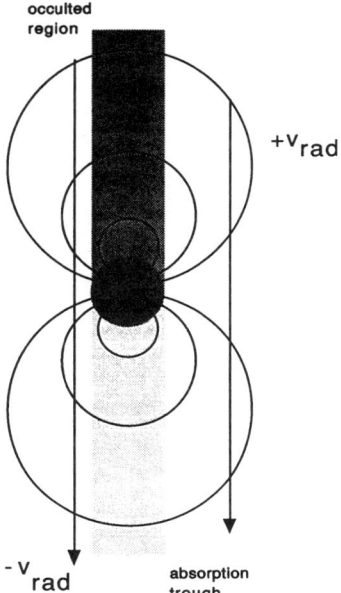

Figure 10.6 Iso-velocity contours for a decelerating mass outflow.

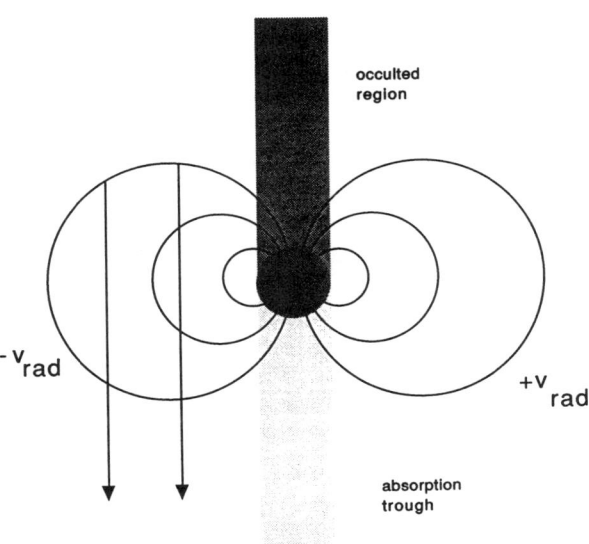

Figure 10.7 Iso-velocity contours for line formation in an accretion disk. Note the degeneracy of the contours across the stellar disk and the occultation of the rearmost portion of the circulation.

surfaces are open and any line of sight intersects the surface at only one point. If the flow decelerates, these surfaces are closed, and a single line of sight intersects the surface at two points. The rearward point is therefore absorbed by the forward one, and there is a more complicated coupling problem. But in terms of the escape probability, we can at least see conceptually what happens, and it isn't that bad. First, we need to know the escape probability for a photon from the rearward portion of the flow. This is multiplied by the probability of absorption in the forward portion of the outflow. Then the total optical depth is essentially the sum of the two, and the combined escape probability becomes

$$\beta' = \int\int (1 - e^{-\tau_{\text{rear}}/\mu}) e^{-\tau_{\text{front}}/\mu'} \, d\mu \, d\mu' \qquad (29)$$

where now the velocity gradient is implicitly included in the definition of the line optical depth.

10.2.7 Continuum Observations of Outflows

In one respect, continuum observations are useful for gauging the strength of an outflow. Thermal radio continuum observations can be used. The key to the analysis is density dependence of the absorption coefficient for thermal bremsstrahlung, also called *free–free* emission. A steady-state outflow at terminal velocity has a power-law density dependence, r^{-2}. This means that as one goes farther from the star, one sees the optical depth increase at progressively lower frequencies, and therefore the shape of the spectrum reflects the mass loss rate. Essentially, the frequency at which the spectral slope changes from being optically thin to optically thick is a measure of the total absorbing column toward the "photosphere." The solid angle subtended by the optically thick surface increases with decreasing frequency and partially offsets the drop in the escape rate for photons. (See Fig. 10.8.) For a thermal (blackbody) source, the continuum at radio frequencies varies like $F_\nu = B_\nu \approx T\nu^2$ and therefore the emissivity varies as $j_\nu \sim k_\nu T_e \nu^2$. For a thermal bremsstrahlung spectrum, the absorption coefficient is given by

$$\kappa_\nu = 3.69 \times 10^8 n_e^2 g_{\text{ff}} T_e^{-1/2} \nu^{-3} (1 - e^{-h\nu/kT_e})$$

$$\approx 8.44 \times 10^{-28} \left(\frac{\nu}{10\,\text{GHz}}\right)^{-2.1} \left(\frac{T_e}{10^4\,\text{K}}\right)^{-1.35} n_e^2 \text{ cm}^2 \qquad (30)$$

where n_e is the electron density, T_e is the electron temperature (which is assumed to be the same as T), and g_{ff} is the Gaunt factor for free–free (thermal bremsstrahlung) emission. For wind with a constant mass loss

10.2 Radiative Transfer on the Cheap

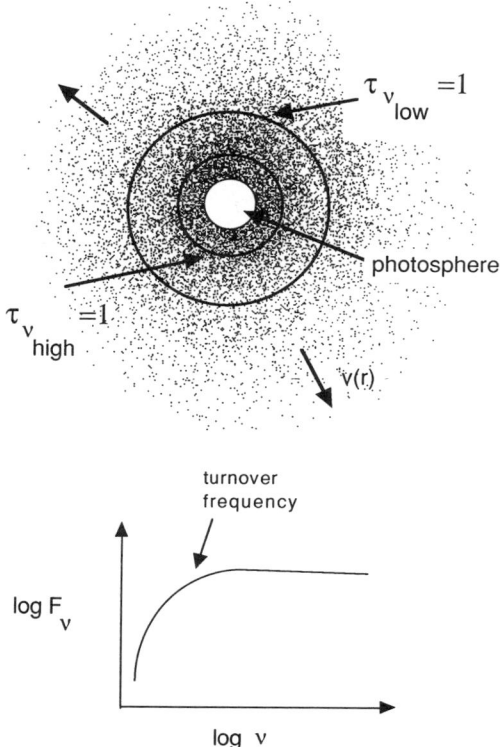

Figure 10.8 Physical picture for the origin of the radio continuum of a stellar wind.

rate expanding at terminal velocity, you recall that the density is given by $n_e \approx \dot{M}/(v_\infty r^2)$. The power-law dependence of the density will change as the wind velocity depends on radius. The optical depth for a completely ionized wind therefore is given by

$$\tau_\nu \sim \left(\frac{\dot{M}}{v_\infty}\right)^2 r^{-3} \tag{31}$$

For the radio continuum, the equation of transfer is especially simple:

$$F_\nu = \kappa_\nu B_\nu(T_e) D^{-2} \int_0^\infty (1 - e^{-\tau_\nu}) 2\pi r \, dr \tag{32}$$

To get this, the optically thick core was assumed to be very small in comparison with the envelope. Finally, using the approximate form for the bremsstrahlung opacity and Planck's law, we get

$$F_\nu \sim \left(\frac{\dot{M}}{v_\infty}\right)^{4/3} \nu^{0.6} T_e^{0.1} \tag{33}$$

The dimensional scaling law that results is

$$F_\nu \approx 130 \left(\frac{\dot{M}/v_\infty}{[M_\odot \text{ yr}^{-1}]/[\text{km s}^{-1}]}\right)^{4/3} T_e^{0.1} \nu^{0.6} D_{\text{kpc}}^{-2} \text{ Jy} \tag{34}$$

Notice that the flux now varies as $\nu^{0.6}$, much less steeply than expected for blackbody emission.[2] The reason is actually quite straightforward after the fact—free–free emission from a diffuse plasma with a density gradient is never Planckian because the emissivity depends on density. Usually we don't know the mass loss in advance; instead the scaling law is inverted to obtain $\dot{M}$ from a measured radio flux. This method can also be applied, with a bit more care, to the far infrared. And there's certainly nothing preventing its application to more general problems than *stellar* flows. The one important piece of information lacking from continuum studies is the dynamics: *you must already know the terminal velocity from some line measurement.*

10.3 A Sampling of Cosmic Flows

10.3.1 Narrow Absorption Features in Hot Stellar Winds

One of the most important results to come out of the newer generation of detectors, the CCD spectrographs, is the discovery of wind line profile variations. The so-called *discrete* or *narrow absorption components* have been discovered in many hot stars, usually but not exclusively in the ultraviolet resonance line profiles. The most important behavior of these features is that they move and change strength over time scales of hours to days. They progress from lower to higher absolute velocities, always moving toward more negative radial velocities. They never seem to appear on the emission portion of the line profile. Typically they are observed between 1500 and 2500 km s^{-1} from the rest wavelength, but recent observations indicate that they may be found at a much lower relative velocity compared with the stellar photosphere.

---

[2] A generalization of the result to nonthermal emission is discussed by Hjellming in Verschuur and Kellermann (1988). It involves a change in the exponents because of the different opacity law. See also Wright and Barlow (1975).

10.3 A Sampling of Cosmic Flows

Well, what are they? Stellar winds are certainly unstable, and the presence of these features is good empirical support for that. The main reason, as we have already discussed, is that radiation pressure is a very unstable driving force. The optical depth in particular lines may change because of the presence of small perturbation. Lines can shadow one another, and the radiative force therefore varies greatly depending on the optical depth in key absorption lines. This means that once a line becomes optically thick, the force imparted through that line to the atmosphere drops. If many such lines should become thick, because of a chance density fluctuation, that part of the atmosphere slows down relative to the optically thin material surrounding it. This material is moving supersonically, as is its environment, and a drop in acceleration produces internal shock waves. In fact, the density and velocity are inversely correlated, so that density fluctuations produce enhanced velocity fluctuations and the result is that the wind breaks up into regions of shocked filaments and blobs. These may persist for some time, it appears, and be advected out in the general flow (they are, after all, still moving outward as a result of the initial driving). Therefore, we see them in absorption against the stellar photosphere and can watch as they move out in the velocity gradient. More detailed discussion is beyond the scope of this work, but there are increasingly many cases in which these features have been observed and analyzed for stars.

However, stars are not alone in possessing time-variable line profiles. Galaxies also show the effect of changes in the environment. Usually, though, this is because of some change in the characteristics of the illuminating source rather than internal dynamics, although that too has been seen. For galaxies, however, we don't have the luxury of observing the phenomenon on time scales of days or hours. Instead, it takes weeks or months. The most dramatic changes are reported for accretion disk–related variations in active galactic nuclei. In two cases, NGC 5548 and NGC 4151, both Seyfert-type galaxies, the line profiles have been observed to show profile variations in the range 500 to 3000 km s^{-1} over periods of weeks to months. Different portions of the line profile vary differently, reflecting the fact that the regions are even more distant and mechanically isolated from each than in the stellar case.

The variations of line profiles in objects other than stars also illustrate a very important problem of interpretation that we have, up to now, been able to avoid. If the source is unresolved, the time scale is the key to the length scale. If the variations are consistent with hydrodynamic properties, then the regions are mechanically coupled to each other and have to be treated self-consistently. This is the situation, for example, in a stellar accretion disk. Velocities of a few thousand kilometers per second and time scales of hours to days provide mechanical coupling on length scales of a few solar radii, about the size of a typical star. On the other hand, for similar velocities, the expected time scale for variability is months, or even years, when discussing galactic-scale phenomena. So you can see that one reason for working on stars is impatience. If you can afford to wait for several decades to possibly see the variations that provide the signature of some special process, then work on galaxies. If you want more immediate gratification, work on stars.

10.3.2 Jets from Young Stars and Herbig–Haro Objects

In a few cases, we are lucky enough to be able to resolve the flows. One of the best examples of this is found in regions of star formation—protostellar jets. These spectacular phenomena have been discovered in the vicinity of recent star-forming complexes and have been instrumental in the understanding of the processes of star formation. In particular, they may help answer the nagging and long-standing question of how a star manages to rid itself of its angular momentum sufficiently to allow gravitational contraction to the main sequence. We won't concern ourselves with that aspect of the problem here. Instead, let's examine what this means for the analysis of the flow.

Atomic and Molecular Jets

For a few regions, like L1551 or the inner region of the Orion Molecular Cloud (OMC), one observes ionized or neutral jets emanating from the cores of the clouds. The flows were initially not recognized by optical imaging. In fact, there were no indications that these objects possessed anything like collimated outflows from their optical spectra. The stars are always deeply imbedded within the environmental dust and gas, and often there is no optical counterpart to a detected infrared source. The flows were initially found during relatively low angular resolution studies of the CO emission from the molecular clouds. The technique is to look at clouds that are strong infrared sources and to look for evidence of molecular material associated with point sources in the deep interior of the cloud. The idea is that by looking for the high-density gas in the vicinity of obviously condensed, very cold, bodies, it should be possible to recognize genuine protostars.

Molecular observations soon showed that instead of merely being more turbulent, or of showing evidence for accretion disks, these sources often showed large-scale bipolar flows. These have a characteristic signature. They show well-separated blue- and redshifted CO rotational lines (usually the $J = 2 \rightarrow 1$ and/or $J = 1 \rightarrow 0$) in emission. The lines often appear to be formed in the same line of sight. Mapping often reveals more

10.3 A Sampling of Cosmic Flows

details, and it is the mapping that has been responsible for much of the revolution in this field in the past decade.

The flows often resolve in two oppositely directed regions. One is always blueshifted with respect to the other. Often, there is no strong velocity gradient, indicating both that the line of sight through the flow is constant in inclination and that there is no compelling evidence for deceleration or acceleration of the flow on the scales of the observation. The flows appear to be oppositely directed, although they sometimes show rather contorted shapes on larger dimensions. (Fig. 10.9.) Here the possibility of imaging *and* studying dynamics underscores the importance of spectral line observations.

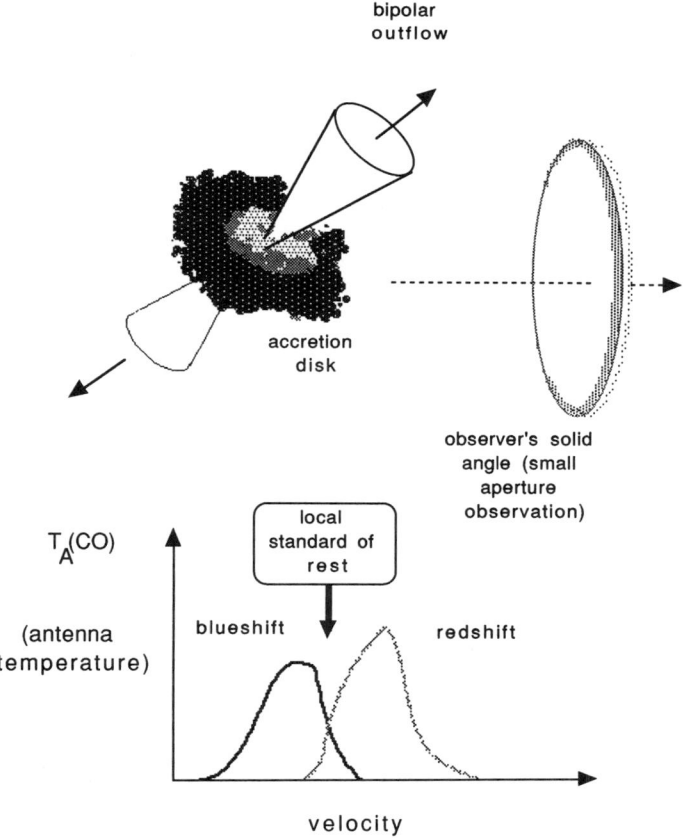

Figure 10.9 Physical picture for the origin of the spectrum of a bipolar outflow. Line profiles and velocity contours for spatially resolved flows are shown.

The same phenomenon has been observed in the optical using $H\alpha$ emission. The difference is that the ionized gas is obviously not tracing precisely the same flow that the molecules are sampling. The flows are usually narrower, but this is probably an artifact of the higher spatial resolution available for optical observations. In addition to the direct observation of accelerations and kinks along the flows, the optical observations are able to probe deeper into the core of the source, going closer to the flow's origin. For the neutral and molecular material, we are obviously looking at much cooler sources or much cooler regions in the flow than we sample with the optical emission lines.

The line profiles do not have a P Cyg form, and in general outflows observed in the millimeter and radio regimes do not. The source function has a lot to do with it. The fact is that the clouds do not have strong thermal continua. If they did, we would see line formation that looks very much like that from a stellar wind. Instead, the line formation takes place because of strongly nonequilibrium conditions within the cold, low-density flow. Collisions between CO and H_2 excite the rotational levels of the CO molecule, causing it to radiate. The transfer of radiation is actually simpler than we encounter in stellar winds, in the sense that the local conditions are reflected almost without alteration by the intervening matter. This condition arises from an important difference between the interstellar medium and a stellar atmosphere.

The lines intrinsically have very small widths, since the kinetic temperatures inferred for either molecular clouds or protostellar outflows are only a few hundreds of meters per second (nearly terrestrial values!). The random motions in the cloud are so much larger than this, and the emission measures so great, that individual regions of the cloud simply produce narrow, local absorption lines. In fact, the appearance of the narrow absorption lines in stellar winds is a very close analog. The only difference is that the ones observed in protostellar flows are stationary. From molecular clouds, we observe the same sort of line profile. The random motions within the cloud appear to be much larger than the local sound speed. Why this is so is an open question, one that has vexed workers for nearly two decades. But the empirical result stands as important in its own right.

The protostellar outflows illustrate something very exciting, something that we have not previously had available to us in astrophysics, the chance to look at different markers in a flow. In the laboratory, one often uses different kinds of tracers to obtain a complete picture of the flow. For instance, to study the dynamics one might use aluminum shavings or some form of insoluble powder to visualize *streamlines*. The density can be studied using some types of tracers, and the temperature can be observed

10.3 A Sampling of Cosmic Flows

either from probes or from temperature-sensitive dyes. Put together, these provide a useful means for probing the full set of physical parameters of the flow. In the protostellar flows, we have all of these, and we simply have to decide what wavelength region to observe for any of them. Unfortunately, one difference from the laboratory is that we can't restart the flow using different initial conditions. What you see is what you get.

The analogy between flow visualization and Herbig–Haro (H-H) objects is striking. These are emission regions, not obviously connected with stars but located in regions of star formation. They are extended (that is, resolved) nebulous patches with emission line spectra indicating moderate excitations. Their spectra are consistent with shock excitation. The [S II] lines are especially strong, indicating collisional rather than photoexcitation, and the typical kinetic temperature derived from their spectra is of order 10^4 K. The most dramatic feature of these objects is their motion. Connected with regions of active star formation, they are often found in close proximity to strong, dust-shrouded infrared sources. Proper motions (motions transverse to the line of sight) have been measured in many cases, precisely like the effect of dropping a fluorescent dye in the path of an emerging jet. The H-H objects are interpreted as the working surfaces of jets impinging on the surrounding interstellar medium.

10.3.3 Analysis of Explosions

Novae

Novae occur in close binary systems in which one of the stars is degenerate (e.g., Starrfield 1989). The canonical model is that mass, transferred from the companion, falls through a shock onto the surface of a compact degenerate object, normally a white dwarf. The accreted layer eventually reaches a temperature high enough for nuclear energy generation to commence. The sudden onset of nucleosynthesis releases an enormous amount of energy, heating the burning layer and, by a positive feedback, further increasing the nuclear reaction rate. The luminosity rises steeply in the outer layers of the white dwarf as a result of this added energy. However, there is initially no structural readjustment because the pressure, which comes from degeneracy, is insensitive to this rise in the temperature. The nuclear luminosity may reach or exceed the L_{Edd}, leading to the equivalent of a massive stellar wind that expands very rapidly away from the accreter. In addition, when the temperature rises above $\approx 10^9$ K, the envelope changes its equation of state and begins to act like an ideal gas, leading to an explosion. These two events, which are not necessarily independent, contribute to a net outward acceleration of the outer layers of the accreter. The characteristic time scale for this initial dynamical stage in the outburst is of order a few days.

Observational Diagnosis of a Nova Outburst

The rise to optical maximum signals that the shell is being ejected since initially radiation escapes longward of the Balmer discontinuity at 3647 Å due to the lower continuum opacity. The rapid expansion of the ejecta drops the temperature. If it falls below 12,000 K, strong absorption line blends of Fe II, Fe III, and other iron peak resonance and low excitation transitions appear in the spectrum, forming a sort of "iron curtain" of opacity and masking the hot dwarf. The peak of the radiation then shifts longward to about 2800 Å. For a classical nova, where the mass of the ejecta may be as large as 10^{-5} $M_\odot$, the ejecta remain optically thick and the nova light curve plateaus for up to several months. For lower-mass ejecta that are typical of recurrent novae, more like 10^{-7} $M_\odot$, this state is severely curtailed and may last only a matter of days if it occurs at all.

As the optical depth of the shell decreases, far ultraviolet radiation from the hot core can leak out. This radiation reionizes the ejecta and anything else in the environment, producing a strong and persistent emission line spectrum. The emission line strength grows as the ionization eats outward through the ejecta, eventually reaching the edge, and thereafter the emission line strength decreases due to the expansion of the ejecta and the decrease in the recombination rate, the physical process producing the emission. Depending on the amount of mass in the ejected shell, the emission line maximum and the onset of the optically thin phase of the expansion may be very slow. But eventually the shell becomes completely optically thin and completely ionized, and at this stage, the so-called nebular phase, it becomes possible to determine the abundances of the elements synthesized during the thermonuclear runaway and mixed into the ejecta in the first seconds to minutes of the outburst.

For a few novae, especially RS Oph, the effects of the environment are readily visible in the evolution of the ultraviolet (UV) line profiles (Figs. 10.10 and 10.11). The wind of the red giant companion is flash ionized by the UV pulses from the explosion, and the low-density wind first produces absorption lines seen against the broad emission lines of the optically thick ejecta and later narrow emission lines which persist as the ejecta fade from view. You will recall that we discussed the problems of precursors in the chapter on shocks. Here is an excellent example of what happens when the environment is dense.

10.3 A Sampling of Cosmic Flows

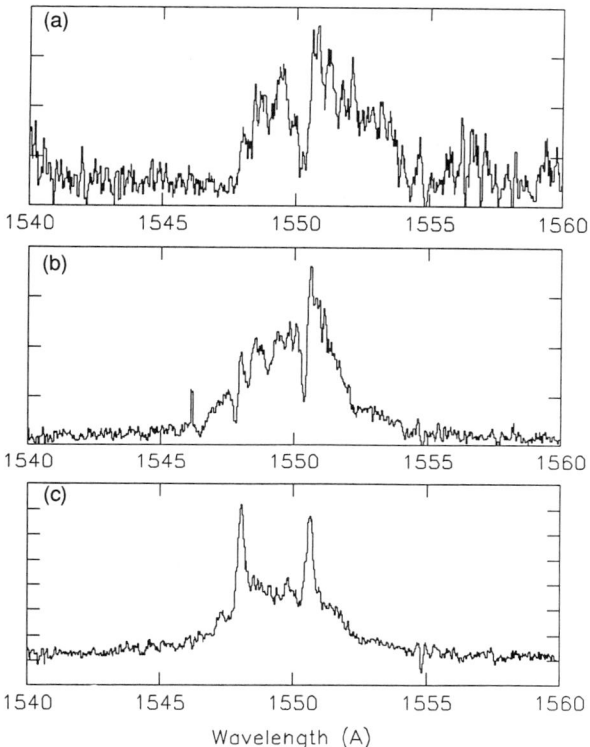

Figure 10.10 IUE high-resolution spectra showing the evolution of the emission line profile of C IV λ1550Å during the early stages of the recurrent nova RS Oph 1985. (a) Day 30, (b) day 45, (c) day 75. The narrow absorption lines are from the red giant wind absorbing against the expanding ejecta.

10.3.4 Novae and Supernovae: Velocity Gradient for Explosions

The fastest material in the outburst gets farthest first. Since both novae and supernovae occur as essentially instantaneous explosions—that is, the layer erupting is geometrically thin compared with the radius of the start—the imposed velocity field is very simple. There is a range of initial velocities going from the top to the bottom of the zone. Because the distance any blob reaches is dependent uniquely on its velocity, this can be turned around to give the velocity gradient. Therefore, the imposed velocity structure is $v(r) \sim r$, often referred to as a *Hubble flow*. The UV resonance lines are the strongest and evince this very fast-moving material.

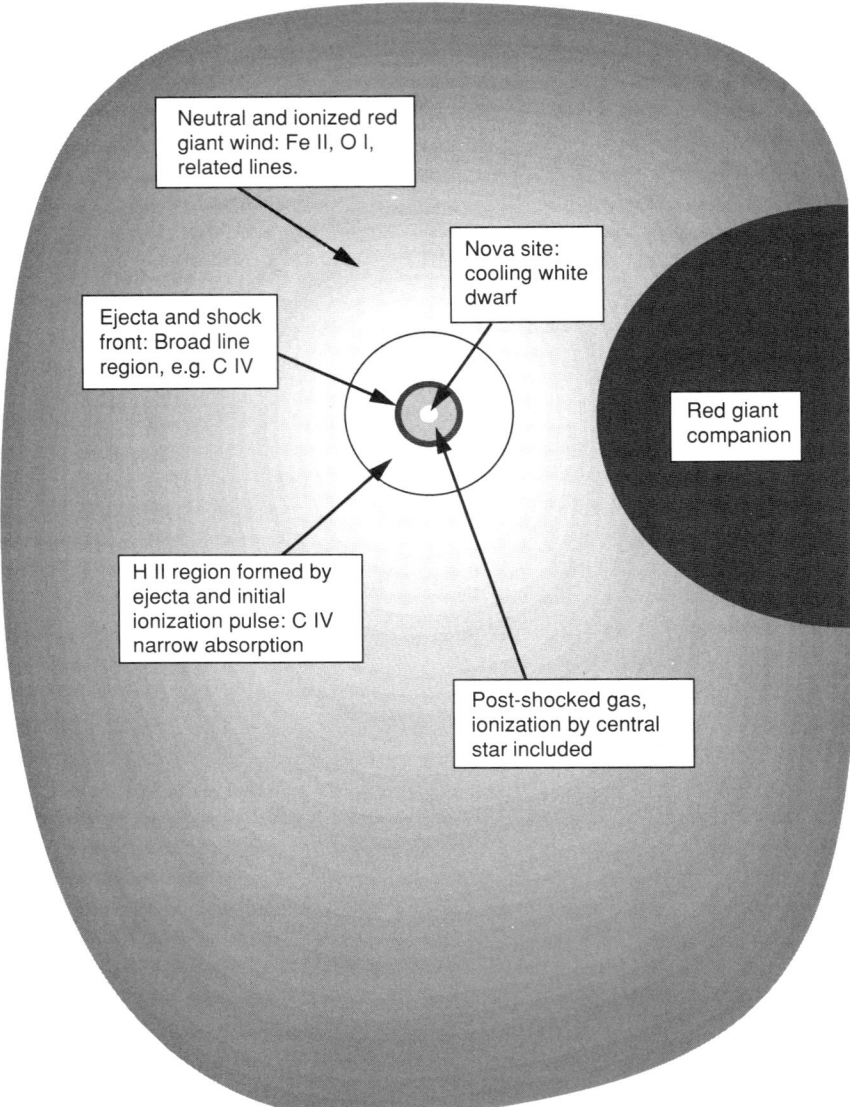

Figure 10.11 Example of the expansion of a nova shell within the environment of a companion's stellar wind.

10.3 A Sampling of Cosmic Flows

Excited state transitions, like $H\alpha$ or $H\beta$, sample only the deeper layers, which have lower expansion velocities.

As the ejecta expand, the P Cygni profiles narrow as the outer layers become optically thin and the fastest-moving matter ceases to contribute to the absorption trough on the line profile. This is most clearly seen in the C IV 1550 Å from LMC 1990 No. 1 resonance line profile. Deeper layers, more slowly moving and coming from lower depths of the processed zone of the white dwarf, can remain hot and optically thick for quite some time. The length of time required for the P Cyg profiles to disappear is also a measure of the mass of the ejecta. However, there are several ways to make the P Cyg profile disappear, or never form. The simplest is for the envelope to be ejected asymmetrically so that the effective covering factor for the core is small. In this case, the optical depth of the shell is always small and coronal lines appear rapidly. An alternative is for the shell to fragment. While observed in many optical spectra of novae in outburst, UV observations are thus far equivocal on this point, lacking the required spectral resolution. The disappearance of P Cyg profiles on the UV resonance lines signals the transition to the optically thin *nebular* phase.

In many ways, supernovae and novae resemble one another. The formation of the Fe forest and especially the formation of strong P Cyg profiles on resonance lines make both type I and type II supernovae extreme versions of the nova problem. The primary difference is in the complexity of the initiating shock and the fact that the imposed velocity field may depart substantially from the simple Hubble flow.

10.3.5 Accretion Disks

Accretion disks were first hypothesized for the nuclei of active galaxies (AGNs) partly on the basis of the observed emission line profiles. These were often extraordinarily broad, of order 10^4 km s^{-1}. Some AGNs show profile variations in which the wings vary differentially from line cores. The idea of the efficient release of gravitational potential in a circulatory accretion flow led to the interpretation of the profiles of almost all active objects as the product of disklike accretion. In fact, the essential picture is probably correct. In an optically thin disk, each annulus contributes throughout the line profile at constant density. The densest parts of the accretion disk being the innermost, near any possible boundary layer, the profile will be weighted toward large radial velocities and will have a characteristically "horned" appearance. On the other hand, if the disk is optically thick, the profiles are not as simple to interpret and depend strongly on the precise scale height as a function of position within the disk and also on the spectrum of the central body. An optically thick disk is essentially a collection of annular stellar atmospheres, each of which has a different "surface gravity" (determined from the distance of the annulus from the central body) and a local flux determined by the local rate of viscous dissipation and external illumination from the central accreter.

For most observational studies, however, the detailed line profiles have been unavailable or ambiguous. The evidence that there are multiple regions in the nucleus makes the line profiles often too complex to disentangle. In such cases, however, continuum emission remains the main tool for understanding the structure and energy balance of the accretion disk. As with outflows, we expect that there should be a characteristic spectral signature if we integrate over the entire surface of the disk. Let's examine this using the fact that the temperature varies as $F^{1/4}$, where F is the local flux generated by dissipation at a distance r from the accreter. Now we know that $T(r) \sim r^{-n}$ (we will come back shortly to fill n in) and we now assume that in an optically thick disk, most of the radiation comes out near the local peak of the local Planck function, $B_\nu(T)$. The consequence of this statement is that we know how to relate ν to T by $\nu_{\max}T =$ constant, the Wien displacement law. In this Rayleigh–Jeans limit, $B_\nu(T) = 2kT\nu^2 c^{-2}$. Now substitute the temperature in for frequency, and use the power law for the temperature to convert between T and r to get the luminosity as a function of frequency:

$$L_\nu = 2\pi \int_{R_{\text{inner}}}^{R_{\text{outer}}} B_\nu(T) r \, dr \sim \nu^{(3n-2)/n} \tag{35}$$

For the α-disk, we found that $n = \frac{3}{4}$, so that $L_\nu \sim \nu^{1/3}$ as our first estimate. Notice that the spectrum increases in intensity, even in this simple approximation, toward higher energy. Remember, the disk is unresolved so it looks like an extremely baroclinic photosphere. This happens for a reason similar to that for the optically thin line profile—the inner part of the disk is more heavily weighted than the periphery. You see that each part of the disk has a characteristic frequency that it is primary contributor to through the Wien law. This is why the formal power law is comparatively simple. But notice that this is in the absence of almost all of the interesting physics. There is no frequency dependence on the opacity (the gray case and completely optically thick continuum formation have been assumed). The complications, though, are left for you to ponder.

10.3.6 Absorption Line Probes: The Diffuse Interstellar Medium

For the observation of strong absorption lines in the interstellar medium (ISM), the best place to go is the ultraviolet. Here you can observe the resonance transitions of a wider range of ionization than in any other part of the spectrum. The difficulty with assessing the dynamical state

10.3 A Sampling of Cosmic Flows

of the gas comes, in part, from the relatively limited resolution available for most of the data. With CO and other molecular observations, the typical velocity resolution is of order 0.1 km s^{-1}; for most optical and UV observations, the resolution is rarely better than 1 km s^{-1}. This means that the spectra do not resolve marginally supersonic small-scale line structure. The detailed dynamics and abundances and optical depths for individual velocity components along any line of sight are probably underestimates. Nonetheless, to continue with the idea of flow visualization, absorption line studies are a very powerful tool for the analysis of the diffuse gas. When using resonance lines, the excitation can be ignored. In the ISM, the line profiles are formed by scattering (the collision frequency is simply too low for the lines to thermalize) and the optical depths for most lines appear small enough that damping is not important.

The spectral line detected along any direction is the integrated profile over the individual clouds that lie along the line of sight. Each cloud is assumed to have an intrinsically narrow Gaussian profile with an internal velocity dispersion $(\sigma^2 + \xi_{turb}^2)^{1/2}$ where ξ_{turb} is the mean turbulent velocity dispersion. This really does not necessarily mean turbulence as we have been discussing it; it may be from homogeneous fully developed dissipative turbulent motions. It may, alternatively, be simply a reflection of the complex scaling of velocity and length scales among interstellar clouds and the fact that there is still considerable unresolved structure in the line profile. Either way, the optical depth for a Gaussian is

$$\tau_\nu = \int_0^\infty P(s|l_{II}, b_{II}) \kappa_0 e^{-(\Delta\nu/\Delta\nu_D)^2} \rho(s)\, ds \approx \langle \tau_0 \rangle \exp{-\left(\frac{\Delta\nu}{\Delta\nu_D}\right)^2} \quad (36)$$

Here $P(s|l_{II}, b_{II})$ is the distribution function for clouds at a specific abundance along the line of sight at distance s, given the galactic longitude l_{II} and latitude b_{II}. The line equivalent width is then given by

$$W = \int_{-\infty}^\infty (1 - e^{-\tau_\nu})\, d\Delta\nu \quad (37)$$

Notice that this measure of the column density also depends on the distribution of the clouds along a line of sight and the turbulent velocity (whatever that means). Ultimately, if the velocity gradient is small and the turbulence is nearly sonic, the line strength ceases to grow linearly with increasing column density. For small mean optical depths, the equivalent width varies as $W \sim \langle \tau_0 \rangle$; for asymptotically black line cores, $W \sim \sqrt{\ln \langle \tau_0 \rangle}$. The resultant curve is called a *curve of growth*, the systematic variation in the core-to-wing ratio for an absorption line as a function of column density, N.

Although we have used a Gaussian profile for the interstellar medium, this is not quite correct. Recall the profile function $y(x) = \int_0^x \phi(x)\,dx$ that we employed in the ray approximation. Because the atomic states have finite lifetimes they have finite widths (by the uncertainty principle). Therefore, the profile has the intrinsic form of a Lorentzian:

$$\kappa_{\nu,\text{intrinsic}} = \mathcal{L} = \frac{\pi e^2}{m_e c} \frac{\Gamma/4\pi^2}{(\nu - \nu_0)^2 + (\Gamma/4\pi)^2} \tag{38}$$

Here $\Gamma = \Sigma_{l<i} A_{il} + \Sigma_{l<j} A_{jl} = \Gamma_{\text{lower}} + \Gamma_{\text{upper}}$ is the sum of the transition probabilities, A_{ji}, for the lower and upper states, in other words the total width of the lower and upper levels. Each atom sees a broader range of frequencies than implied by Γ because it is randomly subjected to turbulent and thermal motions. These produce a Gaussian profile even for a delta function intrinsic profile. The two profiles—$\mathcal{G}$, the Gaussian, and $\mathcal{L}$, the Lorentzian—are convolved together to produce the final line profile $\kappa_\nu \sim \mathcal{G} * \mathcal{L}$ so that

$$\kappa_\nu = \frac{\pi^{1/2} e^2}{m_e c} f \frac{1}{\Delta \nu_D} H(a, v) \tag{39}$$

where the function $H(a, v)$ is called the Voigt profile, defined by

$$H(a, v) = \frac{a}{\pi} \int_{-\infty}^{\infty} \frac{e^{-y^2}\,dy}{(v-y)^2 + a^2} \tag{40}$$

where $a = \Gamma/(4\pi \Delta \nu_D)$ is the damping parameter and $v = \Delta \nu / \Delta \nu_D$, $\Delta \nu_D$ again representing the Doppler width. As a increases, so does the width of the profile, and for large enough column densities these wings are actually observable. (See Fig. 10.12.) They produce departures at the large column density limit from the saturated portion of the curve of growth, causing the equivalent width to grow as $W \sim N^{1/2}$ instead of the saturated result of $W \sim \sqrt{\ln N}$ obtained for a Gaussian profile.[3] Some examples of both satu-

---

[3] The most obvious manifestation of "damped" profiles is observed for Lyα in the interstellar medium, but other very strong resonance lines also show this effect. The most important observation of damped profiles is the Lyα forest, the numerous narrow absorption lines observed in the spectra of many quasars. These lines are formed by intervening galactic disks and halos along the line of sight to the active galaxy, observed at lower redshift relative to the quasar. They are analogous to the diffuse interstellar medium in that they are neither mechanically nor thermally connected with the illuminating source and the same analytical methods can be employed in studying their dynamics and abundances as we have just discussed for galactic diffuse clouds. It must be borne in mind, however, that because of the length of the sight lines through external galactic disks, we will not see the internal dynamics of individual clouds but rather the broadening resulting from the ensemble dynamics. In this sense, we gain insight into the motions within the disk or halo of the intervening galaxy, but lack the detailed dynamical information that we can obtain for individual clouds within our galaxy.

10.3 A Sampling of Cosmic Flows

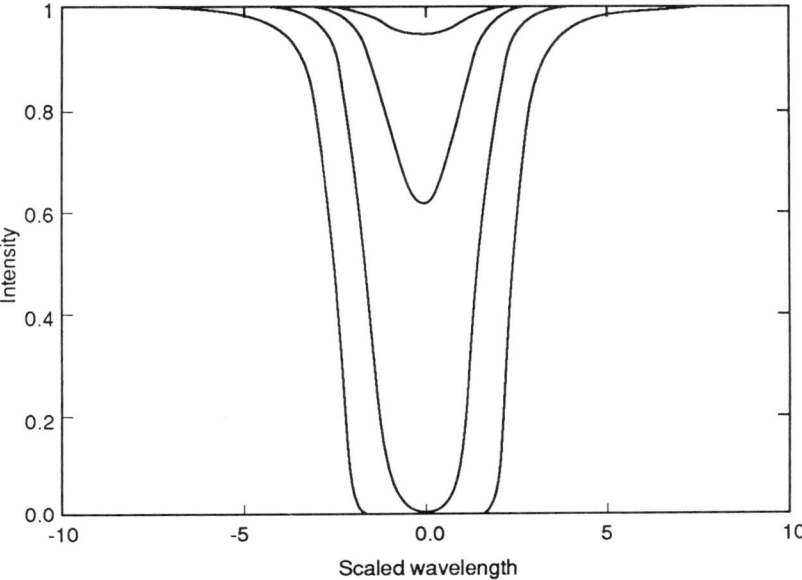

Figure 10.12 Example of line profile changes for increasing column density for a profile with $a = 0.3$. Note the formation of wings on the saturated profile. These profiles are the basis for the curve of growth.

rated and unsaturated profiles are shown in Figure 10.13. These spectra were obtained toward ξ Per, an O star in the Perseus spiral arm (about 1.5 kpc from the sun).

10.3.7 Eclipsing Binaries: Astrophysical Schlieren

Eclipsing binaries offer the best opportunity to determine the velocity structure in stellar atmospheres. The principle works much like that in the interstellar medium problem. A cool star occults a hotter companion as seen from Earth. The energy density in the atmosphere of the cool star is high enough that the radiation from the companion does not affect the state of the atmosphere through which it is viewed. As the eclipse proceeds toward maximum, the companion sweeps out a pencil along the line of sight (see Fig. 10.14a,b). For weak lines, large profile variations may be observable depending on the density contrast in the probed atmosphere and the time scale for motions. If the material is in a net outflow, the broadening of the absorption line formed against the companion will vary as the eclipse proceeds. If due to turbulence, individual convective cells

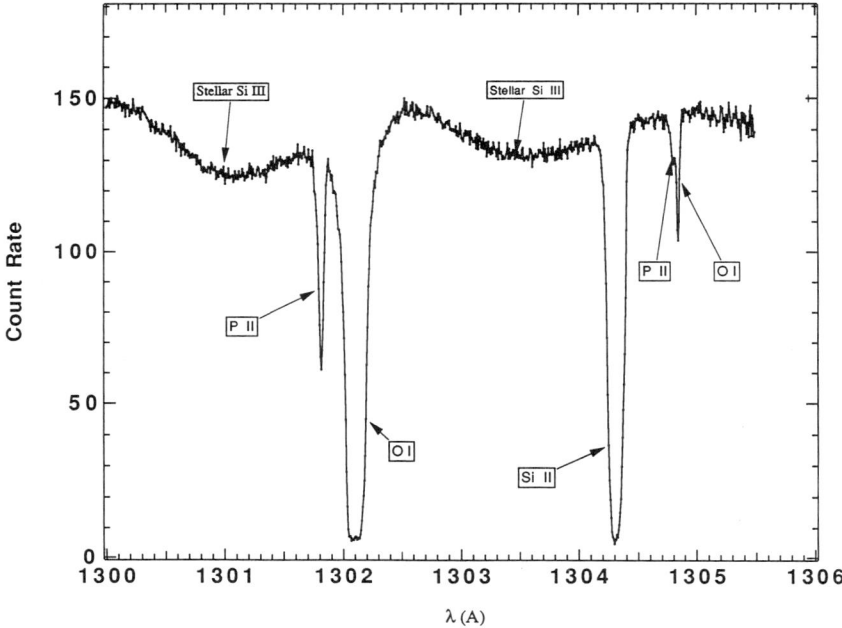

Figure 10.13 GHRS spectrum of the λ1300Å region of the galactic O star ξ Persei, showing the interstellar O I and P II profiles superimposed on the broad photospheric Si III lines. The resolution in this spectrum is approximately 90000 (about 3 km s^{-1}). Notice that the O I λ1302Å resonance line shows several resolved clouds. The base of the profile is due to background that has not been removed to make the cores of the strongest lines more visible.

may absorb temporarily and produce profile variations. The power spectrum (temporal) of these variations is measured by the longitudinal correlation function of the radial velocities.[4]

The picture we have just outlined is observed in several classes of binaries. Most notable for displaying these phenomena are the ζ Aur stars, a group of cool (G and K) giants with hot (B or A) main sequence companions. These stars often have very long orbital periods, of order years or even decades, and can be studied in eclipse for extended periods of time. The most dramatic example is ε Aur, which has a period of about 27 years and whose eclipse lasts for over 1 year. The limitations of the technique point beautifully to the difference between astronomical and laboratory studies of flows by passive probes. The orbital plane must be

---

[4] In effect, we are making use here of the Taylor hypothesis that the spatial and temporal ensemble averages are the same.

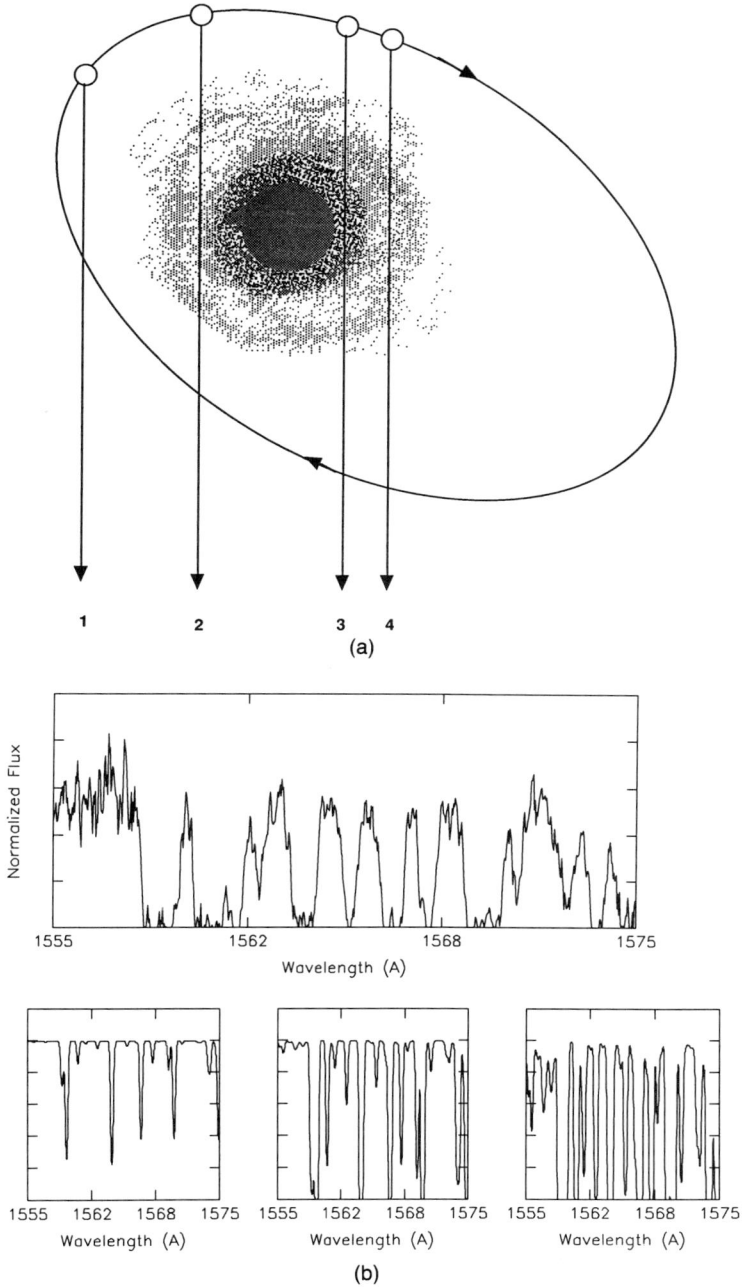

Figure 10.14 Formation of absorption spectrum for a ζ Aurigae–type binary. (a) Different lines of sight (1-4) show different absorption through the atmosphere. (b) Resulting spectrum for the $\lambda 1560$Å region for increasing column densities (10^{19}, 10^{20}, 10^{21} cm^{-2}) compared with an eclipse spectrum of the B star 22 Vul (IUE high-resolution spectrum) (IUE spectrum courtesy of T. B. Ake III.)

suitably oriented relative to the line of sight and the optical depth of the companion's atmosphere must change sufficiently slowly so that it does not appear like a wall. The eclipse period must be long, the period must be well known, and the spectrum of the two stars separately must be well known. Because of these severe limitations, only a few systems have been well studied but the increasing use of high-resolution UV and optical spectrographs promises to greatly extend this method to fainter, shorter-period systems.[5]

10.3.8 Bisector Methods for Absorption Lines: Astrophysical Tomography

Again, we can apply the interstellar analogy to a resolved stellar surface. Of course, at the moment we know of only one that is well resolved.[6] But in the coming decades we should have a larger sample to study as new ground-based and space imagers come on line. The bisector method is ideally suited to the problem of probing the depth dependence of turbulent motions for the solar atmosphere. In essence, the procedure works as follows. (See Fig. 10.15.) Imagine that we have a weak line that is both broadened by turbulence and shifted by velocity fields. If we assume that the line is intrinsically weak, then the depth in the line is proportional to the local optical depth in the atmosphere; this is connected to a physical depth through the intrinsic absorption coefficient. Now determine the centroid velocity as a function of intensity. For a specific portion of the solar surface, this determines whether on average the matter is moving out or going in. If you take a map of the centroid at fixed depth over some surface region, you will have a map of the relative upward and downward velocities as a function of depth. This creates a three-dimensional picture of the velocity structure of the atmosphere. We are doing something like this when, using different molecular probes, we determine the velocity correlation functions as a function of density in galactic clouds. The solar atmosphere is, however, a time-variable medium and we have a reference scale for physical depth in the atmosphere.

---

[5] FF Aqr and V471 Tau, both main sequence K stars with white dwarf companions, have been observed in this way. The eclipses of these systems have extremely rapid onset, of order minutes to hours, but some structural details have been gleaned for the atmospheres of the eclipsing star. The problem, however, is that no dynamical information has yet been extracted from these systems because of the inherent slowness of the available detectors (Shore 1988, Guinan 1990).

[6] The sun.

10.3 A Sampling of Cosmic Flows

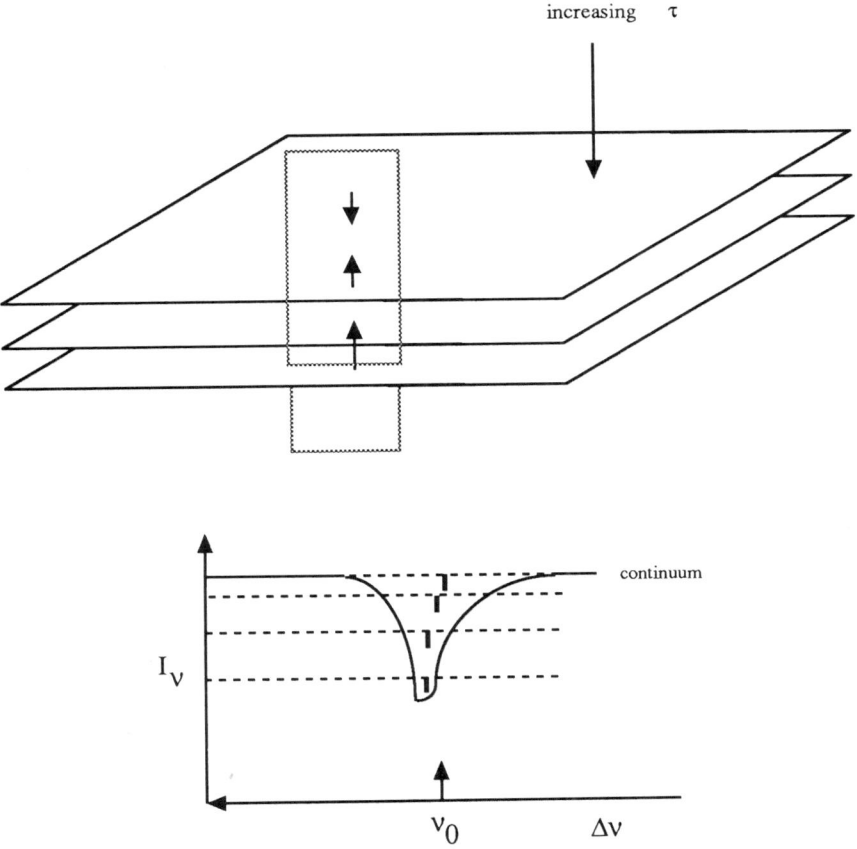

Figure 10.15 The bisector method for the analysis of the solar atmosphere. Velocity in individual layers is shown with the subsequent mapping of the motion into the line profile. The observer sees a slice through the atmosphere and different portions of the line profile, formed at different optical depths, reflect the mean motion in that layer.

10.3.9 Extragalactic Radio Jets

Perhaps the most dramatic objects observed in the universe are the large-scale emission regions surrounding active radio galaxies. These structures, discovered at centimeter wavelengths, are often observed extending to distances of hundreds of kiloparsecs from the centers of their parent galaxies. Dramatic examples have been published by many groups, especially using the Very Large Array (VLA) radio interferometer of the National Radio Astronomy Observatory. Fine structure, on subarcsecond scales, has been observed in many of these sources. Unfortunately, it should be kept in mind that even at the limit of current imaging, using very long baseline interferometry to resolve microarcsecond-scale regions, the structures are still unresolved at the level of many parsecs for most sources. A few have been imaged to the physical scale of less than 1 parsec, and the galactic center has been studied on a scale of about 1 astronomical unit. However, each increase in spatial resolution reveals yet more detailed structures, so the ultimate course for the emission has yet to be determined.

The reason for passing quickly over this class of objects is that we cannot apply the same methods that have been discussed in this chapter for other objects. The time scales and length scales are enormous, and in general we just don't see things move. A few dramatic examples of motion have been demonstrated, perhaps the best one the discovery of subluminal motion in the jet of M87, the central galaxy in the Virgo cluster and one of the closest and best-studied radio galaxies. But the observations are all made in the continuum. The radio jets and lobes appear to have precious little thermal material within them. Their emission in the radio is entirely nonthermal, due to synchrotron radiation by relativistic particles in weak magnetic fields within the lobes. Any changes in morphology may be due to instabilities or excitation waves or some other mechanism that changes structure and intensity without transporting mass. In addition, one of the long-standing problems with these sources is that we observe only the electrons. The bulk of the matter, protons and neutrals, we do not see. So the many heroic attempts to calculate the structure and evolution of these jets and lobes have made many varied and extensive assumptions about the energetics and properties of the transported matter.

10.3.10 Cooling Flows: Diagnosis of a Thermal Instability

One place where the problem of flow diagnosis is especially acute is the observation of large-scale x-ray emission from clusters of galaxies. The argument proceeds like this. The gas is observed to be emitting x-rays, with typical temperatures greater than 10^7 K, at a rate that is very high. Taking $E_{\text{thermal}}/L_{\text{XR}}$ as an estimate of the cooling time for this gas, observations show that it is shorter than H_0^{-1}, the Hubble time given as the inverse of the Hubble constant. This gas has an internal velocity dispersion σ that supports it as a pressure against the cluster gravitational field. If the temperature is dropping steadily, there should be a pressure deficit in the interior of the intracluster medium and the matter should be accreting onto the central galaxy. We have already discussed this class of flow problem, so here let's only comment on the signature of the process.

The assumption that the flow exists comes from the assumption that the cooling is not efficiently counterbalanced by some heating term. It is also assumed that the gas is primordial, having been heated at the time of the cluster's formation, and thence cooling. Observations show that strong metal lines are detected in the x-ray spectra, especially from iron, indicating that the gas has been circulated through stars at least at some time. Therefore it is likely that the assumption of no heating is incorrect, and the assumption that the material is only accreting onto the galaxies and not being recycled is also problematic. One would like to be able to detect some other signature of the flow, some velocity gradient in the gas. This is very hard to do because the galaxies mess things up. Their motion through the gas disturbs it, so it is like watching the flow in a canal constantly being stirred up by barges. We see the evidence for a further thermal instability in action because cool gas, $H\alpha$ filaments, oftentimes accompanies the x-ray–emitting material. But there is still a great deal of work to be done on this problem.

References

Bernes, C. (1979). A Monte Carlo approach to non-LTE radiative transfer problems. *Astron. Astrophys.* **73**, 67.
Bode, M. F., and Evans, A., eds. (1989). *The Classical Novae.* New York: Wiley.
Castor, J. (1970). Spectral line formation in Wolf–Rayet envelopes. *Mon. Not. R. Astron. Soc.* **149**, 111.
Castor, J., and Lamers, H. (1979). An atlas of theoretical P Cygni profiles. *Astrophys. J.* **39**, 481.
Castor, J. I., Latz, J. H., Seaton, M. J. (1981). Ultraviolet spectra of planetary nebulae— III. Mass loss from the central star of UGC 6543. *Mon. Not. R. Astron. Soc.* **194**, 547.
Cowie, L., and Songaila, A. (1986). High resolution optical and ultraviolet absorption line studies of interstellar gas. *Annu. Rev. Astron. Astrophys.* **24**, 499.
Elitzer, M., and Netzer, H. (1985). Line fluorescence in astrophysics. *Astrophys. J.* **291**, 464.
Fransson, C. (1984). Line profiles from supernovae. *Astron. Astrophys.* **132**, 115.
Garmany, C. D., ed. (1990). *Properties of Hot Luminous Stars.* San Francisco: Astron. Soc. Pacific.
Guinan, E. F. (1990). What we learn from eclipsing binaries in the ultraviolet. In *Evolution in Astrophysics* (p. 73). E. Rolfs, ed. ESA SP-310.
Hollenbach, D. J., and Thronson, H. A., eds. (1987). *Interstellar Processes.* Dordrecht, The Netherlands: Reidel.
Hughes, P. A., ed. (1990). *Beams and Jets in Astrophysics.* Cambridge: Cambridge University Press.
Kalkofen, W., ed. (1987). *Numerical Radiative Transfer.* Cambridge: Cambridge University Press.

Kondo, Y., ed. (1989). *Exploring the Universe with the IUE Satellite.* Dordrecht, The Netherlands: Kluwer.

Kudritzki, R. P., Yorke, H. W., and Frisch, H. (1988). *Radiation in Moving Gaseous Media: Lectures from the 18th Advanced Course Saas-Fee.* Y. Chmielewski and T. Lanz, eds. Geneva: Geneva Observatory.

Kwan, J. (1978). Radiative transport and kinematics of molecular clouds. *Astrophys. J.* **223**, 147.

Lada, C. (1985). Cold outflows, energetic winds, and enigmatic jets around young stellar objects. *Annu. Rev. Astron. Astrophys.* **23**, 267.

Lamers, H., Cerruti-Sota, M., and Perinotto, M. (1987). The "SEI" method for accurate and efficient calculation of line profiles in spherically symmetric stellar winds. *Astrophys. J.* **314**, 726.

Lucy, L. B. (1971). The formation of resonance lines in extended and expanding atmospheres. *Astrophys. J.* **163**, 95.

Mihalas, D. (1978). *Stellar Atmospheres.* San Francisco: W. H. Freeman.

Mihalas, D., and Winkler, K.-H. A., eds. (1986). *Radiation Hydrodynamics in Stars and Compact Objects.* Berlin: Springer-Verlag.

Osterbrock, D. (1989). *Astrophysics of Gaseous Nebulae and Active Galactic Nuclei.* Sacramento: University Science Books.

Payne-Gaposchkin, C. (1957). *The Galactic Novae.* New York: Dover.

Shore, S. N. (1988). IUE observations and physical processes in close binary systems. In *A Decade of UV Astronomy with the IUE Satellite* (Vol. 1, p. 67). E. Rolfs, ed. ESA SP-281.

Sobolev, V. (1960). *Moving Envelopes of Stars.* Cambridge, Mass.: Harvard University Press.

Starrfield, S. (1989). Thermonuclear runaways and classical nova outburst. In *The Classical Novae.* M. Bode and A. Evans, eds. New York: Wiley.

Verschuur, G. L., and Kellermann, K. I, eds. (1988). *Galactic and Extragalactic Radio Astronomy.* Berlin: Springer-Verlag. [See especially Astronomical Masers (Reid and Moran), Interstellar Molecules and Astrochemistry (Turner and Ziurys), and Radio Stars (Hjellming).]

Winkler, K.-H. A., and Norman, M. L., eds. (1986). *Astrophysical Radiation Hydrodynamics.* Dordrecht, The Netherlands: Reidel.

Wright, A. E., and Barlow, M. J. (1975). The radio and infrared spectrum of early-type stars undergoing mass loss. *Mon. Not. R. Astron. Soc.* **170**, 41.

Problems and Questions for Further Exploration

> *You can't make an omelette without breaking eggs.*
> Proverb

    1. Estimate the plasma frequency for the following environments: the diffuse interstellar medium, the interplanetary medium near the Earth, and an H II region. How does the plasma frequency relate to scintillation of unresolved extragalactic radio sources?

    2. Calculate the mean atomic collision time in the room you are currently sitting in and compare it to an H II region. How do these compare with typical atomic and molecular transition lifetimes?

    3. Assume a gas of sufficient density that the uncertainty principle governs the pressure. In other words, the phase space is restricted by $\Delta p \, \Delta x = \hbar$, where p here is the momentum of a particle. The condition for a degenerate gas is that its phase space volume permitted by the uncertainty principle exceeds that provided by thermal motion. Show that the equation of state for this gas is $P \sim \rho^{5/3}$ and that there is a finite temperature above which the gas becomes nondegenerate.

    4. The nonlocality of viscous coupling is the origin of the term $\nu \nabla^2$ in the Navier–Stokes equation. Show that this term can also be derived from the assumption that fluid at a position $\mathbf{x}$ in a flow couples to both $\mathbf{x} + \delta \mathbf{x}$ and $\mathbf{x} - \delta \mathbf{x}$.

    5. Show that the equations of motion for an incompressible medium dominated by viscosity (low Reynolds number limit) have only harmonic solutions (see also Purcell, E. (1977). Life at low Reynolds number. *Amer. J. Phys.* **45**, 3.). What are the implications of this for wake formation as the Reynolds number is increased?

6. Suppose a shearing flow has a velocity gradient $dv_\phi(r)/dr$, where v_ϕ is the circulation velocity, and that $v_\phi \sim r^n$ for some $n > \frac{1}{2}$. Find some scaling for the velocity law as a function of n_1 assuming that the Reynolds number is always less than or equal to the critical value Re_c.

7. Derive the velocity profile for viscous flow in a pipe, $v_z(r)$, with a circular cross section and constant ν. Assume Re is low enough that the inertial terms are negligible. This may seem more like engineering, but it is a good way to get a feeling for what a boundary layer means in an astrophysical flow, like a bipolar outflow (see Chapter 10).

8. Show why in a teacup, if the walls rotate faster than the fluid, the Ekman pumping works in the opposite direction to that shown in Figure 3.7.

9. In two-dimensional planar or axisymmetric flow simulations, shear flows often generate turbulence and vorticity. How does the two-dimensional case differ from the full 3-D simulation?

10. Perform the vortex stretching experiment with cream in a coffee cup. Why does the boundary of the cream column begin to oscillate?

11. Derive the functional form of the ratio of the epicyclic frequency to the rotation frequency for a rotation curve with $\Omega = \Omega_0 (r/r_0)^{-n}$. Here Ω_0 and r_0 are constants. Where are the resonances in this rotation curve (those points where κ is a rational multiple of Ω)?

12. Once on an interstate, I noticed that the vertical straps holding down a load on a flatbed truck were flapping at high frequency. These straps were flat-sided and stood some distance out from the load. How is this flapping behavior related to circulation and the von Kármán vortex sheet?

13. Assume that a 1 $M_\odot$ main sequence star has a rotation period of order 6 days. Calculate the Eddington–Sweet circulation timescale and compare it with the evolutionary time for the star.

14. Derive Eq. (4.35) and generalize this to the incident Mach number M_1, using the Rankine–Hugoniot conditions.

15. Plot the polar diagram for the Mach numbers M_2 versus M_1 for an oblique shock. Find the critical value of ϕ for which $M_2 < 1$.

16. Diagram the characteristics for a rarefaction fan. Why does a rarefaction discontinuity propagate at the sound speed?

17. You observe a star behind a shock in the intersellar medium. The shock is traveling at -100 km s^{-1} (that is, toward you) and you see its absorption lines doubled. You assume that the low-velocity material is due to the local interstellar gas in the line of sight. What is the true velocity of the gas and what is its density compression? A typical resonance line has an oscillator strength f of about 1. What is the column density required to see a shock with a width of only 30 km s^{-1} and an equivalent width of about 30 mÅ?

18. Calculate the post-shock temperatures for an adiabatic shock with a velocity of 10 km s^{-1}. Assuming the density of the preshocked gas to be about 10^4 cm^{-3}, calculate the collision frequency for neutral molecular hydrogen in the post-shocked gas. Why might infrared vibrational emission from excitation of H_2 be a signature of such shocks in molecular environments like star-forming regions? You will need to check some references on spectroscopy for the appropriate molecular constants; Herzberg, G. (1945) *Spectra of Diatomic Molecules* (Princeton: Van Nostrand) is an excellent source.

19. Consider two steady-state spherical winds at terminal velocity coming from sources separated by a distance a. The contact surface between these winds forms a surface $\Sigma(x,y)$, where x is along the line of centers and y is perpendicular. Find the differential equation that describes this surface as a function of the ratio of mass loss rates, $\lambda = \dot{M}_2/\dot{M}_1$. Use the condition that there is no net mass flux across Σ, and that the surface is stationary (so that the pressures are balanced across Σ).

20. Compute the profile for an adiabatic shock from Eqs. (5.16–5.18). Assume that the jump conditions are represented by a gas with $\gamma = \frac{5}{3}$ in the strong shock limit. [*Remark*: A Runge–Kutta integration best suits this problem. See Hamming, R. W. (1983) *Numerical Methods for Scientists and Engineers* (New York: Dover Books)].

21. What is the effect of a density gradient on the evolution of a spherical shock? Along these lines, describe qualitatively the effects on its environment of a supernova explosion *inside* a molecular cloud.

22. Calculate the Alfvén Mach number for a diffuse interstellar cloud with $T = 100$ K, $n \approx 10^2$ cm^{-3}, and $B_0 \approx 10$ μG. Compare this with a molecular cloud using a scaling of $B_0 \sim \rho^{2/3}$ with a density of 10^4 cm^{-3} and $T \approx 10$ K.

23. In the Parker instability, the interstellar gas is supported by the galactic magnetic field against collapse toward the midplane. (See also Fig. 6.1.) Describe how this is connected with the Rayleigh–Taylor instability and analyze it in terms of two fluids, one represented by a "relativistic" equation of state (the magnetic field) and the other being the diffuse gas. What is the final configuration of the field–gas system?

24. In the Fermi acceleration mechanism for the origin of cosmic rays, electrons are scattered from magnetic mirrors to accelerate to high energy. What is the role in this mechanism of collisionless magnetic shocks?

25. Combining what you know about vortex shedding and turbulence being pumped by injection of enstrophy, describe why leafy trees might be an important source of turbulence near observatories.

26. Consider the following model. Turbulence in an accretion disk appears at a specific value of the Reynolds number Re$_c$. This provides a relation between a scale length, velocity, and the turbulent viscosity v_t.

Assume that the pressure scale height and/or the shear length provide the length scale and that the sound speed and/or the mean circulation velocity scale the speed. Construct an α-type disk scenario using these (this combines material from Chapters 7 and 8).

27. Derive the Skumanich relation for the spindown of a magnetized star, $\Omega \sim t^{-1/2}$. In particular, assume that the magnetic field depends on the rotation frequency of the star through $B_{\text{surface}} \sim \Omega^n$, leaving n arbitrary. The magnetic field varies as r^{-2}. It constrains the wind out to the Alfvén point, which is the moment arm for angular momentum loss. Take the mass loss rate to be $\dot{M}$ and the terminal velocity to be v_∞. What is the result when you assume that B_0 scales linearly with the astrophysicists's Rossby number?

28. For a plane parallel radiatively driven wind, show that the critical point occurs at $g_{\text{eff}} = 0$. What is the mass loss rate from such a star in terms of the density (assuming an isothermal atmosphere)? Why does the mass loss rate increase with increasing luminosity? (*Hint*: how does the depth at which the effective gravity vanishes depend on the luminosity?) (Lucy and Solomon 1970).

29. Instead of just relying on radiation pressure, assume that there is an additional source of pressure and heating that extends spatially over a considerable radius. For instance, turbulence, Alfvén waves, or sound waves have all been posited for different wind models. Why does this extended momentum input permit a wind that has a very low terminal velocity? You should consider not just the formalism but also the meaning of a slow, nearly breeze solution that gradually crosses the sound speed with distance from the source. [See, e.g., Pijpers, F. P. (1991) *The Dynamics of the Winds of Cool Giants and Supergiants*, Amsterdam: Sternwacht (PhD thesis).]

30. Assume that a blob, perhaps a standing shock, is advected out in a stellar wind with the velocity profile given by $v(r) = v_\infty(1 - R_\star/r)^\beta$, where v_∞ is the terminal velocity, $R_\star$ is the stellar radius, and β is a free parameter (usually of order unity) for the wind law. Calculate the velocity versus acceleration law for this blob as a function of β and show what its motion is through the line profile as a function of time (this is related to the observation of the narrow discrete absorption components observed on UV resonance line profiles in hot luminous stars).

31. Find the specific angular momentum at the L_1 point. Use the approximation for the Roche radius in terms of q and a and assume that the center of the star is at a distance $M_2 a/M$ from the center of mass.

32. What is the effect of including radiation pressure in the calculation of the Roche surface? How does this affect the mass transfer in a close binary system?

33. For the Shakura–Sunyaev α-disk, calculate the dependence of

temperature on radius of two different opacity sources, a Kramers opacity with $\kappa \sim \rho T^{-7/2}$ and a Thomson opacity, $\kappa = constant$, assuming an equation of state for an ideal gas (see, e.g., Shakura and Sunyaev 1973).

34. Calculate the accretion rate for a strongly magnetized star imbedded in a wind on the analogy to the gravitational capture cross section. Remember that the Alfvén radius is important for this problem.

35. For a temperature of 20 K and number density of 10^5 cm^{-3}, typical of a molecular cloud, calculate the Jeans length and the mass of a typical fragment.

36. Assume that opacity stops the collapse of a Jeans fragment. What is the limiting Jeans length, hence mass, for an arbitrary opacity source κ_{rad}?

37. How is the mixing length related to the point of neutral buoyancy for a rising convective blob?

38. Show that the magnetic field replacing the density increases the buoyancy of a blob and calculate the rise time, within mixing length theory, for a magnetic blob in a convectively stable atmosphere.

39. Assume that a gas remains optically thin as a thermal instability develops and calculate the density at which the collapse time from self-gravitation equals the cooling rate.

40. Assume that a Rayleigh–Taylor instability starts in a medium with large density contrast $\Delta\rho \sim \rho$. Find the limiting scale length for the instability, assuming that the growth in the perturbation is limited only by the expansion of the medium (and that the length scale is changing due to $R(t)$, the rate of expansion of the layer).

41. How is the von Kármán vortex sheet related to the Kelvin–Helmholtz instability?

42. Calculate the accretion-generated shear instability for a magnetized neutron star, with a surface field of 10^{12} G, imbedded in an accretion disk. Show that the Alfvén radius causes the shear to be very large at the Alfvén point, generating a strong Kelvin–Helmholtz instability.

43. The terminal velocity on a normal P Cyg profile is especially sharply defined (see Fig. 10.2, for example) and is the blackest part of the line profile. Why? Assume a velocity law of the form $v(r) = v_0 + v_\infty (1 - R_\star/r)^\beta$, where v_∞ is the terminal velocity, v_0 is an initial velocity for the wind, of order the sound speed, $R_\star$ is the stellar radius, and β is a free parameter (usually of order unity) for the wind law. Describe the formation of a line profile in this flow. Plot the isotachs for this flow and look at the nonlocalness of the radiative transfer as you approach the outer parts of the wind.

44. Derive Eq. (10.34). What happens if the velocity law is assumed to vary as $v(r) \sim r^n$, where n is either positive or negative?

45. Calculate the recombination time for a medium ionized by a

strong UV pulse (assuming it is atomic hydrogen only) if the density is about 1000 cm^{-3}. What happens if the central ionizing source is turned on and off on a timescale longer than, or shorter than, the recombination time? What does the finite size of the medium have to do with the observed spectrum variations? (This is especially important in sorting out dynamic from radiative effects in active galactic nuclei.)

46. For a Gaussian profile, compute the depth of line formation as a function of $\Delta\lambda$, the wavelength relative to line center. Now consider the idea behind the bisector method. Assume a velocity field with depth and see how different parts of the line profile map to different depths in the atmosphere.

47. Compute the curve of growth for a set of Voigt profiles with $a = 0.1, 0.3$, and 1.0. How does the damping part of the curve vary as a function of the a parameter at constant column density? How does this relate to the turbulent velocity of a cloud along the line of sight, and how would you be able to determine the b value in practice?

48. This problem supplements the text on similarity solutions and also on shocks. An H II region in the late stages of expansion evolves with constant ionized mass. This is the original condition resulting from the Strömgren sphere mechanism. Assume that the recombination coefficient is α and that the ultraviolet luminosity of the central star is L_{UV}. Show that the similarity transformation for the H II region expansion is $\rho_2^2 R^3(t) =$ constant, where ρ_2 is the ionized gas density and $R(t)$ is the radius of the H II region. Find the constant. The sphere expands like a "stalled" shock, that is, $\rho_1 V_s^2 = \rho_2(v_2^2 + a_{s,2}^2)$, where V_s is the shock velocity, v_2 is the post-ionization gas velocity, ρ_1 is the neutral gas density, and $a_{s,2}^2$ is the sound speed in the H II region. Since $V_s = \dot{R}$, show that the evolution of the ionized region is given by $R(t) = R_0 \left(1 + \dfrac{7 a_{s,2} t}{4 R_0}\right)^{4/7}$, where R_0 is the initial Strömgren radius. Thus the asymptotic evolution of an H II region is given by a similarity law of the form $R(t) \sim t^{4/7}$. State the assumptions you have needed to make to derive the post-shock speed.

Appendix: Some Real Numbers

> *Now the Universe itself keeps on expanding and expanding,*
> *In every direction it can whiz.*
> *As fast as it can go,*
> *The speed of light, you know,*
> *Ten million miles a minute,*
> *That's the fastest speed there is.*
> Monty Python, *The Galaxy Song*

A.1 Physical and Astrophysical Constants

Astrophysicists, as a group, have not yet learned about the SI system, so the constants here will be given in cgs units. They are intended for use in order of magnitude calculations and are not the most accurate ones available in the literature. One of the best quick references for more precise values of constants and physical magnitudes is still Allen, C. W., 1976, *Astrophysical Quantities* (London: Athlone Press).

Constant	Symbol	Magnitude
Astrophysical constants		
Earth mass	$M_\oplus$	5.98×10^{27} g
Earth radius	$R_\oplus$	6.378×10^{9} cm
Solar mass	$M_\odot$	1.989×10^{33} g
Solar radius	$R_\odot$	6.596×10^{10} cm
Solar luminosity	$L_\odot$	3.826×10^{33} erg s^{-1}
Astronomical unit	AU	1.496×10^{13} cm

(*continued*)

Constant	Symbol	Magnitude
Parsec	pc	3.086×10^{18} cm
Galactocentric distance	R_0	8.5 kpc
Physical constants		
Gravitational constant	G	6.670×10^{-8} dyn cm^{2}g^{-2}
Stefan–Boltzmann constant	σ	5.670×10^{-5} erg cm^{-2} K^{-4} s^{-1}
Boltzmann constant	k	1.381×10^{-16} erg K^{-1}
Gas constant	R	8.314×10^{7} erg K^{-1} mol^{-1}
Avogadro number	N_A	6.0221×10^{23} mol^{-1}
Planck constant	h	6.626×10^{-27} erg s
Electron mass	m_e	9.110×10^{-28} g
...	$m_e c^2$	511 keV
Electron charge	e	4.803×10^{-10} esu
Charge-to-mass ratio (electron)	—	5.273×10^{17} esu g^{-1}
Prôton/electron mass ratio	m_p/m_e	1836.1
Electron volt	1 eV	1.602×10^{-12} erg
Temperature in eV	—	11604.8 K eV^{-1}
Electron speed at 1 eV	—	5.931×10^{7} cm s^{-1}
Speed of light	c	2.997×10^{10} cm s^{-1}
Bohr radius	a_0	5.292×10^{-9} cm
Thomson cross section	σ_e	6.652×10^{-25} cm^2

A.2 Astrophysically Useful Orders of Magnitude

Many of the parameters you may need to calculate can be derived from scaling arguments. Many of those will be quite familiar, but it can be useful training to have them available anyway. Remember, most of the time we are scaling results from laboratory to astrophysical environments, so we need to be able to keep the relative variables in mind.

A.2.1 Atomic and Molecular

Characteristic timescales for permitted transitions are 10^{-7} to 10^{-9} sec, so the typical timescale for the complete establishment of LTE is of this order. The radiative timescale should be comparable with the collisional timescale for the medium to be optically thick. Forbidden transitions have typical timescales of 10^{-4} to 10^{14} sec, the latter being about the decay time for the 21-cm transition of H1. Typical energies for fine structure transitions is of order 0.01 to 0.1 eV, while for permitted transitions from the ground state the energies are about 5 to 100 eV. Typical ionization energies are from 5 to 20 eV, depending on the ion, for most species observed in absorption in stellar photospheres. Coronal lines, formed from ions with ionization potentials of order 200 eV or greater, indicate extremely low densities and very hard ionizing continua. Typical molecular dissociation energies are of order a few eV, while the ionization energies are about the same as the atomic values, about 10 eV. Characteristic energies for rotational transitions are 0.01 eV, for vibrational about 0.1 to 1 eV, and for electronic about 5 to 10 eV. Typical absorption cross sections are multiples of the geometric area with the Bohr radius as a length scale. The same is true for the collisional cross sections. The typical minimum for the absorption or scattering cross section is the Thomson or electron scattering cross section, which is also the grey opacity limit for nonrelativistic scattering.

A.2.2 Planetary

The typical scale of mass is $M_\oplus$. The Jovian planets are around 100 $M_\oplus$ and have radii of about 10 $R_\oplus$; the terrestrial planets are all about the same order of magnitude as the Earth. A simple scaling of the mean density of the terrestrial planets is that of a silicate, around 3 or 4 g cm^{-3}, while for the Jovian planets it is around that of water, 1 g cm^{-3}. Comets deviate from this, having typical masses of about 10^{22} g.

The parameters of the interplanetary medium are, of course, dependent on the distance from the sun. But for simple calculations, at the Earth (1 AU), the typical density is about 5 cm^{-3} and the typical flow velocity of the solar wind is about 400 km s^{-1} with essentially complete ionization. The typical velocity in the solar system is the characteristic Keplerian velocity. For the Earth this is about 30 km s^{-1}; the escape velocity from the Earth is about 10 km s^{-1}. The typical scale length is 1 AU.

A.2.3 Stellar

The standard scale for the mass of a star is $M_\odot$. For the surface gravity, useful for estimating the mechanical properties of a star, the standard is, on the main sequence, $\log g \approx 4.0$. For the sun, $\log g = 4.4$. The surface temperature, defined as being the effective temperature $T_{\text{eff}} \equiv (L/4\pi\sigma R^2)^{1/4}$, is about 5800 K for the sun, about 10^4 for a 2 $M_\odot$ star on the main sequence (an A0 V star), and of order 30,000 for a B0 V star of about 10 $M_\odot$. The photosphere corresponds to the surface that is optically thick in the continuum, which, for the sun, has a density of about 10^{16} cm^{-3}. For a quick estimate one can use the Thomson cross section and a length scale of order 1 R$_\odot$.

Main sequence stars range from about 0.1 to 100 M$_\odot$; white dwarfs are confined to a relatively narrow range around 1 M$_\odot$ and 0.01 R$_\odot$; neutron stars are typically 1 M$_\odot$ and about 20 km in radius.

Typical stellar magnetic fields are of order 1 G for global fields but range to higher values in starspot regions and in peculiar stars. Fields have been measured for Ap stars up to a few tens of kG. Fields for white dwarf stars are about 1 MG; for neutron stars they are in excess of 10^4 MG.

A.2.4 Interstellar

There are really three components to the interstellar medium (ISM): the diffuse gas, the warm clouds, and the cold molecular clouds and the complexes they form. The parameters have to be specified separately for each component of the medium.

The typical hydrogen density scale in the diffuse interstellar medium is about 1 cm^{-3}, with a characteristic ionization fraction of anything between 10^{-3} and 1. The magnetic field strength is between 1 and 10 μG in the diffuse medium and as high as a few milligauss in molecular clouds. For molecular clouds the typical densities are about 100 to $>10^4$ cm^{-3}, typical temperatures are about 20 K, and typical masses are about 10^4 M$_\odot$. The characteristic scale length for just about all phases of the ISM is about 1 pc. For the hot interstellar medium, the typical kinetic temperatures are of order 10^6 K with densities less than 10^{-2} cm^{-3}. The very high temperature of this gas is responsible for its spatial extent, several kiloparsecs off the galactic plane, and is the reason for its label as *coronal* gas.

An interesting fact, perhaps coincidence, is that the energy density of all sources in the ISM is about 1 eV cm^{-3}. This includes cosmic rays, photons, turbulence, large-scale motions, and the like. When discussing the energy density, however, it is important to keep in mind that the dark matter contribution is large, even if the nature of its constitution is presently unknown (it may contain as much as 70% of the galactic mass and an even larger contribution to the mass of a galaxy cluster).

A.2.5 Galactic

The typical length scale within a galaxy is about 1 kpc. Between galaxies, the typical scale is about 1 Mpc, especially within a cluster. The typical mass scale is about 10^{12} M$_\odot$, and the characteristic galaxy radius is about 10 kpc, which is about the same as R_0; the peripheral parts of the galaxy may, however, extend to as far as 100 kpc from the galactic center and still be bound. Typical orbital velocities are about 100 to 200 km s^{-1} for spiral galaxies, which are also the same order of magnitude as the velocity dispersions in elliptical galaxies. Recall that for circular orbits, the escape velocity is $\sqrt{2}$ times the local orbital speed. For clusters of galaxies, the typical mass scale is tens to hundreds of galaxies in a region of a few Mpc. Typical velocity dispersions within clusters are about 500 km s^{-1}.

A.2.6 Cosmological

Typical length scales are of order a few tens of Mpc. On this scale, the mass of a structure is of order 10^{14} M$_\odot$.

The characteristic timescale is the inverse of the Hubble constant, which is of order the age of the universe. This constant, H_0, is about 100 km s^{-1} Mpc^{-1}; the timescale is about 10 GYr. There is still considerable uncertainty in this number, so take it as a very approximate value.

Typical differential velocities are about 1000 km s^{-1}.

A.3 Some Important Dimensionless Numbers

An excellent collection of dimensionless numbers is found in the *CRC Handbook of Chemistry and Physics* (Sections F and G). Throughout this book, we have been using such numbers for scaling and here we collect some of the numbers we have introduced.

Number	Symbol and equation	Use
Ekman	$E = l^2 \Omega / \nu$	Boundary layers
Mach	$M = v/a_s$	Fluid speed
Prantdl, Schmidt	$\sigma = \kappa/\nu$	Viscosity
Rayleigh	$R = \alpha g l^3 \Delta T / \kappa \nu$	Thermal buoyancy
Reynolds	$Re = Ul/\nu$	Viscosity
Rossby	$Ro = U/l\Omega \sin \theta$	Geostrophic waves
Taylor	$Ta = 4\Omega^2 l^4 / \nu^2$	Coriolis/viscous

General Bibiliography

> *When I get a little money, I buy books, and if there is any left, I buy food and clothes.*
> Erasmus

1 Monographs

1.1 Classical Fluid Mechanics

The literature on fluid mechanics is quite extensive, and it would be folly to try to be comprehensive in any reference list. Most of the books in this field have a decidedly engineering bent, due mainly to the considerable interest in fluids by workers in aerodynamics and hydraulics. There are, however, a number of important monographs which take a distinctly physical approach and which are of great use in a variety of astrophysical problems.

Abramovich, G. N. (1963). *The Theory of Turbulent Jets.* Cambridge, Mass.: MIT Press.

> The only book available on the subject. Most amusing is the section on use of jets as doors for buildings and the physics of air curtains (a technology used extensively in shopping malls).

Anderson, J. E. (1963). *Magnetohydrodynamic Shock Waves.* Cambridge, Mass: MIT Press.

> A classic reference on the subject and still quite valuable. The treatment of the shock layer is very interesting, despite the annoying habit of the author to put everything in Cartesian coordinates.

Batchelor, G. K. (1959). *The Theory of Homogeneous Turbulence.* Cambridge: Cambridge University Press.

> This is the classic of the applied mathematical literature. Since homogeneous, incompressible turbulence is an astrophysical fiction, it is mainly useful as a

1 Monographs

guide to the mathematics, especially for stochastic integrals. Still, much worth pursuing.

Batchelor, G. K. (1967). *An Introduction to Fluid Dynamics*. Cambridge: Cambridge University Press.

The standard introduction to formal methods in mathematical fluid mechanics.

Birkoff, G. and Zarantonello, E. H. (1957). *Jets, Wakes and Cavities*. New York: Academic Press.

This unique study deals mainly with analytical solutions to subsonic flow problems, but it is an excellent introduction to many of the basic problems associated with disrupted flow environments. It concentrates, as usual, on incompressible fluids but also provides important insights into compressible flows past blunt objects.

Chandrasekhar, S. (1961). *Hydrodynamic and Hydromagnetic Stability*. New York: Dover.

Probably the most interesting book written on the problem of stability, with special emphasis on variational principles. It is most useful for its treatment of MHD stability and for the comprehensive discussion of Rayleigh–Bénard convection.

Chandrasekhar, S. (1967). *Plasma Physics*. Chicago: The University of Chicago Press.

Notes from a course of lectures. Particularly memorable for the discussion of the energy principle in MHD instability calculations.

Cole, G. H. A. (1962). *Fluid Dynamics*. London: Methuen.

A very useful and concise little book, especially notable for Chapter 7 on dimensionless numbers.

Courant, R., and Friedrichs, K. O. (1948). *Supersonic Flows and Shock Waves*. New York: Interscience.

Most useful for the discussion of characteristics and for the informal introduction to theory of supersonic jets.

Cowling, T. G. (1957). *Magnetohydrodynamics* (also 2nd edition). New York: Interscience.

Drazin, P. G., and Reid, W. H. (1981). *Hydrodynamic Stability*. Cambridge: Cambridge University Press.

Covers much the same material as Lin and Chandrasekhar but does it more formally. A useful companion volume. This is an update and extension of Lin's book.

Greenspan, H. P. (1967). *The Theory of Rotating Fluids*. Cambridge: Cambridge University Press.

The only book to deal exclusively with stability problems related to inviscid and viscous rotating fluids. Of special interest for applications to accretion disks and rotating stars, but also to planetary atmospheres.

Hayes, W. D., and Probstein, R. F. (1959). *Hypersonic Flow Theory*. New York: Academic Press.

An excellent monograph for information on compressible supersonic flows, as usual with strong aerodynamic bias.

Hinze, J. O. (1975). *Turbulence*, 2nd edition. New York: McGraw-Hill.

Has a strong engineering orientation, but this actually enhances its utility. There is an excellent review of the methods by which turbulence is measured, and many laboratory results are quoted. Very good companion to Townsend's monograph.

Krall, N. A., and Trivelpiece, A. W. (1973). *Principles of Plasma Physics*. New York: McGraw-Hill.

Recently reprinted by San Francisco Press, this classic is especially useful for plasma wave phenomena.

Lamb, H. (1945). *Hydrodynamics*. New York: Dover.

The classic work in English on fluid mechanics, the chief survivor of the last century. While formal, it contains many topics not treated well elsewhere, including an excellent discussion of potential flows, a good selection of problems, and one of the earliest discussions of solitary waves.

Landau, L. D., and Lifshitz, E. M. (1987). *Fluid Mechanics*, 2nd edition. Oxford: Pergamon.

If it is not here, you should think about whether it is interesting enough to pursue. The discussion of shocks (both one- and two-dimensional treatments) is especially important, and the chapter on superfluids is still the best introduction to the subject.

Landau, L. D., and Lifshitz, E. M. (1960). *Electrodynamics of Continuous Media*. Oxford: Pergamon.

Contains an excellent treatment of plasma problems, as well as their connection with many aspects of kinetic theory.

Leslie, D. C. (1973). *Developments in the Theory of Turbulence*. Oxford: Oxford University Press.

A good introduction to Fourier methods and also to the direct interaction approximation. The DIA method is very similar to renormalization group techniques now being employed, and the exposition is written with a certain apostolic zeal that is very enjoyable.

Liepmann, H. W., and Puckett, A. E. (1947). *Introduction to Aerodynamics of a Compressible Gas*. New York: Wiley.

Mainly interesting as an introduction to supersonic flows for ideal gases. This work is one of the best introductions available in the subject, particularly for sections on hodographic methods, formal solutions to flow problems in the supersonic and transonic regimes, and the discussion of jets and wakes.

Lighthill, M. J. (1986). *An Informal Introduction to Theoretical Fluid Mechanics.* Oxford: Oxford University Press.

Mainly interesting as an introduction to boundary layers and vortex dynamics.

Lin, C. C. (1955). *Theory of Hydrodynamic Stability.* Cambridge: Cambridge University Press.

Still a useful guide and one of the first books on the subject.

Moffatt, H. K. (1978). *Magnetic Field Generation in Electrically Conducting Fluids.* Cambridge: Cambridge University Press.

Still the best introduction to mean field techniques and particularly good as a precursor of topological methods that have recently been used for turbulence.

Monin, A. S., and Yaglom, A. M. (1971). *Statistical Fluid Mechanics* (2 vols.). Cambrige, Mass.: MIT Press.

A massive, mainly formal, guide to the methods required for the study of turbulent flows, this monograph is especially notable for its treatment of topics like turbulent radiative transfer. It also has a wealth of laboratory data related to real shear flows and discusses some of the methods of formal graph-theoretical treatments for turbulence.

Nicholson, D. R. (1983). *Introduction to Plasma Theory.* New York: Wiley.

A good introduction to plasmas; also contains an interesting discussion of solitons and MHD instability.

Pai, S-I. (1954). *Fluid Dynamics of Jets.* New York: van Nostrand.

This is an engineering classic, it is notable for discussions of mixing and supersonic jets. The photographs are also very interesting.

Pedlovsky, J. (1979). *Geophysical Fluid Dynamics.* Berlin: Springer-Verlag.

An excellent introduction, albeit one that presents formidable "linguistic" problems for astrophysics.

Polovin, R. V., and Demutiskii, V. P. (1990). *Fundamentals of Magnetohydrodynamics.* New York: Plenum.

A good guide to the Soviet school of plasma fusion research.

Prandtl, L. (1952). *Essentials of Fluid Dynamics.* New York: Haffner.

Although long out of print, this is one of the best informal introductions to fluids, with an interesting section on supersonic jets that is especially notable for the interesting photographs. It can be considered an update of the lectures recorded in L. Prandtl and O. G. Tietjens (1934), *Fundamentals of*

Hydro- and Aeromechanics and *Applied Hydro- and Aeromechanics* (New York: Dover).

Schlichting, H. (1960). *Boundary Layer Theory*, 4th edition. New York: McGraw-Hill.

> The most comprehensive introduction to the theory of boundary layers; includes a very large amount of engineering and laboratory data.

Sears, W. R., ed. (1954). *General Theory of High Speed Aerodynamics*. Princeton: Princeton University Press.

> The chapter on characteristics (by A. Ferri) is quite good.

Sedov, L. L. (1977). *Similarity Methods in Mechanics*, 5th edition, French. Moscow: Mir.

> The only complete work on the subject of similarity solutions. This work is of special interest for the discussion of shock and blast waves and for the sections on astrophysical flows and stellar models. The only English translation is based on an early edition: L. Sedov (1959), *Similarity and Dimensional Methods in Mechanics* (New York: Academic Press).

Spitzer, L., Jr. (1962). *Physics of Fully Ionized Gases*. New York: Interscience.

> An excellent overview of plasma processes, including an outstanding chapter on collisions in plasmas.

Stix, M. (1963). *Plasma Waves*. New York: McGraw-Hill.

> An excellent introduction to the general theory of plasma oscillations and the most accessible work on the subject in English.

Townsend, A. A. (1976). *The Structure of Turbulent Shear Flows*, 2nd edition. Cambridge: Cambridge University Press.

> The best work available on free and wall-limited turbulent flows.

Tritton, D. J. (1977). *Physical Fluid Dynamics*. London: Van Nostrand Reinhold–UK.

> A delightful introduction to the subject, informal and chatty. The discussion of turbulence is especially good and phenomenological.

Turner, J. S. (1973). *Buoyancy Effects in Fluids*. Cambridge: Cambridge University Press.

> A work notable for its treatment of plumes and jets but also important for its discussion of double-diffusive instabilities and free convection. A very nice companion to Chandrasekhar's work.

van Dyke, M. (1984). *An Album of Fluid Motion*. Stanford: Parabolic Press.

> A unique work, intended to illustrate many classical and beautiful phenomena encountered in fluids. It is most notable for its spectacular collection of shock-related phenomena, especially jets and blunt objects, and for one of the

1 Monographs

outstanding compilations of photographs of vortex sheets. Many of these illustrations are well known from other works, but here they are all collected with very useful commentary and are much enhanced by their juxtaposition. There is also a small Gallery of Fluid Motion published every year in *Physics of Fluids*. However, the principal importance of van Dyke's collection is that it is "from life"; no numerical simulations are included.

Whitham, G. B. (1974). *Linear and Nonlinear Waves*. New York: Wiley.

An outstanding treatment of shock and soliton problems; provides many interesting insights into the mathematics of wave propagation. More formal than most of the other monographs mentioned here.

Zel'dovich, Ya., and Razer, Y. (1965). *High Temperature Flows and Shock Phenomena* (2 vols.). New York: Academic Press.

An excellent introduction to shock wave and blast wave theory, including a wonderful first chapter on multicomponent gas shock physics. Useful for its discussion of shock precursors and radiative effects in shocks and for a chapter on blast waves. Also, one of the more thorough discussions of detonation phenomena. The first two chapters of vol. 1 were reprinted separately as an introduction to shock waves.

1.2 Astrophysical Fluid and Plasma Processes

Alfvén, H., and Falthammer, H. (1963). *Cosmical Electrodynamics*. New York: Oxford University Press.

Aller, L. H. (1987). *Physics of Thermal Gaseous Nebulae*. Dordrecht, The Netherlands: Reidel.

Anile, A. M. (1989). *Relativistic Fluid Mechanics*. Cambridge: Cambrige University Press.

Brandt, J. C. (1970). *An Introduction to the Solar Wind*. San Fransisco: W. H. Freeman.

Chandrasekhar, S. (1967). *Ellipsoidal Figures in Equilibrium*. New Haven, Conn.: Yale University Press; reprinted by Dover.

Cox, J. (1981). *Theory of Stellar Pulsation*. Princeton: Princeton University Press.

The best modern monograph on pulsation theory, one that is heavily motivated by the nonradial aspects of the problem. The background information about stellar evolution is very valuable.

de Jager, C. (1984). *The Brightest Stars*. Dordrecht, The Netherlands: Reidel.

A good introduction to massive stars and mass loss driven by turbulence and radiation pressure.

Dyson, J. E., and Williams, D. A. (1980). *The Physics of the Interstellar Medium*. Manchester, United Kingdom: Manchester University Press.

Foukal, P. A. (1990). *Solar Astrophysics*. New York: Wiley Interscience.

Frank, J., King, A. R., and Raine, D. (1991). *Accretion Power in Astrophysics*. Cambridge: Cambridge University Press.

Kaplan, S., and Pik'lner, S. (1970). *The Interstellar Medium*. Cambridge, Mass: Harvard University Press.

Krause, F., and Rädler, K.-H. (1980). *Mean-Field Electrodynamics and Dynamo Theory*. Oxford: Pergamon.

Melrose, D. (1987). *Instabilities in Astrophysical and Space Plasmas*. Cambridge: Cambridge University Press.

Mihalas, D., and Mihalas, B. W. (1984). *Foundations of Radiation Hydrodynamics*. Oxford: Oxford University Press.

Chapters 2 (idealized fluid flows) and 5 (waves, shocks, and winds) are especially useful. This magnificently encyclopedic book has an overall emphasis on computational methods and radiative transfer. The physics is, however, paramount.

Osterbrock, D. E. (1989). *Astrophysics of Gaseous Nebulae and Active Galactic Nuclei*. Sacramento, Calif.: University Books.

Parker, E. N. (1963). *Interplanetary Dynamical Processes*. New York: Wiley-Interscience.

An introduction to the solar wind, mainly analytical solutions, and still valuable.

Parker, E. N. (1979). *Cosmical Magnetic Fields*. New York: Oxford University Press.

The fundamental work in the field, one that is uniquely personal and contains excellent discussions of the dynamo process and topology of field lines.

Priest, E. R. (1982). *Solar Magnetohydrodynamics*. Dordrecht, The Netherlands: Kluwer.

Sarazin, C. L. (1988). *X-ray Emission from Clusters of Galaxies*. Cambridge: Cambridge University Press.

A good overview of cooling flows and galactic winds.

Scheffler, H., and Elsässer, H. (1987). *Physics of the Galaxy and Interstellar Matter*. Berlin: Springer-Verlag.

Spitzer, L. (1979). *Physical Processes in the Interstellar Medium*. New York: Wiley.

The best overview currently available on the ISM in monograph form.

Stix, M. (1989). *The Sun*. Berlin: Springer-Verlag.

Tassoul, J. (1978). *Theory of Rotating Stars*. Princeton: Princeton University Press.

Zirin, H. (1988). *Astrophysics of the Sun*. Cambridge: Cambridge University Press.

More observationally oriented than the other works, this is a good guide to solar phenomenology.

1.3 Astrophysical Applications: Collections

Belvedere, G., ed. (1989). *Accretion Disks and Magnetic Fields in Astrophysics*. Dordrecht, The Netherlands: Kluwer.

Proceedings of a European Physical Society conference.

Bridle, A. H., and Eilek, J. A., eds. (1984). *Physics of Energy Transport in Extragalactic Radio Sources*. Charlottesville, N. C.: NRAO.

Proceedings of a workshop on astrophysical jets held at Green Bank in 1984.

Cordova, F., ed. (1988). *Multiwavelength Astrophysics*. Cambridge: Cambridge University Press.

A superb cure for wavelength parochialism. The volume consists entirely of review papers. It also summarizes the databases available for multiwavelength research.

Crivellari, L., Hubeny, I., and Hummer, D. G., eds. (1991). *Stellar Atmospheres: Beyond Classical Models*. Dordrecht: Kluwer.

Dalgarno, A. and Layzer, D., eds. (1987). *Spectroscopy of Astrophysical Plasmas*. Cambridge: Cambridge University Press.

See especially Chapter 9, Astrophysical Shocks in Diffuse Gas, by C. McKee.

Gehrels, T. A., ed. (1978). *Protostars and Planets*. Tucson, Ariz.: University of Arizona Press.

This was the first comprehensive collection of reviews on star formation and the formation of the solar system. It was followed by D. Black and M. Matthews, eds. (1985), *Protostars and Planets*, II (Tucson, Ariz.: University of Arizona Press). The two books cannot be seen as replacing one another; they are very complementary and should be used as such. And in the tradition of modern movie-making comes, naturally, D. Black, ed. (1991), *Protostars and Planets*, III (Tucson, Ariz.: University of Arizona Press).

Hollenbach, D., and Thronson, H. A., ed. (1987). *Interstellar Processes*. Dordrecht, The Netherlands: Kluwer.

This is the proceedings of the First Wyoming Conference on interstellar processes. It contains a superb set of review articles and stands as one of the best references available on physical processes in the interstellar medium.

Hughes, P. A., ed. (1990). *Beams and Jets in Astrophysics*. Cambridge: Cambridge University Press.

A good overview of radio jets and large-scale structures observed in active galaxies.

Lucas, R., Omont, A., and Sora, R., eds. (1985). *Birth and Infancy of Stars: Les Houches XLI*. Amsterdam: North-Holland.

The proceedings of the Les Houches 1983 summer school on star formation.

Sellwood, J. A., ed. (1989). *Dynamics of Astrophysical Discs*. Cambridge: Cambridge University Press.

Proceedings of workshop on discs held in 1988. Some excellent review papers and a good summary of the state of the field.

Thronson, H. A., and Shull, J. M., eds. (1990). *The Interstellar Medium in Galaxies*. Dordrecht, The Netherlands: Kluwer.

The invited talks delivered at the Second Wyoming Conference dealing with large-scale problems of the interstellar medium in external galaxies and the galactic intracluster environment.

Zensus, A., and Pearson, T., eds. (1990). *Parsec-Scale Radio Jets*. Cambridge: Cambridge University Press.

Product of a workshop held at NRAO in 1989.

There are also many IAU symposia and colloquia that in some form deal with fluid dynamics–related problems. It is quite impossible to list all of these, many of which contain few reviews. However, periodic listings of meetings and proceedings are available. One series of such symposia should, however, be mentioned. Starting in the late 1940s, there was a series of symposia on astrophysical fluid dynamics held jointly with the IUTAM. The last volume in the series is R. N. Thomas, ed. (1967), *Aerodynamic Phenomena in Stellar Atmospheres* (New York: Academic Press). This volume contains references to the earlier symposia, which brought together workers in all fields of fluid mechanics to discuss astrophysically important questions. The discussions are completely recorded and are still a gold mine for insights and references. I cannot recommend this series too strongly to the researcher and student alike. They are a wonderful legacy.

1.4 Astrophysical Jets

This subject has developed an almost self-contained literature. For this reason, it seemed fitting to list some of the basic papers to which a beginner should turn for information on both the observational and theoretical aspects of astrophysical jets. They are collected here rather than residing

1 Monographs

with the discussions because they would have been spread over too many chapters of this book. The range of these papers reflects the enormous complexity and diversity of the subject of jets. These papers are meant to be representative rather than exhaustive. The papers were chosen because of their treatment, either observational or theoretical, or because they are good guides to the literature. A few monographs dealing with the astrophysics are included. The hydrodynamics monographs related to jets are listed separately above.

Begelman, M. C., Blandford, R. D., and Rees, M. J. (1984). Theory of extragalactic radio sources. *Rev. Mod. Phys.* **56**, 255.

Bicknell, G. V. (1984). A model for the surface brightness of a turbulent low Mach number jet. I. Theoretical development and application to 3C31. *Astrophys. J.* **286**, 68; (1986). A model for the surface brightness of a turbulent low Mach number jet. II. The global energy budget and radiative losses. *Astrophys. J.* **300**, 591.

Blandford, R. D., and Rees, M. J. (1974). A "twin-exhaust" model for double radio sources. *Mon. Not. R. Astron. Soc.*, **169**, 395.

Blondin, J. M., Konigl, A., and Fryxell, B. A. (1989). Herbig–Haro objects as the heads of radiative jets. *Astrophys. J. (Lett.)* **337**, L37.

Bridle, A. H., and Eilek, J. A., eds. (1984). *Physics of Energy Transport in Extragalactic Radio Sources*. Charlottesville, N. C.: NRAO.

Bridle, A., and Perley, R. A. (1984). Extragalactic radio jets. *Annu. Rev. Astron. Astrophys.* **22**, 319.

Buhrke, T., Mundt, R., and Ray, T. P. (1988). A detailed study of HH 34 and its associated jet. *Astron. Astrophys.* **200**, 99.

Canto, J., Tenorio-Tagle, G., and Rozyczka, M. (1988). The formation of interstellar jets by the convergence of supersonic conical flows. *Astron. Astrophys.* **192**, 287.

Curtis, H. C. (1918). Descriptions of 762 nebulae and clusters photographed with the Crossley Reflector. *Publ. Lick Obs,* **13**, 31.

Hartigen, P. (1989). The visibility of the Mach disk and the bow shock of a stellar jet. *Astrophys. J.* **339**, 987.

Hines, D., Owen, F., and Eilek, J. (1989). Filaments in the radio lobes of M87. *Astrophys. J.* **347**, 713.

This contains references to much earlier work on the M87 radio structure.

Hughes, P., ed. (1990). *Beams and Jets in Astrophysics*. Cambridge: Cambridge University Press.

Kahn, F. D. (1982). A turbulent radio jet. *Mon. Not. R. Astron. Soc.* **202**, 553.

Kellermann, K. I., and Owen, F. N. (1988). Radio galaxies in quasars [especially Section 13.4: Extended Structures]. In *Galactic and Extragalactic Radio Astronomy*, 2nd edition (p. 563). G. Verschuur and K. I. Kellermann, eds. Berlin: Springer-Verlag.

Königl, A. (1982). On the nature of bipolar sources in dense molecular clouds. *Astrophys. J.* **261**, 115.

Lelievre, G. Nieto, J. -L., Horville, D., Renard, I., and Servan, B. (1984). Optical structures in the M87 and 3C 273 jets. *Astron. Astrophys.* **138**, 49.

Margon, B. (1984). Observations of SS 433. *Annu. Rev. Astron. Astrophys.* **22**, 507.

Mundt, R. (1986). Jets from protostellar objects. *Can. J. Phys.* **64**, 487.

Norman, M. L., Smarr, L., Winkler, K. -H., and Smith, M. D. (1982). Structure and dynamics of supersonic jets. *Astron. Astrophys.* **113**, 285.

See also an earlier paper by M. L. Norman, L. Smarr, J. R. Wilson, and M. D. Smith (1981). Hydrodynamic formation of twin-exhaust jets. *Astrophys. J.* **247**, 52.

Payne, D. G., and Cohn, H. (1985). The stability of confined radio jets: The role of reflection modes. *Astrophys. J.* **291**, 655.

Reipurth, B. (1989). Observations of Herbig–Haro objects. In *ESO Workshop on Low Mass Star Formation and Pre-Main Sequence Objects* (p. 247). B. Reipurth, ed. Munich: ESO.

Roberts, D. A. (1986). Knots, wiggles, and flaring in ballistic and stochastic jets. *Astrophys. J.* **300**, 568.

Smith, M. D., Norman, M. L., Winkler, K. -H. A., and Smarr, L. (1985). Hotspots in radio galaxies: A comparison with hydrodynamic simulations. *Mon. Not. R. Astron. Soc.* **216**, 67.

Turner, J. S. (1986). Turbulent entrainment: The development of the entrainment assumption and its applications to geophysical flows. *J. Fluid Mech.* **173**, 431.

Walker, R. C., Benson, J. M., and Unwin, S. C. (1987). The radio morphology of 3C 120 on scales from 0.5 parsecs to 400 kiloparsecs. *Astrophys. J.* **316**, 546.

Wilson, M. J., and Falle, S. A. E. G. (1985). Steady jets. *Mon. Not. R. Astron. Soc.* **216**, 971.

2 Journals

2.1 General

The primary journals for fluid mechanics are: *The Journal of Fluid Mechanics* (monthly, each issue constitutes a single volume) and *Physics of Fluids* (originally a monthly, the journal bifurcated in 1988 into parts A, classical fluids, and B, plasma processes). *Geophysical and Astrophysical Fluid Mechanics* is a monthly covering geophysical oceanic and atmospheric phenomena; stellar, galactic, and extragalactic fluid dynamics; and sometimes MHD. This is a unique journal, and while sometimes hard to find, it is often a gold mine for interesting results. *Annual Reviews of Fluid Mechanics* is a very accessible review journal, occasionally including astrophysically slanted articles and also containing good historical overviews. Other review journals that often have astrophysical or hydrodynamics articles are *Reviews of Modern Physics, Reports on the Progress of Physics,* and *Physics Reports*.

A few other journals frequently contain interesting papers that can be metamorphosed into astrophysical applications. These include *Journal of the American Institute of Aeronautics and Astronautics, Journal of Plasma Physics, Proceedings of the Royal Society* (*London*), *Journal de Physique, Journal of Atmospheric Sciences, ASME Journal of Fluid Engineering, AIAA Journal* (especially useful for laboratory work on supersonic flows), *Quarterly Journal of the Royal Meteorological Society* (especially good for review-type papers), and *Reviews of Geophysics*.

2.2 Astrophysics

The standard journals in the astrophysics literature are *The Astrophysical Journal* (*Astrophys. J.*), *Monthly Notices of the Royal Astronomical Society* (*Mon. Not. R. Astron. Soc.*), *Astronomy and Astrophysics* (*Astron. Astrophys.*), *Astronomical Journal* (*Astron. J.*), *Publications of the Astronomical Society of Japan* (*Publ. Astr. Soc. Japan*), and *Soviet Astronomy* (*Sov. Astr.-AJ*), *Solar Physics*, and *Icarus* (the primary journal for solar system research). Other important sources, especially of observational material, are *Australian Journal of Physics, Publications of the Astronomical Society of the Pacific* (*PASP*), and *Journal of Astronomy and Astrophysics* (*J. Astron. Astrophys.*).

The primary astronomical review volume is *Annual Reviews of Astronomy and Astrophysics*; *Annual Reviews of Earth and Planetary Sciences* also often contains much of interest to even traditional astronomers. These are single annual editions, like the other *Annual Reviews*, and are excellent for obtaining a broad perspective and excellent pointers to the literature. In addition, there are several smaller review journals: *Comments on Astrophysics, Reviews in Astronomy and Astrophysics, Fundamentals of Cosmic Physics, Space Science Reviews*, and *Quarterly Journal of the Royal Astronomical Society*.

2.3 When All Else Fails, Some Other Places to Check

Physics journals which often contain papers on fluids, often of astrophysical interest, are *Physical Reviews A, Physica D: Nonlinear Phenomena, Advances in Physics*, and *Nature*. Numerical methods appear in *Journal of Computational Physics* and *Computational Methods in Physics*. *Computers in Physics* and *Physics Today* also sometimes contain relevant reviews.

3 Astronomical Data Centers

There is an excellent discussion of the available databases, primarily from satellite observations, in Cordova (1988). However, for some additional information see C. Jaschek (1989), *Data in Astronomy* (Cambridge: Cambridge University Press).

4 Historical

Currently, issues of *Annual Reviews*, both of astronomy and astrophysics and of fluid mechanics, offer personal reminiscences or historical surveys as the first article of each issue. They are frequently gold mines for insight into why particular problems were important and how certain problems were attacked, and they sometimes uncover methods long buried under the mass of more recent papers. They are an excellent way to start any hike through the mountains of literature.

Collected popular lectures have been published for many of the founders of the field, including Kelvin, Mach, and Helmholtz. Unfortunately, Helmholtz's hydrodynamics papers have not. been separately published. A wonderful source for both biographical information and historical background is the C. G. Gillespie, ed. (1970), *Dictionary of Scientific Biography* (15 vols.), (New York: Scribners). This collection is written at some depth and often contains discussions of important details of a scientist's work, not just a quick skim. It should *always* be consulted as a starting point for further historical research. Historical papers on fluid mechanics and even astrophysical problems can often be found in the following journals: *Journal for the History of Astronomy, Isis, Osiris, British Journal for the History of Science, Historical Studies in the Physical and Biological Sciences, Centaurus, Physics Today, Scientific American*, and the *Archive for the History of the Exact Sciences*.

Abbe, C., ed. (1891). *The Mechanics of the Earth's Atmosphere*. Washington, D.C.: Smithsonian Miscellaneous Collections.

> This rare but uniquely important volume contains translations of nearly all of Helmholtz's fundamental hydrodynamics papers and also Kirchhoff's classic paper on jets. (I thank John Ptak for his invaluable aid in obtaining a copy of this book.)

Bernoulli, J. (1968). *J. Bernoulli: Hydrodynamica and D. Bernoulli: Hydraulica*. New York: Dover.

> The only available translation of the founding work on fluid mechanics.

Beyer, R. T., ed. (1984). *Nonlinear Acoustics in Fluids*. Princeton: van Nostrand Reinhold.

> A very important collection of classical papers, including the Rankine and Hugoniot papers, as well as an English translation of Riemann's paper on nonlinear waves.

Cohen, R. S., and Seeger, R. J., eds. (1970). *Ernst Mach: Physicist and Philosopher*. Dordrecht, The Netherlands: Reidel.

Eddington, A. S. (1927). *The Internal Constitution of the Stars*. Cambridge: Cambridge University Press.

> See also his biography by A. V. Douglass (1957), *Arthur Stanley Eddington* (London: Nelson).

Jeans, J. H. (1923). *Astronomy and Cosmogony*. Cambridge: Cambridge University Press.

> See also his biography by E. Milne (1952), *J. H. Jeans* (Cambridge: Cambridge University Press).

Kelvin (Wm. Thomson Lord Kelvin) (1910). *The Mathematical Papers of Lord Kelvin (Sir William Thomson)*, (6 vols.). J. Larmor, ed. Cambridge: Cambridge University Press.

> Volume IV deals mainly with Kelvin's hydrodynamical studies, especially his treatment of the vortex ring and the instability that bears his name. See also R. Kargon and P. Achinstein, eds., (1987), *Kelvin's Baltimore Lectures and Modern Theoretical Physics: Historical and Philosophical Perspectives* (Cambridge, Mass.: MIT Press). This book contains the full text of the lectures, previously unpublished, and excellent background essays. The discussion of vortex motion is found scattered throughout the first dozen of the 20 lectures. A good

modern biography is H. J. Sharlin (1979), *Lord Kelvin: The Dynamic Victorian* (University Park, Penn.: Penn State).

Landau, L. D. (1965). *The Collected Papers of L. D. Landau.* D. ter Haar, ed. Oxford: Pergamon.

A comprehensive collection of Landau's papers in all fields. Most important for our discussions are the papers related to superfluid dynamics and Landau's classical paper on the transition to turbulence as a bifurcation cascade.

Lyttleton, R. A. (1953). *Equilibrium of Rotating Fluid Masses.* Cambridge: Cambridge University Press.

Now largely superseded by Chandrasekhar (1967), but still of considerable historical interest.

Rayleigh (John William Strutt, Lord Rayleigh) (1964). *Scientific Papers* (6 vols.). Cambridge: Cambridge University Press; reprinted by Dover.

This reprint, which contains some corrections and a brief notice on bibliography, is the complete corpus *sans* monographs of Rayleigh's scientific output. Every volume contains hydrodynamic studies, especially vol. VI, which includes the famous 1916 paper on convection and the work on stability of velocity gradients. See also R. J. Strutt (1968) (reprint), *Life of John William Strutt, Third Baron Rayleigh* (Madison, Wisc.: University of Wisconsin Press), the only available biography of Rayleigh, written by his son (himself a noted physicist).

Riemann, B (1968). *Oeuvres Mathématiques de Riemann.* Paris: Blanchard.

The papers of Riemann have never been translated into English and only the French translation is still in print. This volume contains his papers on characteristics and wave propagation.

Saltzman, B. ed. (1962). *Selected Papers on the Theory of Thermal Convection.* New York: Dover.

A wonderful collection of original papers on convection in all contexts, astrophysical and meteorological as well as laboratory.

Stokes, Sir G. B. (1905). *The Mathematical Papers of Sir George Stokes.* J. Larmor, ed. Cambridge: Cambridge University Press.

Taylor, G. I. (1960). *The Scientific Papers of Sir Geoffrey Ingram Taylor* (4 vols.). G. K. Batchelor, ed. Cambridge: Cambridge University Press.

Volume II contains all of Taylor's papers on statistical theory of turbulence, and vols. III and IV treat papers on fluid mechanics. See also G. Batchelor (1986) *J. Fluid Mech.* **173**, 1 ff, the entire issue of which is devoted to a centennial celebration of Taylor's birth.

Thomson, Sir Wm., and Tait, P. G. (1912). *Principles of Natural Philosophy*, (2 vols.). Cambridge: Cambridge University Press.

4 Historical

The reprint is *Principles of Mechanics and Dynamics* (New York: Dover). Volume 1 contains the best overview of vorticity theory during the nineteenth century and is still a wonderful view of the foundations by the pedagogical masters.

von Kármán, Th. (1958). *Aerodynamics.* Ithaca, N.Y.: Cornell University Press.

A unique and very exciting overview of work on aerodynamics in general and shock theory in particular. These lectures serve to put much of the theory from the first work on hypersonic flows into perspective.

von Neumann, J. (1968). *Collected Papers,* (6 vols.). A. H. Taub, ed. Oxford: Pergamon.

The most important papers on shock. Many War Department reports during the Manhattan Project and not otherwise available are included in vol. 6.

Whittaker, E. (1951). *A History of Theories of Aether and Electricity,* Vol. 1, *The Classical Theories.* New York: Harper.

This work on electromagnetic theory and its foundations also contains many important insights into the hydrodynamic analogies employed by the workers on electrical and magnetic phenomena in the nineteenth century. See also J. Z. Buchwald (1986), *From Maxwell to Microphysics,* (Chicago: The University of Chicago Press), a wonderful update to Whittaker which includes more detail on Helmholtz's theory of vorticity and its development by Larmor and Thomson.

Index

Absolute vorticity, 64
Accretion
  inverse Parker solution, 305–307
  polytropic, 307–309
  spherical, 305–309, 312
  spherical time-dependent, 312
  wind, 309–312
  wind, cross section for, 311
Accretion, cooling flows, 313–314
Accretion disk, 289–305
  α-disk, 298–299
  boundary layer, 303–305
  iso-velocity contours, 393
  Kelvin–Helmholtz, 293
  Shakura–Sunyaev approach (α-disk), 296, 299, 419, 420
  spectrum, continuous, 405–406
  spectrum, line, 393
  thin disk approximation, 295–298
  time-dependent, 299–303
  time-dependent evolution equation, 301
Airfoil, 69–70
Alfvénic Mach number, 184, 277
Alfvén radius, 280
Alfvén velocity, 116, 182, 277
Alfvén wave dispersion relation, 183
Algol paradox, 289
Ambipolar diffusion. *See* Magnetic ambipolar diffusion
Ampère's law, 174

Baroclinic fluid, 87–89
Baroclinic instability. *See* Instability, baroclinic
Barotropic fluid, 67–68
  definition of, 61
Bernoulli equation, 24–31
  irrotational flow, 25
β-plane approximation, 77–78
Bifurcation, and Jeans instability, 323–324
Binary stars
  geostropic approximation, 74–75
  semidetached, 289
  x-ray sources, 293–294
  ζ Aurigae systems, 410–412
Binding energy. *See* Energy, binding
Bipolar flow, molecular, 398–400
Bisector method, line profile analysis, 412–413
Blasius solution. *See* Boundary layer, laminar
Blast waves, 148–149
  breakout phase, 155–157
  snowplow phase, 157–158
Boltzmann equation, 6
Bonnet integral equation, 250
Boundary layer, 46–50
  accretion disk, 303–305
  and astronomical seeing, 49–50
  laminar, 47–49
    Blasius solution, 48–49
  magnetospheric accretion instability, 369

Boundary layer (*continued*)
  similarity transform, 48
  stress, 47
  turbulent, 50
  von Kármán velocity profile, 49–50
Boussinesq approximation, buoyancy in incompressible fluid, 326, 335
Box-counting method, turbulence, 251–253
Broken stick process, 249
Brunt–Vaisala frequency, 369

Centrifugal acceleration, 59
  effective potential, 50
Chandrasekhar number. *See* Hartmann number
Chaos, 253–254
  and convection, 332–333
  and self-inductive dynamos, 211–212
Characteristics
  intersection as shock, 102–104
  sound waves, 97
  Vlasov equation, 7–8
Circulation
  definition of, 68
  lift, 69–70
Circulation currents, 90–93
  speed, 91–92
Coffee, Rayleigh–Taylor instability in, 357–358
Collapse, gravitational. *See* Gravitational collapse
Colliding winds. *See* Winds, colliding
Collision time, mean, from kinetic theory, 51
Conductivity
  and diffusion, 177–178
  Spitzer, 177
Contact discontinuity
  shear flow, 361
  shocks, 116
Continuity equation, 9, 10
  Fokker–Planck equation, 55
  incompressible, 216
  Jeans evaporation, 261
  MHD, 176
Convection
  and dynamos, 196, 199–201
  efficiency, 337
  magnetic, 340–343
  mixing length theory, 333–340
  Rayleigh–Benard problem, 324–333
  in rotating frame, 67
  Schwarzschild criterion, 334–335
  time-dependent accretion disks, 302–303
Convection instability. *See* Instability, convection
Cooling flow, and thermal instability, 414–415
Cooling flows. *See* Accretion, cooling flows
Coriolis force, 63–64
Coriolis parameter, 74
Correlation function, 218–224, 230
  autocorrelation, 218
  evolution of Fourier transform, 231–232
  Fourier transform, 206, 219, 224, 229
  integrated, 218, 222
  longitudinal, 218
  transverse, 218
Correlation tensor, 218
Coulomb parameter, 177
Cowling "antidynamo" theorem, 194–197
Crab Nebula, 149
Critical point
  inner Lagrangian point accretion flow, 291–292
  isothermal wind, 263
  polytropic accretion flow, 308
  radiatively driven wind, 271
Curve of growth method, 407–409
  sample line profiles, 409
Cylindrical coordinates
  equations of motion in, 58–60
  unit vectors in, 58

Damkohler number, 140
Damped line profile, Lyman α forest, 408
Damping parameter, spectral line, 408
Deformation, 37
De Laval nozzle
  Bernoulli flow with confinement, 27–28
  stellar wind analogy, 264, 265–266
Detonation wave, 134–137
D-front, H II region, 137
Dilution factor, 391
Dimensionless numbers
  Alfvénic Mach, 184
  Coulomb parameter, 177
  Damkohler, 140
  dynamo, 201
  Ekman, 79
  Froude, 370

Index

Hartmann, 342–343
Jeans parameter, 321–322
Mach, 107
magnetic Reynolds, 190
Nusselt, 339–340
plasma parameter, 184
Prandtl, 327–328
radiative force parameter, 270
Rayleigh, 327–328
Reynolds, 43–45
Richardson, 370
Rossby, 78–79
rotational parameter, 90
Taylor, 371
Toomre parameter, 322
Dissipation, turbulence, 226, 229–232
Dissipation, viscous, 45–46
  accretion disks, 294–295
  Keplerian motion, 46
  and stress, 45–46
Distribution function, 3
  and Fokker–Planck equation, 53
  Maxwellian, 8
  moments as hydrodynamic variables, 5
  moments of, 5–9, 9–12
  probability and, 4
  symmetry of, 3
Double-diffusive instability. *See* Instability, double-diffusive, magnetic convection as
Dynamo number, 201

Eclipsing binary stars, 409–412
Eddington luminosity, 270
Eddington–Sweet circulation speed, 92
Eddington–Sweet timescale (also Eddington–Vogt timescale), 92
Eddington–Vogt timescale for circulation. *See* Eddington–Sweet timescale
Einstein summation convention, 7
Ekman layer, 79–82
Ekman number, 79, 80
Ekman pumping, 81–82
  spinup time, 81
Ekman spiral, 79–80
Electromotive force, turbulent, 199, 203–204
Energy, binding, 14
Energy conservation, 16–18
  equation of, 16–18, 23
  H II regions and ionization, 137–138

Enstrophy, 62
  and turbulence, 222–223, 240
Enthalpy, 19, 26
Entrainment efficiency, 244–245
Entropy, 17, 18
  and H-theorem, 7
  potential temperature, 65
  rotating bodies, 91
  shocks, 109
Epicyclic frequency, 85, 143, 322
Epicyclic motion, 84
Equation of state
  photon gas, 22
  polytropic, 20–21, 152, 160, 308, 351
Equations of motion, 11
  conservative form, 23
  cylindrical coordinates, 58–60
  density waves, 126
  dynamical mass loss, 262
  rotating frame, 63–65
  self-similar form, 154
  shallow-water approximation, 73–75
  sphere, 75–78
  turbulent, 227–229
Equivalent width, spectral line, 407
Escape probability, 392
  decelerating flows, 392–394
  for photons, 382–383
Eulerian frame, 23–24

Ferraro isorotation theorem, 197, 207
First law of thermodynamics, 17–19
Flux freezing. *See* Magnetic flux freezing
Fokker–Planck equation, 52–55
  current, 55
Force-free magnetic field. *See* Magnetic field, force-free
Fractal, 250–253
Free-fall time, 166–167, 344
Froude number, 370

Geostrophic approximation, 73–82
  and binary stars, 74–75
Gravitational collapse
  isothermal, 167–168
  magnetic flux conservation, 187–188
  pressureless, 166–167
  similarity solution, 166–168
Gravitational potential, effective potential, 66

Hadley circulation, 67
Hartmann number (also Chandrasekhar number), 342–343
Haussdorf–Besicovitch dimension, 251
Hayashi track, 332
Heat flux, 17
Helicity, 62
 conservation of, 63
 energy spectrum, 205
Helicity density, 62
Herbig–Haro object, 398–401
H-theorem, 7
H II region, 106
 expansion, late stages of, 422
 as ionization shock, 132–141
 R- and D-fronts, 137
 Strömgren sphere, 132–133
Hubble flow, and P Cygni line profiles, 382, 403–405
Hurwitz–Routh stability criterion, 349–350
Hyperbolic equations, 101

Induction equation, 174
 magnetic wind, 276
Inner Lagrangian point, 290
Instability
 baroclinic, 87–89
 convection, 324–343
 double-diffusive, magnetic convection as, 340–343
 Jeans, 319–324
 Kelvin–Helmholtz, 293, 361–363
 Parker, 419
 pulsation, 342–353
 Rayleigh criterion for shear flows, 363–365
 Rayleigh–Taylor, 286–288, 357–361
 Richardson criterion, 369–370
 sausage (magnetic pinch), 186
 sound wave as example, 318–319
 Taylor, 370–371
 thermal, 353–357
 thermohaline, 341
Integrated line profile function, 390
Interstellar line profiles, 410
Irrotational flow, around sphere, 70

Jeans evaporation rate, 261
Jeans instability, 319–324
 bifurcation and, 323–324
 virial theorem and, 322–323
Jeans length, 320
Jeans mass, 321
Jeans parameter, 321–322
Jet
 definition, 29
 extragalactic radio, 413–414
 Kelvin–Helmholtz instability for free, 365–368
 light, 31
 P Cygni profile for, 400
 and protostars, 398–401
 rate of momentum transport, 30
 subsonic, 30
 supersonic, 118–119
 supersonic, bending, 144
 turbulent entrainment, 244–246

Kelvin circulation theorem, 68
Kelvin–Helmholtz, 293, 361–363
 for free jet, 365–368
Kelvin–Helmholtz timescale, 91
Keplerian motion, 82
 stability of, 83, 85–86
Kolmogorov cascade. See Kolmogorov theory for turbulence
Kolmogorov length, 225
Kolmogorov spectrum, 227
Kolmogorov theory for turbulence, 224–227

Lagrange multiplier, 190, 193
Lagrangian frame, 24
Lagrangian variation, 350
Landau cascade, 254
Lane–Emden equation, 161
 boundary conditions, 161
 isothermal sphere, 162–163
Langmuir wave, 31–33
Langmuir wave dispersion relation, 33
Large velocity gradient (LVG) approximation, 387
Lift, 70
Lighthill theory for turbulence, 237–239
Line profile
 Lorentzian, 408
 P Cygni, 384, 386
 Voigt, 408

Local thermodynamic equilibrium (LTE), 108, 375
  Planck function as source function, 376
Lorentzian line profile, 408
LTE. *See* Local thermodynamic equilibrium

Mach disk, 95–96, 119
Mach number, 107
Mach stem, 115
Magnetic ambipolar diffusion, 179–181
Magnetic diffusion coefficient, 177
Magnetic dynamo, 194–209
  α, 204–207
  $\alpha-\omega$ dynamo, 198–207
  astrophysical types, 207–209
  Babcock phenomenological model, 195–196
  dependent on rotation, 201–202
  self-inductance, 211–212
Magnetic field
  force-free, 188–193
  force-free, cylindrical solution, 190
  galactic-scale, 209
  planetary, 207–208
  stellar, 201–202, 208–209
Magnetic flux conservation, 186–188
Magnetic flux freezing, 178–179
Magnetic helicity, 189, 192–193
  and Woltjer's theorem, 193
Magnetic pinch equilibrium, 185–188
Magnetic reconnection, 209–210
Magnetic Reynolds number, 191
Magnetohydrodynamics
  condition for applicability, 175–176
  equation for magnetic field evolution, 177–178
Magnus force, 71–73
  and jet bending, 72–73
Markov process, 52
Matching conditions, shocks, 105
Matthew effect, 98
Maxwell equations, 174, 275
Maxwellian distribution. *See* Distribution function, Maxwellian
Mean field electrodynamics, 201–207
Mean intensity, 376
MHD. *See* Magnetohydrodynamics
Mixing length, 333–339
Mixing length parameter, 338–339
μ-barrier, 83

Navier–Stokes equation, 41, 216
  turbulence, 230
Negative specific heat, 15, 19–20
Neutral point
  magnetic, O-type, 195
  magnetic, X-type, 209–210
Nova, 401–405
  P Cygni profiles, 403–405
  RS Ophiuchi 1985, 402–404
Nuclear explosions
  critical overpressure, 155
  precursor, 120
Nusselt number, 339–340

Oblique shock. *See* Shock, oblique
Ohmic dissipation, 177–178
  timescale, 178
Ohm's law, 176
One-zone model, pulsation, 348
Optical depth, 379
  line center, 389
Orbital frequency, and effective potential, 85

P Cygni profile
  examples, Melnick 42, 379–381
  examples, Nova LMC 1990 No. 1, 381–382
  formation, ray approximation, 388–389
  parameters, 384
  schematic, 386
  signature of outflow, 385
  Sobolev approximation, 386–388
Period–luminosity relation, 344
Phase space, 2–3
Picard method, 249
Planar shock. *See* Shock
Planetary nebula, 282–288
  similarity solution, 158–159
Plasma frequency, 32
Plasma parameter, 117, 122, 184
Poisson equation
  charge, 31
  gravity, 160
Polytrope, 159–166
  galaxy clusters, 165
  isothermal sphere, 162–164
  molecular cloud, 165–166
  pressure bounded, 163–164
Potential temperature, 65

Potential vorticity, 65
  conservation of, 64–65
Poynting–Robertson effect, 83
Prandtl number (or Schmidt number), 327–328
Precursor, 120–121
Pressure scale height, 331
Propagator, 53–54
Pseudophotosphere, 382
Pulsation
  Lagrangian approach, 350–353
  one-zone approach, 345–349
Pulsation equation, adiabatic, 352

Radiation pressure, 269
  effect of lines, 271–272
Radiative flux, 376
Radiative force parameter, 270
Radiative temperature gradient, 336
Radiative transfer, $1 \times$, 374–396
  equation of, 376
  equation of, planar, 377, 378, 379
  equation of, spherical, 377, 387
  escape probability, 382–383
  flux, 376
  mean intensity, 376
  optical depth, 379
  and probability, 376
  ray approximation, 388–392
  source function, 376
  thermal equilibrium, 376
Random walk
  equation of motion, 55
  as Markov process, 52–53
Rankine–Hugoniot conditions, 102–109
  colliding stellar winds, 286
  ionization, 136
  Mach number, 106–108, 113
  magnetic shock, 117
  oblique shock, 112
  planar shocks, 102–106
  similarity solutions, 154
Rayleigh frequency, 83
Rayleigh number, 327–328
  critical, 330
Rayleigh stability criterion, 82–86
  on isobaric surfaces, 88
Rayleigh–Taylor instability, 286–288, 357–361
  in colliding winds, 286–288
  instability of cream with coffee, 357–358

Reynolds number, 43–45
  harmonic motion for low values, 417
  and Navier–Stokes equation, 43–44
R-front, H II region, 137
Richardson criterion. *See* Instability, Richardson criterion
Richardson number, 370
Riemann invariants, 99–101
  as characteristics, 101
  piston problem, 101
Roche gravitational potential, 290–292
  binary stars, 74–75
Roche radius, 289. *See also* Inner Lagrangian point
Rossby number, 78–79, 201
  dynamo scaling law, 201
Rossby wave, phase speed, 78
Rotational parameter, 90
Rotation and mixing, 92
Rotation and self-gravitation, 87–93

Schlieren, astrophysical, 409–412
Schwarzschild criterion. *See* Convection, Schwarzschild criterion
Sedov problem
  cylinder, 150
  sphere, 148–150
Seeing, astronomical, 49
Semiconvection, 338
Shakura–Sunyaev approach (α-disk), 296, 299, 419, 420
Shallow-water approximation: *See* Equations of motion, shallow-water approximation
Shear, 37
  fluid yields under, 57
Shock
  blunt body flow, 127–132
  bullets and whips, 95
  collisionless, 122
  collision with incompressible sphere, 131–132
  collision with interstellar clouds, 127–131, 144–145
  curved, 114–115
  density waves, 124–127, 141–144
  entropy jump, 109
  generation by sound wave steepening, 96–101
  H II regions, 106, 132–141
  ionization, 106

magnetic, 116–118
oblique, 111–116
   refraction angle, 113
precursor, 120–121
radiative, 119–122
rarefaction, 112
reflected oblique, 115–116
reflected planar, 110
strong, 105
traffic as example, 123–124
Shock adiabat, 104, 135
Shock tube, 102–104
Similarity methods, 148–169
diffusion equation, 169
general transforms, 150–155
Similarity transformation, 150
Blasius solution, 48
diffusion equation, 169
gravitational collapse, 167–168
H II region, late stages of expansion, 422
Skumanich law for rotational spindown, 281, 420
Snowplow phase, energy loss, 157
Sobolev approximation, 386–388
optical depth, 387
Sound speed, 21, 97, 319
Source function, 376
with scattering, 377, 379, 391–392
Specific heat, 18–19
effect of ionization and dissociation, 130–131
negative, for gravitation, 15, 19–20
polytropic, 20–21
Spherical coordinates, units vectors in, 58
Stagnation pressure, 155–157
Stellar wind bubble, 158
Stochastic differential equations, 249–250
Stochastic functions, and turbulence, 248–249
Streamline, 25
Stress
viscous, 38–43
wall, 47
Stress tensor, 11
diagonal components, 38, 39
kinetic turbulent, 217
pressure, 38
symmetry, 39
viscous, 40, 41–42
Strömgren sphere, 132–133
Superadiabatic temperature gradient, 335

Supernova
expansion and similarity methods, 149–150, 154, 157–158
precursor, 121

Taylor column, 66–67
Taylor hypothesis, 248
Taylor instability. *See* Instability, Taylor
Taylor microscale, 222–223
Taylor number, 371
Taylor–Proudman theorem, 66–67
and Ferraro isorotation theorem, 197
Tea leaves, 81–82
Thermal bremsstrahlung
absorption coefficient, 394
spectrum, wind, 394–396
Thermal instability. *See* Instability, thermal
Thermal instability criterion, 355
Thermal instability dispersion relation, 355
Thermohaline instability. *See* Instability, thermohaline
Tomography, astrophysical, 412–413
Toomre parameter, 322
Topological fluid dynamics, 63
and magnetic fields, 191–194
Total energy, polytrope, 19
Turbulence
accretion disks, 298–299, 419
astrophysical mechanisms, 233–235
astrophysical signatures, 247–248
coherent structures, 243–244
compressible, 237–240
and enstrophy, 222–223
entrainment in jets, 244–246
Lighthill theory, 237–240
and magnetic fields, 203–206
self-gravitating, 242–243
supersonic, 241–242
transition to, 235–236
Turbulent Reynolds stress, 230–231, 245–246

Unified stellar atmosphere, 385–386

Vector potential, solenoidal, 60
Velocity correlation function, 206
Velocity gradient, 387
Verhulst equation, 254

Virial theorem, 12–15
  higher moments, 15
  including pressure, 14
  Jeans instability and, 322–323
  self-gravitation, 22
  temperature for accretion, 304
  total energy with gravity, 14, 19
Viscosity
  accretion disk, 301
  coefficients of, 39, 41–43
  low Reynolds number, harmonic motion, 417
  magnetic analog, 177
  and rotation, 86–87
  turbulent, in accretion disks, 305
Vlasov equation
  characteristics, 7–8
  general, 6
  for plasma, 32
Voigt line profile, 408
Von Kármán vortex sheet, 68–69
Von Zeipl's theorem, 89–90
Vortex stretching, 61–62
  and Taylor–Proudman theorem, 71
Vorticity, 60
  absolute, 64
  and advection, 25
  blunt body shock generation, 131–132
  conserved in incompressible fluids, 61–62
  equations of motion, 60–63
  potential, 65
  and turbulence, 239–240

Wave
  Alfvén, 181–185
  compressive, phase speed, 42
  density, 123–124, 322
  Langmuir, 31–33
  nonlinear sound, 96–98

Rossby, 78
  shear, phase speed, 43
  sound, 318–319
  viscous medium, 42–43
Wind
  breeze solution, 264
  Castor–Abbot–Klein approximation, 271
  critical point, isothermal, 263, 271
  de Laval nozzle analogy, 264, 265–266
  as dynamical mass loss, 262
  effect of temperature gradient, 266–268
  inner Lagrangian point as critical point, 291–292
  iso-velocity contours, accelerating, 389, 420
  iso-velocity contours, decelerating, 393
  magnetic, 274–282
  magnetic, jet-like outflow, 282
  narrow absorption features for hot stellar, 396–397
  Parker solution, 262–265
  radiation pressure driving, 269–273
  rotational spindown, 278, 280–282
  spherical isothermal, 262–265
  thermal bremsstrahlung spectrum, 394–396
  velocity law, accelerating, 421
  Weber–Davis solution for magnetic, 279
Winds, colliding
  in binary star systems, 419
  Rayleigh–Taylor instability, 286–288
  similarity solutions, 283–286
  spherical, 282–288
Woltjer theorem, force-free magnetic fields, 188–190
Working surface, 118–119

Zeeman polarimetry, 187
ζ Aurigae binaries, 410–412